Quantum Liquids

Quantum Liquids

Bose condensation and Cooper pairing in condensed-matter systems

A.J. Leggett

Macarthur Professor and Professor of Physics
University of Illinois at Urbana-Champaign

OXFORD
UNIVERSITY PRESS

Great Clarendon Street, Oxford, OX2 6DP,
United Kingdom

Oxford University Press is a department of the University of Oxford.
It furthers the University's objective of excellence in research, scholarship,
and education by publishing worldwide. Oxford is a registered trade mark of
Oxford University Press in the UK and in certain other countries

First published 2006
Reprinted 2007 (with corrections)
First published in paperback 2022
Impression: 1

Published in the United States of America by Oxford University Press
198 Madison Avenue, New York, NY 10016, United States of America

British Library Cataloguing in Publication Data
Data available

Library of Congress Cataloging in Publication Data
Data available

ISBN 978–0–19–852643–8 (Hbk.)
ISBN 978–0–19–285694–4 (Pbk.)

Printed and bound by
CPI Group (UK) Ltd, Croydon, CR0 4YY

Preface to the Paperback Edition

In the 15 years since the original publication of this book, while some of the topics covered have not seen any particularly noteworthy evolution, others have undergone explosive developments. Were I writing the book from scratch today, I think there is little I would change in chapters 1-3 and 6, and the only major development I would want to note with respect to the subject matter of chapter 5 is that in 2015 superconductivity, believed to be of the "classic" type, was discovered in a class of metallic hydrides at several hundred gigabars pressure, with transition temperatures which now approach room temperature. A fairly up-to-date review is by Pickard et al., *ARCMP** 11, 57 (2020).

With respect to the topic of chapter 4, the situation is very different; there have been so many important developments in the study of ultracold atomic gases over the last 15 years that much of this chapter is severely out of date. Many of these developments are covered, with further references, in the recent book *Ultracold Atomic Physics* by Hui Zhai (Cambridge University Press 2020). The sub-topic of the BEC-BCS crossover (covered here in section 8.4) was already by 2012 the subject of a collection *The BCS-BEC Crossover and the Unitary Fermi Gas*, edited by W. Zwerger (Springer 2012); for more recent developments the reader is recommended to consult the INSPEC index under specific sub-topics.

As to the cuprate (high-temperature) superconductors which are the subject of chapter 7, there have been a number of important developments on the experimental front, particularly as regards the underdoped (pseudogap) regime where various new or improved spectroscopies have revealed that the phase diagram is even more complex than suspected in 2006. I list below a selection of recent reviews (necessarily arbitrary and in no special order) which the reader may find helpful:

Proust and Taiiefer, *ARCMP* 10, 409 (2019)
Robinson et al., *Reps. Prog. Phys.* 22, 126501 (2019)
Tajima, ibid. 75, 097001 (2016)
Frano et al., *J. Phys. Cond. Mat.* 32, 374005 (2020)
Stewart et al., *Advances in Physics* 66, 75 (2017)
Yazdani et al., *ARCMP* 7, 11 (2016)

Despite these important developments, I think my commentary on the theoretical situation would be little changed from the 2006 version: we still don't have a scenario which is endorsed by more than a modest fraction of the relevant community.

Concerning the topics of chapter 8: The BEC-BCS crossover (section 8.4) is briefly discussed above. The field of non-cuprate exotics (section 8.1) has blossomed in recent years, with particular attention being paid to systems which are strongly two-dimensional (indeed in some cases just a single monolayer); see for example the Stewart review cited above. One motivation for this enhanced interest has been the hope that such superconductors may be "topological" in nature and thus be usable for the purpose of topologically protected quantum computing (for these ideas see e.g. Nayak et al., *Revs. Mod. Phys.* 80, 1083 (2008), especially section III. B). However, recent experiments have made it controversial whether the long-favored candidate, Sr_2RuO_4, is indeed a topological superconductor. As regards 3-He in aerogel (section 8.2), there have been some important developments, particularly in uniaxially strained samples, including the realization of the so-called "polar" phase which is unstable in bulk 3-He; for a recent review see Halperin, *ARCMP* 10, 158 (2019). Finally, as regards the topic of section 8.3, the majority belief in the relevant community is now that most of the puzzling phenomena observed in solid 4-He below ~400 mK are not, as originally conjectured, a consequence of "supersolidity" but rather of the unique plasticity of this system; see Beamish and Balibar, *Revs. Mod. Phys.* 92, 045002 (2020). The search for supersolidity however continues in other systems.

In concluding this preface to the paperback edition, I would like to apologize to the many readers who over the last 15 years have notified me of various minor (and in some cases not so minor) typographical and other errors in the 2006 text. For many years I kept a careful consolidated list of these corrigenda, only to have it disappear without trace when my office was reorganized in connection with my retirement. So while I have tried to fix as many errors as possible in this paperback version, I am afraid that the one(s) to which you drew my attention may have gone uncorrected. Sorry!

*ARCMP=Annual Reviews of Condensed Matter Physics.

Preface

When the father of low temperature physics, Heike Kamerlingh Onnes, received the Nobel prize in 1913 for his liquefaction of helium, he concluded his acceptance speech by expressing the hope that progress in cryogenics would "contribute towards lifting the veil which thermal motion at normal temperature spreads over the inner world of atoms and electrons." Speaking only months after Bohr's publication of his atomic model, and long before the advent of modern quantum mechanics, Onnes could not have guessed how prophetic his words would prove. For of all the novelties revealed by the quest for low temperatures, by far the most dramatic have been the phenomena which result from the application of quantum mechanics to systems of many particles – in a word, of quantum statistics. And of these phenomena in turn, the most spectacular by far are those associated with the generic phenomenon which is known, when it occurs in a degenerate system of bosons, as Bose–Einstein condensation, or in a system of degenerate fermions as Cooper pairing. In this phenomenon (which for brevity I shall refer to generically in this Preface as "quantum condensation"), a macroscopic number either of single particles or of pairs of particles – a fraction of order one of the whole – are constrained to behave in exactly the same way, like a well-drilled platoon of soldiers (see the cover of *Science*, December 22, 1995). While the best-known consequence of quantum condensation is superfluidity (in a neutral system like ^{4}He) or superconductivity (in a charged one such as the electrons in metals), this is actually only a special case of a much more general pattern of behavior which has many other spectacular manifestations.

There are many good books on specific quantum condensates (liquid ^{4}He, the alkali Bose gases, superconductors, liquid ^{3}He, etc.); I list a selection below. The present book is in no sense intended as a substitute for these more specialized texts; rather, by giving an overview of the whole range of terrestrial condensates and their characteristic behaviors in what I hope are relatively simple and understandable terms, I aim to put the individual systems in context and motivate the reader to study some of them further.

This book is born of the conviction that it ought to be possible to present the essentials of Bose–Einstein condensation and Cooper pairing, and their principal consequences, without invoking advanced formal techniques, but at the same time without asking the reader to take anything on trust. Thus, the most advanced technique I have introduced is the language of second quantization, which for those not already practised in it is reviewed in a self-contained fashion in Appendix 2A. (However, most of Chapters 1–4 can actually be read even without fluency in the second-quantization language.) While this policy has the drawback of precluding me from introducing the Bogoliubov–de Gennes equations for superfluid Fermi systems (a technique which certainly needs to be learned, eventually, by anyone intending to do serious theoretical research on such systems), I hope it will mean that the book is relatively easily readable by, for example, beginning graduate students in theory or by experimentalists who do not wish to invest the time and effort to cope with more advanced formalism.

It will become clear to the reader from an early stage that I have at least two rather strong convictions about the theory of quantum condensation which are not necessarily shared by a majority of the relevant theoretical community. The first is that it is neither necessary nor desirable to introduce the idea of "spontaneously broken $U(1)$ symmetry," that is to consider (alleged) quantum superpositions of states containing different total numbers of particles; rather, I take from the start the viewpoint first enunciated explicitly by C.N. Yang, namely that one should simply think, in non-technical terms, about the behavior of single particles, or pairs of particles, averaged over the behavior of all the others, or more technically about the one- or two-particle density matrix. At the risk of possibly seeming a bit obsessive about this, I have tried to derive all the standard results not only for Bose but for Fermi systems using this picture; the idea of spontaneous $U(1)$ symmetry breaking is mentioned only to make contact with the bulk of the literature.[1] My second strong conviction is that many existing texts on superconductivity and/or superfluidity do not adequately emphasize the distinction, which to my mind is absolutely crucial, between the equilibrium phenomenon which in the context of a neutral superfluid is known as nonclassical rotational inertia (or the Hess–Fairbank effect) and in a charged system underlies the Meissner effect, and the metastable phenomenon of persistent currents in a neutral or charged system; again, I have tried to place some emphasis on this, see in particular Chapter 1, Section 1.5.

After two introductory chapters on the general phenomenon of quantum condensation, in the remaining chapters I treat in order liquid ^{4}He (Chapter 3), the alkali Bose gases (Chapter 4), "classic" (BCS) superconductivity (Chapter 5), superfluid ^{3}He (Chapter 6), the cuprate superconductors (Chapter 7), and finally, in Chapter 8, a miscellany of mostly recently realized quantum-condensed systems; in particular, Section 8.4 deals with very recent experiments which have made the long-conjectured "crossover" from Bose condensation to Cooper pairing a reality. The flavor of the various chapters is rather diverse, reflecting differences in the history and current status of our understanding of these systems; in particular, while in Chapters 3–6 I try to provide a reasonable theoretical basis for understanding the phenomena described, in Chapters 7 and 8 the treatment is much more descriptive and cautious. It will be noted that in Chapter 5 I have said essentially nothing about a topic which is a major part of most textbooks, namely the way in which properties associated with the normal phase (ultrasound absorption, tunnelling, spin susceptibility, etc.) are modified in the superconducting phase; this is a deliberate policy, so as not to distract attention from the central topic of condensation. Whenever possible, I have tried to provide derivations of standard results which differ somewhat from the conventional ones. While some of these alternative arguments (e.g. that given in Section 5.7 for the Ginzburg–Landau

[1] And in Appendix 5C as a purely formal device to streamline an otherwise cumbrous algebraic step.

formulation of superconductivity theory) may be less rigorous than the standard derivations, I hope they may complement the latter by giving a more physical picture of what is going on.

Two further points of general policy are that in order to keep the focus in the text on the main line of the argument, I have tried wherever possible to relegate cumbrous mathematical derivations to appendices; and that I have for the most part not attempted to trace the detailed history of the various theoretical ideas I discuss. (Any colleagues who feel thereby slighted might want to note that even the original BCS paper on superconductivity does not appear in the list of references!) Generally, I have given specific theoretical references only at points where the discussion in the text needs to be supplemented.

The original delivery date for the manuscript of this book was in the summer of 2004, but unforeseen events forced a postponement of 18 months. This was serendipitous, in the sense that during that period the topics covered in Sections 8.3 and 8.4 of Chapter 8 have undergone explosive experimental expansion, but the down side is that in some other chapters which were written earlier, in particular Chapter 4, the coverage of the most recent developments is not always complete.

It would be pointless to list all the good books which exist on individual quantum-condensed systems. For what it is worth, here are some which I feel readers of this book might find natural further reading:

- On helium-4: Pines and Nozières (1966). Wilks (1967) is a very useful general compendium.
- On the alkali Bose gases: Pethick and Smith (2002) and Pitaevskii and Stringari (2003).
- On "classical" superconductivity: De Gennes (1966) and Tinkham (1975, revised 1996). These two are as useful today as when they were first published.
- On superfluid ^{3}He: Vollhardt and Wölfle (1990) is a very good compendium, but hardly bedtime reading. The more general texts by Tsuneto (1998) and Annett (2004) have useful chapters.
- On cuprate superconductivity: see comments in Chapter 7.

I would like to thank the many colleagues, at UIUC and elsewhere, who have helped my understanding of the various systems treated in this book. Particular thanks are due to Lance Cooper, Russ Giannetta, Laura Greene, Myron Salamon and Charlie Slichter for their comments on a first draft of Chapter 7, Sections 7.3–7.7 and to Man-Hong Yung for proof-reading parts of the manuscript; needless to say, the responsibility for any remaining deficiencies is entirely mine. I am also very grateful to Linda Thorman and Adam D. Smith for their sterling efforts in typing the manuscript against a tight deadline. The writing of this book was supported by the National Science Foundation under grants nos. DMR-99-86199, DMR-03-50842 and PHY-99-07949.

Finally, I am very conscious that there are a number of points in this book where, owing in part to publishers' deadlines, I have not been able to spend all the time and thought that I would ideally have liked.[2] There is no doubt also the usual

[2]This is particularly true of the last few paragraphs of Chapter 5, Section 5.7 and Chapter 7, Section 7.5 and Appendix 7B.

crop of hopefully minor mistakes, typographical errors, etc. In the hope that the book may at some time in the future merit a revised edition, I shall be very grateful to any readers who bring such deficiencies to my attention.

Urbana, IL
30 January 2006

Contents

List of Symbols*

Symbol	Meaning	Page where defined/ introduced
[A, B]	commutator AB − BA	64
{A, B}	anti-commutator AB + BA	66
$\|00>$	doubly-unoccupied state of ($\mathbf{k}\uparrow, -\mathbf{k}\downarrow$)	244
$\|11>$	doubly-occupied state of ($\mathbf{k}\uparrow, -\mathbf{k}\downarrow$)	244
^{4}He*	excited state of ^{4}He atom	20
A	hyperfine interaction constant	114
$\mathbf{A}(\mathbf{r})$	electromagnetic vector potential	25
$\hat{a}$	Bose/Fermi annihilation operator	64
$\hat{a}^{\dagger}$	Bose/Fermi creation operator	64
a_b	bound-state radius	158
a_{bg}	background scattering length	160
a_s	s-wave scattering length	118
a_{zp}	width of single-particle groundstate in harmonic trap	127
B_{1g}	group-theoretic notation for $d_{x^2-y^2}$ symmetry	321
B_{c1}	lower critical field (of superconductor)	197
B_{c2}	upper critical field (of superconductor)	198
B_{hf}	characteristic hyperfine field	7
c_k	coefficient of pair wave function	105, 175, 186
c_s	speed of (hydrodynamic) sound	98
$\boldsymbol{d}$	constant value of $\mathbf{d}(\hat{\mathbf{n}})$ in ^{3}He-A	260
$\hat{\mathbf{d}}(\hat{\mathbf{k}})$ (or $\hat{\mathbf{d}}(\hat{\mathbf{n}})$)	d-vector notation for spin triplet OP	260
$dn/d\varepsilon$	density of states of both spins at Fermi surface of metal	166
$d_{x^2-y^2}$	most popular symmetry of cuprate order parameter	321
e**	electron charge ($\equiv -\|e\|$)	
E_{BP}	broken-pair energy	186
E_{EP}	excited-pair energy	186

*Symbols that are used only close to their definition are not included in the list below; note that some of these duplicate symbols listed here.
**when occurring in expressions for Φ_0, eV etc., $|e|$ is usually written as simply e.

Symbol	Meaning	Page where defined/ introduced
E_{GP}	ground-pair energy	186
E_J	Josephson coupling energy	227
E_k	BCS excitation energy, $\equiv (\varepsilon_k^2 + \|\Delta_k\|^2)^{1/2}$	181
E_R	recoil energy	146
f	filling fraction	149
F	Helmholtz free energy	104
F	Helmholtz free energy density	104
F	total atomic spin	7
$f(\mathbf{pp}'\sigma\sigma')$	Landau interaction function	230
$F(\mathbf{r}\,\sigma\,\mathbf{r}'\sigma' : t)$	order parameter in Fermi systems	51
$f(r_{ij})$	two-particle ingredient of Jastrow function	109
F_k	Fourier transform of Cooper-pair wave function $F(\boldsymbol{\rho}, \mathbf{R})$ with respect to relative coordinate $\boldsymbol{\rho}$	179
F_n	coefficient in expansion of Cooper-pair wave function	217
$f_n(T)$	normal fraction	24
$f_s(T)$	superfluid franction	23
$F^{s,a}$	dimensionless Landau parameters	231
$F_{\alpha\beta}(\mathbf{k})$	matrix form of Cooper pair wave function	258
g_D	nuclear dipolar coupling constant	266
H_D	nuclear dipole energy	265
I_c	critical current of Josephson junction	227
I_{cl}	classical moment of inertia	22
K	total dimer/molecule intrinsic angular momentum	7
k_F	Fermi wave vector	365
k_{FT}	$\equiv q_{TF}$	240
K_{mn}	matrix elements of time-reversal operator	220
ℓ	mean free path	215
$\hat{\boldsymbol{\ell}}$	direction of apparent relative angular momentum of pairs in ^{3}He-A	273
$\mathbf{L}$	orbital angular momentum	22
m*	effective mass	231
m_F	projection of total atomic spin on z-axis	115
$\bar{n}$	time reverse of state n	213
N(0)	density of states of one spin at Fermi surface ($\equiv 1/2(dn/d\varepsilon)$)	176
N_o	condensate number	12, 34
p	doping (of cuprates) (= no. of holes per CuO_2 unit)	290
p_F	Fermi momentum	229
q_{TF}	Thomas-Fermi wave vector	171
$\mathbf{R}$	center-of-mass coordinate	49

Symbol	Meaning	Page where defined/ introduced
R_N	normal state resistance (of Josephson junction)	228
r_o	range of interatomic potential (= van der Waals length)	118
$S(k)$	static structure factor	98
$S(\boldsymbol{r}, t)$	entropy density	83
t	tunneling matrix element [also time, throughout]	147
$T^*(p)$	crossover line (in phase diagram of cuprates)	292
T_c	critical temperature	12
T_F	Fermi temperature	11, 166
T_λ	lambda-temperature of liquid ^{4}He	72
$U(\mathbf{r})$	external potential	25
$u_{\mathbf{k}}$	parameter in BCS wave function	244
U_o	coefficient of δ-function in interparticle potential	41
V	visibility (of interference fringes)	137
$V(r)$	external potential	18
$V(r_i - r_j)$	interparticle potential	47
v_c	Landau critical velocity	101
v_F	Fermi velocity	167
$v_{\mathbf{k}}$	parameter in BCS wave function	244
$V_{\mathbf{kk}'}$	pairing interaction (in BCS problem)	180
V_o	coefficient of BCS contact interaction	174
V_o	height of optical-lattice potential	146
$\mathbf{v}_s(\mathbf{r}t)$	superfluid velocity	35, 191
$\mathbf{v}_s(\boldsymbol{R}t)$	superfluid velocity (in Fermi systems)	52
$Y(T)$	Yosida function	206
Z	atomic number	170
β	$1/k_BT$	9
γ	coefficient of linear term in specific heat (of normal Fermi system)	296
Γ_K	time-reversal-breaking parameter	219
δ	control parameter for interatomic potential (relative to position of resonance)	120
δ_c	characteristic value of δ	123, 162
Δ	constant value of Δ_k	182
Δ	detuning of laser	116
$\Delta_{\mathbf{k}}$	BCS gap parameter	181
$\Delta_{\mathbf{k},\alpha\beta}$	matrix form of gap in spin space	259
ΔN	number imbalance (relative number)	69, 225
$\Delta\phi$	relative phase	44
$\varepsilon(\omega)$	dielectric constant	301
ε_1	real part of dielectric constant	299
ε_2	imaginary part of dielectric constant	299
ε_c	cutoff energy in BCS model	175

Symbol	Meaning	Page where defined/ introduced
ε_F	Fermi energy	10
$\varepsilon_{\mathbf{k}}$	kinetic energy relative to Fermi energy [also used earlier for absolute value of kinetic energy]	175
ζ	$-(k_F a_S(\delta))^{-1}$ [also diluteness parameter $(na^3{}_S)^{1/2}$ in dilute Bose gas, pp. 132–4]	367
ζ	depletion parameter	106
θ_D	Debye temperature	2
κ	circulation of vortex line/ring [also GL parameter, 203: bulk modulus, 238]	91
κ_o	quantum of vorticity h/m [also bulk modulus, 171–5]	95
$\lambda_{ab}(T)$	ab-plane penetration depth (of cuprates)	305
$\lambda_c(T)$	c-axis penetration depth (of cuprates)	307
λ_L	London penetration depth	25
λ_T	thermal de Broglie wavelength	118
μ	chemical potential	9
μ_n	nuclear magnetic moment	15
ξ'	order of magnitude of Cooper pair radius	184
$\xi(T)$	healing length	127, 201
$\xi_{ab}(T)$	in-plane Ginzburg-Landau healing length	324
$\xi_c(T)$	c-axis Ginzburg-Landau healing length	316
ξ_k	absolute value of kinetic energy	372
ξ_o	Cooper pair radius	190
$\boldsymbol{\rho}$	relative coordinate	49
$\rho(E)$	density of single-particle states per unit energy	10
$\rho\ (r)$	single-particle density	44
$\rho(\boldsymbol{R}, \boldsymbol{R}' : \beta)$	many-body density matrix (in coordinate representation)	110
$\rho_1(\mathbf{rr}'t)$	single-particle density matrix	32
$\rho_2(\boldsymbol{r}_1\sigma_1\boldsymbol{r}_2\sigma_2 : \boldsymbol{r}'_1\sigma'_1 \boldsymbol{r}'_2\sigma'_2 : t)$	2-particle density matrix	47
$\rho_n(T)$	normal density	78
$\rho_s(T)$	superfluid density	78
σ	spin projection (in units of $\hbar/2$)	47
τ	relaxation time	167, 234, 297
$\phi(r_1r_2\sigma_1\sigma_2)$	"pseudo-molecular" wave function in BCS problem	178
$\phi(\mathbf{r}t)$	phase of condensate wave function (or order parameter)	35

Symbol	Meaning	Page where defined/ introduced
Φ_o	(superconducting) flux quantum, $\equiv h/2\|e\|$	196
$\chi_0(\mathbf{r}t)$	condensate eigenfunction/wave function	34, 125
$\chi_o(q\omega)$	bare response function	171
$\hat{\psi}(\mathbf{r}t)$	Bose/Fermi field operator	67
$\psi^{\dagger}_{\sigma}(\mathbf{r}t)$	Fermi creation operator	67
$\Psi(\mathbf{R}, t)$ [or $\Psi(\mathbf{r}, t)$]	Ginzburg-Landau order parameter	52
$\Psi(\mathbf{r}_1\mathbf{r}_2\ldots\mathbf{r}_N : t)$	many-body wave function	31
$\Psi(\mathbf{r}t)$	order parameter	34
ω	angular velocity	22
$\boldsymbol{\omega}(\mathbf{r}t)$	vorticity	91
ω_c	quantum unit of angular velocity	22
ω_L	Larmor frequency	266
ω_o	frequency of harmonic trap	19
ω_p	(electron) plasma frequency	167
$\hat{\boldsymbol{\omega}}$	spin-orbit rotation axis (in superfluid ^{3}He)	268
Ω_p	ionic plasma frequency	238

List of Symbols (not defined)

$\|\uparrow>$	state $\sigma = +1$
$\|\downarrow>$	state $\sigma = -1$
c_V	specific heat
g	gravitational acceleration [also interchannel coupling constant, pp. 159–64]
H	Hamiltonian
h	Planck's constant
$\hbar$	Dirac's constant ($\equiv$ h/2π)
$\boldsymbol{J}(\boldsymbol{r})$	current density
$\boldsymbol{k}$	wave vector
k_B	Boltzmann's constant
m	particle mass
n	particle density
P	pressure
t	time
T_1	nuclear spin relaxation time
$\delta n(p\sigma)$	deviation of (quasi)particle occupation number from normal-state value
λ	optical wavelength [also dimensionless ratio Na_s/a_{2p}, pp. 127–28: relative channel weight, pp. 163–64]
μ	micron (10^{-6}m)
μ_B	Bohr magneton
μ_o	permeability of free space
σ	conductivity
χ	magnetic susceptibility
ω	angular frequency

1
Quantum liquids

A "quantum liquid" is, by definition, a many-particle system in whose behavior not only the effects of quantum *mechanics*, but also those of quantum *statistics*, are important. Let us examine the conditions for this to be the case.

Needless to say, if we wish to describe the actual structure of atoms or molecules, then under just about any conditions known on Earth it is essential to use quantum mechanics; a classical description fails to account for even the qualitative properties. However, if we consider the atoms or molecules as themselves simple entities and ask about their dynamics or thermodynamics, we find that classical mechanics is often quite a good approximation. A qualitative explanation of why this should be so goes as follows: The fundamental novelty introduced by the quantum-mechanical description in the motion of particles is that it is necessary to ascribe to the particle wave-like attributes, resulting in phenomena such as interference and diffraction; the quantitative relation between the "wave" and "particle" aspects is given by the de Broglie relation

$$\lambda = h/p \tag{1.0.1}$$

where p is the momentum of the particle and λ the wavelength of the associated wave. However, we know from classical optics that a wave will behave very much like a stream of particles ("physical optics" becomes "geometrical optics") if the wavelength λ is small compared to the characteristic dimension d of whatever is obstructing it ("one cannot see around doors"); the condition to see wave-like effects is, crudely speaking

$$\lambda \gtrsim d \tag{1.0.2}$$

In the case of a many-particle system it is perhaps not immediately clear what we should identify as the length d, but for reasonably closely packed systems, at least, it seems reasonable to take it to be of the order of the interparticle distance, i.e. as $n^{-1/3}$ where n is the density (though see below). On the other hand, the typical value of λ is determined, according to Eqn. (1.0.1), by that of the momentum p, which in thermal equilibrium is determined by the mean thermal energy $k_{\mathrm{B}}T$:

$$p \sim (mk_{\mathrm{B}}T)^{1/2} \tag{1.0.3}$$

Combining Eqns. (1.0.1)–(1.0.3), we find that the conditions for quantum mechanical effects to be important in the (center-of-mass) motion of a set of atoms or molecules is roughly

$$k_{\mathrm{B}}T \lesssim n^{2/3}\hbar^2/m \tag{1.0.4}$$

where m is the mass of the atom or molecule in question.

For a gas in equilibrium with its liquid or solid phase, it is impossible to fulfill the condition (1.0.4), because at low temperatures the density n tends to zero exponentially. Even in the liquid or solid phase (with n say $\sim 10^{23}$ cm^{-3}) it requires temperatures at most[1] of the order of 20 K$/A$ where A is the atomic or molecular weight. By contrast, for electrons, with their much smaller mass, the criterion is well fulfilled in any realistic solid or liquid, so that we would expect them always to behave in a fully quantum-mechanical way.

The upshot of the above argument is that the effects of quantum *mechanics* will always be important for electrons, and will be important also for atoms (and molecules) at temperatures which, although generally lower than room temperature, are nowadays easily attainable. For our system to qualify as a quantum liquid, however, we need more than this: we need the effects of quantum *statistics* to be important. As we shall see in the next section, such statistics are a consequence of the fundamental quantum-mechanical principle of indistinguishability of elementary particles and those of the composites (atoms, molecules, etc.) made out of them. It is crucial to appreciate that the mere fact that a given system of particles shows substantial effects of quantum mechanics such as the quantization of energy is not enough to guarantee that it will automatically show the effects of indistinguishability; it is necessary, in addition, for the particles to be able to "find out" that they are indistinguishable, and they can do this only if they can change places (otherwise, we can as it were "tag" them by their physical location). A good illustration of this principle is the effect (or lack of it) of isotopic identity on (a) the rotational and (b) the vibrational spectra of diatomic molecules: Consider for example a homoatomic molecule such as C_2. Both the rotational and the vibrational spectrum are quantized, and for a heteronuclear molecule (say $^{12}C + ^{14}C$) all possible levels occur for both. If we replace the ^{14}C by ^{12}C, so that the nuclei as well as the electrons of the two atoms are identical, we find that in the rotational motion (in which the atoms "change places" readily) the levels with odd angular momentum (which would correspond to a wave function odd under exchange of the atoms, forbidden for identical bosons) are missing; on the other hand, in the vibrational motion (in which the atoms do not change places) all the original levels still occur.

Thus, for a system of particles to constitute a quantum liquid it is necessary not only that it satisfies the condition (1.0.4), but that the particles should be able to change places fairly readily. In most crystalline or amorphous solids the exchange even of neighboring atoms is so difficult that the effects of indistinguishability, and hence of quantum statistics, are completely negligible; a partial exception is solid helium, but even there the exchange of atoms is in some sense a "weak" effect by comparison with the lattice effects which this system shares with more conventional crystalline solids. (As a quantitative illustration of this state of affairs, the Debye temperature of solid ^{3}He is of the order of 20 K, while the temperature of magnetic ordering, usually associated with the exchange of neighboring atoms, is only about 1 mK.)

[1] This naive argument actually somewhat underestimates the temperature for quantum-mechanical effects to be important in a solid, which is more like $T \lesssim \theta_D$, where θ_D is the Debye temperature. The reason is that in a solid the effective value of d can be considerably less than the interatomic spacing.

Thus, a "quantum liquid" composed of atoms (or molecules) must be in either the gas or the liquid phase (though see Chapter 8, Section 8.3). For a system of electrons the criterion is less stringent, in the sense that provided the electrons can change places fairly readily (i.e. crudely speaking, the system is a metal) it does not matter whether or not they are moving in an atomic background which is solid-like (and indeed, in the most interesting systems they always are).

Collecting the above considerations, we see that the class of (terrestrial) quantum liquids includes (a) electrons in any solid or liquid metal, and (b) any collection of atoms or molecules which simultaneously is in the liquid or gaseous phase and satisfies the criterion (1.0.4). The known members of subclass (b) consist at present of the liquid isotopes of helium[2] (^{4}He, ^{3}He and their mixtures) and the dilute atomic alkali gases (along with the recently stabilized gaseous phase of metastable atomic helium). If one ventures beyond the Earth's surface, then the class of quantum liquids contains also some astrophysical members such as the neutrons in (some regions of) neutron stars. In this book I will concentrate on the terrestrial quantum liquids.

1.1 Indistinguishability and the symmetry of the many-body wave function

In principle, every physical system should be described by quantum field theory, which is by construction fully relativistically covariant. However, for the systems which are the subject of this book the velocity of the (massive) constituents is rarely greater[3] than $\sim$1% of that of light, and indeed usually much smaller; thus, for most purposes nonrelativistic quantum mechanics for the massive particles, augmented when necessary by the standard quantum theory of the electromagnetic field, should be an excellent approximation, and it will be used throughout. Within such a framework it is natural to regard the "elementary" constituents of matter as nucleons and electrons (and the electromagnetic field as composed of photons). As is well known, general principles of quantum mechanics require the angular momentum of any elementary particle, measured in units of $\hbar$, to be either integral or half-odd-integral; in the former case the particle in question is called a boson, in the latter a fermion. It is known that both nucleons and electrons have spin $\frac{1}{2}$ and are thus fermions, while the photon has spin 1 and is a boson. As far as is known, stable "elementary" bosons with nonzero mass do not exist, but for the purposes of the argument of the next paragraph it is convenient to imagine temporarily that they might, in which case they would presumably be describable for our purposes by nonrelativistic quantum mechanics and possibly be characterized by some spin vector $\boldsymbol{\sigma}$ whose eigenvalues would be integral rather than $\frac{1}{2}$.

Consider then the description, in nonrelativistic quantum mechanics, of two elementary particles *of the same species*, each of which is in the general case characterized

[2]Liquid hydrogen is a marginal case; since the freezing temperature is $\sim$20 K, it barely satisfies criterion (1.0.4).

[3]Electrons in a heavy metal (or for that matter a heavy atom) may have velocities greater than this, so that effects which are essentially relativistic in origin, such as the spin–orbit interaction, may not be negligible. However, there are standard ways of handling such effects within the nonrelativistic formalism.

by a coordinate vector $\boldsymbol{r}$ and the projection σ of its spin vector on some specified axis, say the z-axis. Quite generally, the two-particle Schrödinger wave function (probability amplitude) is of the form

$$\Psi_2 \equiv \Psi(\boldsymbol{r}_1, \sigma_1, \boldsymbol{r}_2, \sigma_2) \tag{1.1.1}$$

Now, a fundamental principle of quantum mechanics asserts that any two "elementary" particles of the same species are *indistinguishable*; there is no way of "labeling" them so as to say that it is particle 1 which is at point $\boldsymbol{r}_1$ with spin projection σ_1 and particle 2 at $\boldsymbol{r}_2$ with spin projection σ_2 rather than vice versa. Consequently, all physical properties, and in particular the probability $P(\boldsymbol{r}_1\boldsymbol{r}_2\sigma_2\sigma_2) \equiv |\Psi(\boldsymbol{r}_1\sigma_1\boldsymbol{r}_2\sigma_2)|^2$ of finding particle 1 at position $\boldsymbol{r}_1$ with spin projection σ_1 and particle 2 at $\boldsymbol{r}_2$ with projection σ_2, must be invariant under simultaneous exchange of *all* the variables of 1 and 2:

$$|\Psi(\boldsymbol{r}_1\sigma_1, \boldsymbol{r}_2\sigma_2)|^2 = |\Psi(\boldsymbol{r}_2\sigma_2, \boldsymbol{r}_1\sigma_1)|^2 \tag{1.1.2}$$

Note that there is no requirement that the probability be invariant under exchange of the coordinates alone, without exchange of the spin projections (in that case, the spin projection, if different for the two particles, may be used as it were as a "label"). Equation (1.1.2), which holds independently of the dimensionality of the (coordinate) space, is compatible with the hypothesis that interchange of 1 and 2 multiplies the wave function (1.1.1) itself by an arbitrary phase factor e^{ia} of modulus unity. However, in three or more spatial dimensions[4] (a case which of course includes the physical world we live in) a further restriction follows from the fact that the operation of exchanging two particles of the same spin *twice* is essentially equivalent to moving one completely around the other, and that (for spatial dimension $d \geq 3$) the resulting "loop" can be contracted to a point other than the relative origin and the double interchange must therefore be equivalent to the identity operation. (For a more careful version of this argument, and the inclusion of the spin degree of freedom, see Leinaas and Myrheim (1977).) This requires $\alpha = n\pi$ with n integral, i.e.

$$\Psi(\boldsymbol{r}_1\sigma_1, \boldsymbol{r}_2\sigma_2) = \pm\Psi(\boldsymbol{r}_2\sigma_2, \boldsymbol{r}_1\sigma_1) \tag{1.1.3}$$

To proceed further we need to invoke the celebrated *spin-statistics theorem*, which states that the + sign applies for all elementary particles of integral spin (bosons) and the − sign for all particles of half-odd-integral spin (fermions). The spin-statistics theorem was originally proved (see e.g. Streater and Wightman 1964) within the framework of relativistic quantum field theory, of which the ordinary Schrödinger

[4]In two dimensions the argument fails because there is no way of contracting the relevant "loop" to a point other than the relative origin; that is, the concept of one particle "going around" another is well-defined and not equivalent to the identity. This leads to the possibility of existence of types of particles obeying "fractional statistics," i.e. having a value of the interchange phase α different from $n\pi$ (in modern terminology "anyons"); see the classic paper of Leinaas and Myrheim (1977) on this subject. Although such "anyons" do not as for as we know exist in nature as "elementary" particles, they form an extremely useful way of representing some of the possible states of a system of electrons physically confined to a plane, as in the quantum Hall effect: see e.g. Prange and Girvin 1990. One might indeed say that anyons are as "real" (or as unreal!) as the composite "bosons" such as ^{4}He atoms to be discussed below. Although some very interesting theoretical connections have been made between the theory of quantum liquids (as interpreted here) and the quantum Hall effect, the latter topic deserves (and has generated) books in its own right, and will not be discussed here.

quantum mechanics used in this book may, at least for our purposes, be regarded as the nonrelativistic limit. It is very tempting to think that the theorem should be provable entirely within the framework of Schrödinger quantum mechanics (presumably as a consequence of the single- (double-) valuedness of the probability amplitudes corresponding to integral (half-odd-integral) spin). A particularly interesting attempt in this direction has been made recently by Berry and Robbins (2000; see also Berry and Robbins 1997); for earlier attempts, and experiments which test the spin-statistics relation, see Duck and Sudarshan 1998. For the purposes of this book I shall from now on take this relation as given.

Making the obvious generalization to the case of many particles, we reach the fundamental conclusion that *the many-particle* Schrödinger wave function (probability amplitude) must be invariant under the simultaneous exchange of *all* the coordinates of any two identical particles of integral spin, and must change sign under such an exchange of any two identical particles of half-odd-integral spin. A common, if somewhat inaccurate,[5] shorthand for this state of affairs, specialized to the cases of interest in terrestrial physics, is that nucleons and electrons (with spin $\frac{1}{2}$) obey Fermi–Dirac (or Fermi) statistics, while photons (spin 1) obey Bose–Einstein (or Bose) statistics;[6] hence the names "fermion" and "boson" for the two kinds of particles respectively. A consequence of the occurrence of the minus sign for fermions is the well-known *Pauli exclusion principle*: Consider a set of identical fermions, and expand the many-body wave function in terms of a complete set of orthogonal single-particle wave functions $\phi_i(\boldsymbol{r}, \sigma)$:

$$\Psi(\boldsymbol{r}_1\sigma_1\boldsymbol{r}_2\sigma_2 \ldots \boldsymbol{r}_N\sigma_N) = \sum_{ijkl\ldots s} C_{ijkl\ldots}\, \phi_i(\boldsymbol{r}_1\sigma_1)\phi_j(\boldsymbol{r}_2\sigma_2) \ldots \phi_s(\boldsymbol{r}_N\sigma_N) \tag{1.1.4}$$

Then the minus sign in Eqn. (1.1.3) immediately implies that the coefficient $C_{ijkl\ldots s}$ is completely antisymmetric in its indices, and in particular that it vanishes whenever any pair of indices are equal: occupation of any single-particle state by more than one fermion is forbidden! This principle, and more generally the constraint of antisymmetry is automatically handled by the formalism of "second quantization" described in Appendix 2A; however, it is often helpful to one's intuition to continue to write the many-body wave function in the explicit "Schrödinger" form (1.1.4) (sometimes referred to, somewhat misleadingly, as "first-quantized" notation).

Let us now turn to the wave function of two identical composite objects made up of elementary fermions, such as a pair of ^{87}Rb atoms. Of course, in such a case it is in principle possible to specify the wave function as a function of the coordinates and spins of all the fermions separately (248 degrees of freedom!), and to require the appropriate antisymmetry under interchange of any two protons/neutrons/electrons; however, this

[5]The "spin-statistics" theorem is, as explained, a constraint on the *symmetry* of the many-body wave function with respect to interchange of its arguments; under certain very special circumstances only (see Section 1.3) it leads to a unique distribution of particles between states, i.e. a unique "statistics"

[6]Provided we are prepared to work in a single Lorentz frame, any "Fock" state of the electromagnetic field, that is any state corresponding to a definite total number of photons, can be formally described by a Schrödinger wave function. However for more general states the generalization, while possible, becomes extremely clumsy and the second-quantized formalism (Appendix 2A) is infinitely more convenient.

is obviously extremely clumsy, and moreover unnecessary under most circumstances. To see why, let's consider for definiteness a complex of two ^{4}He atoms, each containing two neutrons, two protons and two electrons. We use the principle (already illustrated for the case of molecular vibration and rotation) that the relative sign of two quantum states can matter only if there is a nonzero probability for a physical transition between them to take place. Thus, the symmetry under a particular exchange process can matter only if there is a nonzero probability amplitude for that process actually to take place (we do not expect the "exchange" of an electron on Sirius with one on Earth to have physical consequences!). Whether it can or not is, of course, a question of energetics. Consider first the interchange of a single neutron (say) on one ^{4}He nucleus with one on the other. Such a process would require each neutron to physically escape from its original ^{4}He nucleus and travel, or tunnel, through space to the other nucleus; provided we restrict ourselves (as we always shall in this book when dealing with terrestrial systems) to states with energies much less than the nuclear binding energy (a few MeV), the WKB probability amplitude for this tunneling process is so tiny that the probability of it occurring is negligible over the age of the Universe.[7] The same applies to the simultaneous exchange of two or three nucleons, so that the formal symmetry of the wave function under such processes can be ignored. However, the simultaneous exchange of all four nucleons, i.e. exchange of the ^{4}He nuclei as a whole, is not similarly forbidden; so we must, at least for the moment, take into account the correct behavior of the many-body wave function under this exchange, namely multiplication by $(-1)^4 = +1$. By the same token, for exchange of two ^{3}He nuclei with identical spin orientations the appropriate factor is $(-1)^3 = -1$. (No particular symmetry is required under exchange of a ^{3}He and ^{4}He nucleus, since this cannot be written as a product of pure exchanges of elementary particles.) More generally, the exchange of any two (zero-spin or identical-spin) even isotopes of the same species must multiply the many-body wave function by $+1$, and the exchange of any two such odd isotopes must multiply it by -1. I postpone for the moment the question of the effect of the spin degree of freedom.

Now consider the electrons. Suppose, first, that the nuclei are separated by a distance large compared to the atomic radius, and that we deal with the electronic ground state. Then the same argument as applied to the nucleons above applies to the electrons: the probability amplitude for a process in which two electrons each physically leave their original atom and migrate to the other is negligibly small. In fact, in this limit, the only relevant process which possesses appreciable probability amplitude is that in which each nucleus carries with it its complete electronic cloud, i.e. the atoms exchange positions as wholes. It is then clear (still neglecting the spin degree of freedom) that the many-body wave function must be multiplied by $(-1)^{N_\mathrm{f}}$ where N_f is the total number of fermions (nucleons plus electrons) per atom. Thus, in particular, under these conditions, all odd-isotope alkali atoms (^{87}Rb, ^{23}Na, ^{7}Li, ...) will behave, as wholes, as bosons, as will ordinary (light) hydrogen, ^{1}H, while all even-isotope ones (^{40}K, ^{6}Li ...) will behave as fermions (as will deuterium, ^{2}H).

[7] Recall that there is a huge Coulomb barrier for the approach of the nuclei to short distances, so that the distance travelled by the neutron must be appreciable.

Although the above consideration actually covers almost all situations currently of practical interest in the real-life quantum liquids, it is useful for completeness to say a word about the appropriate description when the exchange of electrons cannot be neglected, as may be the case when the nuclei approach sufficiently close, and in particular when the atoms form a molecule or dimer. Under these conditions the concept of an "atom" cannot really be defined, and it is more appropriate to imagine the nuclei to be frozen in place and to consider the symmetry of the electronic state with respect to reflection in the plane bisecting their separation. States which are even under this operation are conventionally labeled with a subscript "g" and those which are odd with a "u." Consider the case of a single s-electron on each atom outside closed shells (the case of interest for all alkali-gas BEC systems). Since in this case reflection in the bisector plane is equivalent to exchange of the electrons, it follows that g-states are always even under exchange and thus can only be associated with singlet electronic spin functions (see below), while u-states are always odd and can be associated only with triplet spin configurations. To put it the other way around, if we know that at large interatomic distances, where the atoms behave independently, the electrons are polarized parallel (as is the case e.g. for spin-polarized ^{1}H in a field of 10 T), then they can collide only in u-states and so on. In particular, the electronic ground state of a dimer of spin-polarized atomic hydrogen in a 10 T field at arbitrary nuclear separation is the $^3\sum_\mathrm{u}$ state to a high degree of approximation.

We finally need to consider explicitly the effect of the spin degree of freedom, both electronic and nuclear; the necessary notation is defined in Chapter 4, Section 4.1. At large interatomic distances, the situation is fairly straightforward; I will consider first the limit of most practical interest, $B \lll B_\mathrm{hf}$ when B_hf is the characteristic hyperfine field. In this limit, it is a standard result (see e.g. Chapter 4, Section 4.1, or Woodgate 1970, Chapter 9) that the good quantum numbers for a single atom are the total atomic spin F and its projection m_F on the axis of the external magnetic field. Consider now a pair of such identical atoms. In the absence of appreciable coupling between the hyperfine and (atomic center-of-mass) orbital degrees of freedom, the energy eigenfunctions can always be written as a product of a "spin" function (i.e. a function of the hyperfine degrees of freedom) and an orbital function (i.e. a function of the coordinates of the atomic nuclei). The symmetry of the orbital wave function under exchange of the two atoms is $(-1)^L$ where L is the relative orbital angular momentum. As to the spin wave function, the combination of two atomic (intrinsic) angular momenta each equal to F yields possible values of the total dimer intrinsic angular momentum $K \equiv |F_1 + F_2|$ equal to $0, 1, \ldots, 2F$. Now, it is easy to establish[8] that the symmetry of the spin state under exchange of the two atomic spin coordinates (only) is $(-1)^{K+2F}$. Since the symmetry under simultaneous exchange of both spin and orbital coordinates of the two atoms must be η, where $\eta \equiv +1$ for bosons and -1 for fermions, and since $(-1)^{2F} = \eta$, we obtain a fundamental constraint on the quantum numbers of a dimer composed of two atoms with identical F-value, namely

$$K + L = \text{even} \tag{1.1.5}$$

[8]Start with the state $K = 2F$ and use the orthogonality of states with the same m_K but different K.

irrespective of whether we are dealing with bosons or fermions. Thus, for example, two ^{23}Na atoms each in the $F = 2$ state can undergo a collision in a relative s-state only if K is 0, 2 or 4, not if it has the (allowed) values 1 or 3. Two atoms with different F-values are not subject to any particular symmetry constraint, since the F-value can in effect be used as a "tag."

It is straightforward to adapt the above arguments to the case $B \gg B_{\text{hf}}$, by working in terms of the variables m_s and m_I rather than F and m_F. In particular, spin-polarized ^{1}H in a field of 10 T has $m_{s1} = m_{s2}$, so the minus sign necessary on exchanging the atoms is already provided by exchange of the electrons (remember the electronic state is u, hence odd under interchange). Hence, if we consider only the wave function corresponding to the orbital and nuclear-spin degrees of freedom, *spin-polarized ^{1}H atoms behave as bosons*, and since the nuclear spin is $\frac{1}{2}$, they are sometimes said to be "spin-$\frac{1}{2}$ bosons."

In the intermediate range of B/B_{hf}, or when the atoms are close together, the situation is more complicated since the energy eigenstates are not in general eigenstates of either F or of m_s. However, they can always be expressed as superpositions of such eigenstates, and the fact that the Hamiltonian commutes with the permutation operation guarantees us that all the components must have the same symmetry under interchange. Moreover, since (continuous) adiabatic evolution clearly cannot change the (discrete) symmetry, the latter must be simply that of the state into which the state in question would evolve as the interatomic separation tends to infinity and the magnetic field to zero; this identifies it uniquely.

1.2 The Fermi–Dirac and Bose–Einstein distributions: BEC in a noninteracting gas

In this section I will consider a set of identical *noninteracting* particles whose wave function satisfies Eqn. (1.1.3) with either the $-$ sign (fermions) or the $+$ sign (bosons), and held in thermal equilibrium with a bath at temperature T. The relevant question is: what is the distribution of the particles over the available single-particle states? The material of this section is, with the exception of the last paragraph, standard, and good accounts can be found in many textbooks (e.g. Landau and Lifshitz 1958, Chapter 5; or Huang 1987, Chapters 11 and 12).

Since we eventually wish to describe not only systems which are approximately translation-invariant (such as liquid ^{4}He) but also for example the magnetically trapped atomic alkali gases and the electrons in amorphous (noncrystalline) metals, it is convenient to allow our particles to move in an arbitrary bounded external one-particle potential which in general is a function of both the coordinate $\boldsymbol{r}$ and the spin projection[9] σ. However complicated this potential, we can always find a complete set of orthonormal eigenfunctions $\varphi_i(\boldsymbol{r}, \sigma)$ of the single-particle Hamiltonian and the corresponding eigenvalues ϵ_i; it will be convenient to choose the zero of energy to coincide with the minimum value of ϵ_i. Note that the eigenfunctions $\varphi_i(\boldsymbol{r}, \sigma)$ need not be simple products of a coordinate-space and a spin-space function (this is true only if

[9]In principle it could be a matrix with respect to σ and even depend on velocity; this in no way affects the argument.

the potential is a sum of terms which depend only on coordinates and only on spin respectively). For the moment I will assume that the only conservation law which is relevant to the thermodynamics is that of the total particle number N; see below.

It follows from the basic principles of quantum statistical mechanics that for a set of N particles under the above conditions the density matrix of the many-body system is diagonal in the energy representation and given by the standard macrocanonical Gibbs distribution; that is, the probability of occurrence of a many-body state E_α is

$$p_\alpha = Z^{-1} \exp -\beta E_\alpha, \quad Z \equiv \sum_\alpha \exp -\beta E_\alpha \tag{1.2.1}$$

when we introduce the standard notation β for $1/k_{\mathrm{B}}T$, a convention which will be used throughout this book. Since the total energy E_α is given by

$$E_\alpha = \sum_i n_i \epsilon_i \tag{1.2.2}$$

and thc total particle number N is constrained to satisfy the condition

$$N = \sum_i n_i \tag{1.2.3}$$

it follows that we can rewrite (1.2.1) in the form

$$p\{n_i\} = Z^{-1} \exp -\beta \sum_i n_i \epsilon_i, \quad Z = \sum_{\{n_i\}} \exp -\beta \sum_i n_i \epsilon_i \tag{1.2.4}$$

provided that the values of n_i are constrained by the condition (1.2.3). This constraint prevents the n_i from being regarded as completely independent, and is very inconvenient. As is well known, the standard solution is to relax the constraint that the total particle number N is fixed, replacing it by the condition that the system is held at a constant chemical potential μ which is then obtained from the condition that the calculated quantity $\sum_i \langle n_i \rangle$ is equal to the actual number N of particles in the system. The effect is to replace (1.2.1) by the grand canonical distribution

$$p_\alpha = \tilde{Z}^{-1} \exp -(E_\alpha - \mu N_\alpha), \quad \tilde{Z} \equiv \sum_\alpha \exp -\beta(E_\alpha - \mu N_\alpha) \tag{1.2.5}$$

where the quantity N_α is the value of $\sum_i n_i$ in the state α. One can now write p_α in the form

$$p_\alpha\{n_i\} = \tilde{Z}^{-1} \exp -\beta \sum_i n_i(\epsilon_i - \mu), \quad \tilde{Z} \equiv \sum_{\{n_i\}} \exp -\beta \sum_i n_i(\epsilon_i - \mu) \tag{1.2.6}$$

where the n_i may now be taken as *independent* variables. This means that we can rewrite $p\{n_i\}$, which no longer depends explicitly on α as a product:

$$p\{n_i\} = \prod_i p_i(n_i), \quad p_i(n_i) = \exp -\beta(\epsilon_i - \mu) n_i \Big/ \sum_i \exp -\beta(\epsilon_i - \mu) \tag{1.2.7}$$

where the allowed values of n_i are 0 and 1 for fermions, and $0, 1, 2, \ldots$ for bosons. It is worth noting explicitly that the fact that the grand canonical distribution factorizes in this simple way is a consequence of the fact that in each case there is one and only one many-particle state which corresponds to any given set of values $\{n_i\}$, and that in the Bose case this is itself a consequence of the requirement that the wave particle be totally symmetric under the exchange of any pair of particles. From (1.2.7) we immediately obtain the standard Fermi–Dirac (or Fermi) and Bose–Einstein (Bose) formula for the average number $\langle n_i(T)\rangle$ of fermions and bosons respectively in the single particle state i at temperature T:

$$\langle n_i(T)\rangle = (\exp\beta(\epsilon_i - \mu) + 1)^{-1} \qquad \text{fermions} \tag{1.2.8}$$

$$\langle n_i(T)\rangle = (\exp\beta(\epsilon_i - \mu) - 1)^{-1} \qquad \text{bosons} \tag{1.2.9}$$

where in each case the chemical potential $\mu(T : N)$ is implicitly defined by the condition

$$\sum_i \langle n_i(\mu, T)\rangle = N \tag{1.2.10}$$

Note that in the Bose case the formula (1.2.9) makes sense only if the chemical potential μ is negative: we return below to the question of what happens if and when this condition becomes inconsistent with the constraint (1.2.10).[10] Also note that in the limit $T \to \infty(\beta \to 0)$ the condition (1.2.9) can be fulfilled, for any physically reasonable density of single particle energy levels ϵ_i, only by taking $\mu \to -\infty$, so that the expressions (1.2.8) and (1.2.9) both reduce to the classical Maxwell distribution (which is what we would have got by treating the particles as distinguishable).

Let us now examine some characteristic features of the Fermi and Bose distributions (1.2.8) and (1.2.9), starting with the Fermi case. At zero temperature the RHS of Eqn. (1.2.8) reduces to the Heaviside step function $\theta(\mu - \epsilon_i)$, so the chemical potential at zero temperature (the Fermi energy ϵ_F) is simply given by the condition

$$\sum_{i(\epsilon_i<\mu)} = N \tag{1.2.11}$$

or equivalently, introducing the single-particle density of states[11] $\rho(\epsilon)$ by

$$\rho(\epsilon) \equiv \sum_i \delta(\epsilon - \epsilon_i) \tag{1.2.12}$$

by the formula

$$\int_0^{\mu} \rho(\epsilon)d\epsilon = N \tag{1.2.13}$$

The numerical value of $\mu(T = 0)$ depends on the density of states $\rho(\epsilon)$, which in turn depends on the nature of the confining potential (if any) and the number of available

[10] In the case of photons the total number N is not conserved and we must therefore set $\mu = 0$ from the start: this gives of course the standard Planck distribution, which is then trivially consistent with (1.2.9).

[11] I have assumed that N is very large compared to unity, so that the relevant levels may be taken as continuously distributed.

values of the spin projection, g_s (cf. below). I quote two useful special cases: For a set of particles of mass m moving freely in a volume Ω we have

$$\rho(\epsilon) = \Omega \frac{g_s}{4\pi^2} \frac{(2m)^{3/2}}{\hbar^3} \epsilon^{1/2}$$

and hence

$$\mu(T = 0)(\equiv \epsilon_{\mathrm{F}}) \equiv k_{\mathrm{B}}T_{\mathrm{F}} = (\hbar^2/2m)(6\pi^2 n/g_s)^{2/3} \; (n \equiv N/\Omega) \tag{1.2.14}$$

while in the case of fermions confined in a three-dimensional isotropic harmonic trap with frequency ω_0 we have $\rho(\epsilon) = g_s\epsilon^2/2(\hbar\omega_0)^3$ and hence

$$\mu(T = 0) = (6N/g_s)^{1/3}\hbar\omega_0 \tag{1.2.15}$$

At nonzero temperature an analytic expression for $\mu(T)$ is usually not available. However, provided $\rho(\epsilon)$ is not pathologically varying in the neighborhood of ϵ_{F}, it is easy to see that the first correction to μ is of relative order $(\rho'(\epsilon_{\mathrm{F}})(k_{\mathrm{B}}T)^2/(\rho\epsilon_{\mathrm{F}}))$, and for any power-law dependence $\rho(\epsilon) \sim \epsilon^n$ with $n > 0$ is negative and of relative order $(k_{\mathrm{B}}T/\epsilon_{\mathrm{F}})^2$. Extrapolating this behavior, we find that the condition for the effect of the quantum (Fermi) statistics, i.e. of indistinguishability, to be visible (which is very crudely $\mu > 0$) is, to an order of magnitude, $k_{\mathrm{B}}T \lesssim \epsilon_{\mathrm{F}}$. Putting in the value of ϵ_{F} from (1.2.15), we see that this is just the criterion we got by a "hand-waving" argument in Section 1.1. Similar considerations hold for the case of a harmonic trap, although in that case the density must itself be determined self-consistently.

It should be emphasized that all the above formulae are based on the assumption that the only conservation law relevant to the thermodynamics is that of total particle number; in particular, in the case of more than one spin (hyperfine) species it is assumed that interspecies conversion can take place on a time scale fast compared to that of the relevant experiments. This is overwhelmingly so for (e.g.) electrons in metals, but need not be so for the alkali gases, where interspecies conversion is a very slow process. In such a case it may be necessary to treat the different spin species as effectively separate physical systems (just as one would different chemical species) and to assign to each species α its own chemical potential μ_α, which is then fixed by Eqn. (1.2.15) with $N \to N_\alpha$ and $g_s \to 1$.

I now turn to the Bose case described by (1.2.9). As we saw, that equation makes sense only if the chemical potential μ is negative. Now, μ is fixed as a function of N and T by the combination of (1.2.9) and (1.2.10), that is by the implicit equation

$$\sum_i (\exp\beta(\epsilon_i - \mu) - 1)^{-1} = N \tag{1.2.16}$$

Imagine for the moment that we could treat the energy levels ϵ_i as forming a continuum and thus define a single-particle density of states $\rho(\epsilon)$ as we did in the Fermi case. Then (1.2.16) becomes

$$\int_0^\infty \frac{\rho(\epsilon)d\epsilon}{\exp\beta(\epsilon - \mu) - 1} = N \tag{1.2.17}$$

Now, the LHS of Eqn. (1.2.17) is a uniformly increasing function of μ, and we know that μ must be <0; consequently, the equation implies the condition

$$\int_0^\infty \frac{\rho(\epsilon)d\epsilon}{\exp(\beta\epsilon)-1} \geqslant N \tag{1.2.18}$$

At sufficiently high temperatures ($\beta \to 0$) this criterion is certainly fulfilled. However, whether or not it is fulfilled for all nonzero temperatures depends crucially on the form of $\rho(\epsilon)$ in the limit $\epsilon \to 0$, i.e. on the low-energy density of single-particle states. If $\rho(\epsilon)$ is constant or proportional to a negative power of ϵ (as is the case of a 1D system in free space) then the LHS of (1.2.18) is infrared divergent for any finite β and hence the condition is automatically fulfilled. If on the other hand $\rho(\epsilon)$ is proportional to a positive power of ϵ, then (1.2.18) cannot be fulfilled below a critical temperature T_c defined by the implicit equation ($\beta_c \equiv 1/k_B T_c$)

$$\int_0^\infty \frac{\rho(\epsilon)d\epsilon}{\exp(\beta_c\epsilon)-1} = N \tag{1.2.19}$$

As in the Fermi case, the actual value of T_c for a given system depends on the form of the potential. Two important cases are, again, three-dimensional free space and an isotropic 3D harmonic trap, for which $\rho(\epsilon)$ is given by the two forms cited above respectively. Inserting these forms into (1.2.19) we find

$$k_B T_c = \frac{2\pi}{[\zeta(\frac{3}{2})]^{2/3}}\frac{(n/g_s)\hbar^2}{m} \approx 3.31\frac{(n/g_s)^{2/3}\hbar^2}{m} \quad \text{3D free space} \tag{1.2.20}$$

$$= (N/g_s)^{1/3}\frac{\hbar\omega_0}{[\zeta(3)^{1/3}]} \approx 0.94(N/g_s)^{1/3}\hbar\omega_0 \quad \text{3D harmonic trap} \tag{1.2.21}$$

It is interesting that in both these cases T_c is just of the order of the "Fermi temperature" $T_F \equiv \epsilon_F/k_B$ of the corresponding Fermi system with the same mass m, spin multiplicity g_s and total particle number N; this is not surprising, since except for the "marginal" case $\rho(\epsilon) \sim \epsilon^n, n \to 0$, the criterion (1.2.18) says, to an order of magnitude, that the number of single-particle states with energies less than $k_B T_c$ is of the order of the total number N of particles which must be accommodated, and it is exactly this criterion which defines T_F in the Fermi case.

What happens for $T < T_c$? As shown by Einstein in his original paper, it is now necessary *not* to treat the single-particle energy spectrum as a continuum, but to single out for special treatment the lowest state 0, which we recall has energy zero according to our convention. (The remaining states may still be treated as a continuum without appreciable error.) Equation (1.2.17) is still formally valid, but the chemical potential μ tends to a very small ($\sim N^{-1}$) negative value, and as a result a macroscopic number of particles (i.e. a number of order N) occupy the lowest single-particle state, with the rest distributed over the excited states according to (1.2.17) with μ set equal to 0. This is the simplest manifestation of the phenomenon known as *Bose–Einstein condensation* (hereafter abbreviated as BEC); as we shall see, in a more general form this phenomenon is at the root of most of the effects which we shall discuss in this book. We will denote the total number of particles in the lowest single-particle state (the state labeled 0) by N_0 and refer to those particles as the "condensate"; then, taking into account that for $i \neq 0$ the energy levels may without appreciable error be replaced by a continuum and the chemical potential μ be taken equal to zero, we find that in thermodynamic equilibrium for $T < T_c$ the equation fixing the condensate number $N_0(T)$ is the condition that the mean total number of particles should equal N, which reads

$$N_0(T) + \int_0^\infty \frac{\rho(\epsilon)d\epsilon}{\exp\beta\epsilon - 1} = N \tag{1.2.22}$$

If the single-particle density of states $\rho(\epsilon)$ is proportional to ϵ^{n-1}, then the integral is proportional to T^n; thus, comparing with (1.2.19) we have the simple formula

$$N_0(T) = N(1 - (T/T_c)^n) \quad T < T_c \tag{1.2.23}$$

For a free 3D gas n is $\frac{3}{2}$, while for a 3D harmonic trap it is 3.

The reader might be legitimately concerned that the derivation just given of the fundamental phenomenon of BEC rests on the use of the grand canonical ensemble, i.e. on the relaxation of the condition of conservation of total particle number, and yet it is precisely this condition which is invoked "on average" in Eqn. (1.2.10). Is this consistent? That is, would a calculation using the macrocanonical ensemble give essentially the same results? Provided N is sufficiently large, the answer is yes: see e.g. Politzer 1996. There are in fact some features of the free Bose gas as treated by the grand canonical distribution which are "pathological," for example the fluctuations in N are of order N rather than of order $N^{1/2}$ as for a standard extensive system, and some of these features can be eliminated by the use of the macrocanonical distribution; however, it turns out that they are already eliminated by the introduction of interactions, and it is not necessary to concern ourselves with them here.

As in the Fermi case, the above derivation can be generalized to the case where there are several species α which cannot interconvert on the relevant experimental time scales; in such a case each species is described by its own chemical potential μ, and undergoes BEC at a different temperature. However, in the Bose case this is not necessarily the end of the story; as we shall see in the next chapter, even for a free gas, if a certain kind of quite physical conservation law is imposed, the "simple" kind of BEC described above cannot occur at all! It is remarkable that it took more than 75 years from Einstein's original paper for this to be clearly appreciated.

1.3 Cooper pairing

As we have seen in Section 2, a gas of *noninteracting* fermions is described in thermal equilibrium by the Fermi–Dirac distribution (1.2.8), which allows a maximum of one particle per state; thus there is no question of BEC or anything similar occurring. Consider, however, a very dilute system of fermions (say for definiteness for fermionic atoms, e.g. D) with an attractive interaction sufficient to bind two atoms into a molecule; for the moment let us assume for simplicity that there is only one molecular bound state and that it is s-wave, and (perhaps less physically) that its radius is large compared to the atomic size. By the arguments of Section 1.2, such a molecular complex should behave as a boson, and if the gas is sufficiently dilute we should be able to neglect intermolecular interactions to a first approximation; thus, applying the results of the previous section, we expect BEC *of the molecules* to set in below a critical temperature given by Eqn. (1.2.20), where n is now the density of the molecules and m their mass; note that this temperature is, up to a factor of order unity, just the Fermi temperature which would have characterized the atoms in the absence of interactions. Note however that in the formation of the individual molecules themselves the "Fermi statistics" played no role.[12]

Now imagine that we gradually increase the density, while keeping the attractive interaction for the moment constant. When the molecules start to overlap (i.e. when the intermolecular distance becomes comparable to the molecular radius) we can of course no longer ignore the molecule–molecule interactions. However, equally importantly, we can no longer ignore the effects of the underlying Fermi statistics. A simple argument regarding this point goes as follows: If a molecule has a radius r_0 (defined e.g. by the root-mean-square value of the interatomic distance in the bound state), then by the uncertainty principle the kinetic energy of the single-particle states involved in forming it is of order $\hbar^2/mr_0^2 \equiv E_0$. If in work in volume Ω, then the total number of single-particle states available in this energy range is of order $\Omega(2mE_0/\hbar^2)^{3/2} \sim \Omega r_0^{-3}$. Thus for N molecules the average occupation f of a given single-particle state is $\sim nr_0^3$, where $n \equiv N/\Omega$ is the density. When f is small compared to unity (the dilute limit) the veto imposed by Fermi statistics on double occupation is essentially irrelevant; however, when $f \sim nr_0^3$ becomes comparable to one (i.e. the "molecules" begin to overlap appreciably) then it is essential to take it into account. Indeed, under those conditions the concept of an identifiable "molecule" loses its meaning; it is impossible to say which atom is paired with which.

In general we do not have much prior knowledge of how a given system of Fermi atoms will behave under these "dense" conditions; we cannot for example necessarily exclude a priori the possibility that it will form a solid. However, it is certainly not absurd to imagine that a feature which is qualitatively similar to the formation of diatomic molecules and their BEC which occurs in the dilute limit might persist as the density is increased. In Chapter 2 we will define this "feature" more quantitatively and explore some of its consequences. It turns out, perhaps contrary to intuition, that the problem stated, which is difficult in the intermediate regime, actually becomes easier to analyze again in the extremely dense limit when the radius of the (putative) "molecules" is much larger than the interparticle spacing. This is the so-called BCS limit, and was analyzed by Bardeen et al. in their classic 1957 paper; in this limit the "molecules" are called Cooper pairs. We will discuss the BCS theory in detail in Chapter 5. Whether one thinks of the process of Cooper pairing as a kind of BEC or as something completely different is perhaps a matter of taste; however, it is important to appreciate that it differs qualitatively from the BEC of dilute di-fermionic molecules in at least two respects. In the first place, it turns out (again perhaps somewhat contrary to intuition) that the Fermi degeneracy which occurs in the high-density limit actually *helps* the process of pair formation, so that even a two-particle attraction which is too weak to bind a molecule in free space can nevertheless induce Cooper pairing. Secondly, while in the dilute-gas limit the process of formation of the diatomic molecules can be thought of as quite different from that of their BEC (so that the typical temperature at which dissociation of the molecules into two free atoms occurs is orders of magnitude

[12]Except in so far as they restrict the spin state (e.g. for two spin-$\frac{1}{2}$ particles only the singlet state is allowed).

larger than T_c for BEC), in the BCS limit the processes of formation of the pairs and of their "condensation" are essentially identical – it turns out to be thermodynamically unfavorable, in this limit, to form pairs which are not condensed.

Despite these differences, it is important to appreciate that there is no clear qualitative distinction between the processes of BEC in the dilute-gas limit and of Cooper pairing in the ultradense one, and indeed it is possible to construct an ansatz for the many-body ground state wave function which interpolates smoothly between these limits: see Chapter 8, Section 8.4. Indeed, in that section we shall see that it is now becoming possible to probe this "BEC-BCS crossover" experimentally.

1.4 The experimental systems

In this section I shall give a brief overview of the various different physical systems I shall be discussing in this book, with emphasis on the differences in their physical properties and in the experimental techniques necessary to manipulate and diagnose them, rather than on the commonality of behavior, associated with the occurrence of (pseudo-) BEC, which will be the focus of subsequent chapters.

1.4.1 The helium liquids

The liquid phases of the two stable isotopes of helium, ^{4}He and ^{3}He, are in many ways the simplest quantum liquids. Under normal low-temperature conditions and in the absence of irradiation, we may assume that the helium atom is always in its electronic ground state, $(1s)^2\ {}^1S_0$; thus the ^{4}He atom, with zero nuclear spin, has no internal degrees of freedom, and the ^{3}He atom has a single binary degree of freedom corresponding to the nuclear spin of $\frac{1}{2}$. Two helium atoms, of whatever species, interact via a potential which is crudely of the Lennard-Jones type, with a minimum of ~11 K at a nuclear separation of ~2.3 Å. Two ^{4}He atoms in free space can form a diatomic molecule (dimer), with however a very tiny binding energy (~1 mK); this would correspond to an enormous s-wave scattering length (see Chapter 4), but this is unlikely to be relevant at liquid-state densities. Because of their lighter mass, two ^{3}He atoms should not be bound (nor should a ^{3}He–^{4}He pair). It is, of course, the combination of weak attraction and small mass which prevents either isotope of helium forming a solid under its own vapor pressure; in fact, the solid phase of ^{4}He is stable only above ~26 atm and that of ^{3}He only above 34 atm.

The number density of liquid ^{4}He varies from ~0.02 Å^{-3} at svp to ~0.023 Å^{-3} at melting pressure, and that of liquid ^{3}He similarly from ~0.014 Å^{-3} to ~0.02 Å^{-3}; thus in either case the typical interatomic separation is of the same order as the position of the potential minimum.

In addition to the Lennard-Jones potential, which is of course rotationally invariant, two ^{3}He atoms interact via their nuclear magnetic dipole moments. This interaction is of the standard form

$$\hat{H}_{\text{dip}} = \frac{\mu_0 \mu_n^2}{4\pi r^3}(\boldsymbol{\sigma}_1 \cdot \boldsymbol{\sigma}_2 - 3\boldsymbol{\sigma}_1 \cdot \hat{\boldsymbol{r}}\boldsymbol{\sigma}_2 \cdot \hat{\boldsymbol{r}}) \tag{1.4.1}$$

where $\boldsymbol{r} \equiv r\hat{\boldsymbol{r}}$ is the nuclear separation, $\boldsymbol{\sigma}$ the Pauli-matrix components and μ_n the nuclear moment, which for ^{3}He is ~(−)2.13 nuclear magnetons (~$10^{-3}\ \mu_{\text{B}}$). Since the Lennard-Jones potential prevents the nuclear separation from ever being much less than 2 Å, the maximum magnitude of the expression (1.4.1) is about of order 10^{-7} K; and since no temperature even close to this has yet been attained in liquid ^{3}He one might think that the nuclear dipole interaction would have negligible consequences. On the contrary, we shall see in Chapter 6 that in the superfluid phases of liquid ^{3}He it has an unexpectedly large, indeed dominant effect.

Although it has proved possible in recent years to produce droplets (up to $\sim 10^7$ atoms) of both ^{4}He and ^{3}He in vacuum which appear to be liquid, and also droplets of liquid ^{3}He in solid ^{4}He of sizes up to ~ 2 μ, the overwhelming majority of experiments have of course been done on bulk samples enclosed in some kind of macroscopic solid container. The effects of such a container may be crudely represented by a potential $V(\boldsymbol{r})$ which is zero except within a few Å of the container walls, where it is first attractive (because of van der Waals effects) and then, as we approach the wall itself, strongly repulsive. As a result of the van der Waals attraction, the density of the helium increases as we approach the wall, and in fact the first couple of layers are generally believed to be solid-like; in addition, if the liquid is at saturated vapor pressure and has a free surface, a liquid film will coat the walls up to a considerable height above the surface, and as we will see in Chapter 3 this produces spectacular effects in the superfluid phase of ^{4}He. Apart from this the nature of the container appears to have very little effect on the behavior of either helium isotope; a reason for this, apart from the very short range of the potential, is that while liquid helium is very compressible compared to a typical solid (the velocity of sound is only ~250–350 m/sec), the interactions are still strong enough to resist any substantial change in the density profile from a constant. In particular, the direct effect of the onset of BEC in ^{4}He on the density profile is to this day unobserved.[13]

I now turn to the diagnostic techniques used to study liquid helium. First, one can do a variety of thermodynamic measurements (specific heat, sound velocity, thermal expansion,...) and transport experiments (viscosity, thermal conductivity, sound attenuation, ...). The static microscopic density correlations can be studied by elastic x-ray scattering, and the dynamic correlations by inelastic x-rays; however, in the case of ^{4}He an even more useful experimental probe is thermal neutron scattering, since thermal neutrons are well matched to both the energy and the momentum scales characteristic of liquid ^{4}He. For ^{3}He this technique is much less useful, because the ^{3}He nucleus has a very large cross-section for the absorption of thermal neutrons in the charge-exchange reaction $n + {}^3\text{He} \to {}^3\text{H} + {}^1\text{H} + 746$ keV; however, this feature has itself been exploited to create intense localized heating in the liquid (Baüerle et al. 1996, Ruutu et al. 1996). Finally, in the case of ^{3}He a very large amount of information, particularly in the superfluid phases, can be obtained from nuclear magnetic resonance (NMR) experiments. The above list is far from exhaustive; there is certainly no shortage of diagnostic techniques available.

1.4.2 Electrons in metals

In this book I shall not be very much concerned with the properties of electrons in the normal phase of metals in their own right, but more in the ways in which they affect the behavior in the superconducting state, if and when it occurs. In fact, for those

[13]Other than as a weak singularity in the thermal expansion coefficient.

characteristics of the superconducting state in which I shall be primarily interested, it seems that it is only some rather generic properties of the normal state which are relevant. With this in mind, it is convenient to divide the known superconductors into two classes: (a) "classic" superconductors, that is, crudely speaking, those whose normal-state behavior conforms at least approximately to the traditional textbook account (a class which includes essentially all superconductors known before 1975, and most though not all of those known before 1986), and (b) "exotic" superconductors, a class which includes the heavy-fermion superconductors, the organics, the fullerides, various assorted systems such as Sr_2RuO_4, and above all the large class of cuprate (high-temperature) superconductors. I will discuss the normal phases of classes (a) and (b) separately.

As regards the normal phase of the classic superconductors, by our definition this is at least approximately a "textbook" metal, and it is necessary to remind the reader of only a few basic points (a more extended discussion is given in Chapter 5, Section 5.1). First, it is important to remember that the normal metal in question need not be crystalline; there are many amorphous metals and alloys which become superconducting, some with transition temperatures which are quite high by classical standards (e.g. Nb-Zr). The crystalline members of this class generally have relatively simple and fairly isotropic structures. Secondly, in the crystalline case a textbook metal is (by definition!) described at least qualitatively by the Bloch–Sommerfeld picture, that is, the model of independent electrons obeying Fermi statistics and moving in the periodic potential of the atomic ions, and thus described by the familiar energy bands; similarly, in the amorphous case a picture of independent electrons moving in a static random potential is qualitatively valid. (A partial explanation of how this model can continue to work in the presence of the strong Coulomb interaction between the conduction electrons, and of dynamic effects associated with ionic motion, is given by the Landau–Silin approach, see Chapter 5, Section 5.1). Thirdly, we can define a "Fermi energy" (chemical potential at $T = 0$) which in essentially all cases of interest turns out to be of the order of 1–10 eV; thus, at all temperatures below melting the electrons are strongly degenerate and their low-energy behavior is determined entirely by the properties of the states in a narrow energy region (width $\sim k_B T$) close to the Fermi surface. Thus the experimentally measured quantities have the standard temperature-dependences: specific heat $\propto T$, resistivity independent of T for alloys and at sufficiently low temperature for (slightly impure) crystals, Hall coefficient independent of T, etc.

Turning now to class (b), we find that the members of this class seem to have rather few properties in the normal state in common (other than being crystalline). One feature which is shown by most though not all is that the crystal structure is complicated and often highly anisotropic; this is of course spectacularly so in the cuprates, where the characteristic CuO_2 planes (or pairs or triples thereof) are well separated from one another in the perpendicular direction. Generally speaking the evidence is that for these systems electron interactions in the normal phase cannot be accommodated by any simple modification of the Landau–Silin picture, and indeed the characteristic temperature-dependences predicted by that model are often not observed (e.g. in the cuprates the Hall coefficient is strongly T-dependent). A more extended discussion is given in Chapter 7, Section 7.8 and Chapter 8, Section 8.1.

1.4.3 The atomic alkali gases[14]

Over the last two decades a brilliant program of experimental research[15] has achieved the cooling of dilute gases of atoms, usually though not exclusively of an alkali element, into a regime of density and temperature where the onset of BEC is expected. In the present context we need not be concerned with the details of the preparation and cooling techniques (on which see e.g. Inguscio et al. 1999), and rather focus on the final state achieved. Further details are given in Chapter 4, Section 4.1.

Typically, the system consists of anywhere between $\sim 10^3$ and $\sim 10^9$ alkali atoms, all of the same isotopic species (e.g. ^{87}Rb, ^{40}K, ...) confined by either laser or magnetic trapping.

In the case of laser trapping two experimental arrangements are of particular interest: In the first case one uses a single red-detuned laser and focuses the beam so that the intensity is maximum in a particular region of space, which then acts as a trap for the atoms; near the bottom of the trap the potential is that of an anisotropic harmonic oscillator. In the second arrangement, one uses three pairs of lasers (usually red-detuned) propagating in the directions $\pm x$, $\pm y$, $\pm z$ with each pair (e.g. the "$\pm x$" one) mutually phase-locked and with identical intensities and frequencies. In this set-up, the alternating constructive and deconstructive interference of the electric fields of each pair gives rise to a three-dimensionally periodic attractive potential for the atom.

$$V(\boldsymbol{r}) = \text{const.}\{I_x \cos 2k_z x + I_y \cos 2k_y y + I_z \cos 2k_z z\} \tag{1.4.2}$$

where I_i and k_j are respectively the intensity and wave vector of the pair propagating in the $\pm i$ direction, $i = x, y, z$. (The most common case is that the $I's$ and $k's$ are chosen identical.) This arrangement is known as an "optical lattice," and as we shall see in Chapter 4 has been recently used for analog simulation of certain solid-state systems. Of course, hybrids of the two arrangements described (1D and 2D optical lattices) are also possible, and one can also, by a slight relative detuning of the lasers of a pair, translate the lattice in time. A characteristic of laser trapping which has been fundamental to many recent experiments on the alkali gases is that the time scale over which the potential (1.4.2) can be applied is essentially limited only by the time required to turn on the laser, and can thus be made many orders of magnitude smaller than the "intrinsic" time scales of the system; in this respect the alkali gases are unique among the systems considered in this book.

The principle of magnetic trapping will be explained in Chapter 4, Section 4.1; for present purposes we need the result that only certain hyperfine species can be trapped this way, and that for those the effective potential is typically of the form

$$V(\boldsymbol{r}) = C_0 + \lambda |B(\boldsymbol{r})| \qquad (\lambda > 0) \tag{1.4.3}$$

where the constants C_0 and λ may depend on the hyperfine index of the atomic state in question. Since in a typical magnetic trap the field magnitude near its minimum

[14] I will treat H as an "alkali," although this is somewhat unconventional.

[15] Recognized by the award of the 1997 Nobel Prize in physics to S. Chu, C. Cohen-Tannoudji and W.D. Phillips.

(taken as the origin) has the form

$$|B(r)| = B_0 + \alpha z^2 + \beta(x^2 + y^2) \tag{1.4.4}$$

the potential is generally of an anisotropic harmonic-oscillator form.

Drawing together the results of the last two paragraphs, and setting aside the optical-lattice case, we see that provided the extent of the trapped atomic cloud is small on the scale of overall variation of the laser and/or magnetic field (a condition usually satisfied in practice) the generic form of the potential in which the atoms move is, up to an irrelevant constant, that of an anisotropic harmonic oscillator:

$$V(\boldsymbol{r}) = \tfrac{1}{2}m\omega_0^2(z^2 + K(x^2 + y^2)) \tag{1.4.5}$$

where the axial oscillation frequency ω_0 may be a function of the hyperfine index (but the anisotropy parameter K is not). Experimentalists commonly specify the trap in terms of the parameters ω_0 and K (or $\omega_r \equiv K^{1/2}\omega_{ax}$); note that K may, depending on the nature of the trap, be much larger, much smaller or of order unity. It is useful to note that the zero-point width $a_{Ho} \equiv (\hbar/m\omega_0)^{1/2}$ is typically of order 1μ; while for a completely noninteracting gas this would also be the width of the condensate cloud, we shall see in Chapter 4 that interactions normally broaden the latter to a considerably greater value. I quote at this point some further results, namely that a typical value of the maximum atomic density in the BEC phase is[16] $\sim 10^{11}$–10^{15} cm^{-3}, of the velocity an amazingly low value of $\sim$1 mm/sec and of the transition temperature to the BEC phase a few hundred nK.

It might at first sight seem surprising that a gas of alkali atoms under these conditions would remain stable for long enough to permit meaningful experiments on it. In the first place, the thermodynamic equilibrium state at nK temperatures of a collection of (say) 10^8 ^{87}Rb atoms is certainly not a gas – it is a crystalline solid! One might indeed think that the atoms of the gas phase would rapidly recombine to form molecules, and the molecules would then promptly aggregate to form a liquid and eventually a solid. Even were this not to happen, one might think that in a magnetic trap the trappable hyperfine states could gain energy by flipping their spins, after which they would be expelled from the trap in short order.

Fortunately, neither of these considerations poses a major problem in real life (at least at the densities, etc., explored so far). In the case of magnetic trapping, spin flip by a single atom with emission of a photon is astronomically improbable, because of the very low density of photon states for the transitions involved. Two-atom collisions, even those involving central forces, can lead in the general case to spin flip followed by expulsion of one or more of the atoms from the trap, but it is straightforward to show that such "dangerous" collisions cannot occur if one works with one of the so-called "maximally stretched" states (see Chapter 4, Section 4.1). The residual noncentral (electromagnetic dipole) forces are too weak to be a serious problem. As to recombination into molecules and eventually into a solid, the first stage, namely the process of recombination of two atoms in free space into a diatomic molecule, is forbidden by so many conservation laws that it is astronomically improbable. "Three-body" recombination (i.e. two atoms recombining into a molecule in the presence of a third atom

[16]For comparison, the density of air at STP is about 2.5×10^{19} molecules/cm^3.

which can carry away the surplus energy, etc.) is not completely negligible, but the rate per atom is proportional to the square of the density and at the maximum densities currently achieved leads (for most alkali gases – Cs is an exception) to lifetimes of the gas of the order of seconds or minutes. (An exception is the behavior of bosonic atoms very close to the "Feshbach resonances" to be discussed in Chapter 4 where the life time may be much shorter). Thus, nature has been kind to us, and most of the atomic alkali-gas systems are sufficiently stable that we can carry out a wide variety of meaningful experiments on them.

The above considerations refer as they stand to the heavier alkalis (Li, Na, K, Cs, Rb, ...). The case of hydrogen is somewhat special; while most of the statements made above apply to it, the temperature at which BEC is expected is for attainable densities of the order of 50 μK rather than hundreds of nK. In addition, because of the very much higher energy of the principal dipole transition ($\sim$10 eV, as opposed to 1–2 eV for the heavier alkalis) it has not so far proved possible to trap ^{1}H by laser methods, and magnetic traps commonly use fields $\sim$10 T, so that the electron spins are completely polarized; ^{1}H under these conditions has the peculiar property of being a "spin-$\frac{1}{2}$ Bose system," see Section 1.2.

Finally, I should mention that very recently BEC has been attained in a gas of ^{4}He atoms excited to the $(1s)(2s)^3 \sum_2$, state (conventionally denoted ^{4}He*). The lifetime of this excited state is $\sim$2 sec, so that the stability of this system is not much worse than that of the alkalis. This system is unique among gaseous systems so far cooled into the BEC state in having nuclear spin zero, so that in a strong magnetic field the "internal" state is essentially unique.

If the alkali-gas systems are in many ways very rich compared to liquid helium, there is one aspect in which they are much poorer, namely the diagnostic techniques available. Because of the very low density, not only conventional[17] NMR or ESR but also scattering experiments with massive particles such as neutrons would almost certainly be totally infeasible, and direct measurements of mass flow also seem impossible. Indeed, all the experimental information we have (and all that we are likely to get in the foreseeable future) comes directly or indirectly from optical measurements of either the total density or the density of a particular hyperfine species (which can be targeted by appropriate choice of the laser frequency and/or polarization). The two principal techniques are absorption imaging and phase-contrast (Zernike) imaging; for details of these techniques and their limitations, see e.g. Ketterle et al. (1999). It is fortunate that for a very dilute and compressible system such as the alkali gases (in strong contrast to the case of liquid helium) a study of the density alone gives almost all the information we want.

To conclude this section I give in Table 1.1 the various quantum liquids which will be discussed in this book.

1.5 Superconductivity and superfluidity: basic phenomenology

The two fundamental premises on which this book is based – premises which by now command almost universal assent in the community of condensed-matter

[17]While NMR or ESR can be used to excite the system, the power absorption is too weak to be measured directly and the excited atoms have to be detected optically.

Table 1.1 Terrestrial quantum liquids (in order of discovery).

System	Statistics	Density (cm^{-3})	T_c (K)	Comments
Electrons in classical metals	Fermi	$\sim 10^{23}$	1–25*	Superconducting state well described by BCS theory
Liquid ^{4}He	Bose	$\sim 10^{22}$	2.17	Only known dense Bose superfluid
Liquid ^{3}He	Fermi	$\sim 10^{22}$	2×10^{-3}	Pairing anisotropic
Cuprates and other exotics	Fermi	$\sim 10^{21}$	1–160	Pairing often believed to be anisotropic
Bose alkali gases	Bose	$\sim 10^{15}$	10^{-7}–10^{-5}	First really dilute BEC system
Fermi alkali gases	Fermi/Bose	$\sim 10^{12}$	10^{-6}	Laboratory for "BEC-BCS crossover"

*In 2022 T_c's close to room temperature have been reached in some metallic hydrides under several hundred gigabar pressure.

physicists – are (a) that "superconductivity" and "superfluidity" are essentially the same phenomenon, occurring respectively in a charged and in an electrically neutral system, and (b) that this phenomenon is the result, in a Bose system, of BEC or, in a Fermi system, of the analogous phenomenon, Cooper pairing ("pseudo-BEC"). Arguments in favor of premise (b) will be presented throughout the book, in particular in Chapter 2, Section 2.5; here I briefly sketch the reasons for believing (a).

It is necessary first to define exactly what is meant by "superfluidity" and "superconductivity." Actually, in practice the word is applied in each case to describe not a single phenomenon, but a *complex* of phenomena which normally occur in conjunction (but, as we shall see in Chapter 2, do not inevitably do so under all possible circumstances). The most fundamental ingredients in each case relate to the flow properties. Historically, the first experimental manifestations of both superconductivity and superfluidity were frictionless flow in a "simple" geometry – in the former case, flow of electric current through a mercury wire between "normal" (nonsuperconducting) leads accompanied by zero voltage drop across the wire; in the latter, mass flow of liquid ^{4}He between closely spaced plates under zero pressure drop. However, it turns out that those experiments are in some sense of a hybrid nature, and that the true nature of the phenomenon emerges more clearly if we consider a multiply connected (ring) geometry and distinguish two superficially similar but conceptually very different effects.

Consider then a system of either neutral particles (e.g. ^{4}He atoms) or charged ones (electrons) moving in a toroidal ("doughnut") geometry with thickness d much less than its mean radius R (see Fig. 1.1); in the following I shall neglect effects which are of relative order d/R or smaller. In the case of electrons it is convenient for a technical reason to require d to be even small compared to the London penetration depth λ_L (on which see below); this will guarantee that any current flow will be homogeneous across the cross-section of the torus. It is natural in the case of ^{4}He to think of the torus as a hollow container with solid walls, while in the case of electrons the physically realistic implementation would be a solid ring of (e.g.) crystalline material; however,

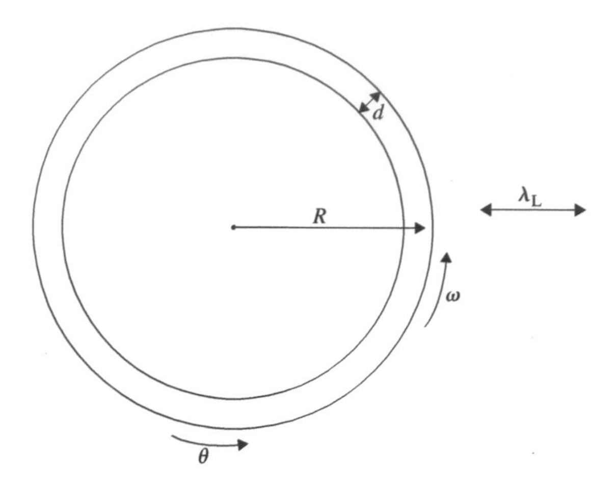

Fig. 1.1 The toroidal geometry used in the discussion of Section 1.5. The container may be stationary, or rotating with angular velocity ω.

all that really matters in the ensuing argument is that the "container," whatever its detailed nature, should when at rest provide some potential $V(\boldsymbol{r}) \equiv V(r, z, \theta)$ (see Fig. 1.1), and that when the container is rotated at angular frequency ω the potential transforms according to the prescription

$$V(r, z, \theta) \to V(r, z, \theta - \omega t) \tag{1.5.1}$$

It is furthermore essential to some of the conclusions to be drawn below that the interaction between the particles in question is a function only of the magnitude $|\boldsymbol{r}_i - \boldsymbol{r}_j|$ of their relative separation, and is thus invariant under rotation of the orbital coordinates alone. (This condition is not always exactly fulfilled in real life, e.g. by the spin–orbit interaction in metals, but it is usually fulfilled to an adequate approximation.) It is convenient to define a characteristic quantum unit of angular velocity, ω_c, by the prescription $\omega_c \equiv \hbar/mR^2$ where m is the mass of the particles in question; for a container of radius $R \sim 1$ cm, ω_c is of the order of 1 sec^{-1} for electrons but only $\sim 10^{-4}$ sec^{-1} for ^{4}He atoms. It is clear that if the system has angular momentum $L = I_{cl}\omega$ where I_{cl} is the classical moment of inertia NmR^2, then $L = N\hbar(\omega/\omega_c)$.

In this geometry we can now consider two superficially similar but conceptually very different kinds of experiment. In the first, we produce at $t = 0$ a state in which the system is rotating at some angular velocity $\omega \gg \omega_c$ while the container is stationary. In the charged case, this could be done by starting with the electrons in equilibrium with the static lattice and a finite magnetic flux through the loop, then reducing the latter to zero over a sufficiently short period. In the neutral case, we can start at a sufficiently high temperature (where everything behaves completely normally), rotate the container at angular velocity ω, wait for the system to come into equilibrium with it, then cool (while still rotating) into the temperature regime $T < T_c$ (say for simplicity to $T = 0$), then stop the container. It is an experimental fact that such fast-rotating ($\omega \gg \omega_c$) states can be produced for any system, charged or neutral, normal or superfluid. However, for a normal system (i.e. one which is not superconducting or superfluid) these states are very transient; the circulating current will die away to zero over a time scale which is set in the charged case by the electrical circuit parameters (resistance and self-inductance of the loop) and in the neutral case by the viscosity and sample dimensions. What is peculiar to the superfluid (superconducting) state is that provided the temperature is maintained at its initial value, the initial circulating current will persist indefinitely.[18] What is even more remarkable is that by heating and cooling the system in the superfluid (superconducting) phase we can change the angular momentum carried by this persistent current *reversibly*; in fact, provided that we do not vary the conditions too fast and never exceed the critical temperature T_c, the angular momentum is uniquely given by the formula

$$L = f_s(T) I_{cl} \tilde{\omega} \tag{1.5.2}$$

where $f_s(T)$ is for any given system a definite function of T which tends to 1 for $T \to 0$ and to 0 for $T \to T_c$ (the "superfluid fraction," see Chapter 3), and $\tilde{\omega}$ is the integral multiple of ω_c closest to the value of the angular velocity which the container possessed at the point when the system originally cooled through T_c. Note that in the above discussion, once the circulating-current state has been set up we do not need to distinguish between the charged and the neutral case; the analogy between "superconductivity" and "superfluidity" as regards this phenomenon is quite obvious.

It is intuitively clear that since the container is at rest, the circulating-current state cannot be the thermodynamic equilibrium state of the system. In fact, it is easy to demonstrate that no state with angular momentum larger than $N\hbar/2$ can be the equilibrium state; the proof proceeds by reductio ad absurdum, by multiplying the alleged ground state wave function[19] by the factor $\exp -i\sum_i \theta_i$ where θ_i is the angular coordinate of the ith particle. This operation leaves the potential energy unchanged and adds to the kinetic energy a term $-\omega_c L + \frac{1}{2} I_{cl}\omega_c^2$; thus for $L > N\hbar/2$ the total energy is decreased, contrary to our original assumption. Since (1.5.2) and the assumption that $\tilde{\omega} \gg \omega_c$ together imply that $L/N\hbar \gg \frac{1}{2}$ (except possibly very close to T_c when the proof needs to be generalized), it follows that the circulating-current state cannot be the ground state but must be an astronomically long-lived *metastable* (excited) state. In future I shall refer to this phenomenon, indifferently for the neutral and charged cases, as "metastability of supercurrents."

The second major ingredient in the complex of phenomena which we call superfluidity or superconductivity is, by contrast, a *thermodynamic equilibrium* phenomenon. With respect to this phenomenon, the analogy between the behavior in the charged and neutral cases is less trivially obvious then in the case of persistent-current experiments. To discuss it for the neutral case, we need to consider the thermodynamic equilibrium of the system when the *container* is rotating at some nonzero angular velocity ω (the case of immediate interest is when $\omega < \frac{1}{2}\omega_c$, but we may as well consider the general case). A standard result of statistical mechanics, derived in Appendix 1A,

[18]In the superconducting case a lower limit of $\sim 10^{15}$ years has been established for the relaxation time of the current in a particular superconducting loop.

[19]For simplicity I give the proof explicitly only for the case $T = 0$. However, the generalization to finite T is completely obvious since the operation specified does not change the entropy.

says that under these conditions[20] the equilibrium state is to be found by minimizing the effective free energy $F_{\text{eff}} \equiv \langle H_{\text{eff}} \rangle - TS$, where the time-independent operator H_{eff} is defined by the prescription

$$\hat{H}_{\text{eff}} = \hat{H} - \boldsymbol{\omega} \cdot \hat{\boldsymbol{L}} \equiv \hat{H} - \boldsymbol{\omega} \cdot \sum_i \hat{\boldsymbol{r}}_i \times \hat{\boldsymbol{p}}_i \tag{1.5.3}$$

Here $\hat{H}$ is the Hamiltonian of the system when the container is at rest, and $\hat{p}_i$ is the *laboratory-frame* momentum of the ith particle.

Now, it is a fact of everyday observation that in a "normal" system such as water or He-I (liquid ^{4}He above T_c) the effect of the second term in (1.5.3) is to induce an angular momentum which is (very nearly) equal to $I_{\text{cl}}\omega$ (where $I_{\text{cl}} \equiv NmR^2$ is as above the classical moment of inertia). However, it is characteristic of liquid ^{4}He in its superfluid phase (and more generally of neutral superfluids) that under the condition stated ($\omega < \frac{1}{2}\omega_c$) this is no longer so; in fact, the formula for the angular momentum is

$$\boldsymbol{L} = f_n(T) I_{\text{cl}} \boldsymbol{\omega} \tag{1.5.4}$$

where the quantity $f_n(T)$ tends to zero for $T \to 0$ and to 1 for $T \to T_c$. Indeed, $f_n(T)$ is simply related to the quantity $f_s(T)$ introduced in Eqn. (1.5.2).

$$f_n(T) = 1 - f_s(T) \tag{1.5.5}$$

$f_n(T)$ is called the "normal fraction" (see Chapter 3). The effect described by Eqn. (1.5.4) is sometimes called "nonclassical rotational inertia" or the "Hess–Fairbank" (HF) effect (Hess and Fairbank established its existence experimentally following its theoretical prediction by F. London); I stress again that it is a *thermodynamic equilibrium* effect.

It is possible to discuss the direct analog of the HF effect for a charged system, e.g. by considering the response of the electrons in a metallic ring when the latter is rotated (the so-called "London moment"). However, the real-life situation is complicated by the fact that in this case the potential exerted by the crystalline lattice is not necessarily small, and thus, as mentioned above, the thermodynamic equilibrium state of the system (electrons) is not time-independent in the laboratory frame. What is usually regarded as the analog of the HF effect in a charged system is the **Meissner effect**, or more precisely the fundamental effect which when combined with the general laws of electromagnetism gives rise to the latter. I now discuss this effect.

Consider our standard toroidal geometry (with $d \ll \lambda_L$), keep the "container" (the crystal lattice) at rest, and apply to the ring a small "Aharonov–Bohm" flux Φ (that is, a magnetic field configuration such that a finite flux threads the ring but the external field on the metal itself is zero). It is convenient to assume the condition $\Phi < \frac{1}{2} \cdot (h/2|e|)$ (which corresponds to the condition $\omega < \frac{1}{2}\omega_c$ in the neutral case). Under these conditions the effect of the flux is completely described by the vector

[20] It is an important implicit assumption in the derivation (see Appendix 1A) that the interaction with the container is so weak that it need be taken into account only for the establishment of thermal equilibrium, and can otherwise be neglected. (In the more general case, it is clear that the state of the system cannot be time-independent in the laboratory frame.)

potential on the ring, $\boldsymbol{A}(\boldsymbol{r}) = (\Phi/2\pi R)\hat{\boldsymbol{\theta}}$ when $\hat{\boldsymbol{\theta}}$ is a unit vector tangential to the ring circumference, and the thermodynamic equilibrium state of the system is given by minimizing the quantity $F_{\text{eff}} \equiv \langle \hat{H}_{\text{eff}} \rangle - TS$, where the Hamiltonian $\hat{H}_{\text{eff}}$ is given by the standard "gauge-coupling" prescription $\hat{\boldsymbol{p}} \to \hat{\boldsymbol{p}} - e\boldsymbol{A}(r)$, i.e.[21]

$$\hat{H}_{\text{eff}} = \frac{1}{2m} \sum_i (\hat{\boldsymbol{p}}_i - e\boldsymbol{A}(\hat{\boldsymbol{r}}_i))^2 + \sum_i U(\boldsymbol{r}_i) + \frac{1}{2} \sum_{i,j} V(\boldsymbol{r}_i - \boldsymbol{r}_j) \tag{1.5.6}$$

It is essential to appreciate that in this formula $\hat{\boldsymbol{p}}_i$ is the **canonical** momentum, so that the electric current is given (apart from symmetrization of the operators) by the expression

$$\boldsymbol{j}_{\text{cl}}(\boldsymbol{r}) = e/m \sum_i (\hat{\boldsymbol{p}}_i - e\boldsymbol{A}(\boldsymbol{r}_i))\delta(\hat{\boldsymbol{r}} - \hat{\boldsymbol{r}}_i) \tag{1.5.7}$$

Now, it is again an experimentally observed fact that when applied to a "normal" ring an Aharonov–Bohm flux does not generate any circulating current. By contrast, if the ring is superconducting the flux induces a nonzero current density in a "diamagnetic" sense and given by the formula

$$\boldsymbol{j}_{\ell}(\boldsymbol{r}) = -\Lambda(T)\boldsymbol{A}(\boldsymbol{r}) \tag{1.5.8}$$

where the constant $\Lambda(T)$ is given by the expression

$$\Lambda(T) = (ne^2/m^*) f_{\text{s}}(T) \tag{1.5.9}$$

where m^* is a quantity (for the moment regarded as defined by (1.5.9)) which turns out to have the order of magnitude of the electron mass, n is the electron density and $f_{\text{s}}(T)$ is the same function as appears in Eqn. (1.5.2).

If now we relax the condition $d \ll \lambda_L$ and assume, as seems very natural, that the result (1.5.2) is insensitive to the origin of the vector potential $\boldsymbol{A}(\boldsymbol{r})$ (in particular, to whether or not the field which produces it satisfies the Aharonov–Bohm condition or not), then we can combine (1.5.8) with the generally valid Maxwell formula curl $\boldsymbol{H}(\boldsymbol{r}) = \boldsymbol{j}(\boldsymbol{r})$ to obtain the well-known result that the magnetic field falls off to zero within the body of a superconductor (whether multiply or simply connected) over a distance of the order of the "London penetration depth" $\lambda_L(T) \equiv [\Lambda(T)]^{-1/2} \equiv (m^*/Ne^2)^{-1/2}(f_{\text{s}}(T))^{-1/2}$, which is typically of the order of a few hundred Å (for details see Chapter 5, Section 5.6). Thus, if we consider a macroscopic (dimension $\gg \lambda_L$) sample of a metal which becomes superconducting at some temperature T_{c} placed in a weak magnetic field, we find that for $T > T_{\text{c}}$ the field penetrates the sample essentially completely, but that the moment T fall below T_{c} the field appears in effect to be immediately[22] expelled from the interior – the sample becomes a perfect diamagnet. It is this effect which is usually called the Meissner (or Meissner–Ochsenfeld) effect; however, it should be emphasized that the fundamental phenomenon is the diamagnetism described by Eqn. (1.5.8).

[21] In this case (unlike the neutral case) it does not actually matter whether $V(\boldsymbol{r}_i - \boldsymbol{r}_j)$ is rotationally invariant (or indeed whether it is a function of $\boldsymbol{r}_i$ and $\boldsymbol{r}_j$ separately).

[22] Very precise observation should establish that the expulsion is actually a continuous process, according to (1.5.9), but it is completed within a very small range of $(T - T_{\text{c}})/T_{\text{c}}$.

While it is clear that there is a strong qualitative similarity between the phenomenon described by (1.5.8) and the HF effect for neutral superfluids,the precise correspondence is somewhat obscured by the presence of a quadratic term (in $\boldsymbol{A}(r)$) in (1.5.6) which has no analog in (1.5.3). We can clarify the situation by viewing it in the neutral case not from the laboratory frame but from the frame of reference rotating with the container: as shown in Appendix 1A, the expressions both for the effective Hamiltonian, apart from a relatively uninteresting "centrifugal" term, and for the mass current then exactly coincide for the charged and neutral cases, with the correspondence

$$e\boldsymbol{A}(\boldsymbol{r}) \rightleftarrows m\boldsymbol{\omega} \times \boldsymbol{r} \tag{1.5.10}$$

Thus, the HF effect as viewed from the *rotating* frame corresponds exactly (apart from the centrifugal effect) to the effect described by Eqn. (1.5.8) as viewed from the *laboratory* frame. I will return in Chapter 5 to the details of the behavior of a superconducting ring in an applied Aharonov–Bohm flux, and in footnote 22 of that chapter will briefly address the question of whether there is a direct analog, in neutral systems, of the Meissner effect as such.

Appendix

1A Statistical mechanics in a rotating container

In order to apply the standard prescriptions of equilibrium statistical mechanics, we need to work in a frame of reference in which the Hamiltonian is time-independent. In the case of a uniformly rotating container, the only such frame is that which rotates with the container (hereafter called simply the "rotating frame"). The question then arises, what is the appropriate form of the Hamiltonian to use in this frame?[23]

For pedagogical simplicity I shall attack this problem in three stages. First, consider a single classical particle of mass m in a container which for convenience I take to be approximately but not exactly cylindrically symmetric. If the container is stationary in the laboratory frame of reference,[24] then the (classical) Hamiltonian is

$$H = \boldsymbol{p}^2/2m + V(\boldsymbol{r}) \equiv H_0 \tag{1.A.1}$$

where $V(\boldsymbol{r})$ is the potential due to the container. Suppose now that the container rotates with angular velocity ω about some axis which we take as the z-axis of an appropriately chosen cylindrical polar coordinate system, and define $\boldsymbol{\omega} \equiv \omega\hat{\boldsymbol{z}}$. Then the form of the kinetic-energy term in (1.A.1) is unchanged, but the potential term becomes $V(\boldsymbol{r}'(t))$, where $\boldsymbol{r}'(t)$, the coordinate viewed from the rotating frame, is specified in terms of its cylindrical polar components by

$$r' = r, \quad z' = z, \quad \theta' = \theta - \omega t \tag{1.A.2}$$

[23]In the following, it is important to distinguish between the statement that the probability distribution is "stationary," that is, time-independent, as viewed from a given frame, and the statement that the system is at rest in that frame; the latter statement implies the former but not vice versa.

[24]Which for the purposes of this discussion I take to be an inertial frame, although strictly speaking, owing to the Earth's rotation, etc., it is not.

Note that the time derivatives of $\boldsymbol{r}$ and $\boldsymbol{r}'$ are related by

$$\dot{\boldsymbol{r}}(t) = \dot{\boldsymbol{r}}'(t) + \boldsymbol{\omega} \times \boldsymbol{r}' \tag{1.A.3}$$

How to find the form of Hamiltonian appropriate for use in the rotating frame? It is tempting but incorrect to simply take the lab-frame Hamiltonian [i.e. (1.A.1) with $V(\boldsymbol{r}) \to V(\boldsymbol{r}'(t))$] and express it in terms of the rotating-frame variables. Rather, we should follow the canonical prescription for doing Hamiltonian mechanics in an arbitrary coordinate system: see e.g. Goldstein (1980), Chapter 8. That is, we start by writing down the Lagrangian and expressing it in rotating-frame coordinates using (1.A.3):

$$\begin{aligned}\mathcal{L}(\boldsymbol{r}, \dot{\boldsymbol{r}}, t) &= \tfrac{1}{2} m \dot{\boldsymbol{r}}^2 - V(\boldsymbol{r}'[\boldsymbol{r}, t]) \\ &= \tfrac{1}{2} m (\dot{\boldsymbol{r}}' + \boldsymbol{\omega} \times \boldsymbol{r}')^2 - V(\boldsymbol{r}') \equiv \mathcal{L}(\boldsymbol{r}', \dot{\boldsymbol{r}}')\end{aligned} \tag{1.A.4}$$

Next, we obtain the canonical momentum in the rotating frame by the standard prescription:

$$\boldsymbol{p}' = \frac{\partial \mathcal{L}(\boldsymbol{r}', \dot{\boldsymbol{r}}')}{\partial \dot{\boldsymbol{r}}'} = m(\dot{\boldsymbol{r}}' + \boldsymbol{\omega} \times \boldsymbol{r}') \tag{1.A.5}$$

Note that $\boldsymbol{p}'$ is not the kinematic momentum $m\dot{\boldsymbol{r}}'$ as viewed from the rotating frame. The final step is to define the rotating-frame Hamiltonian $H'(\boldsymbol{r}', \boldsymbol{p}')$ in the standard way, by

$$H'(\boldsymbol{r}', \boldsymbol{p}') = \dot{\boldsymbol{r}}' \cdot \boldsymbol{p}' - \mathcal{L} \tag{1.A.6}$$

Expressing $\mathcal{L}$ in terms of $\boldsymbol{r}'$ and $\boldsymbol{p}'$, and rearranging the terms in the triple product, we find

$$H'(\boldsymbol{r}', \boldsymbol{p}') = \boldsymbol{p}'^2/2m - \boldsymbol{\omega} \cdot \boldsymbol{r}' \times \boldsymbol{p}' + V(\boldsymbol{r}') \tag{1.A.7}$$

If now the particle is in equilibrium with a thermal bath which is itself stationary in the rotating frame (e.g. the phonons in the container walls) then we can go ahead and apply all the usual rules of equilibrium statistical mechanics: e.g. the probability of finding the particle with coordinate $\boldsymbol{r}'$ and $\boldsymbol{p}'$ will be proportional to the Gibbs factor $\exp[-\beta H'(\boldsymbol{r}', \boldsymbol{p}')]$, and will be (trivially) stationary in the rotating frame and thus in general not stationary in the lab frame. However, we are free to choose a "special" time $t = 2n\pi/\omega$ at which the rotating and lab frames coincide, and express H' as a function of the lab-frame coordinate $\boldsymbol{r}$ and momentum $\boldsymbol{p}(\equiv m\dot{\boldsymbol{r}})$: using (1.A.5) and (1.A.3) we find that at such times

$$H'(\boldsymbol{r}', \boldsymbol{p}') = \frac{p^2}{2m} - \boldsymbol{\omega} \cdot \boldsymbol{r} \times \boldsymbol{p} + V(r) \equiv H_0 - \boldsymbol{\omega} \cdot \boldsymbol{L} \equiv H_{\text{eff}}(\boldsymbol{r}, \boldsymbol{p}) \tag{1.A.8}$$

Thus we know that at such special times the distribution is determined in the lab frame by H_{eff} (and in general, if $V(\boldsymbol{r})$ lacks cylindrical symmetry, will itself not be cylindrically symmetric). At general times, since we know that the distribution is stationary in the rotating frame, its anisotropy will simply rotate, as viewed from the lab frame, with angular velocity ω.

Actually it is often the case that we wish to take the limit of a cylindrically symmetric $V(\boldsymbol{r})$.[25] In that case it is clear that we can simply use the effective Hamiltonian (1.A.8) at all times, and the distribution is then stationary (i.e. time-independent) also as viewed from the lab frame (but may of course correspond to a finite angular velocity as viewed from that frame).

Before proceeding to the many-body case, let's make the analogy with electromagnetism explicit. To do so we go back to (1.A.7) and rewrite it in the equivalent form

$$\begin{aligned} H'(\boldsymbol{r}', \boldsymbol{p}') &= (\boldsymbol{p}' - m\boldsymbol{\omega} \times \boldsymbol{r}')^2/2m + \tilde{V}(\boldsymbol{r}') \\ \tilde{V}(\boldsymbol{r}') &\equiv V(\boldsymbol{r}') - \tfrac{1}{2}m(\boldsymbol{\omega} \times \boldsymbol{r}')^2 \end{aligned} \tag{1.A.9}$$

This is the Hamiltonian appropriate to the rotating frame. As viewed from this frame the mass current (kinematic momentum) is

$$\boldsymbol{j}' = m\dot{\boldsymbol{r}}' = \boldsymbol{p}' - m\boldsymbol{\omega} \times \boldsymbol{r}' \tag{1.A.10}$$

Let's now consider a charged system and write down the standard expressions, in the lab frame but in the presence of an electromagnetic vector potential $A(\boldsymbol{r})$, for the Hamiltonian and for the particle current:

$$H = (\boldsymbol{p} - e\boldsymbol{A}(\boldsymbol{r}))^2/2m + V(\boldsymbol{r}) \tag{1.A.11}$$

$$\boldsymbol{j} = m\dot{\boldsymbol{r}} = \boldsymbol{p} - e\boldsymbol{A}(\boldsymbol{r}) \tag{1.A.12}$$

From a comparison of (1.A.9) and (1.A.10) with (1.A.11) and (1.A.12) we see that the problem of the neutral system viewed from the rotating frame (with a potential which includes the "centrifugal" contribution $-\frac{1}{2}m(\boldsymbol{\omega} \times \boldsymbol{r}')^2$) is formally identical to the problem of the charged system viewed from the rest frame, with the correspondence $e\boldsymbol{A}(\boldsymbol{r}) \rightleftharpoons m(\boldsymbol{\omega} \times \boldsymbol{r})$. In the case of a constant magnetic field $\boldsymbol{B}$ the vector potential $\boldsymbol{A}(\boldsymbol{r})$ can be written, by a suitable choice of gauge, in the form $\frac{1}{2}(\boldsymbol{B} \times \boldsymbol{r})$, so we have the correspondence $\boldsymbol{\omega} \rightleftharpoons e\boldsymbol{B}/2m$.

It is now straightforward to generalize the above results to the many-body case of interest, *provided* the two-body interatomic-potential $U(\boldsymbol{r}_i - \boldsymbol{r}_j)$ is a function only of the distance $|\boldsymbol{r}_i - \boldsymbol{r}_j| \equiv |\boldsymbol{r}_i' - \boldsymbol{r}_j'|$ and hence invariant under the rotation. (If this condition is not fulfilled, as is likely to be the case (e.g.) in the case of appreciable spin–orbit interaction, then there exists in general no frame of reference in which the Hamiltonian is time-independent, so we cannot do equilibrium statistical mechanics and the problem becomes similar to that of Couette flow (see e.g. Chossat and Iooss 1994).) With this premise the argument is an obvious generalization of the above one for the single case: in the rotating frame the Hamiltonian is

$$H'\{\boldsymbol{r}_i', \boldsymbol{p}_i'\} = \sum_i (\boldsymbol{p}_i' - m(\boldsymbol{\omega} \times \boldsymbol{r}_i'))^2/2m + \sum_i \tilde{V}(\boldsymbol{r}_i) + \frac{1}{2}\sum_{ij} U(|\boldsymbol{r}_i' - \boldsymbol{r}_j'|) \tag{1.A.13}$$

(where $\tilde{V}$ contains the centrifugal term), and in the lab frame we have

$$H'\{\boldsymbol{r}_i' . \boldsymbol{p}_i'\} = H_0\{\boldsymbol{r}_i, \boldsymbol{p}_i\} - \boldsymbol{\omega} \cdot \boldsymbol{L}, \quad \boldsymbol{L} \equiv \sum_i \boldsymbol{r}_i \times \boldsymbol{p}_i \tag{1.A.14}$$

[25] For consistency we must take this limit only *after* taking the limit of infinite time, in order to establish thermal equilibrium with the rotating container.

Equation (1.A.13) leads, in the classical case, to an interesting general result. Consider the particle current as viewed from the rotating frame

$$\boldsymbol{J}_{\text{rot}}(\boldsymbol{r}') = \sum_i \delta(\boldsymbol{r}' - \boldsymbol{r}'_i)(\boldsymbol{p}'_i - \boldsymbol{\omega} \times \boldsymbol{r}'_i)$$

The expectation (average) value of this quantity is

$$\langle \boldsymbol{J}(\boldsymbol{r}')\rangle_{\text{rot}} = \int d\{r'_i\}d\{p'_i\}\delta(\boldsymbol{r}' - \boldsymbol{r}'_i)(p'_i - \boldsymbol{\omega} \times \boldsymbol{r}'_i) \exp -\beta H'\{\boldsymbol{r}'_i, \boldsymbol{p}'_i\} \qquad (1.\text{A}.15)$$

Now in classical statistical mechanics $\boldsymbol{p}'_i$ and $\boldsymbol{r}'_i$ are independent variables, and we can perform the change of variables $\boldsymbol{r}'_i \to \boldsymbol{r}_i \equiv \boldsymbol{r}'_i, \boldsymbol{p}'_i \to \boldsymbol{p}_i \equiv p'_i - m(\boldsymbol{\omega} \times \boldsymbol{r}'_i)$. Then since $\boldsymbol{p}_i$ occurs only quadratically in the exponent in the integral of (1.A.15) and only linearly in the prefactor, it immediately follows that $\langle \boldsymbol{J}(\boldsymbol{r}')\rangle_{\text{rot}} = 0$. In words: in equilibrium classical statistical mechanics, *the system must be at rest as viewed from the frame of the container*, whether the latter is at rest or rotating. This is the exact analog of the famous Bohr–Van Leeuwen theorem on the absence, in classical statistical mechanics, of diamagnetism (no equilibrium current in the rest frame even when $\boldsymbol{A}(\boldsymbol{r}) \neq 0$).

We finally go over to the quantum-mechanical case. Both (1.A.13) and (1.A.14) remain true, with all quantities now treated as operators. However, the above theorem now in general fails (as does the Bohr–Van Leeuwen theorem) because the operators $\hat{p}_i - m\boldsymbol{\omega} \times \hat{\boldsymbol{r}}_i$ do not obey the same commutation relations as do the operators $\hat{p}_i$ (the analogous consideration for the electromagnetic case is standard in the theory of the quantum Hall effect). Thus, even in the absence of observable centrifugal effects, in quantum mechanics the system "knows the difference" between a "lab" (inertial) and a rotating frame!

2

BEC: Its definition, origin, occurrence, and consequences

In Section 1.2 of Chapter 1 we saw, following Einstein's original 1925 argument, that in a gas of spinless noninteracting bosons there is a temperature T_c below which a macroscopic number of particles occupy the lowest-energy single-particle state. It seems probable that not all of Einstein's contemporaries were persuaded that this observation had any real physical relevance; they could well have argued that the introduction of even weak interactions would destroy the phenomenon. As we shall see, from a modern point of view such skepticism is not entirely without foundation; however, ironically, it turns out that under the circumstances most often met in real life the effect of interactions is actually to *reinforce* the tendency to BEC. In this chapter I will present some general considerations concerning BEC in an interacting system, starting with its definition.

2.1 Definition of BEC in an interacting system

The original argument of Einstein not only assumes the absence of interactions but is restricted to thermal equilibrium; we would like to be able to generalize the definition of BEC to an interacting system which is not necessarily in equilibrium. The obvious generalization is, in words, "a macroscopic number of particles occupies a single one-particle state"; let us try to make this notion more precise and quantitative, starting with the "spinless" case (the particles have no internal degrees of freedom, as e.g. in the case of liquid ^{4}He).

For such a system by far the most useful and generally applicable definition of BEC is, to my mind, one given nearly 50 years ago by Penrose and Onsager (1956). Consider a system consisting of a large number N of bosons characterized by coordinates $\boldsymbol{r}_i (i = 1, 2, \ldots, N)$. It may have arbitrary interparticle interactions and in addition be subject to an arbitrary single-particle potential, in general time-dependent. Any pure many-body state s of the system at time t can be written in the form

$$\Psi_N^s(t) \equiv \Psi_s(\boldsymbol{r}_1, \boldsymbol{r}_2, \ldots, \boldsymbol{r}_N : t). \tag{2.1.1}$$

where the function Ψ_s is symmetric under exchange of any pair i, j (definition of "boson"!). Similarly, the most general state of the system can be written as a mixture of different such normalized and mutually orthogonal pure states s with weights p_s.

Consider now the *single-particle density matrix* $\rho_1(\boldsymbol{r},\boldsymbol{r}')$ which is defined by the prescription

$$\rho_1(\boldsymbol{r},\boldsymbol{r}':t) \equiv N\sum_s p_s \int d\boldsymbol{r}_2 d\boldsymbol{r}_3 \ldots d\boldsymbol{r}_N \Psi_s^*(\boldsymbol{r},\boldsymbol{r}_2,\ldots,\boldsymbol{r}_N:t)\Psi_s(\boldsymbol{r}',\boldsymbol{r}_2,\ldots,\boldsymbol{r}_N:t)$$
$$\equiv \langle\hat{\psi}^\dagger(\boldsymbol{r}t)\hat{\psi}(\boldsymbol{r}'t)\rangle \quad (\boldsymbol{r}\equiv\boldsymbol{r}_1,\boldsymbol{r}'\equiv\boldsymbol{r}_1') \tag{2.1.2}$$

where in the last expression we have introduced the Bose field operator $\hat{\psi}(\boldsymbol{r}t)$ defined in Appendix 2A. Note that in the explicit many-body expression for ρ_1 there is no integration over $\boldsymbol{r}_1$ (which we picked out arbitrarily because of the symmetry of the wave function; $\boldsymbol{r}_j$ with any $j \leq N$ would do just as well); thus, the physical significance of $\rho_1(\boldsymbol{r},\boldsymbol{r}':t)$ is, crudely speaking, the probability amplitude, at time t, to find a particular particle at $\boldsymbol{r}$ times the amplitude to find it at $\boldsymbol{r}'$, averaged over the behavior of all the other $N-1$ particles. ρ_1 may be regarded as a matrix with respect to its spatial index $\boldsymbol{r}$.

Now, it follows directly from the definition (2.1.2) that the matrix $\rho_1(\boldsymbol{r},\boldsymbol{r}':t)$ is Hermitian:

$$\rho_1(\boldsymbol{r},\boldsymbol{r}':t) \equiv \rho_1^*(\boldsymbol{r}',\boldsymbol{r}:t) \tag{2.1.3}$$

It then follows by a well-known theorem that ρ_1 can be diagonalized, i.e. written in the form

$$\rho_1(\boldsymbol{r},\boldsymbol{r}':t) = \sum_i n_i(t)\chi_i^*(\boldsymbol{r}t)\chi_i(\boldsymbol{r}'t) \tag{2.1.4}$$

where the functions $\chi_i(\boldsymbol{r}t)$ form, for any given t, a complete orthogonal set. Note that in general both the eigenfunctions χ_i of ρ_1 and its eigenvalues n_i are functions of time, and that the χ_i certainly need not be eigenfunctions of the single-particle Hamiltonian or indeed of any simple operator, other than $\hat{\rho}_1$ itself.

We now state the following definitions:

1. If all the eigenvalues $n_i(t)$ of ρ_1 at time t are of order unity (none of order N) then we say the system is "normal" at time t (i.e. not Bose-condensed).
2. If there is exactly one eigenvalue of order N, with the rest all of order unity, then we say the system exhibits "simple BEC".
3. If there are two or more eigenvalues of order N, with the rest of order unity, then we say it exhibits "fragmented BEC".

It is clear that with these definitions the spinless noninteracting Bose gas discussed in Chapter 1, Section 1.3, is characterized below T_c by simple BEC.

Strictly speaking, of course, phrases like "of order N" and "of order unity" are not quantitatively defined. For certain kinds of systems we can make the definition rigorous by taking an appropriate thermodynamic limit; e.g. for a gas of atoms moving freely in space in volume Ω we can consider the limit $N \to \infty, \Omega \to \infty, N/\Omega = \text{const.}$, $T = \text{const.}$; then the statement that some n_i is "of order N" may be defined as equivalent to the statement that the limiting value of n_i/N under these conditions is a constant, while the statement that "n_i is of order unity" means that lim $(n_i/N) \to 0$. If for example we consider a 3D noninteracting gas, impose periodic boundary conditions and let $i = 0$ and $i = 1$ denote respectively the single-particle ground state ($\boldsymbol{k} = 0$ state) and the first excited single-particle state ($k \sim \Omega^{-1/3}$), then according to the results of Chapter 1, Section 1.3, below $T_c \lim_{N\to\infty}(N_0/N) \to (1-(T/T_c)^{3/2}) \sim$ const., while $\lim_{N\to\infty} n_1/N \sim N^{-1/3} \to 0$; thus simple BEC has indeed occurred according to our definition.

In various real-life situations, and in particular for an atomic alkali gas in a trap, there may be no physically meaningful way of taking the thermodynamic limit and hence no rigorous definition of the phrase "of order N". However, in such cases there is little difficulty in applying our definitions in a "rule-of-thumb" sense. Consider for example a set of 10^8 atoms in a trap, and suppose that one eigenvalue of ρ_1 is 5×10^7 and none of the others is greater than 10^3. Under these conditions few people would deny that (simple) BEC is occurring; arguments about the precise definition in hypothetical marginal cases are, at least in this kind of context, of little practical consequence. There is however one kind of situation when it pays to re-examine the definition, namely when BEC according to the literal definition is forbidden because of the reduced dimensionality of the system (see Chapter 1, Section 3, and below Section 3).

The definition of BEC given above for the spinless case may be straightforwardly extended to the case of particles possessing an internal (e.g. hyperfine) degree of freedom α. The pure-state many-body wave functions Ψ_N^s are then of the form

$$\Psi_N^s = \Psi_s(\boldsymbol{r}_1\alpha_1, \boldsymbol{r}_2\alpha_2, \ldots, \boldsymbol{r}_N\alpha_N) \tag{2.1.5}$$

and it is worth noticing explicitly that they are required to be symmetric only under the simultaneous exchange $\boldsymbol{r}_i \rightleftarrows \boldsymbol{r}_j, \alpha_i \rightleftarrows \alpha_j$, not necessarily under (e.g.) exchange of $\boldsymbol{r}_i$ and $\boldsymbol{r}_j$ alone. The single-particle density matrix is now defined by

$$\begin{aligned}\rho_1(\boldsymbol{r}\alpha, \boldsymbol{r}'\alpha' : t) &\equiv N \sum_s p_s \sum_{\alpha_2 \ldots \alpha_N} \int d\boldsymbol{r}_2 \ldots d\boldsymbol{r}_N \Psi_s^*(\boldsymbol{r}\alpha, \boldsymbol{r}_2\alpha_2, \ldots, \boldsymbol{r}_N\alpha_N) \\ &\quad \times \Psi_s(\boldsymbol{r}_1'\alpha_1', \boldsymbol{r}_2\alpha_2, \ldots, \boldsymbol{r}_N\alpha_N) \\ &\equiv \langle \hat{\psi}_\alpha^\dagger(\boldsymbol{r})\hat{\psi}_{\alpha'}(\boldsymbol{r}')\rangle \quad (\boldsymbol{r} \equiv \boldsymbol{r}_1, \alpha \equiv \alpha_1, \text{etc.})\end{aligned} \tag{2.1.6}$$

and is a matrix with respect to the pair of indices $\boldsymbol{r}, \alpha$. Again it is Hermitian and can be diagonalized:

$$\rho_1(\boldsymbol{r}\alpha, \boldsymbol{r}'\alpha' : t) = \sum_i n_i(t)\chi_i^*(\boldsymbol{r}\alpha : t)\chi_i(\boldsymbol{r}'\alpha' : t) \tag{2.1.7}$$

where the eigenfunctions $\chi_i(\boldsymbol{r}\alpha : t)$ are spinors;[1] note that they need not be simple products of a space and a spin (internal) function[2]. We can now define the concepts of a "normal" state and of simple and fragmented BEC just as in the spinless case. We shall see in Section 2.6 that the concept of "fragmentation," which is of limited interest in the spinless case, comes into its own in the more general one.

It is worth noting that it is an immediate consequences of our definitions that no mixture of many-body states which do not individually possess simple BEC can itself

[1] I follow the loose and somewhat inaccurate usage, which has become conventional in the BEC field, of calling any multicomponent wave function a "spinor".

[2] For example, in a typical magnetic trap the spin orientation of an atom depends on its position in space and such a factorization is therefore impossible.

possess simple BEC. However, the converse is not true: it is perfectly possible to have a mixture of states, each possessing simple BEC, which is fragmented or even normal (if the number of states in the mixture is itself of order N). Indeed, it turns out that by generalizing the notion of "mixture" somewhat it is possible to express an *arbitrary* state of the N-boson system in terms of fully (simply) Bose-condensed states: see Steel et al. 1999. We will not need this generalization in this book.

When we are dealing with simple BEC, it is convenient to denote the single macroscopic eigenvalue of the density matrix by the symbol $N_0(t)$ and call it the "condensate number" (so that the ratio N_0/N is the "condensate fraction"). In the general case it is important to distinguish the "condensate fraction" defined in this way from the "superfluid fraction" which will later be defined (cf. also Chapter 1, Section 1.6) in terms of the flow properties (though in a few special limits they are numerically almost the same). We also pick out the normalized eigenfunction corresponding to this single macroscopic eigenvalue, denote it $\chi_0(\boldsymbol{r}t)$ (or in the "spinful"[3] case $\chi_0(\boldsymbol{r}, \alpha : t)$) and call it the "wave function of the condensate" (or, when differently normalized, the "order parameter": cf. next section). Note that while $\chi_0(\boldsymbol{r}t)$ is not in general an eigenfunction of the single-particle part of the Hamiltonian, it behaves in all other respects just like a single Schrödinger one-particle wave function, and in particular the multiplication of χ_0 by a *uniform* (in space and time) phase factor $e^{i\varphi}$ has no physical significance (this property is sometimes called "$U(1)$ gauge symmetry", though it should be remembered that it is a symmetry of the Hamiltonian only in the rather trivial sense that the latter conserves the particle number; see next section).

It is clear also that the above definitions of N_0 and χ_0 fail when the system is "fragmented", and in that case we need to start again from scratch.

To conclude this section, I should note that while I personally believe that the Penrose–Onsager definition of (simple) BEC used above is by far the most conceptually satisfactory, one can find in the literature various alternative definitions; I will discuss a couple of these in the next section.

2.2 The order parameter and the superfluid velocity; alternative definitions of BEC

In the main body of this section I will consider exclusively the case of simple BEC; the fragmented case will be mentioned briefly at various points. It is convenient to define a single quantity which will give information simultaneously about the degree of BEC and the single-particle state in which it occurs. Let us work for simplicity with the spinless case. An appropriate quantity is the *order parameter*, conventionally denoted $\Psi(\boldsymbol{r}, t)$ and defined by the equation

$$\Psi(\boldsymbol{r}t) \equiv \sqrt{N_0(t)}\chi_0(\boldsymbol{r}, t) \tag{2.2.1}$$

Since by definition the single-particle state χ_0 is normalized to unity, the normalization of Ψ is

$$\int |\Psi(\boldsymbol{r}t)|^2 d\boldsymbol{r} = N_0(t) \tag{2.2.2}$$

[3]There appears to be no simple term in the (existing) English language to characterize particles which possess a nontrivial internal (e.g. hyperfine) degree of freedom, and with some reluctance I therefore use from now on the neologism "spinful" in this sense.

It is clear from the definition (2.2.1) that there is nothing in the least mysterious about the order parameter, and in particular that its existence does not require "spontaneous breaking of the $U(1)$ gauge symmetry" (cf. below); in fact, the *overall* phase of $\Psi(\boldsymbol{r}t)$, like that of $\chi_0(\boldsymbol{r}t)$, is physically meaningless.

The generalization of (2.2.1) to the spinful case is obvious: the quantity Ψ, like χ_0, is now a spinor, defined by

$$\Psi(\boldsymbol{r}, \alpha : t) \equiv \sqrt{N_0(t)}\chi_0(\boldsymbol{r}, \alpha : t) \tag{2.2.3}$$

with the normalization

$$\sum_\alpha \int |\Psi(\boldsymbol{r}, \alpha : t)|^2 d\boldsymbol{r} = N_0(t) \tag{2.2.4}$$

Note that even when all the condensed particles occupy the same spatial state, it is also necessary for the definition (2.2.3) to make sense that they occupy a single spin state (which can of course be a linear combination of the basis states of our chosen representation).

Returning to the spinless case, let us define the phase $\varphi(\boldsymbol{r}\,t)$ of the condensate wave function (or equivalently of the order parameter):

$$\chi_0(\boldsymbol{r}, t) \equiv |\chi_0(\boldsymbol{r}t)| \exp i\varphi(\boldsymbol{r}t) \quad (\text{or } \Psi(\boldsymbol{r}t) \equiv |\Psi(\boldsymbol{r}t)| \exp i\varphi(\boldsymbol{r}t)) \tag{2.2.5}$$

The "density of the condensate" $\rho_c(\boldsymbol{r}t)$ and the current carried *directly* by the condensate particles, $j_c(\boldsymbol{r}t)$, are then given by the expressions

$$\begin{aligned} \rho_c(\boldsymbol{r}t) &= N_0(t)|\chi_0(\boldsymbol{r}t)|^2 \equiv |\Psi(\boldsymbol{r}t)|^2 \\ \boldsymbol{j}_c(\boldsymbol{r}t) &= N_o(t)\left(-\frac{i\hbar}{2m}\chi_0^*(\boldsymbol{r}t)\boldsymbol{\nabla}\chi_0(\boldsymbol{r}t) + c.c.\right) \\ &\equiv N_0(t)|\chi_0(\boldsymbol{r}t)|^2\frac{\hbar}{m}\boldsymbol{\nabla}\varphi(\boldsymbol{r}t) \quad \left(\equiv |\Psi(\boldsymbol{r}t)|^2 \cdot \frac{\hbar}{m}\boldsymbol{\nabla}\varphi(\boldsymbol{r}t)\right) \end{aligned} \tag{2.2.6}$$

Let us define the ratio $\boldsymbol{j}_c(\boldsymbol{r}t)/\rho_c(\boldsymbol{r}t)$, which has the dimensions of velocity, as the *superfluid velocity*,[4] and denote it as $\boldsymbol{v}_s(\boldsymbol{r}t)$: we see that in its definition the magnitude of the order parameter drops out and we have

$$\boldsymbol{v}_s(\boldsymbol{r}, t) \equiv \frac{\hbar}{m}\boldsymbol{\nabla}\varphi(\boldsymbol{r}t) \tag{2.2.7}$$

Equation (2.2.7) will turn out to play a very important role in the theory of superfluidity. We see that it has two immediate consequences: In any region of space throughout which $\chi_0(\boldsymbol{r}t)$ is nonzero (so that $\boldsymbol{v}_s$ is everywhere defined) we have

$$\operatorname{curl} \boldsymbol{v}_s(\boldsymbol{r}, t) = 0 \tag{2.2.8}$$

while if we consider a closed path C on which χ_0 is everywhere finite but which envelops a region in which the latter vanishes (e.g. a path taken around the torus

[4]With hindsight it would probably be more logical to call this quantity the "condensate velocity" and to denote it $\boldsymbol{v}_c$. The conventional terminology reflects the history of the subject.

of Chapter 1, Section 1.6), then since $\varphi(\boldsymbol{r}t)$ is defined only modulo 2π we have the celebrated *Onsager–Feynman quantization condition*

$$\oint \boldsymbol{v}_s \cdot d\boldsymbol{\ell} = nh/m \tag{2.2.9}$$

The explicit occurrence of h in this formula reminds us that $\boldsymbol{v}_s$ is an essentially quantum-mechanical object.

It is interesting to contrast the superfluid velocity $\boldsymbol{v}_s(\boldsymbol{r}t)$ with two other "velocities" which we can define for a quantum-mechanical system. First, consider a single particle with Schrödinger wave function $\chi_1(\boldsymbol{r}t)$. We can of course define the phase $\varphi_1(\boldsymbol{r}t)$ of the latter, and then define

$$\boldsymbol{v}_1(\boldsymbol{r}t) \equiv \frac{\hbar}{m}\boldsymbol{\nabla}\varphi_1(\boldsymbol{r}t) \tag{2.2.10}$$

The quantity $\boldsymbol{v}_1(\boldsymbol{r}t)$ evidently satisfies the conditions (2.2.8) and (2.2.9). However, in the standard interpretations[5] of quantum mechanics it is not usually regarded as possessing any great physical significance. The reason is that in general an actual "measurement" of the particle velocity will not yield anything which has much to do with the quantity defined by (2.2.10). As an example, consider a particle in the ground state of a one-dimensional square well of length L, so that the time-independent Schrödinger wave function is of the familiar form

$$\Psi_1(x) = \sqrt{\frac{2}{L}}\sin\left(\frac{\pi x}{L}\right) \equiv \sqrt{\frac{2}{L}}\left(-\frac{i}{2}(\exp i\pi x/L - \exp -i\pi x/L)\right) \tag{2.2.11}$$

Clearly, for such a state $v_1(x)$ as defined by (2.2.10) is zero. On the other hand, by the standard quantum theory of measurement we expect that any attempt to "measure the velocity"(i.e. the momentum operator $i\hbar\partial/\partial x$ divided by m)[6] will give us one of the two eigenvalues, namely $\pm\pi\hbar/mL$, with probability 50% each. More generally the fluctuations in the measured value of the "velocity" of a single particle generally turn out to be quite comparable to the mean value (except in the semiclassical limit), so that the quantity defined by (2.2.10) does not have much meaning in any situation when the effects of quantum mechanics are important.

The second quantity with dimensions of velocity which we might want to introduce is the hydrodynamic velocity $\boldsymbol{v}_h(\boldsymbol{r}t) \equiv \boldsymbol{j}(\boldsymbol{r}t)/\rho(\boldsymbol{r}t)$, where $\rho(\boldsymbol{r}t)$ and $\boldsymbol{j}(\boldsymbol{r}t)$ are respectively the *total* density and current of all the particles in the system. Expressing those quantities in the terms of the complete set of eigenfunctions $\chi_i(\boldsymbol{r}t)$ and eigenvalues $n_i(t)$ of the single-particle density matrix ρ_1 (see Section 2.1), we find that the quantity $\boldsymbol{v}_h(\boldsymbol{r}t)$ is given by the expression (where $\varphi_i(\boldsymbol{r}t)$ is the phase of the eigenfunction $\chi_i(\boldsymbol{r}t)$)

$$\boldsymbol{v}_h(\boldsymbol{r}t) = \frac{\hbar}{m}\left\{\frac{\sum_i n_i(t)|\chi_i(\boldsymbol{r},t)|^2\overrightarrow{\nabla}\varphi_i(\boldsymbol{r}t)}{\sum_i n_i(t)|\chi_i(\boldsymbol{r}t)|^2}\right\} \tag{2.2.12}$$

[5] In the Bohm–de Broglie interpretation $\boldsymbol{v}_1(\boldsymbol{r}t)$ is regarded as the actual velocity of the particle in question.

[6] e.g. (in principle!) by bringing up an ultrasensitive magnetometer which will measure the electric current density $e\boldsymbol{v}(x)$.

While this expression formally contains $\hbar$, we see that the quantity $\boldsymbol{v}_h(\boldsymbol{r}t)$ in general satisfies neither (2.2.8) nor (2.2.9) (except of course in the special case of complete BEC, when it is identical to $\boldsymbol{v}_s(\boldsymbol{r}t)$). Thus it does not reflect the specifically quantum-mechanical nature of the system. On the other hand, the fluctuations in the measured value of $\boldsymbol{v}_h$ are of relative order $N^{-1/2}$ (see below).

We now see that the superfluid velocity $\boldsymbol{v}_s(\boldsymbol{r}t)$ defined by Eqn. (2.2.7) has in a sense the best of both worlds: On the one hand, it directly reflects the quantum-mechanical nature of the system (Eqn. 2.2.5). On the other hand, since it refers to the common state of a large number N_0 of particles, it is not subject to the large fluctuations characteristic of the single-particle velocity $\boldsymbol{v}_1(\boldsymbol{r}t)$, even in situations where quantum-mechanical effects dominate; for example, if we have BEC in the state (2.2.11) and we measure the *average* velocity of the particles,[7] then the result is peaked[8] around zero (the value of $\boldsymbol{v}_s(x)$ according to (2.2.7)) with a spread of relative order $N_0^{-1/2}$. Thus, the superfluid velocity may be regarded as an essentially classical quantity which nevertheless reflects the intrinsic effects of quantum mechanics.

The extension of the definition (2.2.7) to the spinful case is straightforward only in the special case when the condensate spinor $\chi(\boldsymbol{r}\alpha : t)$ can be factorized into a product of a space function $\chi_0(\boldsymbol{r}t)$ and a fixed (in space) spinor $\Psi_\alpha(t)$; in that case the definition (2.2.7) goes through as it stands with $\chi_0(\boldsymbol{r}t)$ having the above meaning. In the general case, when the spin orientation of the condensate is position-dependent, we need to specify a definite choice of axes and find various "Berry-phase" type effects; in particular, there is no simple theorem corresponding to either (2.2.8) or (2.2.9). (Needless to say, those complications arise equally in the case of a single particle and do not have much to do with BEC as such.) See for example Ho and Shenoy (1996).

It is clear that the definition of "order parameter", given by Eqn. (2.2.1), and a fortiori the definition of $\boldsymbol{v}_s$, rests fundamentally on the assumption that one and *only* one eigenvalue of the single particle density matrix is of order N, and thus by construction excludes the possible occurrence of fragmentation. If and when such fragmentation occurs, it is necessary to consider from scratch what "one-particle" like quantities it will be convenient to introduce.

I now turn to the question of alternative definitions of the "order parameter" and of the phenomenon of BEC. For simplicity I will confine myself to the spinless case. A definition of $\Psi(\boldsymbol{r}t)$ (Yang, 1962) which is closely related but not identical to that given above goes as follows: Consider, just as above, the single-particle density matrix $\rho_1(\boldsymbol{r}, \boldsymbol{r}' : t)$, but instead of asking explicitly for its eigenvalues, focus on its behavior in the limit $|\boldsymbol{r} - \boldsymbol{r}'| \to \infty$. In that case we postulate that ρ_1 can be written in the form

$$\lim_{|\boldsymbol{r}-\boldsymbol{r}'|\to\infty} \rho_1(\boldsymbol{r}, \boldsymbol{r}' : t) = f^*(\boldsymbol{r}t)f(\boldsymbol{r}'t) + \tilde{\rho}_1(\boldsymbol{r}\boldsymbol{r}' : t) \tag{2.2.13}$$

where $\tilde{\rho}_1(\boldsymbol{r}\boldsymbol{r}'t)$ tends to zero for $|\boldsymbol{r} - \boldsymbol{r}'| \to \infty$ and where the function $f(\boldsymbol{r}t)$ may or may not be zero. If $f(\boldsymbol{r}t)$ turns out to be nonzero, then we identify it with the order parameter $\Psi(\boldsymbol{r}t)$ and say the system is Bose-condensed; otherwise the system is

[7] As in footnote 6.

[8] This result is most simply obtained from a transcription to this case of the familiar results on coin-tossing.

normal. A nonzero value of $f(\boldsymbol{r}t)$ is sometimes said to correspond to the presence in the system of "off-diagonal long-range order" (ODLRO).

For the case of simple BEC in a uniform system, where the thermodynamic limit can be taken, it is clear that the above definition of $\Psi(\boldsymbol{r}t)$ is, if not precisely identical to the one given in (2.2.1), at least very closely related to it. To see this, we simply use the definition (2.2.1) to write expression (2.1.4) in the (generally valid) form

$$\rho_1(\boldsymbol{r}\cdot\boldsymbol{r}' : t) = \Psi^*(\boldsymbol{r}t)\Psi(\boldsymbol{r}'t) + \sum_{i\neq 0} n_i(t)\chi_i^*(\boldsymbol{r}t)\chi_i(\boldsymbol{r}'t) \tag{2.2.14}$$

Let us now consider the behavior of this expression when we take the limit $|\boldsymbol{r}-\boldsymbol{r}'| \to \infty$. Unless the behavior of $n_i(t)$ is "pathological", then destructive interference of the different states i should lead to the vanishing of the second term in this limit[9] so that we get agreement with (2.2.13) with the identification of $f(\boldsymbol{r}t)$ with $\Psi(\boldsymbol{r}t)$. Thus, we may loosely think of the two definitions of the order parameter, and the two criteria for BEC, as effectively equivalent in this case. However, it is nontrivial in any given case to specify exactly what constitutes "pathological" behavior, and in addition the definition based on ODLRO obviously cannot be applied in cases such as that of a trapped atomic alkali gas when the limit $|\boldsymbol{r}-\boldsymbol{r}'| \to \infty$ cannot sensibly be taken. Finally, this definition clearly cannot cope with the case of fragmented BEC. For these reasons I personally prefer the approach of Section 2.1.

There is one more definition of the order parameter, and an associated criterion for the occurrence of BEC, which is very commonly used in the literature, particularly that on the alkali gases. Namely, one defines $\Psi(\boldsymbol{r}t)$ by the prescription

$$\Psi(\boldsymbol{r}t) \equiv \langle\hat{\psi}(\boldsymbol{r}t)\rangle \tag{2.2.15}$$

where $\hat{\psi}(\boldsymbol{r}t)$ is the Bose field operator defined in Appendix 2A, and says that the occurrence of BEC is characterized by a nonzero value of the right-hand side of (2.2.15). This is sometimes said to be analogous to the technique used in defining a "classical" electric (or magnetic) field by the prescription

$$\mathcal{E}_{\text{class}}(\boldsymbol{r}t) = \langle\hat{\mathcal{E}}(\boldsymbol{r}t)\rangle \tag{2.2.16}$$

where $\hat{\mathcal{E}}(\boldsymbol{r}t)$ is the quantum-mechanical electric field operator, which can be expressed in terms of the standard annihilation and creation operators. This analogy has been very fruitful in suggesting experiments on the BEC alkali gases which are analogs of some already well-known in laser physics; see Chapter 4.

However, it is necessary to note a crucial difference between the photons which constitute the electromagnetic field on the one hand, and on the other, massive atoms such as those of ^{4}He or ^{87}Rb: Photons can be singly created and destroyed, atoms cannot! (and it is, of course, precisely this difference which allows BEC to occur in thermal equilibrium for atoms but forbids it for photons – in the latter case, "BEC" requires human intervention, in the form of an antenna or a laser). Thus, while nothing forbids the state of the electromagnetic field to be a superposition of states corresponding to different values of the total photon number (and indeed the coherent

[9]The reader may find it instructive to do the calculation explicitly for the case of thermal equilibrium below T_c in a homogeneous noninteracting 3D gas.

states widely discussed in the quantum-optics literature are precisely of that nature), a strong superselection rule forbids the occurrence in nature of superpositions of states with different total number N of atoms of a given species. Since the Bose field operator automatically decreases N by one, this means that in any physically allowed state of the system the right-hand side of Eqn. (2.2.15) is identically zero.[10]

One possible way around the problem relies on the idea[11] of "spontaneously broken $U(1)$ gauge symmetry". An analogy frequently used to explain this idea is that of an ideal Heisenberg ferromagnet whose Hamiltonian has a strict $O(3)$ rotation symmetry. Above the Curie temperature T_c, the spins of this system are effectively pointing in random directions and the net magnetization in zero external field is zero (more strictly, it is of order $N^{1/2}$ in zero field and of order $N\mathcal{H}$ in a small field $\mathcal{H}$). Below T_c, the spin-dependent interaction constrains a finite fraction of the spins (at $T = 0$ all of them) to point parallel, but does not specify in which direction they should point. Hence in strictly zero field the quantum ground state is a superposition of all possible directions and the total average is zero. However, a very small ($o(N^{-1})$) field is now adequate to orient all the spins along it, so that at $T = 0 \lim_{\mathcal{H}\to 0} \lim_{N\to\infty} \boldsymbol{M} \neq 0$ (in distinction to the situation above T_c).[12] The $O(3)$ symmetry of the Hamiltonian is said to be "spontaneously broken" by the thermodynamic equilibrium state of the system below T_c. In just such a way, it is argued, the $U(1)$ gauge symmetry of the Hamiltonian of the Bose system, which ensures conservation of total particle number N, is spontaneously broken below the onset temperature of BEC, and any perturbation of the form (the analog of a uniform magnetic field)

$$\hat{H}' = -\left\{\lambda \int \hat{\psi}^\dagger(\boldsymbol{r})d\boldsymbol{r} + h.c.\right\} \tag{2.2.17}$$

will lead to a finite ($O(1)$) value of $\langle\hat{\psi}(\boldsymbol{r})\rangle$, provided only that λ, while small compared to $k_B T$, is large compared to $k_B T/N$. Thus, just as we never in practice worry about assigning a Heisenberg ferromagnet a definite nonzero value of $\boldsymbol{M}$ despite the rotational invariance of the Hamiltonian, so we may as well say in the BEC case that the operator $\psi(\boldsymbol{r}t)$ has below T_c a finite expectation value, which can then be identified with the order parameter $\Psi(\boldsymbol{r}t)$ as in (2.2.15).

The question of the validity, at a fundamental level, of the analogy between ferromagnetism and the phenomenon of BEC is a very delicate one. To see why, we first note a prima facie difference between the two cases: whereas a great many different physical systems in the universe can act as sources of a magnetic field, the "conservation of ^{87}Rb-ness" implies that the only possible source of the quantity λ in Eqn. (2.2.17) for a system of ^{87}Rb atoms is another set of ^{87}Rb atoms, and thus the *total* number of

[10]It is tempting to think we can get around this difficulty by allowing our system to be "open," that is, to exchange atoms with a reservoir. However, this is not so: as shown in Eqn. (1.3.5) for the special case of the noninteracting gas, the density matrix of the "system," although permitting different values of N, is still diagonal in the N-representation, so that the RHS of Eqn. (2.2.15) is still zero). This result holds quite generally, since provided the total number of atoms in the "universe" (system + reservoir) is conserved the trace over the reservoir implicit in the definition of the system density matrix automatically makes the off-diagonal elements in the N-representation zero.

[11]This idea has been extensively promoted by P.W. Anderson, see e.g. Anderson 1966.

[12]At finite temperature T the limit must be taken so that $\mathcal{H} \ll k_B T/\mu$ but $\gg k_B T/N\mu$.

^{87}Rb atoms (in the system plus the reservoir) must still be conserved. To this, however, it may be replied that actually the Universe as a whole must also satisfy $O(3)$ symmetry, so that our common prescription of treating the external magnetic field on the ferromagnet as a classical quantity is itself dubious at this level. Eventually, if we follow this line of argument to its logical conclusion, we arrive at some very fundamental (and, at least in my opinion, currently not well understood) questions about the validity of describing a universe which is essentially quantum-mechanical in nature[13] by classical ideas at all.[14]

However, even if we conclude that the definition of the order parameter (and hence implicitly of BEC) by Eqn. (2.2.15) is technically valid, I believe it is quite unnecessary and has on more than one occasion in the history of the subject generated various pseudoproblems. Moreover, it is clear that the definition of BEC used in (2.2.15) cannot handle the case of fragmentation. I would claim that, at least in the Bose case, the definition given by (2.2.1) is perfectly satisfactory for all (legitimate) purposes for which one needs the concept of "order parameter", and that it automatically avoids the generation of pseudoproblems. The reader should be warned that this opinion, while strongly held by the present author, is probably not at present the majority view in the relevant community.

2.3 Why should BEC occur in an interacting system? When does it (not)?

The original argument of Einstein which is reproduced in Chapter 1, Section 1.2, tells us only that BEC will occur, below a critical temperature T_c, in a noninteracting system in thermal equilibrium. However, all the real-life systems in which BEC is believed to occur are interacting to some degree or other, and many of the phenomena we shall study (e.g. metastability of superflow, observation of interference fringes), occur well away from (global) thermal equilibrium. What can we say about the occurrence and stability of BEC under these more general conditions?

It is obvious that there cannot be a general theorem to the effect that BEC will always occur in an interacting Bose system, even at $T = 0$, since the crystalline-solid phase of ^{4}He is a counter example (at least assuming that it is anything like a textbook crystal[15]). It is an interesting question, which to the best of my knowledge is currently unresolved, whether any *liquid* phase of a Bose system at $T = 0$ must inevitably be Bose-condensed. At any rate, it can be safely said that at the time of writing no rigorous proof exists of the occurrence of BEC in any realistic[16] interacting system of bosons, even at $T = 0$. Nevertheless, there are strong qualitative arguments which suggest that it is likely to occur, and moreover to be "locally" stable, when the interparticle interactions are repulsive and not too strong, and I now sketch these.

[13] Or at least is believed to be so by most physicists in A.D. 2005.

[14] A question somewhat related to the above is that of a possible "phase standard", see Leggett 1995, 2000: Dunningham and Burnett 2000.

[15] In a "textbook" crystal, each lattice point is occupied by a single atom, and all the eigenvalues of the single-particle density matrix are thus unity. Weak exchange processes should not affect this result qualitatively (though cf. Chapter 8, Section 8.3).

[16] A proof has been given for a lattice gas at exactly half filling at $T = 0$ (Kennedy et al. 1988).

The first argument, which has already been used implicitly in the standard derivation of BEC in a noninteracting system in Chapter 1, Section 1.3, is purely statistical in nature. Consider the problem of distributing N "objects" (particles) between s different "boxes" (states) labeled $i = 1, 2, \dots s$). Start with the trivial case $s = 2$; then if the objects are distinguishable (particles obeying "classical" statistics) there are 2^N states in all and the number of ways $W(m)$ of distributing the objects so that m ends up in one box and $N - m$ in the other is given by the standard binomial ("coin-tossing") formula

$$W(m) = \frac{N!}{m!(N-m)!} \tag{2.3.1}$$

This function is strongly peaked around $m = N/2$; the point I want to stress is that the "extreme" values (m close to zero or to N) are strongly weighted down (by a factor $\sim 2^{-N}$ relative to $m = N/2$). If on the other hand the objects are indistinguishable bosons, then each value of m corresponds only to a *single* state of the system, and $W(m)$ is simply N^{-1} (N states in all); the "extreme" states now have equal weight to the rest. The situation is similar for general values of s: for distinguishable objects the weight of the configuration $\{m_i\}$ in which there are m_i objects in box i is given by the multinomial formula

$$W\{m_i\} = \frac{N!}{\prod_{i=1}^{s}(m_i!)} \tag{2.3.2}$$

and is strongly peaked about configurations when each m_i is close to N/s, while for indistinguishable bosons $W\{m_i\}$ is equal to 1 independently of the configuration; again, the "extreme" states (e.g. all N objects in a single box) have equal weight to the rest. Thus, if there is some factor (e.g. lower energy ϵ_i) which favors one or a few boxes (states) over the rest, it will have a relatively small effect in the distinguishable case but may have a very large influence in the Bose case. It is intuitively clear that the difference between classical and Bose statistics will be important only for $s \lesssim N$; note that if in the real thermodynamic situation we take s to be of the order of the number of single-particle states with energies $< k_B T$, then this rule of thumb just reproduces, to an order of magnitude, the criterion for the onset of BEC obtained in Chapter 1, Section 1.3.

It is important to appreciate that statistical considerations, while they as it were level the entropic playing field between "extreme" states (i.e. crudely speaking, those with BEC) and the rest, do nothing in general to enhance the stability of these states. To illustrate this, let us go back to the case $s = 2$, take $T = 0$ (in which case entropic considerations are irrelevant) and apply an energy bias ϵ, to one single-particle state (say "L") relative to the other ("R"). If m denotes the number of particles in L, then evidently the system energy $E(m)$ is simply $m\epsilon$, so the ground state corresponds to $m = N$ for $\epsilon < 0$ and to $m = 0$ for $\epsilon > 0$. Imagine now that we tune ϵ as a function of time: $\epsilon(t) = At$, $A > 0$, and that there are weak kinetic processes which allow the transfer of individual particles between states. Under these conditions we expect that the system, which for $t < 0$ had $m = N$, will pass for small positive t through decreasing values of m (corresponding to "fragmented" states) and at large positive t will settle into its true ground state ($m = 0$). Thus in this case the (simple) BEC state has no thermodynamic stability with respect to fragmental states. As we shall see below, this is not the state of affairs observed in (e.g.) real superfluid ^{4}He, and to account for this we need to invoke *energetic* considerations, as follows.

Let us imagine for definiteness that we are dealing with a dilute system of spinless particles, and that the interparticle forces may be adequately modelled by a short-range pseudopotential of the form

$$U(\boldsymbol{r}) = U_0\delta(\boldsymbol{r}) \tag{2.3.3}$$

As we shall see in Chapter 4, the above assumptions are a fairly good approximation for an atomic alkali gas such as ^{87}Rb which occupies a single hyperfine state; however, we do not need to specialize to that case here. Consider an orthonormal set of single-particle states $\chi_i(\boldsymbol{r})$ (which may, but need not be eigenfunctions of the single-particle Hamiltonian), which are reasonably extended in space, and suppose for the moment that there are just two atoms present. Let us first neglect the interatomic interaction and suppose that the two-particle wave function has some given form $\Psi(\boldsymbol{r}_1, \boldsymbol{r}_2)$. Now let us evaluate the interaction energy in the Hartree–Fock approximation, i.e. by calculating the expectation value of the operator (2.3.3) in the state $\Psi(\boldsymbol{r}_1\boldsymbol{r}_2)$. We obtain the result

$$\begin{aligned}\langle E_{\text{int}}\rangle &= U_0 \int d\boldsymbol{r}_1 d\boldsymbol{r}_2 \delta(\boldsymbol{r}_1 - \boldsymbol{r}_2)|\Psi(\boldsymbol{r}_1, \boldsymbol{r}_2)|^2 \\ &\equiv U_0 \int d\boldsymbol{r}|\Psi(\boldsymbol{r}, \boldsymbol{r})|^2 \end{aligned}\tag{2.3.4}$$

(where $\Psi(\boldsymbol{r}, \boldsymbol{r}) \equiv \Psi(\boldsymbol{r}_1, \boldsymbol{r}_2)_{\boldsymbol{r}_1=\boldsymbol{r}_2=\boldsymbol{r}}$). Now consider two specific possibilities for $\Psi(\boldsymbol{r}_1, \boldsymbol{r}_2)$:

(a) The two particles are in the same state, i.e. $\Psi(\boldsymbol{r}_1, \boldsymbol{r}_2) = \chi(\boldsymbol{r}_1)\chi(\boldsymbol{r}_2)$. Then we evidently have

$$\langle E_{\text{int}}\rangle_{\text{same}} = U_0 \int d\boldsymbol{r}|\chi(\boldsymbol{r})|^4 \tag{2.3.5}$$

(b) The particles are in different (mutually orthogonal) states χ_1, χ_2. Then the correctly symmetrized and normalized two-particle wave function is

$$\Psi(\boldsymbol{r}_1, \boldsymbol{r}_2) = 2^{-1/2}\{\chi_1(\boldsymbol{r}_1)\chi_2(\boldsymbol{r}_2) + \chi_2(\boldsymbol{r}_1)\chi_1(\boldsymbol{r}_2)\} \tag{2.3.6}$$

and

$$|\Psi(\boldsymbol{r}, \boldsymbol{r})|^2 = 2|\chi_1(\boldsymbol{r})|^2 \cdot |\chi_2(\boldsymbol{r})|^2 \tag{2.3.7}$$

so that the probability to find the two particles at their relative origin is just twice what is would have been for distinguishable particles. Substituting (2.3.7) into (2.3.4), we find[17]

$$\langle E_{\text{int}}\rangle_{\text{diff}} = 2U_0 \int |\chi_1(\boldsymbol{r})|^2 \cdot |\chi_2(\boldsymbol{r})|^2 d\boldsymbol{r} \tag{2.3.8}$$

Thus, even when $|\chi_1|^2$ and $|\chi_2|^2$ coincide (as is the case, e.g. for plane waves) the interaction energy is a factor of 2 larger when the particles are in different states than when they are in the same state.[17] (If $|\chi_1|^2$ and $|\chi_2|^2$ do not coincide, the factor is less

[17] The extra term is of course nothing but a special case of the "Fock" term in the interaction energy, which for bosons adds to the "Hartree" term with a positive sign (cf. below).

than 2 but still may be greater than 1). The result is straight forwardly generalized to the many-body case (see Appendix 2A); in the Hartree–Fock approximation we have quite generally for the expectation value of the interaction (2.3.3).

$$\langle E_{\text{int}}\rangle = \frac{1}{2}U_0 \sum_{ij} n_i n_j (2 - \delta_{ij}) \int |\chi_i(\boldsymbol{r})|^2 |\chi_j(\boldsymbol{r})|^2 d\boldsymbol{r} \tag{2.3.9}$$

where n_i is the average occupation number of single-particle states χ_i.

While the model (namely Eqn. (2.3.3)) of the interparticle interaction used in deriving Eqn. (2.3.9) is too specific for this latter equation to be universally quantitatively applicable, it is an extremely useful guide to the qualitative behavior of a Bose system in a variety of different physical situations. I emphasize again that it is not necessary for the application of (2.3.9) that the single-particle eigenstates χ_i be eigenstates of the single-particle Hamiltonian.

Let us assume for the present that the set of basis states relevant to our problem is such that the probability densities $|\chi_i(\boldsymbol{r})|^2$ for the different χ_i overlap strongly in space; this is the case, in particular, when we consider thermodynamic equilibrium in a uniform system, since then the relevant χ_i's are the plane-wave states. Then it is clear from (2.3.9) that the effect of a repulsive interparticle interaction ($U_0 > 0$) will be to give an energy advantage, other things being equal, to states with simple BEC over both the normal state and states with fragmented BEC; conversely, an attractive interaction ($U_0 < 0$) will prima facie favor fragmentation and/or the normal state.

However, there is actually a further complication which we need to address at this point: Suppose that for some reason we find it advantageous to have a macroscopic occupation of two different single-particle states χ_1 and χ_2, so that (say) $\langle n_1\rangle = N|\alpha|^2$ and $\langle n_2\rangle = N|\beta|^2$ with $|\alpha|^2 + |\beta|^2 = 1$. (Such a situation might arise when we discuss the decay of superflow (see Chapter 3) or in considering a gas with attractive interactions.) Now, there are at least two obvious ways of doing this.

1. We can simply put (approximately) $N|\alpha|^2$ particles into state 1 and $N|\beta|^2$ particles independently into state 2, i.e. the many body wave function is, apart from normalization, of the form

$$\Psi_N = (a_1^+)^{n_1}(a_2^+)^{n_2}|\text{vac}\rangle \text{ with } n_1 \cong N|\alpha|^2, \quad n_2 \cong N|\beta|^2. \tag{2.3.10}$$

   where $|\text{vac}\rangle$ denotes, here and subsequently, the vacuum (no particles present). A state of the form (2.3.10) will be denoted throughout this book as a "(relative) Fock state"; it is clearly fragmented, since the single-particle density matrix may be easily verified to have two eigenvalues n_1 and n_2 which are already each macroscopic.
2. We can put all N particles in a linear combination of states 1 and 2: again apart from normalization,

$$\begin{aligned}\Psi_N &= (\alpha a_1^+ + \beta a_2^+)^N|\text{vac}\rangle \\ &\equiv (|\alpha|e^{i\Delta\varphi/2}a_2^+ + |\beta|e^{-i\Delta\varphi/2}a_2^+)^N|\text{vac}\rangle \\ &\equiv \Psi_N(\Delta\varphi).\end{aligned} \tag{2.3.11}$$

where in writing the second expression I have taken into account that the overall phase of the single-particle wave function has no physical meaning. A state of the form (2.3.11) will be denoted throughout this book as a "(relative) coherent state" or "Gross–Pitaevskii (GP) state"; it is evidently *not* fragmented but is an example of simple BEC.

Now, the average single-particle density $\langle\rho(\boldsymbol{r})\rangle(\equiv \rho_1(\boldsymbol{r},\boldsymbol{r}))$ is different for the Fock and GP states, being given to leading order in N in the former case by

$$\langle\rho(\boldsymbol{r})\rangle_{\text{Fock}} = N\{|\alpha|^2|\chi_1(\boldsymbol{r})|^2 + |\beta|^2|\chi_2(\boldsymbol{r})|^2\} \tag{2.3.12}$$

and in the latter by

$$\langle\rho(\boldsymbol{r})\rangle_{\text{GP}} = N\{|\alpha|^2|\chi_1(\boldsymbol{r})|^2 + |\beta|^2|\chi_2(\boldsymbol{r})|^2 + 2|\alpha|\cdot|\beta|\cdot\text{Re}\{e^{i\Delta\varphi}\chi_1(\boldsymbol{r})\chi_2^*(\boldsymbol{r})\} \tag{2.3.13}$$

Suppose now that we introduce a small perturbation of the form

$$H' = \int V(\boldsymbol{r})\rho(\boldsymbol{r})d\boldsymbol{r} \tag{2.3.14}$$

The difference between the ground-state energy of the GP state $\Psi_N(\Delta\varphi)$ and that of the Fock state (2.3.10) is then given by

$$E_{\text{GP}}(\Delta\varphi) - E_{\text{Fock}} = 2N|\alpha|\cdot|\beta|\text{Re}\left\{e^{i\Delta\varphi}\int V(\boldsymbol{r})\chi_1(\boldsymbol{r})\chi_2^*(\boldsymbol{r})d\boldsymbol{r}\right\} \tag{2.3.15}$$

and it is clear that barring pathology it will always be possible to make the RHS negative by an appropriate choice of $\Delta\varphi$. Thus, a perturbation of the form (2.3.14) will generally pick out a particular GP state (namely that which minimizes (2.3.15)) rather than the Fock state. Note that the expression (2.3.15) vanishes when averaged over $\Delta\varphi$. This is not a surprise, if we anticipate the result (to be derived and discussed in Appendix 2B) that the Fock state (2.3.10) is actually nothing but a superposition of GP states of different $\Delta\varphi$ with equal weight for $r_1 = r_2$; schematically, and neglecting normalization,

$$\Psi_{\text{Fock}}^{(N)} = \int \frac{d(\Delta\varphi)}{2\pi}\Psi_{\text{GP}}^{(N)}(\Delta\varphi) \tag{2.3.16}$$

Of course, similar considerations are already present at the level of a single particle, where generally speaking a particular superposition of two states is energetically favored over a mixture.

Even if no perturbation of the form (2.3.14) exists, it is quite likely that the interaction terms will favor a GP state over the Fock one. For the contact interaction (2.3.3), the difference between the interaction energies in the two states turns out to be

$$E_{\text{GP}}^{(\text{int})}(\Delta\varphi) - E_{\text{Fock}}^{(\text{int})} = NU_0\text{Re}\{Ae^{i\Delta\varphi} + Be^{2i\Delta\varphi}\}$$

where

$$A \equiv 2\int |\alpha|\cdot|\beta|\cdot(|\alpha|^2|\chi_1(\boldsymbol{r})|^2 + |\beta|^2|\chi_2(\boldsymbol{r})|^2)\chi_1(\boldsymbol{r})\chi_2^*(\boldsymbol{r})d\boldsymbol{r}$$

$$B \equiv 4|\alpha|^2\cdot|\beta|^2\int \chi_1^2(\boldsymbol{r})\chi_2^*(\boldsymbol{r})^2 d\boldsymbol{r} \tag{2.3.17}$$

From Eqn. (2.3.17) it is clear that unless A and B are both zero (which would be the case for plane-wave states but is otherwise unlikely), or there is a pathological cancellation, a particular GP state will again be favored over the Fock state; note that this conclusion is independent of the sign of the interaction.

The upshot of the above argument is that *simple BEC tends to be favored* even under many conditions when one would at first might think fragmentation might occur. We will study in Section 2.6 some unusual situations where a fragmented state actually wins out.

Let us now return to the case of thermodynamic equilibrium in a uniform system, when the natural choices of eigenfunctions $\chi_i(\boldsymbol{r})$ in Eqn. (2.3.9) are the standard plane-wave states. Consider first the case of repulsive interparticle interactions ($U_0 > 0$). It is clear that other things being equal, the interaction energy (2.3.9) will tend to advantage the (simple) BEC state not only over possible fragmented states but also over the normal state. Hence we should expect that to the extent that the interaction is well described by Eqn. (2.3.3) (and the density n is not too large, see below) the effect of the repulsion would be to increase the BEC transition temperature T_c over its value T_{co} for a noninteracting gas of the same density. Indeed, in the last few years a number of microscopic calculations have reached this conclusion; see in particular Kashurnikov et al. (2001), who find (as do others) that for a gas with interactions adequately described by (2.3.3), the relative increase of T_c is proportional to $n^{1/3}U_0$ for small values of this parameter; for larger values T_c passes through a maximum and then decreases sharply, dropping below T_{co} (Grüter et al. 1997). This latter behavior may be tentatively attributed to the "hindering" behavior of neighboring atoms on the motion of a given atom, which one might try to take into account by replacing the true mass m by some effective mass $m^* > m$; however, such effects cannot be adequately handled by the considerations of this section.

In the case of an attractive interaction ($U_0 < 0$), Eqn. (2.3.9) suggests that a state with simple BEC might be unstable with respect to fragmentation and/or the normal state. However, application of the considerations explored above about the relation between GP and Fock states indicates that it would be at least as attractive for the system to remain simply Bose-condensed, but in a state which is a superposition of different plane-wave states and hence has a spatially varying density in real space (cf. Eqn. (2.3.15)). Indeed, a very simple dimensional argument indicates that the uniform BEC state is unstable with respect to "collapse" in real space, a process whose end point certainly depends on details of the interatomic interactions which are not adequately encapsulated in (2.3.3). Thus, BEC in Einstein's original sense is indeed unstable against an arbitrary weak interaction, if the sign of the latter is attractive: any 1920s-era skepticism was not entirely wrong!

It must be strongly emphasized that the above considerations, and in particular Eqn. (2.3.9), apply to a spinless system which has the choice of two (or more) different *orbital* states. There is no direct analog of the result (2.3.9) when the labels i and j represent different internal (hyperfine) states, and indeed fragmentation with respect to this degree of freedom is not necessarily excluded (see Section 2.6). In fact, the effects of interactions in a "spinful" system need a separate discussion: see e.g. Leggett (2001), Section IV.E). The above results are however adequate as they stand to discuss a truly spin-zero Bose system (e.g. liquid ^{4}He) or a system where the spin degree of freedom, though nontrivial, is fixed[18] (such as the $F = 1$, $m_F = -1$ state of ^{87}Rb in a magnetic trap).

[18]Though the orientation may depend on position in space.

To conclude this section, let us inquire whether we can say anything rigorous about the occurrence or not of (simple) BEC in an extended system in thermodynamic equilibrium, or failing that at least about the condensate fraction as a function of the system parameters (density n, temperature $T, \ldots$). Actually, results in this area are rather few. More than 40 years ago, Gavoret and Nozières (1964) showed in a classic paper that BEC persists in an interacting gas in three-dimensional free space at $T = 0$ *provided* that perturbation theory in the interaction converges (a premise which excludes solid ^{4}He as a counterexample); however, this method cannot be used to set a nontrivial limit on the transition temperature T_c, nor on the condensate fraction at $T = 0$. More recently Kennedy et al. (1988) proved the existence, at $T = 0$, of BEC in a "lattice gas" at half filling without relying on perturbation theory. However, the existence of BEC has not been proven even at $T = 0$, to my knowledge, for any other extended system with short-range interactions.

In the opposite direction, the best-known "no-go" theorem is associated with the name of Hohenberg, and states that BEC cannot occur in equilibrium at finite temperature is any system moving freely in space in $d \leq 2$ dimensions. (The standard thermodynamic limit is assumed.) This is a fairly strong result, since it is completely insensitive to the existence or sign of interparticle interactions. (we have already seen, in Chapter 1, Section 1.3, that it is true in the noninteracting case). In 3D a lemma proved by Hohenberg en route to his theorem can be used to set an upper bound on $N_0(T)$, and a rather different kind of limit can be obtained for the model described by (2.3.3): see Leggett (2003).

2.4 Pseudo-BEC in a Fermi system (Cooper pairing)[19]

If for a Fermi system we define the single-particle density matrix by Eqn. (2.1.2), and diagonalize it as in (2.1.4), then the Pauli principle immediately tells us that none of the eigenvalues can exceed unity; thus, trivially, BEC in the literal sense cannot occur. On the other hand, even for a completely noninteracting Fermi gas, there is nothing in the *kinematics* which forbids us considering as in Chapter 1, Section 1.4, a state in which each pair of fermions is combined into a "diatomic molecule"[20] whose radius is small compared to the mean interparticle spacing, and imagining those "molecules" to be themselves Bose-condensed. It is intuitively plausible (and in fact true) that in this rather artificial example the appropriately normalized *two-particle* density matrix, or rather an appropriately defined element of it, will have a maximum eigenvalue equal to N. Thus, there is no kinematic restriction on having an eigenvalue of the two-particle density matrix "of order N", and it therefore makes sense, in the case of a Fermi system, to focus on this quantity. In the following I shall consider explicitly the case of spin $\frac{1}{2}$, which is appropriate both to electrons in metals and to ^{3}He; however, as I will indicate in Chapter 8, Section 8.4, the generalization to higher spin is straightforward in the extreme.[21]

[19]The discussion of this section is largely based on that of Yang (1962).

[20]Which would of course not be energetically stable in the absence of interactions.

[21]Provided only two different hyperfine states are involved. In principle, for $F \geq 3/2$, it is possible to envisage formation of (e.g.) quadruples rather than pairs without violating the Pauli principle; I shall not consider this possibility in this book.

Consider then a system of fermions whose single-particle states are specified by a coordinate vector $\boldsymbol{r}$ and a spin label $\sigma = \pm 1$ which describes the spin projection (in units of $\hbar/2$) on some chosen axis. Any pure many-body state s of the system at time t can be written in the form

$$\Psi_N^s(t) \equiv \Psi_s(\boldsymbol{r}_1\sigma_1, \boldsymbol{r}_2\sigma_2, \ldots, \boldsymbol{r}_N\sigma_N : t) \tag{2.4.1}$$

where the function Ψ_s is antisymmetric under the simultaneous exchange $\boldsymbol{r}_i \rightleftarrows \boldsymbol{r}_j$, $\sigma_i \rightleftarrows \sigma_j$. The most general state of the system can be written as a mixture of different such normalized and mutually orthogonal states s with weight p_s. So far, everything is exactly parallel to what we did in Section 2.1 for the Bose system. Consider now the *two-particle density matrix* $\rho_2(\boldsymbol{r}_1\sigma_1, \boldsymbol{r}_2\sigma_2 : \boldsymbol{r}_1'\sigma_1', \boldsymbol{r}_2'\sigma_2')$ defined by

$$\begin{aligned}
&\rho_2(\boldsymbol{r}_1\sigma_1, \boldsymbol{r}_2\sigma_2 : \boldsymbol{r}_1'\sigma_1', \boldsymbol{r}_2'\sigma_2' : t) \\
&\quad \equiv N(N-1)\sum_s p_s \cdot \sum_{\sigma_3\sigma_4\ldots\sigma_N} \int d\boldsymbol{r}_3 d\boldsymbol{r}_4 \ldots d\boldsymbol{r}_N \\
&\quad \times \Psi_s^*(\boldsymbol{r}_1\sigma_1\boldsymbol{r}_2\sigma_2\boldsymbol{r}_3\sigma_3\ldots\boldsymbol{r}_N\sigma_N : t) \cdot \Psi_s(\boldsymbol{r}_1'\sigma_1'\boldsymbol{r}_2'\sigma_2'\boldsymbol{r}_3\sigma_3\ldots\boldsymbol{r}_N\sigma_N : t) \\
&\quad \equiv \langle \hat{\psi}_{\sigma 1}^\dagger(\boldsymbol{r}_1 t)\hat{\psi}_{\sigma 2}^\dagger(\boldsymbol{r}_2 t)\hat{\psi}_{\sigma' 2}(\boldsymbol{r}_2' t)\hat{\psi}_{\sigma' 1}(\boldsymbol{r}_1' t)\rangle
\end{aligned} \tag{2.4.2}$$

which must satisfy the condition

$$\rho_2(\boldsymbol{r}_1\sigma_1, \boldsymbol{r}_2\sigma_2 : \boldsymbol{r}_1'\sigma_1'\boldsymbol{r}_2'\sigma_2' : t) \equiv -\rho_2(\boldsymbol{r}_2\sigma_2, \boldsymbol{r}_1\sigma_1 : \boldsymbol{r}_1'\sigma_1', \boldsymbol{r}_2'\sigma_2' : t) \text{ (etc.)} \tag{2.4.3}$$

The two-particle density matrix is a particularly convenient way of characterizing the properties of a Fermi system, since the complicated nodal structure (due to the antisymmetry) which appears in the many-body wave function $\Psi_s(\boldsymbol{r}_1\sigma_1 \ldots \boldsymbol{r}_N\sigma_N : t)$ is averaged over and appears in ρ_2 only in a much more manageable form (see below). Note that just as the expectation values of one-particle properties such as the kinetic energy are naturally expressed in terms of ρ_1, so the expectation value of any two-particle property can be expressed in terms of ρ_2; for example if the interparticle potential energy operator is spin-independent and local, i.e.

$$\hat{V} = \frac{1}{2}\sum_{ij} V(\boldsymbol{r}_i - \boldsymbol{r}_j) \tag{2.4.4}$$

then its expectation value is given by the formula

$$\langle V\rangle(t) = \sum_{\sigma_1\sigma_2} \iint d\boldsymbol{r}_1 d\boldsymbol{r}_2 V(\boldsymbol{r}_1 - \boldsymbol{r}_2)\rho_2(\boldsymbol{r}_1\sigma_1, \boldsymbol{r}_2\sigma_2 : \boldsymbol{r}_1\sigma_1, \boldsymbol{r}_2\sigma_2 : t) \tag{2.4.5}$$

Note also that a knowledge of ρ_2 is sufficient to give ρ_1:

$$\rho_1(\boldsymbol{r}\sigma, \boldsymbol{r}'\sigma') = N^{-1}\sum_{\sigma_2} \int d\boldsymbol{r}_2 \rho_2(\boldsymbol{r}\sigma, \boldsymbol{r}_2\sigma_2 : \boldsymbol{r}'\sigma', \boldsymbol{r}_2\sigma_2 : t) + o(N^{-2}) \tag{2.4.6}$$

Despite the apparent generality of the description of the many-body state in terms of ρ_2, one caveat is in order: While any form of the many-body density matrix must correspond according to (2.4.2) to a definite form of ρ_2 (and ρ_1), *the converse is not true*; one cannot necessarily assume that any arbitrary form of ρ_2 (or for that matter of ρ_1) which one may choose to write down corresponds to a possible many-body density matrix, and in case of doubt this should really be checked by an explicit construction of the latter.

The quantity ρ_2 as defined by (2.4.2), when regarded as a function of the four indices $\boldsymbol{r}_1\sigma_1$, $\boldsymbol{r}_2\sigma_2$, satisfies the condition of Hermiticity, i.e.

$$\rho_2(\boldsymbol{r}_1\sigma_1, \boldsymbol{r}_2\sigma_2 : \boldsymbol{r}_1'\sigma_1', \boldsymbol{r}_2'\sigma_2' : t) = \rho_2^*(\boldsymbol{r}_1'\sigma_1', \boldsymbol{r}_2'\sigma_2' : \boldsymbol{r}_1\sigma_1, \boldsymbol{r}_2\sigma_2 : t) \tag{2.4.7}$$

and therefore can be diagonalized, i.e. written in the form

$$\rho_2(\boldsymbol{r}_1\sigma_1\boldsymbol{r}_2\sigma_2 : \boldsymbol{r}_1'\sigma_1', \boldsymbol{r}_2'\sigma_2' : t) = \sum_i n_i(t)\chi_i^*(\boldsymbol{r}_1\sigma_1, \boldsymbol{r}_2\sigma_2 : t)\chi_i(\boldsymbol{r}_1'\sigma_1', \boldsymbol{r}_2'\sigma_2' : t) \tag{2.4.8}$$

Because of the condition (2.4.3), the functions χ_i must themselves be antisymmetric under the exchange $\boldsymbol{r}_1, \rightleftarrows \boldsymbol{r}_2, \sigma_1 \rightleftarrows \sigma_2$:

$$\chi_i(\boldsymbol{r}_1\sigma_1, \boldsymbol{r}_2\sigma_2 : t) = -\chi_i(\boldsymbol{r}_2\sigma_2 : \boldsymbol{r}_1\sigma_1 : t) \tag{2.4.9}$$

The different eigenfunctions χ_i are mutually orthogonal (and can be chosen normalized), in the sense that

$$\sum_{\sigma_1\sigma_2} \iint d\boldsymbol{r}_1 d\boldsymbol{r}_2 \chi_i^*(\boldsymbol{r}_1\sigma_1 : \boldsymbol{r}_2\sigma_2 : t)\chi_j(\boldsymbol{r}_1\sigma_1 : \boldsymbol{r}_2\sigma_2 : t) = \delta_{ij} \tag{2.4.10}$$

By inserting the definition (2.4.2) into (2.4.8), integrating the diagonal elements over $\boldsymbol{r}_1, \boldsymbol{r}_2(\equiv \boldsymbol{r}_1'\boldsymbol{r}_2')$ and summing over $\sigma_1\sigma_2(\equiv \sigma_1'\sigma_2')$, we find that the eigenvalues $n_i(t)$ must satisfy the condition

$$\sum_i n_i(t) = N(N-1) \tag{2.4.11}$$

Equation (2.4.11) is in itself compatible with ρ_2 having one or more eigenvalues of order N^2. However, as demonstrated by Yang, the Fermi statistics forbids this; the maximum possible eigenvalue of ρ_2 is in fact[22] N. In analogy with the Bose case, we should distinguish three cases:

1. The "normal" case: no eigenvalue of ρ_2 is of order N (all are of order 1).
2. "Simple" (pseudo-) BEC: one and only one eigenvalue is of order N, the rest of order 1.
3. "Fragmented" (pseudo-) BEC: more than one eigenvalue is of order N.

In the literature, the onset of pseudo-BEC in a Fermi system is usually called the occurrence of *Cooper pairing*, and it is almost invariably assumed without argument to be of the "simple" type (as we shall see, there are good reasons for this).

As in the Bose case, we may raise the question: Why should Cooper pairing ever occur in a Fermi system, and why should the pseudo-BEC which it describes be, overwhelmingly, of the "simple" type? We first note that for the case of a *noninteracting* Fermi system in equilibrium at $T = 0$ it is very straightforward to write down the eigenvalues and eigenfunctions[23] of the two-particle density matrix: Since the many-body ground state is an eigenfunction of the plane-wave occupation numbers $\hat{n}_{\boldsymbol{k}}$, with

[22] Actually it is $N(M - N + 2)/M$ where M is the number of available single-particle states (see Yang, ref. cit. Eqn. (9)).

[23] Or more precisely a possible (and natural) choice of eigenfunctions; because of the extreme degeneracy of the eigenvalues, this choice is not unique.

eigenvalue 1 for $k < k_F$ and 0 for $k > k_F$, the eigenvalues of $\hat{\rho}_2$ are also 0 and 1, and the eigenfunctions with eigenvalue 1 can be labeled by a "center-of-mass" momentum $\boldsymbol{K}$ and a "relative" momentum $\boldsymbol{q}$ such that both $|(\boldsymbol{K}+\boldsymbol{q})/2|$ and $|(\boldsymbol{K}-\boldsymbol{q})/2|$ are less than k_F, and by a total spin $S = 0$ or 1 and a spin projection S_z. Explicitly, when chosen in this way, the normalized eigenfunctions corresponding to nonzero (unit) eigenvalues are

$$\chi_{\text{singlet}}(\boldsymbol{r}_1\sigma_1\boldsymbol{r}_2\sigma_2) = \frac{1}{\sqrt{2\Omega}} \exp i\boldsymbol{K}\cdot\boldsymbol{R} \cos \boldsymbol{q}\cdot\boldsymbol{\rho} \times \frac{1}{\sqrt{2}}(\delta_{\sigma_1+}\delta_{\sigma_2-} - \delta_{\sigma_1-}\delta_{\sigma_2+}) \tag{2.4.12a}$$

$$\chi_{\text{triplet}}(\boldsymbol{r}_1\sigma_1\boldsymbol{r}_2\sigma_2) = \frac{1}{\sqrt{2\Omega}} \exp i\boldsymbol{K}\cdot\boldsymbol{R} \sin \boldsymbol{q}\cdot\boldsymbol{\rho} \times \begin{cases} \delta_{\sigma_1+}\delta_{\sigma_2+} \\ \frac{1}{\sqrt{2}}(\delta_{\sigma_1+}\delta_{\sigma_2-} + \delta_{\sigma_1-}\delta_{\sigma_2+}) \\ \delta_{\sigma_1-}\delta_{\sigma_2-} \end{cases} \tag{2.4.12b}$$

(where $\boldsymbol{R} \equiv \frac{1}{2}(\boldsymbol{r}_1 + \boldsymbol{r}_2)$ $\boldsymbol{\rho} \equiv \boldsymbol{r}_1 - \boldsymbol{r}_2$), with as above the restriction $|(\boldsymbol{K}+\boldsymbol{q})/2|$, $|(\boldsymbol{K}-\boldsymbol{q})/2| \leq k_F$. Note that the value $\boldsymbol{K}= 0$ is in no way singled out (except in so far as it permits the maximum number of choices of relative momentum $\boldsymbol{q}$). Evidently, the ground state of the noninteracting gas (the "Fermi sea") is "normal", as we should intuitively expect.

The most obvious factor which can lead to the onset of Cooper pairing is an attractive interparticle interaction. Consider for definiteness the case of a contact potential, $V(\boldsymbol{r}) = -V_0\delta(\boldsymbol{r})$, where $V_0 > 0$. According to Eqn. (2.4.2), the expectation value of this potential is proportional to the quantity

$$\begin{aligned} P &\equiv \sum_{\sigma_1\sigma_2} \int d\boldsymbol{r}\, \rho_2(\boldsymbol{r}\sigma_1\boldsymbol{r}\sigma_2 : \boldsymbol{r}\sigma_1\boldsymbol{r}\sigma_2) \\ &= \sum_{\sigma_1\sigma_2}\sum_i n_i \int |\chi_i(\boldsymbol{r}\sigma_1\boldsymbol{r}\sigma_2)|^2 d\boldsymbol{r} \end{aligned} \tag{2.4.13}$$

which is the probability density for two particles to be at the same point in space. In the normal ground state described by (2.4.12a), (2.4.12b) the triplet states do not contribute to P (since in such a state the particles never reach the relative origin), and the singlet states contribute an amount which is of order $\Omega^{-1}(\sim N^{-1})$ for each i and hence of order N in total; this is just the standard Hartree term.[24] Suppose now that we consider the possibility of simple pseudo-BEC, so that there exists one eigenvalue (call it N_0) of ρ_2 which is of order N; and suppose, moreover that the corresponding eigenfunction χ_0 is bounded in the relative coordinate $\boldsymbol{\rho}$ (so that the normalization factor is $\sim\Omega^{-1/2}$, not $\sim\Omega^{-1}$). Then the integral over $\boldsymbol{r}$ cancels the square of the normalization factor, and if we suppose for simplicity that $|\chi_0|^2$ is independent of the center-of-mass variable $\boldsymbol{R}$ (so that $\chi_0 \sim \Omega^{-1/2}\chi_{\text{rel}}(\boldsymbol{\rho})$, $\chi_{\text{rel}}(0) \sim 1$), we find that the contribution of this single eigenfunction to the expectation value of the interparticle potential energy,

[24]There is no Fock term in the case of a contact potential.

which I will call $\langle V\rangle_{\text{pairing}}$, is

$$\langle V\rangle_{\text{pairing}} = -N_0V_0 \sum_{\sigma_1\sigma_2} |\chi_{\text{rel}}(\sigma_1, \sigma_2, \rho = 0)|^2 \sim N \tag{2.4.14}$$

Thus, provided that the formation of the "pseudocondensate" (Cooper pairs) does not cost too much in kinetic or other (e.g. Hartree) energy, it will be energetically advantageous to form it. We shall see in Chapter 5 that for a Fermi gas with a weakly attractive s-wave interaction, the normal ground state is always unstable against Cooper pairing, no matter how weak the attraction.

While a simple interparticle attraction is the most obvious source of Cooper pairing, it should be emphasized that it is not necessarily the only one. Indeed, while it seems consistent to apply the above picture, at least qualitatively, to the classical superconductors (where the normal state is not too different from that of a free gas, and the attraction, as we shall see in Chapter 5, is generated by the exchange of virtual phonons), it is not at all clear that it bears much relation to what happens in the high-temperature (cuprate) superconductors, where the "normal" state bears little resemblance to a noninteracting gas; the mechanism of pair formation in these systems has been speculated to be a reduction of kinetic energy, a reduction in the (repulsive) Coulomb interaction or something even more indirect, see Chapter 7.

Whatever the origin of the pairing, can we understand the fact that (as far as we know) fragmentation seems to be disfavored? Let us consider for definiteness the competition between two specific eigenfunctions χ_0 and $\chi_{\boldsymbol{K}}$, which correspond to the same (unspecified) dependence of $\chi(\boldsymbol{r}_1\sigma_1\boldsymbol{r}_2\sigma_2 : t)$ on the spins and on the relative coordinate $\boldsymbol{\rho} \equiv \boldsymbol{r}_1 - \boldsymbol{r}_2$ but to different dependences on the center-of-mass condensate $\boldsymbol{R} \equiv \frac{1}{2}(\boldsymbol{r}_1 + \boldsymbol{r}_2)$, with $\chi_0(\boldsymbol{R}) \sim$ const., $\chi_K(\boldsymbol{R}) \sim \exp i\boldsymbol{K}\cdot\boldsymbol{R}$. Evidently, the value of the quantity P defined by Eqn. (2.4.14) is identical for the two eigenfunctions and thus, if either alone is occupied with eigenvalue N_0, the expectation value of the potential energy (for the contact potential above) is given by (2.4.14). Now, it is tempting to think that one could shuffle pairs of particles at will between χ_0 and $\chi_{\boldsymbol{K}}$ in such a way that the sum of the eigenvalues $n_0 + n_{\boldsymbol{K}}$ is conserved, and that in view of the equality of $|\chi_0|^2$ and $|\chi_K|^2$ the potential energy would be independent of the relative fraction. Indeed, in the "Bose limit", where the χ_i represent infinite tightly bound and dilute molecules and the Fermi statistics is irrelevant, this would be correct. However, once the Fermi statistics become important the χ_i's cannot be considered as independent, and it turns out that if one wishes to obtain a macroscopic occupation of the two χ_i's simultaneously while expending no extra kinetic energy, the expectation value of P and hence of the (negative of the) potential energy is decreased relative to the case of simple BEC; conversely, if we try to perform the double occupation so as to conserve the potential energy, the kinetic energy must increase. This is probably the trickiest point in the whole of the theory of Cooper pairing, and to discuss it adequately we will need the microscopic considerations to be developed in Chapter 5, so I postpone it until then and for now simply quote the result of the discussion, namely that in a Fermi system *attractive interactions favor simple BEC* over fragmented states.

Once we have decided to confine ourselves to the case of simple (pseudo-) BEC, a possible definition of the order parameter follows at once. In analogy to the Bose case, we denote the single macroscopic eigenvalue of ρ_2 as $N_0(t)$ and the corresponding eigenfunction (a *two*-particle quantity!) as $\chi_0(\boldsymbol{r}_1\sigma_1, \boldsymbol{r}_2\sigma_2 : t) \equiv \chi_0(\boldsymbol{r}\sigma, \boldsymbol{r}'\sigma' : t)$. Then we define the order parameter Ψ, or as it has become conventional to denote it in the literature, F, by

$$F(\boldsymbol{r}\sigma, \boldsymbol{r}'\sigma' : t) \equiv \sqrt{N_0(t)}\chi_0(\boldsymbol{r}\sigma, \boldsymbol{r}'\sigma' : t) \tag{2.4.15}$$

Since χ_0 is by definition normalized, with the convention (2.4.15) we have for the "condensate number" $N_0(t)$ the expression

$$N_0(t) = \sum_{\sigma\sigma'} \iint d\boldsymbol{r}\, d\boldsymbol{r}' |F(\boldsymbol{r}\sigma, \boldsymbol{r}'\sigma' : t)|^2 \tag{2.4.16}$$

In the Fermi systems which have been of experimental interest until very recently, the condensate fraction N_0/N has been quite small even at $T = 0$ (typically $\sim 10^{-2}$–10^{-4}). See however Chapter 8, Section 8.4.

An important difference between the Cooper-pair order parameter (2.4.15) and the Bose one (2.2.3) is that while in the latter case any internal degrees of freedom are those of the original atoms, in the Cooper-pair case F may have interesting internal structure due to the pairing process itself. To investigate this, let us rewrite F in terms of the center-of-mass coordinate $\boldsymbol{R} \equiv \frac{1}{2}(\boldsymbol{r} + \boldsymbol{r}')$ and the relative coordinate $\boldsymbol{\rho} \equiv \boldsymbol{r} - \boldsymbol{r}'$, and for the moment fix the center-of-mass coordinate at some definite value, say zero:

$$F = F(\boldsymbol{R}, \boldsymbol{\rho}, \sigma\sigma' : t)|_{\boldsymbol{R}=0} \equiv F(\boldsymbol{\rho}, \sigma\sigma' : t) \tag{2.4.17}$$

Suppose that the state of the many-body system to which (2.4.17) refers is one of thermal equilibrium, or not far from it. There then are two possibilities.

1. The form of $F(\boldsymbol{\rho}, \sigma\sigma')$ is essentially dictated by the energetics of the pairing process, so that in free space under fixed thermodynamic parameters (density, temperature, external magnetic field) it is unique up to an overall phase. This is the case for the simple s-wave pairing which is believed to characterize the classical superconductors (Chapter 5); in this case the quantity F has the form (up to a phase)

$$F(\boldsymbol{\rho}, \sigma\sigma') = f(|\boldsymbol{\rho}|) \times \frac{1}{\sqrt{2}}(\delta_{\sigma+}\delta_{\sigma'-} - \delta_{\sigma-}\delta_{\sigma'+}) \tag{2.4.18}$$

   where the function $f(|\boldsymbol{\rho}|)$ is uniquely fixed as a function of density and temperature. It may also be the case for "exotic" pairing (see below) if the residual invariance is broken by some external factor such as the crystal lattice (this is the case for the $d_{x^2-y^2}$ pairing generally believed to describe the cuprates, see Chapter 7). However, in either case both the phase and the amplitude of the parameter may vary as a function of the center-of-mass variable $\boldsymbol{R}$ (the latter because, e.g. of the necessity to satisfy boundary or topological constraints). Since we are not interested in the internal structure, it is convenient to chose specific values of σ, σ' and $\boldsymbol{\rho}$ (for s-wave pairing a convenient choice[25] is $\sigma = -\sigma' = +1, \boldsymbol{\rho} = 0$) and define a "macroscopic" order parameter which is

[25] An alternative choice would be to integrate $F(\boldsymbol{R}, \boldsymbol{\rho})$ over $\boldsymbol{\rho}$.

a function only of $\boldsymbol{R}$. That is, taking the case of s-wave pairing for definiteness, we define the quantity (also for nonequilibrium conditions)

$$\Psi(\boldsymbol{R}:t) \equiv F(\boldsymbol{R},\boldsymbol{\rho}:\sigma\sigma')_{\substack{\sigma=-\sigma'=+1,\\ \rho=0}} \tag{2.4.19}$$

The quantity defined by (2.4.19) is up to the normalization, which is a matter of convention, the celebrated *Ginzburg–Landau order parameter*, introduced by them into the theory of superconductivity on the basis of phenomenological arguments before the microscopic basis of the phenomenon was understood. Like F itself, it is a complex quantity, and we can immediately use it to define a "superfluid velocity" in analogy to the Bose case[26]

$$\Psi(\boldsymbol{R},t) \equiv |\Psi(\boldsymbol{R},t)| \exp i\varphi(\boldsymbol{R},t) \tag{2.4.20}$$

$$\boldsymbol{v}_s(\boldsymbol{R},t) \equiv \frac{\hbar}{2m}\boldsymbol{\nabla}_R\varphi(\boldsymbol{R},t) \tag{2.4.21}$$

We have used a factor $2m$ rather than m in the definition on the grounds that the object described by F and hence by Ψ is a *pair* of electrons: $\boldsymbol{v}_s$ is the velocity of their center of mass. Evidently, as in the Bose case, $\boldsymbol{v}_s$ satisfies the relations (2.2.8) and (2.2.9) (the latter with $m \to 2m$), and its physical significance as a "classical quantity reflecting the effects of quantum mechanics" is similar to that in the Bose case.

2. The more complicated possibility is that the quantity $F(\boldsymbol{\rho},\sigma\sigma')$ is not uniquely determined by the energetics in thermal equilibrium. Typically, this is because the form of F as a function of $\boldsymbol{\rho}$ and/or of the spin degree of freedom breaks some symmetry of the Hamiltonian, and the resulting "orientational" degree of freedom is not pinned by any external factor. In such a case we need to generalize the definition (2.4.19) of the "Ginzburg–Landau" order parameter so as to specify not only the magnitude and phase, but also the "orientation" of the pair wave function as a function of the center-of-mass coordinate $\boldsymbol{R}$. (Compare the case of a Bose system with a nontrivial internal degree of freedom.) I will postpone the formalism necessary to handle this complication to Chapter 6.

As in the Bose case, definitions of the order parameter alternative to (2.4.19) (and hence of the idea of Cooper pairing) are possible and to be found in the literature. For an extended system, (2.4.19) is effectively equivalent to the existence of off-diagonal long-range order in the two-particle density matrix, that is to the assertion that in the limit $|(\boldsymbol{r}_1+\boldsymbol{r}_2)/2-(\boldsymbol{r}'_1+\boldsymbol{r}'_2)/2| \to \infty$, but $|\boldsymbol{r}_1-\boldsymbol{r}_2|, |\boldsymbol{r}'_1-\boldsymbol{r}'_2|$ finite,

$$\rho(\boldsymbol{r}_1\sigma_1,\, \boldsymbol{r}_2\sigma_2 : \boldsymbol{r}'_1\sigma'_1\boldsymbol{r}'_2\sigma'_2 : t) \to F^*(\boldsymbol{r}_1\sigma_1, \boldsymbol{r}_2\sigma_2 : t)F(\boldsymbol{r}'_1\sigma'_1, \boldsymbol{r}'_2\sigma'_2 : t) \tag{2.4.22}$$

a statement which gives an alternative definition of F. With this definition, the onset of Cooper pairing is equivalent to that of ODLRO. Alternatively one can define F as

[26] For electrons in the presence of an external or self-consistent vector potential we will need to generalize this definition somewhat, see Chapter 5.

an "anomalous average" which (allegedly) arises as a result of spontaneous breaking of the $U(1)$ gauge symmetry:

$$F(\boldsymbol{r}\sigma, \boldsymbol{r}'\sigma' : t) \equiv \langle \hat{\psi}_\sigma(\boldsymbol{r}t)\hat{\psi}_{\sigma'}(\boldsymbol{r}'t)\rangle \tag{2.4.23}$$

so that (for s-wave pairing) the Ginzburg–Landau order parameter is defined by

$$\Psi(\boldsymbol{R} : t) \equiv \langle \hat{\psi}_\uparrow(\boldsymbol{R}t)\hat{\psi}_\downarrow(\boldsymbol{R}t)\rangle \tag{2.4.24}$$

While the conceptual issues related to the definition (2.4.23) are identical to those arising in the Bose case and discussed in Section 2.2, the trick of relaxation of conservation of total particle number, which in the Bose case (at least in the present author's opinion) brings no special advantages, in the Fermi case is an essential ingredient in the original microscopic calculation of Bardeen et al., and has been very widely used in the literature. However, we shall see in Chapter 5 that even in the Fermi case it can be avoided.

We may finally enquire, as in the Bose case, about possible rigorous limits on T_c and/or on N_0. There does in fact exist an application of Hohenberg's theorem (Section 2.3) to the Fermi case, but as applied to existing Cooper-pairing systems it is so weak as to be of little practical interest. (It may be of interest in the context of the Fermi alkali gases discussed in Chapter 8, Section 8.4). A more interesting bound is given by theorem (6) of Yang's original paper, which in our notation reads

$$\frac{N_0}{N} \leq \frac{M - N + 2}{M} \tag{2.4.25}$$

where M is the number of single-particle states available for occupation. Although for a realistic system the concept of "available for occupation" is not well-defined, we can use the result (2.4.25) qualitatively by assuming (e.g.) that the maximum energy (relative to the Fermi energy ε_F) of the states involved in pairing in a classical superconductor is of order of the Debye energy $\hbar\omega_D$ (see Chapter 5); then $M \sim N(1 + \hbar\omega_D/\varepsilon_F)$, so we predict that N_0/N is certainly no greater than $\sim\hbar\omega_D/\varepsilon_F$ (a correct, if somewhat weak conclusion).

2.5 The consequences of BEC: preview of coming attractions

The onset in a quantum liquid of BEC, or of its Fermi analog Cooper pairing, leads to a number of spectacular and unique consequences which include, but are by no means restricted to, the complex of phenomena known as superfluidity or superconductivity. Some of these properties are generic to all BEC systems, while others require in addition to BEC itself that the order parameter describing the condensate should be of a particular type (e.g. a complex scalar, or a spinor). For this and other reasons, not all of these phenomena occur (or are easy to detect) in any particular BEC system, and they will therefore appear scattered through out this book, in the chapter(s) devoted to the system(s) when they are most easily or spectacularly seen. It may be useful, therefore, to give in this section a brief and qualitative preview of the various types of effects we shall be seeing, and to indicate why they are natural consequences of BEC.

To recapitulate the conclusions of Sections 2.2–2.4, when BEC or Cooper pairing takes place in a many-body system, it almost invariably occurs in "simple" form, i.e. at any given time one *and only one* eigenvalue of the one-particle density matrix (for a Bose system) or of the two-particle density matrix (for a Fermi system) is of order N, the rest of order unity. The reasons for this state of affairs are basically energetic, as sketched in Sections 2.2 and 2.4 (for the few exceptions, see the next section). In other words, the state of the condensate (or in the Fermi case of the Cooper pairs) is *unique*, and can be characterized by an *order parameter* which in the most general case is given up to normalization by the appropriate eigenfunction of the density matrix in question, i.e. by (2.2.3) in the Bose case and by (2.4.15) in the Fermi case. Actually, in the Fermi case we are usually not interested in all the details of the dependence of $F(\boldsymbol{r}\sigma, \boldsymbol{r}'\sigma : t)$ on the relative coordinate, and it is therefore convenient to define a "macroscopic" (coarse-grained) order parameter which depends only on the center-of-mass variable (cf. 2.4.19) and possibly on the "orientation" in orbital and/or spin space. If we do this, then we have a unified description of Bose-condensed Bose systems and Cooper-paired Fermi systems: in each case the description is by a complex quantity $\Psi(\boldsymbol{r}, \alpha)$, where $\boldsymbol{r}$ denotes the center-of-mass variable of the condensed object(s) and α labels one or more internal degrees of freedom. In the simplest case (spinless Bose system, or a superconductor with s-wave pairing) the argument α is absent and the order parameter $\Psi(\boldsymbol{r})$ is a simple complex scalar function of position; in the case of superconductivity it is associated with the names of Ginzburg and Landau (GL) and in ^{4}He (or a spinless alkali gas) with those of Gross and Pitaevskii (GP).

A consideration which goes somewhat beyond the mere existence of an order parameter is that in most cases of practical interest it is *locally stable*. That is, in the absence of certain special kinds of perturbation[27] energetic considerations tend to constrain the order parameter to be spatially uniform in both magnitude and phase (and also in orientation, when that is relevant). In the simple GP–GL case of a complex scalar order parameter, this consideration is often expressed phenomenologically by incorporating in the free energy density, expressed as a function of $\Psi(\boldsymbol{r})$, a term proportional to $|\Psi|^4$ and a term proportional to $|\boldsymbol{\nabla}\Psi|^2$, both with positive coefficients: the function of the first is to disfavor spatial variations of the magnitude of the order parameter, while the second, as well as reinforcing this tendency, also suppresses variations of the phase. In the case of a dilute spinless Bose system such as the alkali gases, the origin of those terms is rather obvious: the gradient term is just the kinetic-energy term $(\hbar^2/2m)|(\nabla\psi)|^2$ which occurs as a function of the single-particle Schrödinger wave function $\psi(\boldsymbol{r})$ in the energy of the macroscopically occupied single-particle state, while the quartic term arises directly from Eqn. (2.3.5): see Chapter 4, Section 4.3. In the case of a more strongly interacting Bose system such as ^{4}He, or in the Fermi case, both the gradient and the quartic term can be rigorously justified only close[28] to the transition temperature T_c; elsewhere they must be regarded as phenomenological ansätze which encapsulate the principal physical effects of the kinetic energy and the interparticle interactions respectively. I will return to this question, in the Fermi case, in Chapter 5, Section 5.7 (see also Chapter 3, Section 3.7.2).

The spectacular consequences of BEC are all in some sense consequences of the fact that every particle (or pair of particles) of the condensate must, at any time, be

[27]e.g. a magnetic field, for a charged system.

[28]But not too close, because of critical-fluctuation effects: see e.g. Goldenfeld 1992, Chapter 6.

doing exactly the same thing; they are like a well-drilled platoon of soldiers marching in lockstep.[29] By contrast, the particles in a normal system (or in the "normal" (noncondensed) fraction of a BEC system) are, in general, all doing *different* things. Let us consider some specific examples, each of which will be discussed in more detail later in the book in the context of the specific system(s) in which it occurs:

1. Direct visualization of the BEC phenomenon: The very definition of (simple) BEC is the macroscopic occupation of a single-particle state. Is it possible to confirm directly that this is happening? For a free gas in thermodynamic equilibrium in uniform space, the macroscopically occupied state should, according to the considerations of Chapter 1, Section 1.3, be the zero-momentum plane-wave state, and it seems almost certain that this is also the case in real liquid ^{4}He. In principle, there exists ways of verifying that below T_c the $\boldsymbol{k} = 0$ state is indeed macroscopically occupied, but they are rather circumstantial: see Chapter 3, Section 2. By contrast, were we to deal with a Bose system which is noninteracting but subject to an external potential (e.g. the harmonic-oscillator potential of a magnetic trap) so that the single-particle energy eigenstates are spatially localized, the effect would be dramatic: Above T_c the particles should be distributed between the different harmonic oscillator states according to formula (1.2.9), with no one state particularly favored, and the density distribution may then be shown to be approximately Gaussian, with a width R_{th} of order $(k_B T/m\omega_0^2)^{1/2}$ where ω_0 is the harmonic frequency. By contrast, below T_c a finite fraction (of order 1) of the particles occupy the harmonic-oscillator ground state, whose width $R_Q \sim (\hbar/m\omega_0)^{1/2}$ is very much less than R_{th} for $N \gg 1$ (see Chapter 4, Section 4.3); thus we should see a very sharp spike in the density appearing above the "thermal" background. We shall see in Chapter 4, Section 4.3 that although the effects of repulsive interparticle interactions is to broaden the width of the condensate distribution, typically to several times R_Q, it is still narrow compared to R_{th}. This phenomenon was seen in an early experiment on ^{23}Na. (Andrews et al. 1996).[30]
2. Detection of "latent" BEC by measurement: Although the most spectacular example of this phenomenon actually demonstrated experimentally to date may be the celebrated MIT interference experiment on ^{23}Na (see Chapter 4, Section 4.5), it may be easier to understand the principle involved if we consider a different (thought-) experiment, of which only half has actually been done to date. Consider a set of spin-$\frac{1}{2}$ Bose particles which start at $t < 0$ with all their spins aligned (say in the $s_z = -\frac{1}{2}$ state), but may or may not have a single *orbital* state macroscopically occupied (i.e. be Bose-condensed). Let us imagine that at $t = 0$ we apply a $\pi/2$ pulse so as to flip all the spins into the xy-plane, so that (say) $S_x(t = 0+) \sim N/2$. For $t > 0$, suppose the spins precess under a fluctuating magnetic field in the z-direction which is a random function not only of time but also of position in space; this field may be of either classical

[29]cf. the cover of *Science*, 22 December 1995.

[30]The spectacular images shown in the first paper to detect BEC in an alkali gas (Anderson et al. 1995) are actually taken after expansion, and therefore are a measure of the original (in-trap) distribution in velocity rather than position space.

or quantum origin (the latter case is in some ways more interesting). In either case, let us wait an appropriate time T and then examine the spin state of the atoms in a given small region of space. Since the spin of any atom, and hence the total spin of the sample, has precessed through an unknown angle, and we assume that we have waited long enough for the uncertainty to be $\gg 2\pi$, averaging over this unknown angle will give $\langle S_x\rangle(T) = \langle S_y(T)\rangle = 0$. This is true whether or not the original state of the system was Bose-condensed. However, the significance of the above statement is profoundly different in the two cases: In the "normal" case, each atom is (crudely speaking) in a different orbital state and will therefore have sampled over the time interval $0 < t < T$, a different sequence of magnetic fields and hence undergone a different precession angle. The destructive interference between the different spins will then give, for the whole sample, not only $\langle S_x\rangle = \langle S_y\rangle \cong 0$ but also $\langle \boldsymbol{S}^2\rangle \cong 0$, i.e. the *magnitude* of the total spin polarization is zero. Hence, if we "measure" (say) $S_x(T)$ e.g. by applying a second $\pi/2$ pulse and measuring the resulting z-component, we will always get a zero value (or more precisely a value $\ll N$, when N is the total number of atoms in the sample). (This half of the experiment had not been done at the time of writing).

For a BEC system the situation is very different: While each of the atoms of the condensate has seen a "random" sequence of fields over the interval $0 < t < T$, since they are all in the same orbital state they have all experienced *the same* sequence and therefore undergone the same (unknown) precession angle. There is no destructive interference between different atoms! Thus the N-particle density matrix as regards the spin degree of freedom corresponds to a statistical mixture (in the case of classical fluctuating fields) or an entangled quantum superposition (in the quantum case) of different states with random values of S_x and S_y but the same large value, $\sim(N/2)^2$, of the *magnitude* of total spin. Technically, the density matrix corresponds to fragmentation in spin space (the two eigenvalues of the single-particle density matrix are simply $\pm\frac{1}{2}$, corresponding to our total ignorance of the spin direction). If now we "measure" (say) $S_x(T)$ as above, what we will find is that on each individual run of the experiment we get a value of magnitude $\sim N$, but varying from $-N/2$ to $N/2$ at random from shot to shot. (This half of the experiment has been done, with precisely the predicted result: see (Hall et al. 1998).) Thus, although the spin state of the system before measurement was technically fragmented, the act of measurement forces it to behave "as if" it had been in a definite but unknown coherent state, i.e. possessed "simple" BEC. We will give a more detailed discussion in Chapter 4, Section 4.5, where we will see that similar considerations apply to the MIT interference experiment.

3. Nonclassical rotation inertia (Hess–Fairbank effect): When an annulus containing a system of (electrically neutral) particles such as ^{4}He atoms rotates at angular velocity ω, then according to the considerations of Chapter 1, Section 1.6, the "effective" single-particle energy levels are shifted according to the prescription $\varepsilon_n \to \varepsilon_n - \hbar\omega\ell_n$, where ℓ_n is the angular momentum quantum number associated with state n. In a normal system, the distribution function $f(\varepsilon_n : T)$ shifts so as to accommodate this energy shift, and the resulting angular momentum L

of the system is then just $I_{\rm cl}\omega$ where $I_{\rm cl}$ is the classical moment of inertia. In a BEC system, by contrast, *all* the N_0 atoms in the condensate must be in the *same* state n, and in particular have the same value of l_n; hence the only allowed values of their contribution to L are $N_0 n\hbar$ where n is integral, and in particular for small enough ω the value $n = 0$ is realized and the condensate remains at rest in the laboratory frame. Although the above argument uses the language appropriate to a noninteracting gas, it is easily generalized to the interacting case: see Chapter 3. As we have seen in Chapter 1, Section 1.6, this effect can be shown to be impossible in classical statistical mechanics, and it is often regarded as the fundamental definition of "superfluidity".

As indicated in Chapter 1, Section 1.4, the diamagnetic effect which is the direct analog of the HF effect in a charged system (metal), when combined with the standard laws of electromagnetism, leads to the very spectacular *Meissner effect* and to related phenomena such as flux quantization. See Chapter 5, Section 5.6.

4. The Josephson effect: Consider a single electron moving in a geometry composed of two bulk metals separated by an insulating barrier (Josephson junction). In general the electron will be described by a wave function which is, schematically, a superposition of amplitudes to be in one bulk metal or the other, with some relative phase $\Delta\varphi_{\rm el}$. Changes in $\Delta\varphi_{\rm el}$ in time will in general induce currents across the junction; however, the energy which tends to favor one value of $\Delta\varphi_{\rm el}$ over another is of the order of the single-electron tunneling matrix element t across the barrier, a (very) microscopic quantity, so at any reasonable temperature the different electrons behave independently and their first-order contribution (in t) to the current vanishes, leaving an "incoherent" second-order term which just represents the usual ohmic conductance of the junction. In the Cooper-paired (superconducting) phase the *Cooper pairs* are in general represented by a wave function (order parameter) which is a superposition of amplitudes to be in one bulk metal or the other with a definite relative phase $\Delta\varphi_{\rm CP}$. However, by contrast with the single-electron case, *all* the Cooper pairs must possess *the same* value of $\Delta\varphi_{\rm CP}$; moreover, the energy which prefers one value to another is of the order of the pair tunneling matrix element (which, because two electrons must tunnel, is proportional to t^2 not t, but contains no explicit factors of N^{-1}), *multiplied* by the number of pairs ($\sim N$). Thus the value of $\Delta\varphi_{\rm CP}$ is stabilized against thermal (and quantum) fluctuations,[31] and moreover when it is induced to change, e.g. by applying a potential difference across the junction, the current carried by the different pairs is exactly in phase and sums to a macroscopic value.
5. "Bose amplification" of ultraweak effects: The Josephson effect is actually a special case (probably the only one realizable in a system with a simple complex order parameter, such as the classical superconductors) of a much more general effect in BEC systems, namely the amplification of effects which in a normal system would be completely drowned out by thermal noise. In a BEC system with

[31] This argument fails if the "bulk" metals are too small (the "mesoscopic" case discussed in e.g. Tinkham 1996, Section 7.3): cf. Section 2.6.

one or more "orientational" degrees of freedom, such as superfluid ^{3}He, these effects can be even more spectacular. As an example, consider the case of the nuclear dipole–dipole interaction (to be discussed in detail in Chapter 6) If we consider two small magnetic moments (spins), let us say for definiteness oriented parallel, then the energy associated with the standard dipole–dipole interaction prefers a configuration when they are on end-to-end rather than side-by-side; correspondingly, if they have a given magnitude of relative orbital angular momentum, then the energy is lower when the (vector) angular momentum is perpendicular to the spin direction than when it is parallel. However, if the spins in question are the nuclear spins in liquid ^{3}He, their interaction energy is never greater than $\sim 10^{-7}$ K, several orders of magnitude smaller than the thermal energy at currently attainable temperatures. Thus, in normal ^{3}He the relative orbital angular momentum of two ^{3}He atoms with parallel spins is to all intents and purposes random. However, the Cooper pairs which form in the superfluid phase must, for given spin, all have *exactly the same* value of the (vector) relative orbital angular momentum; the factor favoring the perpendicular over the parallel orientation is now not 10^{-7} K, but $\sim 10^{23} \times 10^{-7}$ K $\sim 10^{16}$ K, many orders of magnitude *greater* than the thermal energy! In fact, under most conditions this "tiny" nuclear dipole energy turns out to be the dominant effect in determining the orientation of the Cooper pairs. We will see in Chapter 6 that when this orientation is disturbed from its equilibrium value (for example by application of an rf magnetic field) the dipole energy gives rise to a unique spin dynamics.

6. Metastability of superflow: It is somewhat ironic that while the indefinite persistence of currents circulating in a ring is the most spectacular and easily demonstrated manifestation of "superfluidity" in liquid ^{4}He and "superconductivity" in a metal, it is actually not an automatic consequence of BEC as such but requires the condensate to be of a certain type. We have seen above that the free energy of a general BEC system contains two types of term which tend to stabilize the uniform state: a term, associated with the kinetic energy, proportional to $|\nabla\Psi|^2$, and a term, normally associated with interparticle interactions, proportional to $|\Psi|^4$. Now the $|\Psi|^4$ term constraints only the overall magnitude of the order parameter, not its phase or orientation; and the gradient term is a monotonic function of the circulating current. Thus the mere existence of these terms does not in itself in any way inhibit the relaxation of the circulating current from a finite value to zero. What may inhibit the latter is *topological* considerations. Suppose that the order parameter of the system in question is a simple complex scalar $\Psi(\boldsymbol{r})$, as it is for superfluid ^{4}He and for the classical superconductors. Then we can define a superfluid velocity $\boldsymbol{v}_\mathrm{s}$ by Eqn. (2.2.7), and the current circulating in a ring will turn out to be proportional to the quantity $\oint^C \boldsymbol{v}_\mathrm{s} \cdot d\boldsymbol{\ell}$ (where the contour C is taken around the ring). Now, according to Eqn. (2.2.9) this quantity is quantized in units of h/m – this is just the statement that "winding number", that is number of "turns" made in the Argand diagram by the phase φ as we go around the ring, is integral. It is clear that so long as the magnitude of Ψ is finite (so that its phase can be defined) the winding number is topologically conserved, and hence so is the circulating current.

What is remarkable is that an argument of the above type generally *fails* as soon as the order parameter is of a type more complicated than a simple complex order (e.g. if it is a spinor): even though it may be possible to define a unique "phase" (or equivalently a "superfluid velocity") for the initial state (with finite circulating current) and for the final state (with zero current), it is not in general possible to define it for all the possible states which interpolate between those, and thus there is no guarantee of the existence of a topological conservation law analogous to that in the complex scalar case. Consequently, *metastability of superflow is not a generic consequence of BEC.* We return to this issue in Chapter 4, Section 7.2, and Chapter 6, Section 6.5.

7. Topological defects: Closely associated with the above considerations related to a ring geometry is the existence, in a bulk BEC system, of highly metastable topological singularities (defects) of the order parameter. For a simple complex scalar order parameter the bulk singularity which corresponds to a metastable circulating supercurrent in the ring geometry is a simple vortex line (with "winding number" ± 1, see Chapter 3); in a normal system such a flow configuration would rapidly dissipate its flow energy and disappear, but in the BEC system the same topological considerations which guarantee the metastability of the circulating supercurrent make the vortex line also highly metastable. For a more complicated form of order parameter, we have seen that there is in general no topological conservation law associated with the overall phase, but there may well be other such conservation laws, and if so these will be reflected in the possible existence in the bulk of more "exotic" types of topological singularities.
8. "Exotic" collective excitations: Generally speaking, the possible long-wavelength collective excitations of a many-body system are determined by the conservation laws satisfied by various macroscopic quantities. In a normal system the relevant conserved quantities are normally only particle number, energy, momentum and possibly spin. However, in a BEC system we can also get conservation laws, of a rather different nature, which are associated specifically with the condensate and are usually of topological character, as described under (6) and (7) above. This circumstance gives rise to the possibility of types of collective excitation which are peculiar to the BEC phase (second and fourth sound in ^{4}He, superfluid spin waves in ^{3}He etc.).

Although the above list of consequences of BEC hopefully includes most of the exotic phenomena observed in existing BEC systems, it should be emphasized that it may not be exhaustive: it is entirely possible that just as Josephson in 1962 discovered an effect which with hindsight is clearly implicit in the idea of Cooper pairing but had not previously been explicitly exhibited, so tomorrow someone may discover some spectacular and heretofore unknown consequence of the BEC phenomenon!

In addition to the above list of phenomena which can be regarded as arising directly from the existence of BEC as such, there are many important effects which arise, in a Fermi system, from the modification of the *normal*-component excitation spectrum by the onset of Cooper pairing (similar but less dramatic effects occur in a Bose system): well-known examples include the behavior in the superconducting phase of a metal of quantities such as the spin susceptibility, NMR relaxation state, thermal conductivity, etc. While I will refer to such effects occasionally in Chapter 5, Appendix 5B, they are not central to the theme of this book. However, there is one related remark which is worth making at this point: As we will see in Chapter 5, one consequence of the onset of (s-wave) pairing in a classical superconductor is that the single-electron energy spectrum acquires a gap Δ, which for $T \ll T_c$ is of the order of the transition temperature T_c. Consequently, in this limit the fraction of excited, i.e. unpaired, electrons is of order $\exp[-(\text{const.}T_c/T]$; as a striking quantitative illustration, in a 1 cm^3 cube of Nb ($T_c \sim 9\,\text{K}$) at 10 mK, there is on average *not even one* unpaired electron! The relevance of this remark to the central theme of this book is that normal fluid-condensate interactions, which under most circumstances tend (e.g.) to damp the collective excitations of the condensate and in other ways make some of the consequences of BEC more difficult to see, are under these circumstances totally negligible: Nb at 10 mK is essentially pure condensate. (Were it not for this circumstance, the prospect of using macroscopic superconducting systems or elements in a quantum computer (see e.g. Makhlin et al. 2001) would never have got off the ground.) A qualitatively similar situation occurs in the B phase of superfluid ^{3}He, which also has a finite gap, though the currently attainable lower limit on T/T_c is considerably higher than in Nb; other Cooper-paired Fermi systems, such as ^{3}He-A and the cuprate superconductors, show only a much weaker version of the effect, since as described in Chapters 6 and 7 the energy gap tends to zero at some points on the Fermi surface and the density of unpaired electrons therefore declines with temperature only as a power law, not exponentially.

2.6 Fragmented BEC

With the exception of a small amount of material on spin coherence in Chapter 4, Section 4.5, the rest of this book will be predicated on the assumption that the BEC we are dealing with is "simple", that is, not fragmented. As a sort of appendix to this chapter, therefore, I will briefly list in this section some of the (relatively rare) circumstances in which fragmentation does or may occur.

1. Bosons with attractive interactions. Consider a dilute gas of spinless bosons with an interaction potential of the "contact" form, Eqn. (2.3.3). We saw in Section 2.3 that if the constant U_0 is positive (repulsive interaction) then the effect is to make fragmentation energetically unfavorable and thus to reinforce simple BEC. By the same token, a negative value of U_0 (attractive interaction) should favor fragmentation over a simple uniform BEC. This case is not entirely unphysical, as certain of the alkali gases (^7Li, and ^{85}Rb in a certain range of magnetic field) appear to have an attractive interparticle interaction. However, we also saw in Section 2.3 that rather than fragment, the system is at least as likely to choose simple BEC in a linear combination of uniform states which correspond to a spatially varying density, i.e. to "collapse" in real space. The relatively small amount of evidence currently available on ^{7}Li and ^{85}Rb suggest that indeed something like this happens and results in massive recombination; thus, even if fragmentation occurs at all, it appears to be highly transient. I will not discuss this case further here.

2. "Coulomb blockade" and related effects. One kind of generic situation where fragmentation may be energetically favorable is when the system of bosons (or Cooper pairs) in question has available to it two states which are rather well separated in space; a typical example is two mesoscopic superconducting grains L and R of limited capacitance C separated by some kind of Josephson junction. In these circumstances the term in the Hamiltonian which tends to prefer simple BEC (that is, condensation of all N pairs into a linear combination of states localized in L and R), namely the Josephson tunneling energy, E_{J}, may be quite small, while the capacitance energy $Q^2/2C \equiv E_{\mathrm{c}}$ induced by a charge imbalance Q between L and R may be substantial; this effect is known as the "Coulomb blockade". If the ratio $E_{\mathrm{c}}/E_{\mathrm{J}}$ is sufficiently large then the ground state is not a coherent superposition but rather corresponds to a definite number of pairs in each grain; such a state is by our definition clearly fragmented (cf. Tinkham 1996, Section 7.3).

The above considerations refer to a pair of states which are separated in real (orbital) space. We have already met (in Section 2.4, point (2)) a rather trivial example of fragmentation in spin space, namely the case in which a perturbation connecting the two spin species (the r.f. magnetic field, or in practice a pair of Raman lasers) is turned on, so as to produce a coherent superposition of the two states, but subsequently turned off: the "dephasing" due to random fields then leads to a fragmented state. It is intuitively plausible, by analogy with the Coulomb-blockade situation, that if the interspecies repulsion is larger than the intraspecies one then the fragmented state may be energetically stable even in the presence of d.c. "tunneling" processes which connect the two species. Such a situation could almost certainly be engineered, if desired, in the alkali gases, and may occur in superfluid ^{3}He-A under sufficiently extreme conditions: see Leggett (2004).

3. *Spin-1 bosons: the LPB state.* Consider a system of bosons with spin 1, moving freely in space in zero external magnetic field. If we neglect interactions, the ground state of such a system can be either of the coherent (simple BEC) type, that is a spinor which is constant in space in magnitude, phase and direction (e.g. all atoms in the state $\boldsymbol{k} = 0$, $S_z = +1$), or fragmented, e.g. all particles in the orbital state $\boldsymbol{k} = 0$ but with one-third each in the states $S_z = +1, 0, -1$; in the absence of interactions these states (and many intermediate ones) are degenerate. Now imagine that we introduce a weak interparticle interaction which is isotropic in spin space. Barring pathology, when applied to the collision of two atoms in an s-state the interaction will favor either a spin configuration corresponding to the total spin K of the colliding atoms equal to 2, or that corresponding to $K = 0$ (note that according to the considerations of Chapter 1, Section 1.2, $K = 1$ is forbidden for two bosons in a relative s-state). In the former case it is intuitively clear that the effect is to favor a coherent state with $\boldsymbol{k} = 0$, $S_z = +1$ (or some rotation of it) over the fragmented state, since the latter will generally give some probability for both $K = 2$ and $K = 0$, while the former guarantees $K = 2$. In the case when $K = 0$ is favored, however, it was pointed out by Law et al. (1998) that the ground state is not a coherent state of the type (2.6.1), but rather a *correlated* state of the form (apart from normalization)

$$\Psi_{\mathrm{LPB}} = (a_{01}^+ a_{0,-1}^+ + a_{0,-1}^+ a_{0,1}^+ - a_{0,0}^+ a_{0,0}^+)^{N/2} |\mathrm{vac}\rangle \tag{2.6.1}$$

(where the first subscript refers to the $\boldsymbol{k}$-vector and the second to the spin projection). While the *two-particle* density matrix corresponding to this state has a single eigenvalue of order N, the one-particle density matrix has three different degenerate eigenvalues of $N/3$, so that the LPB state is by our definition "fragmented".

Unfortunately, while it seems probable that the LPB state, or something close to it, is indeed the ground state of the idealized spin-1 system described, it has been shown (Ho and Yip (2000) that it is stable over a coherent state only by an energy of relative order N^{-1} and would thus in practice very likely be destabilized by any small perturbation which favored the latter, e.g. a weak magnetic field. It is interesting to ask whether we can find examples of fragmented states in an extended system which are more stable than this. As we shall see in the next subsection, the answer is yes.

4. Spin-$\frac{1}{2}$ bosons: the KSA state. For a system of bosons of spin $\frac{1}{2}$ moving freely in space in zero magnetic field, no analog of the LPB state exists, since the Bose statistics forbid two particles in a relative s-state to have a spin singlet wave function. However, it turns out that this very consideration (or something closely related to it) can in certain circumstances produce fragmentation *even in the absence of any interparticle interaction.*

Consider the following thought-experiment, which was originally proposed by Siggia and Ruckenstein in the context of spin-polarized hydrogen more than two decades ago. We take a set of N spin-$\frac{1}{2}$ *mutually noninteracting* Bose particles in zero magnetic field and allow them to come to complete equilibrium (including the spin degree of freedom) with an environment which is at a temperature T much greater than the Bose condensation temperature T_c calculated from formula (1.3.20). It is clear that in the thermodynamic equilibrium state the net spin magnetization is zero (more precisely, the mean-square magnetization is $O(N)$ not $O(N^2)$). Next, we isolate the system from its environment and cool it down through T_c, and in fact to $T = 0$, by a process (such as evaporation, or adiabatic expansion)[32] *which conserves all components of the total spin.* What is the final state of the system?

Remarkably, it is only within the last four years that this simply posed problem has been definitively solved, by Kuklov and Svistunov (2002) and independently by Ashhab (2002). The answer is: Under the stated conditions macroscopic occupation of a single orbital state (i.e. "simple BEC in coordinate space only") is impossible, and the ground state involves macroscopic occupation of the *two* lowest orbital states ("fragmentation"). The argument to this conclusion is in retrospect spectacularly simple: If the ground state were to have all N particles in the same uniform orbital state (whether the lowest eigenfunction of the single-particle Hamiltonian or some other), then the Bose statistics require that the spin function be totally symmetric. But the most general totally symmetric pure state of N particles of spin $\frac{1}{2}$ corresponds to a total spin $\hat{S}^2 = N/2(N/2+1)$.[33] On the other hand, since all components of $\mathbf{S}$ were conserved in the cooling process, so is S^2, so that according to our earlier considerations S^2 is only $O(N)$, not $O(N^2)$! From this it follows at once that simple BEC

[32] It may be objected that any such process, if it is to result in a final "temperature", must involve interparticle collisions and hence finite interactions. However, we can make these arbitrarily weak, and more importantly, invariant under spin rotation, so the conclusion is not affected.

[33] Probably the easiest way of seeing this is to note that of the 2^N possible states of N spins $\frac{1}{2}$, exactly $N+1$ are totally symmetric, with $\hat{S}^2 = N/2(N/2+1), S_z = -N/2, , \ldots, N/2$.

is impossible. In fact, the above authors showed that the true "constrained" ground state has the form (apart from normalization and for N even)

$$\Psi_{\mathrm{KSA}} = (a^+_{0\uparrow}a^+_{1\downarrow} - a^+_{0\downarrow}a^+_{1\uparrow})^{N/2}|\mathrm{vac}\rangle \tag{2.6.2}$$

where 0, 1 denote the lowest two orbital eigenstates of the single-particle Hamiltonian and $\uparrow$, $\downarrow$ indicates $S_z = \pm\frac{1}{2}$. The state (2.6.2) is clearly antisymmetric under the exchange of the spins or of the orbital coordinates of any two atoms separately, but symmetric under the total exchange as it should be. As in the LPB case, ρ_2 has a single large eigenvalue, while ρ_1 has two degenerate eigenvalues. It seems likely that a state of the form (2.6.2) will remain the approximate ground state in the presence of weak repulsive interactions, and if so the system then has some very amusing properties; in particular, in a toroidal geometry it should develop a spontaneous *orbital* angular momentum, see Ashhab and Leggett (2003).

For completeness, it should be mentioned that the argument given in the last paragraph cannot as it stands exclude the possibility of simple BEC in a single-particle state such that the spin orientation depends on position. However, in the presence of an arbitrarily weak repulsive interaction this state appears to be unstable with respect to the KSA state (2.6.2).

It appears entirely possible that the above examples fail to exhaust the opportunities for obtaining fragmented states. However, in the rest of the book we shall see that most of the currently observed "exotic" properties of BEC systems do not require us to involve such states, and from now on I shall assume unless explicitly otherwise stated that any BEC (or pseudo-BEC) we encounter is "simple".

Appendices

2A The second-quantization formalism

In the context of relativistic quantum field theory, where it was originally invented, the formalism of "second quantization" embodies new physics, such as the possibility of creation of particle–antiparticle pairs. In a condensed-matter context, by contrast, it is essentially just a book-keeping device, albeit an extremely convenient one, which allows one to keep track in an automatic way of the effects of the relevant (Bose–Einstein or Fermi–Dirac) "statistics". I hope that most readers of this book will have already acquired some familiarity with the formalism by use; for any who have not, this appendix is intended as a concise and self-contained introduction.

Consider a set of N identical bosons, with each of which is associated an identical single-particle Hilbert space spanned by a complete orthonormal set of basis vector $|i\rangle$; the Hilbert space of the many-body system is the subspace of the tensor product of those N spaces which is completely symmetric under interchange of any two particles. It is convenient to introduce a definite basis for the many-body system, namely those states in which exactly n_i of the N bosons occupy state i, where n_i may be any positive integer $\leq N$; we denote such a state by

$$|\{n_i\}\rangle \equiv |\ldots, n_i, n_j, n_k, \ldots\rangle \tag{2.A.1}$$

It is convenient at this stage to relax the restriction on N and allow the latter to take any positive definite integral value; then, since the values of the n_i for different i are mutually independent, the total Hilbert space of the many-body system may be regarded as a tensor product of spaces associated with the different i separately, and any particular state $\{n_i\}$ may be written symbolically

$$|\{n_i\}\rangle = \prod_i |n_i\rangle \tag{2.A.2}$$

Let us now define a set of "annihilation" operators $\hat{a}_i$ and their Hermitian conjugate "creation" operators $\hat{a}_i^\dagger$ by their matrix elements in the representation (2.A.2), as follows:

$$\langle\{n_i'\}|\hat{a}_i|\{n_i\}\rangle = \sqrt{n_i}\; \delta_{n_i-1,n_i'} \prod_{j\neq i} \delta_{n_j,n_j'} \tag{2.A.3a}$$

$$\langle\{n_i'\}|\hat{a}_i^\dagger|\{n_i\}\rangle = \sqrt{n_i+1}\; \delta_{n_i+1,n_i'} \prod_{j\neq i} \delta_{n_j,n_j'} \tag{2.A.3b}$$

In words: $\hat{a}_i$ decreases the number of particles in state i by one, leaving all the other n_j unchanged, and multiplies the resulting (normalized) state by $\sqrt{n_i}$; similarly, $\hat{a}_i^\dagger$ increases the number in state i by one, leaving all the other n_j unchanged, and multiplies the resulting normalized state by $\sqrt{n_i+1}$. It is easy to check that $\hat{a}_i^\dagger$ is indeed the Hermitian conjugate of $\hat{a}_i$; moreover, it is clear from (2.A.3a) that acting on any state with $n_i = 0$ the operator $\hat{a}_i$ must give zero identically. The unique vacuum state (corresponding to $N = 0$) is $|\ldots 0, 0, 0, \ldots\rangle \equiv |\text{vac}\rangle$, and it is easy to check from (2.A.3b) that the normalized state $|\{n_i\}\rangle$ may be written in the form

$$|\{n_i\}\rangle = \prod_i \{(n_i!)^{-1/2}(a_i^\dagger)^{n_i}\}|\text{vac}\rangle \tag{2.A.4}$$

(where as usual the symbol 0! is interpreted as 1). The "number operator" $\hat{n}_i$, whose eigenvalues are the positive integers n_i, is given by

$$\hat{n}_i \equiv \hat{a}_i^\dagger \hat{a}_i \tag{2.A.5}$$

From their definitions (2.A.3), the operators $\hat{a}_i$ and $\hat{a}_i^\dagger$ satisfy the fundamental commutation relations

$$[\hat{a}_i, \hat{a}_j] = [\hat{a}_i^\dagger, \hat{a}_j^\dagger] = 0 \quad \forall i, j \tag{2.A.6a}$$

$$[\hat{a}_i, \hat{a}_j^\dagger] = \delta_{ij} \tag{2.A.6b}$$

Consider now a transformation of the single-particle basis $|i\rangle$ to a new complete orthonormal basis $|\mu\rangle$, such that any single-particle state $|\psi\rangle$ may be written

$$|\psi\rangle \equiv \sum_i c_i|i\rangle \equiv \sum_\mu c_\mu|\mu\rangle \tag{2.A.7}$$

The transformation must be unitary, and we define a unitary matrix $\hat{U}$ such that

$$|\mu\rangle \equiv \sum_i U_{\mu i}^*|i\rangle \tag{2.A.8a}$$

so that

$$c_\mu = \sum_i U_{\mu i} c_i \tag{2.A.8b}$$

How do the $\hat{a}_i$ and $\hat{a}_i^\dagger$ behave under this transformation? When they act on the vacuum, the answer is clear: since a state with a single particle in $|i\rangle$ is generated by $\hat{a}_i^\dagger|\text{vac}\rangle$, and similarly for $|\mu\rangle$, it immediately follows that when acting on $|\text{vac}\rangle$

$$\hat{a}_\mu^\dagger = \sum_i U_{\mu i}^* \hat{a}_i^\dagger \tag{2.A.9a}$$

It is not immediately obvious (but can be shown with a little labor[34]) that (2.A.9a) is actually correct independently of the state to the right, i.e. as a general operator relation, and thus that

$$\hat{a}_\mu = \sum_i U_{\mu i}\ \hat{a}_i \tag{2.A.9b}$$

i.e. the annihilation operator transforms like the coefficients in the single-particle basis. It is clear from the unitarity of $\hat{U}$ that the transformations (2.A.9) preserve the form of the commutation relations, i.e.

$$[\hat{a}_\mu, \hat{a}_\nu] = [\hat{a}_\mu^\dagger, \hat{a}_\nu^\dagger] = 0 \tag{2.A.10a}$$

$$[\hat{a}_\mu, \hat{a}_\nu^\dagger] = \delta_{\mu\nu} \tag{2.A.10b}$$

Consider now an arbitrary Hermitian single-particle operator $\hat{\Omega}$ with matrix elements $\Omega_{\mu\nu}$ in the μ-basis. We can certainly find a basis (call it i) which diagonalizes $\hat{\Omega}$ with eigenvalues Ω_i. We are interested in the operator

$$\hat{\Omega} \equiv \sum_{s=1}^{N} \hat{\Omega}^{(s)} \equiv \sum_{s=1}^{N} \hat{\Omega}(\hat{\xi}_s) \tag{2.A.11}$$

where the sum over s is over the N different *particles* in the system. It is clear that in the i-representation the expression for $\hat{\Omega}$ is simply

$$\hat{\Omega} = \sum_i \hat{n}_i \Omega_i \equiv \sum_i \Omega_i \hat{a}_i^\dagger \hat{a}_i \tag{2.A.12}$$

Then from Eqns. (2.A.9) it is clear that in a general (μ-) representation we have (since $\hat{a}_i = U_{i\mu}^\dagger \hat{a}_\mu$)

$$\hat{\Omega} = \sum_{\mu\nu} \Big(\sum_i U_{\mu i} \Omega_i U_{i\nu}^\dagger \Big) \hat{a}_\mu^\dagger \hat{a}_\nu \equiv \sum_{\mu\nu} \Omega_{\mu\nu} \hat{a}_\mu^\dagger \hat{a}_\nu \tag{2.A.13}$$

This simple result is the key payoff of the second-quantized formalism.

It is now very straightforward to derive representations similar to (2.A.13) for operators involving two(or more) particles. Consider a two-particle operator[35] $\hat{V}$ with matrix elements

$$V_{\kappa\lambda,\mu\nu} \equiv \int d1 \int d2 \phi_\kappa^*(1) \phi_\lambda^*(2) \hat{\Omega}_2 \phi_\mu(1) \phi_\nu(2) \tag{2.A.14}$$

[34] A quick argument is that the transformation must have the property that if we choose the initial basis i so that $\langle \hat{a}_i a_j \rangle = n_i \delta_{ij}$, then the quantity $\langle n_\mu \rangle \equiv \langle \hat{a}_\mu^\dagger a_\mu \rangle$ must be given, for all $\{n_i\}$, by $\sum_i |U_{ki}|^2 n_i$. Equations (2.A.9) are obviously a sufficient condition for this, although it needs a little more argument to show that they are also necessary.

[35] The most familiar example of such an operator is the two-body potential energy, hence the choice of notation.

where "1" and "2" are shorthand for the variables describing particles 1 and 2 respectively. In this case we are interested in the quantity (though see below)

$$\hat{V} \equiv \frac{1}{2}\sum_{s=1}^{N}\sum_{s'=2}^{N}\hat{V}(s,s') \equiv \frac{1}{2}\sum_{ss'}\hat{V}(\hat{\xi}_s,\hat{\xi}_{s'}) \tag{2.A.15}$$

Now, we can always write $\hat{V}$ in the form of a sum of products of two single-particle operators referring only to particles 1 and 2 respectively, i.e.

$$\hat{V} = \sum_{lm}\lambda_{lm}\hat{\Omega}_1^{(l)}(1)\hat{\Omega}_1^{(m)}(2) \tag{2.A.16}$$

Then, applying the result (2.A.13) to the $\hat{\Omega}_1^{(l)}$ separately and reconstituting the result, we have

$$\hat{V} = \frac{1}{2}\sum_{\kappa\lambda\mu\nu}V_{\kappa\lambda,\mu\nu}\hat{a}_\kappa^\dagger\hat{a}_\mu\hat{a}_\lambda^\dagger\hat{a}_\nu \tag{2.A.17}$$

There is one slight complication here: Very often, we wish to exclude from the sum over s and s' in (2.A.15) the terms with[36] $s = s'$. The appropriate modification turns out to be to "normal-order" the expression (2.A.17), that is to rearrange it so that all creation operators stand ahead of all annihilation operators:

$$\hat{V}_{\text{mod}} = \frac{1}{2}\sum_{\kappa\lambda\mu\nu}V_{\kappa\lambda\mu\nu}\hat{a}_\kappa^\dagger\hat{a}_\lambda^\dagger\hat{a}_\nu\hat{a}_\mu \tag{2.A.18}$$

The reason for the relative ordering of a_μ and a_ν will be explained below. Since the easiest proof uses the coordinate representation of the operators $\hat{a}$, $\hat{a}^\dagger$, I postpone it until we have discussed this.

All the above analysis applies to a system of identical bosons. In the case of N identical fermions, the crucial difference is that the allowed values of each of the n_i are only zero and one; thus, the only non-zero matrix elements of the annihilation and creation operators $\hat{a}_i$, $\hat{a}_i^\dagger$ are

$$\langle\{n_i'\}|\hat{a}_i|\{n_i\}\rangle = \delta_{n_i',0}\delta_{n_i,1}\prod_{j\neq i}\delta_{n_j,n_j'} \tag{2.A.19a}$$

$$\langle\{n_i'\}|\hat{a}_i^\dagger|\{n_i\}\rangle = \delta_{n_i',1}\delta_{n_i,0}\prod_{j\neq i}\delta_{n_j,n_j'} \tag{2.A.19b}$$

Equations (2.A.19) imply that when acting on the allowed states, $\hat{a}_i$ and $\hat{a}_i^\dagger$ satisfy the *anti*commutation relation

$$\{\hat{a}_i,\hat{a}_i^\dagger\} \equiv \hat{a}_i\hat{a}_i^\dagger + \hat{a}_i^\dagger\hat{a}_i = 1 \tag{2.A.20}$$

Moreover, to avoid generating unphysical states (e.g. $n_i = 2$) from the physical ones, we must have

$$(\hat{a}_i^\dagger)^2 = (\hat{a}_i)^2 \equiv 0 \tag{2.A.21}$$

Now by the argument given above (which in the fermion case needs no supplementation) the transformation properties of the $\hat{a}$'s and $\hat{a}^\dagger$'s must be given as in the boson

[36]For example, in the physically realistic case that $\hat{V}$ is the two-particle interaction, we wish to exclude any interaction of a particle with itself.

case by Eqns. (2.A.9). If then Eqns. (2.A.20) and (2.A.21) are to be invariant under unitary transformation, i.e. valid for any basis μ, we must postulate that the operators for *different* $i(\mu)$ also anticommute. In other words we must postulate that for an arbitrary complete orthonormal basis $\{|\mu\rangle\}$ we have

$$\{\hat{a}_\mu, \hat{a}_\nu\} = \{\hat{a}_\mu^\dagger, \hat{a}_\nu^\dagger\} = 0. \tag{2.A.22a}$$

$$\{\hat{a}_\mu, \hat{a}_\nu^\dagger\} = \delta_{\mu\nu} \tag{2.A.22b}$$

(where as above $\{\hat{A}, \hat{B}\} \equiv \hat{A}\hat{B} + \hat{B}\hat{A}$). Given these results, the arguments leading to Eqns. (2.A.13) and (2.A.17) go through exactly in parallel to the boson case. (On Eqn. (2.A.18), see below.)

A particularly useful basis is the coordinate one, obtained by identifying the basis vectors $|\mu\rangle$ with the eigenstates $|\boldsymbol{r}\rangle$ of the coordinate operator. (Recall that the one-particle Schrödinger wave function $\psi(\boldsymbol{r})$ is simply the set of coefficients c_μ in this basis.) When we use this basis, it is conventional to change the notation and denote the annihilation and creation operators a_μ, $a_\mu^\dagger$ by $\hat{\psi}(\boldsymbol{r})$ and $\hat{\psi}^\dagger(\boldsymbol{r})$ respectively; $\hat{\psi}$ and $\hat{\psi}^\dagger$ are often called "field operators". They satisfy, in the case of bosons, the commutation relations

$$[\hat{\psi}(\boldsymbol{r}), \hat{\psi}(\boldsymbol{r}')] = [\psi^\dagger(\boldsymbol{r}), \psi^\dagger(\boldsymbol{r}')] - 0, \quad \forall \boldsymbol{r}, \boldsymbol{r}' \tag{2.A.23a}$$

$$[\hat{\psi}(\boldsymbol{r}), \hat{\psi}^\dagger(\boldsymbol{r}')] = \delta(\boldsymbol{r} - \boldsymbol{r}') \tag{2.A.23b}$$

and for fermions similar relations with the commutators replaced by anticommutators. The appropriate transcription of the prescriptions (2.A.13) and (2.A.17) to this notation is that if the matrix elements of a single-particle operator $\hat{\Omega}$ in the coordinate representation are $\langle \boldsymbol{r}|\Omega|\boldsymbol{r}'\rangle \equiv \Omega(\boldsymbol{r}, \boldsymbol{r}')$, then

$$\sum_{s=1}^{N} \hat{\Omega}(r_s) = \iint dr\, dr'\, \hat{\psi}^\dagger(r)\Omega(r, r')\hat{\psi}(r') \tag{2.A.24}$$

Thus, for example, the second-quantized expressions for the total external potential energy $\hat{U}$ and kinetic energy $\hat{K}$ of the system are respectively[37]

$$\hat{U} \equiv \sum_i U(\boldsymbol{r}_i) = \int d\boldsymbol{r} U(\boldsymbol{r})\hat{\psi}^\dagger(\boldsymbol{r})\hat{\psi}(\boldsymbol{r}) \tag{2.A.25}$$

$$\hat{K} \equiv -\frac{\hbar^2}{2m}\sum_i \boldsymbol{\nabla}_i^2 = -\frac{\hbar^2}{2m}\int \hat{\psi}^\dagger(\boldsymbol{r})\boldsymbol{\nabla}^2\hat{\psi}(\boldsymbol{r})\mathrm{d}\boldsymbol{r} \tag{2.A.26}$$

A similar prescription, corresponding to (2.A.17), follows for an arbitrary two-particle operator: since the case of overwhelming interest is when $\hat{V}$ is diagonal in the coordinate representation, I quote here only the result for that case:

$$\hat{V} \equiv \frac{1}{2}\sum_{i=1}^{N}\sum_{j'=1}^{N} V(\boldsymbol{r}_i, \boldsymbol{r}_j') = \frac{1}{2}\iint d\boldsymbol{r} d\boldsymbol{r}'\, V(\boldsymbol{r}, \boldsymbol{r}')\, \hat{\psi}^\dagger(\boldsymbol{r})\hat{\psi}(\boldsymbol{r})\hat{\psi}^\dagger(\boldsymbol{r}')\hat{\psi}(\boldsymbol{r}') \tag{2.A.27}$$

[37] At this point I revert to the more familiar index i to sum over particles, since there is now no risk of confusion with the basis vectors.

At this stage it is convenient to return to the point raised above, namely that if $\hat{V}$ represents the physical interaction energy then we need to subtract off from (2.A.27) the terms with $i = j'$. According to (2.A.24), these correspond to the expression

$$\frac{1}{2}\sum_{i=1}^{N} V(r_i, r_j) = \frac{1}{2}\int d\boldsymbol{r}\, V(\boldsymbol{r}, \boldsymbol{r})\hat{\psi}^\dagger(\boldsymbol{r})\hat{\psi}(\boldsymbol{r}) \tag{2.A.28}$$

When we subtract (2.A.28) from (2.A.27) and use the (anti)commutation relations (2.A.23), we find that the physical interaction term is given by

$$\hat{V} \equiv \frac{1}{2}\sum_{i\neq j} V(\boldsymbol{r}_i, \boldsymbol{r}_j) = \frac{1}{2}\int d\boldsymbol{r} \int d\boldsymbol{r}'\, \hat{\psi}^\dagger(\boldsymbol{r})\hat{\psi}^\dagger(\boldsymbol{r}')V(\boldsymbol{r}, \boldsymbol{r}')\hat{\psi}(\boldsymbol{r}')\hat{\psi}(\boldsymbol{r}) \tag{2.A.29}$$

and when transcribed into an arbitrary basis μ this then yields (2.A.18). Note carefully the order of the two annihilation operators in (2.A.29) (and (2.A.18)), which makes no difference in the boson case but in the fermion case is essential to get the right overall sign.

To conclude this appendix, let us consider the simplest (Hartree–Fock) scheme for decoupling the expectation values of operators of the form (2.A.10), namely the quantity

$$\langle V\rangle \equiv \frac{1}{2}\sum_{\kappa\lambda\mu\nu} V_{\kappa\lambda,\mu\nu}\langle a_\kappa^\dagger a_\lambda^\dagger a_\nu a_\mu\rangle \tag{2.A.30}$$

We know (simply because $\sum_\nu \langle\hat{a}_\nu^\dagger \hat{a}_\nu\rangle \equiv \sum_\nu \langle n_\nu\rangle = N \neq 0$) that independently of the basis some of the quantities $\langle\hat{a}_\nu^\dagger\hat{a}_\nu\rangle$ must be finite, while, depending on the basis chosen, quantities of the form $\langle\hat{a}_\mu^\dagger\hat{a}_\nu\rangle$ ($\mu \neq \nu$ may or may not be. Consequently it makes some sense to choose a basis in which we have reason to expect the latter quantities to be small (e.g. for a weakly interacting Bose gas in free space, the plane-wave states) and make the ansatz ($\kappa \neq \lambda$)

$$\langle a_\kappa^\dagger a_\lambda^\dagger a_\nu a_\mu\rangle \cong \langle a_\kappa^\dagger a_\mu\rangle\langle a_\lambda^\dagger a_\nu\rangle + \eta\langle a_\kappa^\dagger a_\nu\rangle\langle a_\lambda^\dagger a_\mu\rangle \cong \delta_{\kappa\mu}\delta_{\lambda\nu}\langle n_\kappa\rangle\langle n_\lambda\rangle + \eta\delta_{\kappa\nu}\delta_{\lambda\mu}\langle n_\kappa\rangle\langle n_\lambda\rangle \tag{2.A.31}$$

where $\eta \equiv +1$ for bosons and -1 for fermions. Then the expectation value of the potential energy (2.A.30) is

$$\langle V\rangle \cong \frac{1}{2}\sum_{\kappa\lambda}(V_{\kappa\lambda,\kappa\lambda} + \eta V_{\kappa\lambda,\lambda\kappa})\langle n_\kappa\rangle\langle n_\lambda\rangle \tag{2.A.32}$$

where the first term is the Hartree term and the second the Fock term. Note that for a contact potential of the form (2.3.3), if the basis is plane-wave states, and any internal quantum numbers are identical, $V_{\kappa\lambda,\lambda\kappa} \equiv V_{\kappa\lambda,\kappa\lambda}$, so the total expectation value for $\kappa \neq \lambda$ is twice the Hartree value for bosons and zero for fermions. If $\kappa = \lambda$, the expectation value is still zero for fermions (exclusion principle!) but for bosons we find the *single* term $V_{\kappa\kappa,\kappa\kappa}\langle n_\kappa\rangle\langle n_\kappa - 1\rangle \cong V_{\kappa\kappa,\kappa\kappa}\langle n_\kappa\rangle^2$ if $\langle n_\kappa\rangle$ is large.

2B Relative number and phase states

Consider a system of N identical bosons, each of which is restricted to occupy a Hilbert space spanned by two orthonormal eigenvectors $|1\rangle$ and $|2\rangle$, which for the moment we choose arbitrarily; we denote the corresponding single-particle creation operators

a_1^+, a_2^+. The Hilbert space of the N-particle system is thus $N+1$ dimensional and spanned by the eigenvectors in the relative Fock basis cf. Eqn. (2.A.4)

$$
\begin{aligned}
|M\rangle \equiv & \left[\left(\frac{N}{2}+M\right)!\left(\frac{N}{2}-M\right)!\right]^{-1/2} \\
& \times (a_1^+)^{N/2+M}(a_2^+)^{N/2-M}|\text{vac}\rangle \\
M = & -N/2, -N/2+1 \cdots N/2, \quad N \quad \text{even} \\
= & -N/2-\frac{1}{2}, -N/2+\frac{1}{2} \cdots N/2-\frac{1}{2}, \quad N \quad \text{odd}
\end{aligned} \tag{2.B.1}
$$

where $|\text{vac}\rangle$ denotes the vacuum; it is clear that M is the (nearest integer to)[38] half the difference ΔN in the numbers of particles in states 1 and 2. This model is then isomorphic to the $(N+1)$-dimensional representation of the group SU(2). It is convenient to define operators $\hat{M}$, $\Delta\hat{N}$ by

$$\hat{M} \equiv \tfrac{1}{2}(\hat{a}_1^+\hat{a}_1 - \hat{a}_2^+\hat{a}_2) \equiv \tfrac{1}{2}\Delta\hat{N} \tag{2.B.2}$$

Consider now the "relative GP state"

$$\Psi_N(\Delta\varphi : \alpha, \beta) = (N!)^{-1/2}(\alpha e^{i\Delta\varphi/2}\hat{a}_1^+ + \beta e^{-i\Delta\varphi/2}\hat{a}_2^+)^N|\text{vac}\rangle$$

where we choose α, β to be real and positive and such that

$$\alpha^2 + \beta^2 = 1 \tag{2.B.3}$$

We can expand this state in the Fock basis (2.B.1):

$$\Psi_N = \sum_M |C_M|e^{iM\Delta\varphi}|M\rangle \tag{2.B.4}$$

$$|C_M| \equiv \frac{\alpha^{N/2-M}\beta^{N/2+M}}{[(N/2-M)!(N/2+M)!]^{1/2}} \tag{2.B.5}$$

It is clear that for large N the probability distribution $P_M = |C_M|^2$ is strongly peaked around a value $\bar{M}$ close to $N(\alpha^2-\beta^2)$.

From Eqn. (2.B.4) we see that up to normalization the Fock state $|M\rangle$ can be expressed as (e.g.) a linear superposition of "symmetric"[39] GP states $\Psi(\Delta\varphi)$:

$$|M\rangle = \text{const.} \int d(\Delta\varphi)\exp(-iM\Delta\varphi)\Psi(\Delta\varphi : \alpha = \beta = 2^{-1/2}) \tag{2.B.6}$$

This result is often helpful in trying to visualize the relation between "relative number" and "relative phase" eigenstates.

It is interesting to enquire whether it is possible to define a "relative phase operator" whose eigenstates correspond to the relative GP states (2.B.3), and which would be canonically conjugate to $\Delta\hat{N}$. The answer is that while this is not rigorously possible,

[38] This choice of definition for M, while technically convenient, means that some of the formulas below, which are written explicitly for even N, need correction by factors $\sim(N^{-1})$ for odd N. This affects none of the conclusions and will not be noted explicitly below.

[39] It is of course also possible to express it in terms of asymmetric GP states, but this is not necessary for our present purposes.

an approximate solution can be found for large N, provided that we never have to deal with any states which have large projections on the "extreme" states $|M = \pm N/2\rangle$; see Leggett 2001, Section VII.C. The appropriate definition under these conditions is (ref. cit.)

$$\Delta\hat{\varphi} \equiv -i \cdot \arg\{[(N/2 - \hat{M})(N/2 + \hat{M} + 1)]^{-1/2} \hat{a}_1^+ \hat{a}_2\} \tag{2.B.7}$$

and satisfies (approximately)

$$[\hat{M}, \Delta\hat{\varphi}] \equiv \frac{1}{2}[\Delta\hat{N}, \Delta\hat{\varphi}] = -i \tag{2.B.8}$$

3

Liquid ^{4}He

From the thermodynamic point of view condensed ^{4}He is a very simple system indeed: the ^{4}He atom has an electronic closed shell and zero nuclear spin, thus no internal degrees of freedom, and while one would expect that external electric and magnetic fields should in principle change the atomic structure (in the latter case via diamagnetic effects) and thus the interatomic potentials, and thereby modify the behavior, simple order-of-magnitude estimates indicate (cf. Leggett 1992) that at commonly attainable field values these effects are very tiny. Thus the only relevant thermodynamic variables are temperature (T) and pressure (P).

The phase diagram of ^{4}He in the P-T plane is shown in Fig. 3.1. In this book I shall not be much interested in the solid phase (though some remarks about some intriguing recent experiments are made in Chapter 8, Section 8.3). There are two different liquid phases, conventionally known as He-I and He-II, separated by a line of second-order phase transitions in the P-T plane known as the λ-line. The λ-line is marked by a sharp peak in the specific heat plotted as a function of temperature (this gives the λ-line its name, since the shape of the anomaly resembles the Greek letter λ), and by weaker singularities in other thermodynamic properties such as the thermal expansion

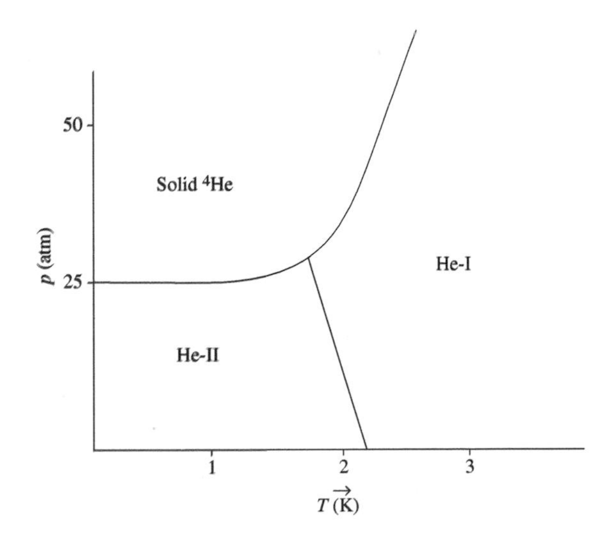

Fig. 3.1 The phase diagram of ^{4}He below 4 K.

coefficient, but appears to have no observable effect on the density profile of the liquid, which as already noted (Chapter 1, Section 1.5) is flat up to a few atomic diameters from the walls. While in the higher-temperature He-I phase the liquid appears to behave qualitatively like any other simple liquid, the He-II phase below the λ-line is characterized by a whole range of exotic properties of which the most dramatic is superfluidity; thus, "superfluid ^{4}He" is used synonymously for the He-II phase and "normal ^{4}He" for the He-I phase. In the next section I review these exotic properties. For a more complete account, see e.g. Wilks (1967).

3.1 Anomalous properties of the He-II phase

In 1938 two groups (Allen and Misener in Cambridge and Kapitza in Moscow) simultaneously conducted experiments in which liquid ^{4}He was driven between two reservoirs, maintained at given pressure difference ΔP, through a narrow plate-shaped channel. By measuring the ratio of mass flow $\dot{m}$ to ΔP and applying standard hydrodynamic theory, one should be able to infer the viscosity η of the liquid. What both groups found was that while the value of η in the He-I phase was not particularly anomalous, the value inferred in this way appeared to jump by a factor of at least 1500 the moment the temperature fell below the λ-line. (We now know that in such a geometry the ratio $\dot{m}/\Delta P$ diverges as $\Delta P \to 0$, so that the very concept of "viscosity" in the usual sense is inapplicable in the He-II phase.) It was for this striking behavior that Kapitza coined the name "superfluidity"; the liquid behaves, in the He-II phase, as if it had zero viscosity (just as in the "superconducting" phase a metal behaves as if it had zero resistance).

The frictionless flow characteristic of the superfluid phase manifests itself in other exotic phenomena. If we take a (closed) beaker which is leak-proof for any ordinary liquid (including He-I) and fill it with liquid helium, we find that below T_λ it may develop "superleaks," that is, the helium may flow out through pores which are so narrow that any ordinary liquid would be completely clamped by its viscosity. An even more spectacular effect is that if the beaker is open, the liquid may flow up over the sides and drip off the bottom! This is explained by the fact that even in the normal phase the sides of the beaker above the bulk liquid surface are covered, because of the Van der Waals attraction, by a thin film of liquid (the "Rollin film"), but this film is clamped, like the liquid in the narrow pores, by its viscosity. When we cool through the λ-temperature, the vanishing of the viscosity allows the film to move freely. This phenomenon is known as "film creep".

Although the phenomenon of "vanishing viscosity" is the most dramatic manifestation of "superfluidity" in He-II, we have already seen in Chapter 1, Section 1.6, that that phenomenon is actually better regarded as the conjunction of two more "elementary" effects, namely "nonclassical rotational inertia" (the Hess–Fairbank effect), in which the liquid ceases to rotate with a rotating container, and "metastability of supercurrents" in which, in an annular geometry, the liquid continues to circulate indefinitely despite the fact that the walls are at rest. Once we have explored the basis of these two "elementary" effects in Section 3.3, I will return to the question of the relationship between them and the original 1938 experiments.

If the liquid (or at least part of it – see below) has zero friction vis-a-vis the container walls (metastability of supercurrents), one might perhaps intuitively expect that an object moving through the (stationary) liquid would also experience reduced or zero friction. One manifestation of this is the Andronikashvili oscillating-disk experiment: if a pile of disks is suspended in liquid helium on a torsional fiber and made to perform torsional oscillations, it is found that above T_λ the whole of the liquid between the disks follows their oscillation, while below T_λ only a fraction (equal in fact to the "normal fraction" defined in Chapter 1, Section 1.6) does so. A somewhat related effect is that of increased ion mobility – provided they are traveling below a critical velocity (see below, Section 3.6) ions traveling in liquid He-II appear to experience a reduced viscosity (which in fact vanishes in the limit $T \to 0$) and hence to possess an increased mobility in the presence of an external electric field.

Just as the ratio of mass flow $\dot{m}$ to pressure head ΔP diverges in the He-II phase as $\Delta P \to 0$, so does the ratio of heat flow $\dot{Q}$ to temperature drop ΔT; thus, the thermal conductivity κ, like the viscosity η, is not defined, and as we shall see the basic mechanism of heat flow in He-II is a peculiar kind of convection. Associated with this phenomenon are various other exotic effects (some of them predicted ahead of their experimental discovery, on the basis of the two-fluid model– see below) such as the "fountain effect", in which application of a temperature difference across a superleak leads to a sudden pressure increase, sufficient to produce a fountain-like spout of the liquid, and "second sound", an (almost) pure temperature wave.

A rather different kind of manifestation of exotic behavior in the He-II phase is the existence in it of vortex lines and rings with circulation quantized in units of h/m. An array of vortex lines is believed to be the stable configuration of bulk He-II in a (simply connected) rotating cylinder, and has been directly imaged by "decoration" with electrons, while vortex rings have not been directly imaged, but are believed to be responsible for much of the behavior of ions in the liquid, and in addition it is believed that in most geometries they are a determining factor in the flow properties. Of course, vortex lines and rings are not unique to He-II, but in normal systems they are not quantized and not metastable.

Finally, following the observation and microscopic understanding of the Josephson effect in superconductors, a number of experiments have been conducted to search for this and related phenomena in ^{4}He. Although the results are less spectacular than in the superconducting case, there is little doubt that the effect exists.

3.2 Direct evidence for BEC in He-II

Although London's 1938 proposal, namely that the anomalous experimental properties of the He-II phase were due to the existence in that phase of BEC, won widespread acceptance from the start and thus forms the basis of most theoretical work on that phase for the last half-century, it has actually proved exceedingly difficult to provide direct experimental evidence for his hypothesis. As we shall see in Chapter 4, this state of affairs is in stark contrast to that in the Bose alkali gases, where the onset of BEC is reflected, inter alia, in a dramatic change in the density profile. Why is no such effect seen in liquid ^{4}He? As already indicated in Chapter 1, Section 1.5, the reason has to do with the "crude" form of the potential exerted by a typical container, and more importantly with the strong and mostly repulsive interatomic interactions, which resist any appreciable change in the density. This consideration also explains the absence, at an observable level, of another effect which one might think could signal the onset of BEC, namely a decrease in the static density correlations $|\Psi(\boldsymbol{r},\boldsymbol{r}')|^2$ at short distances, due to the fact that the factor of 2 in Eqn. (2.3.7) does not occur when both particles are in the condensate (cf. 2.3.5); numerical simulations (Ceperley and Pollock 1986) indicate that while such an effect exists, it is so tiny that it is almost certainly currently undetectable in neutron or X-ray scattering experiments (see Fig. 2a of the cited reference).[1]

Thus, it seems that the only feasible way of detecting the onset of BEC in He-II is to look for it in the most direct possible way, that is to seek evidence that a single one-particle state, namely (in bulk) the $\boldsymbol{k} = 0$ state, is macroscopically occupied. The favored way of doing this (see Sokol 1994) is based on the following idea, which is due to Hohenberg and Platzman (1966): Imagine for a moment that we could ignore interactions completely (the "impulse approximation"), and consider the scattering from the system of a massive particle such as a neutron. Suppose that the initial momentum of the atom doing the scattering is $\hbar\boldsymbol{k}$. The energy and momentum loss of the neutron are denoted respectively $\hbar\omega$ and $\hbar\boldsymbol{Q}$, and we define the quantities[2]

$$\hbar\omega_R \equiv \frac{\hbar^2 Q^2}{2m}, \quad \boldsymbol{v}_R \equiv \frac{\hbar\boldsymbol{Q}}{m} \tag{3.2.1}$$

where m is the atomic mass. Then the laws of conservation of energy and momentum imply the result

$$\omega = \omega_R + \boldsymbol{v}_R \cdot \boldsymbol{k} \tag{3.2.2}$$

and since the atomic scattering cross-section is a function only of $\boldsymbol{Q}$, this implies that for fixed $\boldsymbol{Q}$ the number of neutrons scattered with energy loss $\hbar\omega$ is proportional to the quantity

$$S_{IA}(\boldsymbol{Q},\omega) \equiv \sum_{\boldsymbol{k}} n_{\boldsymbol{k}} \delta(\omega - \omega_R - \boldsymbol{v}_R \cdot \boldsymbol{k}) \tag{3.2.3}$$

where n_k is the number of atoms with momentum $\hbar\boldsymbol{k}$. If $n_{\boldsymbol{k}}$ is rotationally invariant and moreover smoothly decreasing as a function of $|\boldsymbol{k}|$, it is clear that when plotted (for fixed $\boldsymbol{Q}$) as a function of the variable $y \equiv (\omega - \omega_R)/v_R$ the expression (3.2.3) is a smooth symmetric function with a maximum at $y = 0$. If on the other hand the $\boldsymbol{k} = 0$ state has an occupation N_0 of order N (while all other $n_{\boldsymbol{k}}$ are of order unity) then scattering by these "condensate" atoms should give rise to a δ-function peak at $y \equiv 0$, superimposed on the smooth background due to the noncondensate ($\boldsymbol{k} \neq 0$) atoms.

Ignoring for a moment any practical difficulties in isolating such a δ-function peak, let us ask: Can we apply the above argument, or some variant of it, to real liquid ^{4}He, where the interatomic interactions are certainly not negligible? The problem lies in the so-called "final-state effects": as the atom which has scattered the neutron recoils, it may transfer part or all of its momentum and energy to other atoms, so that the simple formula (3.2.2), and hence (3.2.3), no longer applies. It is known that

[1] One might perhaps regard the change in sign of $d\rho/dT$ close to T_λ (see Donnelly 1967, Fig. 1.4) as qualitative evidence for this effect.

[2] Sometimes known (perhaps slightly confusingly) as the "free atom recoil energy and velocity".

these effects vanish in the limit $Q \to \infty$, but at the highest values of Q currently experimentally realistic ($\sim$30 Å^{-1}) they still have a nonnegligible influence on the observed scattering. In the literature various techniques for taking this complication into account have been proposed, the most sophisticated probably being that of Glyde (1994). Recently, Glyde et al. (2000) conducted neutron scattering measurements on liquid ^{4}He at saturated vapor pressure and at various temperatures (0.5, 1.5, 1.6, 2.3 and 3.5 K), using Q-values in the range $15 < Q < 29$ Å^{-1}. Analyzing their data by the above technique and taking into account finite instrumental resolution, etc., they inferred that the temperature-dependent value of the condensate fraction $f_0(T) \equiv N_0(T)/N$ is consistent[3] with the expression

$$f_0(T) = A\left(1 - \left(\frac{T}{T_\lambda}\right)^\gamma\right) \tag{3.2.4}$$

$$A = (7.25 \pm 0.75) \times 10^{-2}, \quad \gamma = 5.5 \pm 1.0 \tag{3.2.5}$$

The inferred value of the zero-temperature condensate fraction A is similar to but somewhat less than the value ($\sim$10%) inferred in earlier work (Snow and Sokol 1995). It is interesting that the most recent numerical simulations (Moroni et al. 1997) are consistent with the value (3.2.5) of A, while earlier calculations of the temperature-dependence of $f_0(T)$ (Ceperley and Pollock 1986) appear to be consistent with the form (3.2.4) (though the predicted value of A is rather larger than that in (3.2.5)).

While neutron scattering experiments have certainly provided the most quantitative evidence concerning the condensate in He-II, recently its existence has also been inferred from a different though related experiment, namely the evaporation of ^{4}He atoms from the free surface of the liquid under the impact of a phonon beam of well-defined wavenumber $\boldsymbol{q}$ (Wyatt 1998). The evaporation of a single atom is believed to be accomplished by the absorption of a single phonon, and in this process not only the energy but the component of $\boldsymbol{q}$ parallel to the surface should be conserved. On re-analyzing his earlier data, Wyatt found the remarkable result that while the magnitude of momentum of the evaporated atoms could not be measured, the distribution with respect to angle in the plane of the surface always has a sharp and narrow peak which is closely parallel to the angle of incidence (in this plane) of the phonon beam. Since a continuous distribution of the initial atomic momenta in the surface plane (with a width of the order of that inferred from X-ray data) could not give rise to such a peak, it seemed difficult to avoid the conclusion that the latter is due to scattering by condensate atoms – since the latter have zero in-plane momentum, the $\boldsymbol{q}_\parallel$-value of the evaporated atom should be just equal to that of the incident phonon, giving the observed effect. Such an interpretation would require that the condensate have appreciable density in the surface layers of the liquid, a condition which is certainly consistent with theoretical considerations (Griffin and Stringari 1996, Draeger and Ceperley 2002).

In sum, both experiment and (numerical) theory seem consistent with the hypothesis that in the He-II phase of liquid ^{4}He a finite fraction $f_0(T)$ of the atoms are

[3]Since there are only three data points in the He-II region, the fit to a power law is of course somewhat speculative. (In fact, theoretical considerations suggest a different T-dependence near T_λ.)

condensed into the $\boldsymbol{k} = 0$ state, with $f_0(T)$ having a maximum value in the range 6–8% at $T = 0$ and declining to zero as $T \to T_\lambda$ according to a formula reasonably approximated by (3.2.4). In the remainder of this chapter I shall assume that such a condensate indeed exists, although the quantitative details will not be particularly important.

3.3 The two-fluid model of He-II: static effects

A very large fraction of the exotic properties of the He-II phase of liquid ^{4}He can be satisfactorily explained by the *two-fluid model.* The modern quantitative form of this model was conceived by Landau in a stunning feat of phenomenological intuition, without invoking the notion of BEC;[4] however, given the evidence presented in the last section that this phenomenon is indeed occurring in ^{4}He, it seems overwhelmingly natural to frame the model in terms related to it. We start then from the considerations of Chapter 2, Section 2.2, and in particular from the definition (2.2.7) of the "superfluid velocity" $\boldsymbol{v}_\mathrm{s}(\boldsymbol{r}t)$. In this section and the next I shall present those consequences of the model which do not rely explicitly on the fact that the quantity $\varphi(\boldsymbol{r}, t)$ which occurs in (2.2.7) is the phase of the condensate wave function (i.e. crudely speaking, those effects where Planck's constant does not enter explicitly);[5] in Section 2.5 I turn to "explicitly quantum" effects, such as vortex lines and the Josephson effect.

As we have seen, the superfluid velocity (2.2.7) is uniquely defined in any region of space in which the condensate wave function $\chi_0(\boldsymbol{r}, t)$ is nonvanishing, whether or not the system is in equilibrium, and satisfies the irrotationality condition (2.2.8). By taking the time derivative of (2.2.8) and interchanging time and space derivatives, it follows that $\nabla \times \partial \boldsymbol{v}_\mathrm{s}/\partial t = 0$, and that $\partial \boldsymbol{v}_\mathrm{s}/\partial t$ can therefore be expressed as the gradient of some scalar quantity $\Omega(\boldsymbol{r}, t)$. I shall return in Section 3.4 to the identification of $\Omega(\boldsymbol{r}, t)$; what is important for the moment is to notice that in a system which is uniform in space and in (stable or metastable) thermal equilibrium with the boundary conditions, $\Omega(\boldsymbol{r}, t)$ *must be constant in space*, and consequently $\boldsymbol{v}_\mathrm{s}(\boldsymbol{r}, t)$ must be constant in time. The crucial point is that such constancy *is quite compatible with a nonzero value* of $\boldsymbol{v}_\mathrm{s}(\boldsymbol{r})$; for example, as we saw in Chapter 2, Section 2.5 (point (4)), in an annular geometry the circulation $\oint \boldsymbol{v}_\mathrm{s} \cdot \mathbf{d}\boldsymbol{l}$ can be nonzero (in fact equal to nh/m, but we do not need to know this at the present stage) even though the system has had plenty of time to equilibrate with the walls. In other words, $\boldsymbol{v}_\mathrm{s}(\boldsymbol{r}, t)$ *must be treated as an independent thermodynamic variable.*[6]

Consider now a state of the system in which it has come to thermodynamic equilibrium subject to a given form of $\boldsymbol{v}_\mathrm{s}(\boldsymbol{r})$. We would now like to introduce a second "velocity" characterizing the system, namely the *normal velocity* $\boldsymbol{v}_\mathrm{n}(\boldsymbol{r})$. It should be emphasized right away that $\boldsymbol{v}_\mathrm{n}$ is *not* (except in the trivial case of a noninteracting gas) the "mean velocity" of the noncondensed atoms; it is a more subtle concept,

[4]About which it indeed seems that Landau was skeptical to the end of his days.

[5]Or more precisely enters only via the "quantum unit of rotation" ω_c defined in Chapter 1 section 1.6 and below.

[6]We may compare the status of $\boldsymbol{v}_\mathrm{s}$ with that of the total number of atoms of a given type in (say) a binary alloy, which we routinely treat as conserved even in cases where (very rare) radioactive processes mean that strictly speaking it is not.

which may be expressed intuitively as "the velocity of the boundary conditions with which the system is (or could be) in equilibrium". Perhaps it is easiest to visualize what this means with the help of a concrete example: In Chapter 1, Section 1.6, we saw that the condition for equilibrium in a container rotating uniformly with angular velocity $\boldsymbol{\omega}$ is the minimization of an "effective" free energy F_{eff} given by Eqn. (1.5.3)

$$F_{\text{eff}} = F - \boldsymbol{\omega} \cdot \langle \hat{\boldsymbol{L}} \rangle \tag{3.3.1}$$

(where F is the free energy associated with the Hamiltonian $\hat{H}$ in the stationary container). Using a standard vector identity, we can write the second term in the form

$$\Delta F = -\int d\boldsymbol{r}\, \boldsymbol{v}_{\text{n}}(\boldsymbol{r}) \cdot \langle \hat{\boldsymbol{J}}(\boldsymbol{r}) \rangle \tag{3.3.2}$$

where $\hat{J}(\boldsymbol{r})$ is the mass current (momentum density) operator and $\boldsymbol{v}_{\text{n}}(\boldsymbol{r})$ is given by the simple formula

$$\boldsymbol{v}_{\text{n}}(\boldsymbol{r}) = \boldsymbol{\omega} \times \boldsymbol{r} \tag{3.3.3}$$

Although in real life the physical object (the container) with which the liquid is in equilibrium is present only at the walls, it is clear that (for example) the introduction of a very flimsy wire mesh, extending throughout the liquid, which is attached to the container and rotates with it ($\boldsymbol{v} = \boldsymbol{\omega} \times \boldsymbol{r}$), would not affect the equilibrium. Thus, in this example $\boldsymbol{v}_{\text{n}}(\boldsymbol{r})$ can indeed be regarded as "the velocity of a boundary with which (were it present) the system would be in equilibrium". A more formal definition, which follows from (3.3.2), is

$$\boldsymbol{v}_{\text{n}}(\boldsymbol{r}) \equiv \frac{\delta F}{\delta \langle \hat{\boldsymbol{J}}(\boldsymbol{r}) \rangle} \tag{3.3.4}$$

where the derivative is to be evaluated in thermal equilibrium *at constant* $\boldsymbol{v}_{\text{s}}(\boldsymbol{r})$. The definition (3.3.4) is somewhat more general than indicated by (3.3.2), e.g. it can apply to the case of an infinite moving straight tube. Note carefully that in contrast to $\boldsymbol{v}_{\text{s}}(\boldsymbol{r})$, which is defined for an arbitrary state of the system, the normal velocity $\boldsymbol{v}_{\text{n}}(\boldsymbol{r})$ is defined only under conditions of (possibly metastable) thermodynamic equilibrium; it is thus a typical *hydrodynamic* concept. We will see below that (as with more familiar hydrodynamic concepts) this definition can be generalized to cover cases where the equilibrium is only local rather than global. However, it should be noted that there are plenty of examples (e.g. Couette flow, which is steady-state but not equilibrium) where a sensible definition of $\boldsymbol{v}_{\text{n}}$ is not possible or, at least, not trivially obvious. In general there are no particular constraints on the space-dependence of $\boldsymbol{v}_{\text{n}}(\boldsymbol{r})$ (in particular, it does not have to obey any irrotationality condition analogous to (2.2.8)), and it is convenient for the following argument to imagine that it can be subjected to an arbitrary (but sufficiently slow) spatial variation.

Consider then the quantity (in general a tensor)

$$K(\boldsymbol{r}, \boldsymbol{r}') \equiv \frac{\delta \langle \hat{\boldsymbol{J}}(\boldsymbol{r}) \rangle}{\delta \boldsymbol{v}_{\text{n}}(\boldsymbol{r}')} \tag{3.3.5}$$

the derivative once more being taken in thermal equilibrium at constant $\boldsymbol{v}_{\text{s}}(\boldsymbol{r})$. In general, $K(\boldsymbol{r}, \boldsymbol{r}')$ will have a finite range in $|\boldsymbol{r} - \boldsymbol{r}'|$, which we might intuitively guess to be of the order of the interatomic spacing. However, in the limit where the variation of $\boldsymbol{v}_n$ is on a scale large compared to this range, we may coarse-grain Eqn. (3.3.5) so as to write (introducing now the standard notation)

$$\rho_n \equiv \int K(\boldsymbol{r}\boldsymbol{r}')d\boldsymbol{r}' \equiv \frac{\partial\overline{\langle\hat{\boldsymbol{J}}(\boldsymbol{r})\rangle}}{\partial \boldsymbol{v}_n(\boldsymbol{r})} \tag{3.3.6}$$

when the bar indicates coarse-graining over an appropriate region around $\boldsymbol{r}$. The quantity ρ_n is known as the *normal density*; in the most general case it may be a (tensor) function of $\boldsymbol{v}_s(\boldsymbol{r})$ and $\boldsymbol{v}_n(\boldsymbol{r})$ as well as of pressure and temperature, though for most practical purposes the dependence on $\boldsymbol{v}_s$ and $\boldsymbol{v}_n$ can be ignored (we shall mostly do this in what follows). We note that combining (3.3.4) and (3.3.6) gives an alternative definition of ρ_n, namely

$$\rho_n(\boldsymbol{r}) \equiv \left(\frac{\partial^2 F}{\partial^2\langle \boldsymbol{J}\rangle}\right)^{-1}_{\boldsymbol{v}_s} \tag{3.3.7}$$

Finally, we define the *superfluid density* ρ_s by the implicit equation

$$\boldsymbol{J}(\boldsymbol{r}) = \rho_s(\boldsymbol{r})\boldsymbol{v}_s(\boldsymbol{r}) + \rho_n(\boldsymbol{r})\boldsymbol{v}_n(\boldsymbol{r}) \tag{3.3.8}$$

Actually, Galilean invariance implies a simple relation between ρ_s and ρ_n. Consider a transformation from the original (inertial) reference frame to a second (inertial) frame moving with velocity $-\boldsymbol{u}$ with respect to the original one. Under this transformation the single-particle states $\chi_\nu(\boldsymbol{r})$ (including the condensate state $\chi_0(\boldsymbol{r})$) transform according to the standard prescription (see e.g. Schwabl 1992, p. 301)

$$\chi_\nu(\boldsymbol{r}) \to \chi'_\nu(\boldsymbol{r}) \equiv \chi_\nu(\boldsymbol{r}) \exp im\boldsymbol{u}\cdot\boldsymbol{r}/\hbar \tag{3.3.9}$$

and consequently we find, from (2.2.7) and (2.2.12) respectively[7]

$$\boldsymbol{v}_s(\boldsymbol{r}) \to \boldsymbol{v}_s(\boldsymbol{r}) + \boldsymbol{u} \tag{3.3.10}$$

$$\boldsymbol{J}(\boldsymbol{r}) \to \boldsymbol{J}(\boldsymbol{r}) + \rho(\boldsymbol{r})\boldsymbol{u} \tag{3.3.11}$$

In addition, the normal velocity $\boldsymbol{v}_n(\boldsymbol{r})$, which as discussed is just the "velocity of the boundary conditions" evidently also transforms in a simple way:

$$\boldsymbol{v}_n(\boldsymbol{r}) \to \boldsymbol{v}_n(\boldsymbol{r}) + \boldsymbol{u} \tag{3.3.12}$$

Combining Eqns. (3.3.10) and (3.3.12), inserting into (3.3.8) and comparing with (3.3.11), we find the simple result

$$\rho_s(\boldsymbol{r}) + \rho_n(\boldsymbol{r}) \equiv \rho(\boldsymbol{r}) \tag{3.3.13}$$

It should be emphasized that in deriving this result we have not used any assumption of translation-invariance, only the (much more general) postulate of Galilean invariance.

[7]The numerator of the RHS of (2.2.12) is just $\boldsymbol{J}(\boldsymbol{r})$.

Finally, we enquire what is the extra kinetic energy of flow associated with finite values of $\boldsymbol{v}_{\mathrm{s}}$ and $\boldsymbol{v}_{\mathrm{n}}$. If $\mathcal{E}_0$ is the energy density of the state corresponding to $\boldsymbol{v}_{\mathrm{s}} = \boldsymbol{v}_{\mathrm{n}} = 0$, then rotational invariance implies that we can write to lowest order in $\boldsymbol{v}_{\mathrm{s}}$ and $\boldsymbol{v}_{\mathrm{n}}$

$$Q(\boldsymbol{r}) \equiv \mathcal{E}(\boldsymbol{v}_{\mathrm{s}}, \boldsymbol{v}_{\mathrm{n}}) - \mathcal{E}_0 = \frac{1}{2}(Av_{\mathrm{s}}^2 + Bv_{\mathrm{n}}^2 + C\boldsymbol{v}_{\mathrm{s}} \cdot \boldsymbol{v}_{\mathrm{n}}) \tag{3.3.14}$$

Once more, Galilean invariance allows us to fix the coefficients in (3.3.14); under the transformation (3.3.9), the energy density transforms according to the prescription

$$\mathcal{E}(\boldsymbol{r}) \rightarrow \mathcal{E}'(\boldsymbol{r}) = \mathcal{E}(\boldsymbol{r}) + \boldsymbol{u} \cdot \boldsymbol{J}(\boldsymbol{r}) + \frac{1}{2}\rho(\boldsymbol{r})\boldsymbol{u}^2 \tag{3.3.15}$$

Substituting (3.3.10) and (3.3.12) in (3.3.14), comparing with (3.3.15) and using (3.3.11), we find for the coefficients in (3.3.14)

$$A = \rho_{\mathrm{s}}(\boldsymbol{r}), B = \rho_{\mathrm{n}}(\boldsymbol{r}), C = 0. \tag{3.3.16}$$

Recapitulating, we have found that the (stable or metastable) thermodynamic equilibrium states of He-II are described by two independent "velocities" $\boldsymbol{v}_{\mathrm{s}}(\boldsymbol{r})$ and $\boldsymbol{v}_{\mathrm{n}}(\boldsymbol{r})$ and associated "densities" $\rho_{\mathrm{s}}(\boldsymbol{r})$ and $\rho_{\mathrm{n}}(\boldsymbol{r})$, such that the densities of mass current $\boldsymbol{J}(\boldsymbol{r})$ and of kinetic energy of flow $Q(\boldsymbol{r})$ are given by the expressions

$$\boldsymbol{J}(\boldsymbol{r}) = \rho_{\mathrm{s}}(\boldsymbol{r})\boldsymbol{v}_{\mathrm{s}}(\boldsymbol{r}) + \rho_{\mathrm{n}}(\boldsymbol{r})\boldsymbol{v}_{\mathrm{n}}(\boldsymbol{r}) \tag{3.3.17a}$$

$$Q(\boldsymbol{r}) = \frac{1}{2}\rho_{\mathrm{s}}(\boldsymbol{r})v_{\mathrm{s}}^2(\boldsymbol{r}) + \frac{1}{2}\rho_{\mathrm{n}}(\boldsymbol{r})v_{\mathrm{n}}^2(\boldsymbol{r}) \tag{3.3.17b}$$

where, for example, $\rho_{\mathrm{s}}(\boldsymbol{r}) \equiv \rho_{\mathrm{s}}(T(\boldsymbol{r}), \rho(\boldsymbol{r}))$. Equation (3.3.17) lend themselves nicely to an intuitive picture of He-II as a mixture of two independent, interpenetrating "fluids" or "components", the "superfluid" and "normal" components, each associated with its own current and flow energy; this is the "two-fluid" model. However, this picture should not be taken too literally, and in particular it is important not to identify the superfluid and normal components with the condensate and noncondensate atoms respectively.

To illustrate this last point, let's consider briefly the way in which ρ_{s} and ρ_{n} depend on temperature. Since $\boldsymbol{v}_{\mathrm{s}}$ loses its meaning in the He-I phase above T_λ, we should expect that $\rho_{\mathrm{s}}(T)$ approaches zero (and thus $\rho_{\mathrm{n}}(T)$ approaches ρ) continuously as T rises to T_λ from below, and experiment confirms this expectation (see below). The limit $T \rightarrow 0$ is a bit more tricky: for a single-component, translation-invariant system such as bulk liquid ^{4}He, there are theoretical arguments (see below, Section 3.4) which suggest that in this limit ρ_{s} should be equal to the total density ρ (and hence ρ_{n} to zero), and experiment certainly seems to confirm this. Thus, at $T = 0$ the whole liquid is "superfluid" despite the fact that only a few percent of the atoms are "condensed" (cf. eqn. (3.2.5))! In some sense, the atoms of the condensate constrain the much more numerous noncondensed ones, in a way that may become clearer in the next chapter when we discuss the Bogoliubov approximation.

Armed with the two-fluid Eqns. (3.3.17) and the Onsager–Feynman constraint (2.2.9) on the superfluid velocity $\boldsymbol{v}_{\mathrm{s}}$, we are now in a position to explain the two fundamental ingredients of "superfluidity", namely the Hess–Fairbank effect (nonclassical rotational inertia) and the metastability of supercurrents. For simplicity we consider the geometry of Chapter 1, Section 1.6, namely an annulus (toroid) of transverse dimension d negligible in comparison with its mean radius R; then, if the annulus (container) rotates with angular velocity ω, the normal velocity $\boldsymbol{v}_\mathrm{n}(\boldsymbol{r})$ is just $\omega R\hat{\boldsymbol{\theta}}$, where $\hat{\theta}$ is a unit vector tangent to the annulus at point $\boldsymbol{r}$. It is convenient to define, as in Chapter 1, Section 1.6, a characteristic quantum unit of rotation ω_c by $\omega_\mathrm{c} \equiv \hbar/mR^2$. I denote the cross-sectional area of the annulus by A.

Consider first the HF effect. As in Chapter 1, Section 1.6, we cool the sample through T_λ while rotating the container at an angular velocity ω which for the moment we allow to be arbitrary. In the He-I phase above T_λ ρ_n is trivially equal to ρ, so that in equilibrium the circulating current density is just $\rho\boldsymbol{v}_\mathrm{n}$, and the total angular momentum of the helium is thus $2\pi RA\rho v_\mathrm{n}R = Mv_\mathrm{n}R = MR^2\omega \equiv I_\mathrm{cl}\omega$, where M is the total mass of the helium and $I_\mathrm{cl} \equiv MR^2$ its classical moment of inertia. Now when we cool into the He-II phase, we need to decide the behavior of $\boldsymbol{v}_\mathrm{s}$ (i.e. of the condensate state). From the symmetry of the situation we expect that its magnitude v_s is constant around the ring, and the OF quantization condition then implies that the allowed values are only the discrete ones

$$v_\mathrm{s}^{(n)} = n\omega_\mathrm{c}R \tag{3.3.18}$$

(n is the "winding number" introduced in section 2.5) Which value of n will the system prefer? We may actually pose this question for general values of ω. According to Chapter 1, Appendix 1A we must minimize the effective free energy (1.6.3), and by substituting (3.3.18) into (3.3.17b) we see that the terms in this expression which depend on n are of the form

$$\mathcal{E}_\mathrm{sup} = \rho_\mathrm{s}(T)R^2\left(\frac{1}{2}n^2\omega^2 - n\omega\omega_\mathrm{c}\right) \tag{3.3.19}$$

It is clear that the expression (3.3.19) is minimized by choosing n equal to the nearest integer to ω/ω_c. In particular, if $\omega < \frac{1}{2}\omega_\mathrm{c}$, as in the case of the HF effect, then on passing through T_λ the system will fix n to be zero (and thereafter keep it at that value). Thus $\boldsymbol{v}_\mathrm{s} = 0$, and the superfluid contributes nothing to the current (3.3.17a), which is given simply by $\rho_\mathrm{n}\boldsymbol{v}_\mathrm{n}$. The total angular momentum is thus reduced by a factor $\rho_\mathrm{n}(T)/\rho \equiv f_\mathrm{n}(T)$:

$$L(T) = \frac{\rho_\mathrm{n}(T)}{\rho}I_\mathrm{cl}\omega \equiv f_\mathrm{n}(T)I_\mathrm{cl}\omega \tag{3.3.20}$$

in agreement with Eqn. (1.6.4).

It is clear that the argument can be extended to the case $\omega > \frac{1}{2}\omega_\mathrm{c}$. Generally, the condensate will choose the n-value closest to ω/ω_c; its contribution to the current will then be $\rho_\mathrm{s}(T)n\omega_\mathrm{c}R$, and its contribution to the total angular momentum $(\rho_\mathrm{s}(T)/\rho)I_\mathrm{cl}\cdot n\omega_\mathrm{c} \equiv f_\mathrm{s}(T)I_\mathrm{cl}\cdot n\omega_\mathrm{c}$. This adds to the contribution (3.3.20) of the normal component, so the total angular momentum is

$$L(T) = I_\mathrm{cl}(f_\mathrm{n}(T)\omega + f_\mathrm{s}(T)n\omega_\mathrm{c}) \tag{3.3.21}$$

$$n \equiv \mathrm{int}\left[\frac{\omega}{\omega_\mathrm{c}} + \frac{1}{2}\right] \tag{3.3.22}$$

This result is illustrated in Fig. 3.2. Note that when int $[\omega/\omega_\mathrm{c}] > \frac{1}{2}$, the angular momentum is *greater* than the classical value.

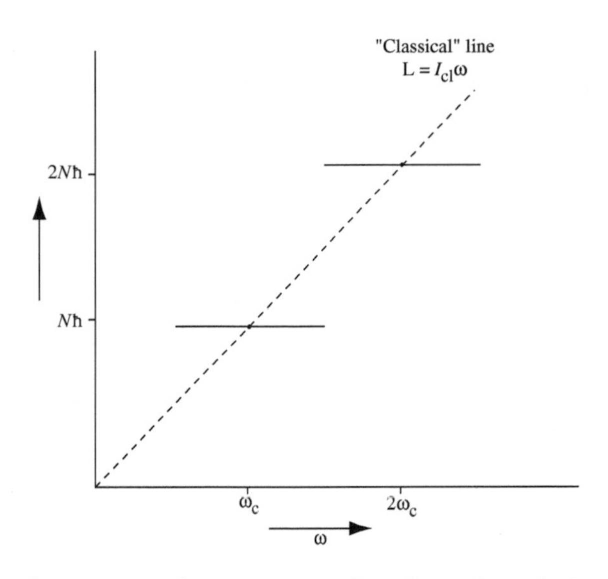

Fig. 3.2 Relation between angular momentum L and angular velocity ω for a superfluid at $T = 0$. For nonzero T the horizontal plateaux tilt upwards, until at $T = T_c$ they coincide with the "classical" line.

I now turn to the other main defining phenomenon of superfluidity, namely the metastability of circulating supercurrents. Imagine that as in Chapter 1, Section 1.6 we cool the system through T_λ while rotating the container at a *large* angular velocity, $\omega \equiv \omega_0 \gg \omega_c$. The analysis goes through as above, and the value of n chosen by the condensate will now be large (typically 10^4, in a realistic experiment). In view of this it is difficult in practice to distinguish between (3.3.21) and the "normal" result $L = I_{cl}\omega_0$, so it looks very much as if the whole liquid is rotating with the container. Now we stop the rotation of the latter and allow the system to come, as far as possible, into equilibrium with the stationary walls. The normal velocity $\boldsymbol{v}_n$ is now zero. However, because of the topological considerations discussed in Chapter 2, Section 2.5, the system cannot change its winding number n, and the angular momentum is therefore given by (3.3.21), with $\omega = 0$ but $n = \text{int}[\omega_0/\omega_c + \frac{1}{2}]$. Thus, we find approximately

$$L(T) \cong \frac{\rho_s(T)}{\rho} I_{cl}\omega_0 \equiv f_s(T) I_{cl}\omega_0 \tag{3.3.23}$$

in agreement with Eqn. (1.6.2). Further, from (3.3.13) we see that the $f_s(T)$ which occurs in (3.3.21) and the $f_n(T)$ which occurs in (3.3.20) sum to unity, as found experimentally (see Eqn. (1.6.5)). Note that by varying T (but staying always below T_λ) we can *reversibly* increase or decrease $L(T)$!

While the "thin" toroidal (annular) geometry is particularly simple from a conceptual point of view, many experiments on liquid helium are not conducted in such a geometry, or indeed in any which is multiply connected. In particular, the original experiments of Kapitza and of Allen and Misener which discovered the phenomenon of superfluidity simply monitored the flow of the liquid through a narrow orifice between two bulk reservoirs. How do the above considerations, or similar, apply in this kind of geometry? Since the walls are at rest (and the capillary is too narrow to permit Poiseuille flow) we can set $\boldsymbol{v}_\mathrm{n} = 0$, so that any current must be carried entirely by the superfluid. In general, the condensate wave function will have a nonzero amplitude, and hence a well-defined (in general time-dependent) phase in each of the bulk reservoirs; call these phases $\phi_\mathrm{L}(t)$, $\phi_\mathrm{R}(t)$, and define the phase difference

$$\Delta\phi(t) \equiv \phi_\mathrm{L}(t) - \phi_\mathrm{R}(t) \tag{3.3.24}$$

Note that $\Delta\phi(t)$ can be either larger or smaller than π. Although one is used to thinking of phase differences as defined only modulo 2π, in this situation the absolute value does matter, as is clear from integrating (2.2.7) along the tube: in fact, assuming the flow to be uniform, we find

$$\boldsymbol{v}_\mathrm{s}(t) = \frac{\hbar}{m}\frac{\Delta\phi}{L} \tag{3.3.25}$$

where L is the length of the tube. If we assume that $\Delta\phi$ is independent of time (the conditions for this will be established in the next section), then we can assume that the system has come to (stable or metastable) thermal equilibrium, so that the current is proportional to $\boldsymbol{v}_\mathrm{s}$ according to (3.3.17a). Similarly, the flow energy is, by (3.3.17b), proportional to $(\Delta\phi)^2$.

Now an amusing point arises:[8] Consider the possibility of a "jump" in which $\Delta\phi \to \Delta\phi - 2n\pi$ with n integral; how such a jump might occur in practice is discussed in Section 3.5 below. Such a jump can lower the value of the flow energy only if $|\Delta\phi| > \pi$. Hence, for $|\Delta\phi| < \pi$ the capillary flow is "absolutely" stable, and the phenomenon we are seeing can be regarded as a version of the HF effect. On the other hand, for $|\Delta\phi| > \pi$ the flow is only metastable, and can be regarded as a variant of the phenomenon of metastability of supercurrents in a toroidal geometry!

So, was the effect seen by Kapitza and Allen and Misener "really" the HF effect, or the metastability of supercurrents, or both? The answer is: we don't know! (insufficient details are given in the original papers of the capillary geometry, magnitude of current observed, etc., to work it out). My guess is that as they varied the parameters of the experiment they saw both effects. But in any case, the essential point to keep in mind is that, as emphasized in Chapter 1, Section 1.6, and Chapter 2, Section 2.5, the two effects are conceptually quite distinct, and while they always seem to occur together in (bulk) liquid ^{4}He, this is actually a consequence of the particularly simple nature of the order parameter and not a generic property of BEC systems.

The above considerations allow us to explain some other experimental flow properties of He-II in a relatively straightforward way. In particular, the Andronikashvili oscillating-disk experiment (see section 1) can be explained by the assumption that the value of n, which starts equal to zero, always remains so, i.e. the superfluid component always remains (metastably) at rest; on the other hand, provided that the oscillation period is much greater than the time taken to establish thermal equilibrium, the normal component should follow the boundary conditions, i.e. the disks, and we expect the effective moment of inertia of the liquid, as observed in the experiment, to be proportional to $f_\mathrm{n}(T)$, as indeed it is. Discussion of the ion mobility experiments is best postponed until we have looked at the excitation spectrum (Section 3.6).

[8] Cf. also the discussion in Chapter 5, Section 10.

3.4 The two-fluid model: dynamical effects

The discussion of the last section is restricted by the assumption that the liquid has come into complete thermal equilibrium with the boundary conditions, subject only to the topological constraints on the relaxation of the superfluid velocity $\boldsymbol{v}_\mathrm{s}$. It is thus incapable as it stands of discussing time-dependent phenomena, such as the collective oscillations of the liquid. In the present section I will indicate the generalization to this case; however, I shall always assume that the time scale of any phenomenon considered is long compared to the relaxation time to local thermal equilibrium (i.e. crudely speaking, the time needed to reach a situation where the local state of the normal component can be uniquely characterized by a single velocity $\boldsymbol{v}_\mathrm{n}(\boldsymbol{r})$). Under these conditions we expect the local state of the liquid to be characterized by eight thermodynamic variables, which we can choose as (for example) the pressure $P(\boldsymbol{r}, t)$, the temperature $T(\boldsymbol{r}, t)$, and the three components of the superfluid velocity $\boldsymbol{v}_\mathrm{s}(\boldsymbol{r}, t)$ and of the normal velocity $\boldsymbol{v}_\mathrm{n}(\boldsymbol{r}, t)$. A complete two-fluid hydrodynamics then consists of a set of equations which will entirely determine the behavior of these quantities given their values at $t = 0$.

To formulate such a two-fluid hydrodynamics for the general case, it is necessary to keep, in expressions for the energy density, etc., terms up to bilinear order in the velocities $\boldsymbol{v}_\mathrm{s}$ and $\boldsymbol{v}_\mathrm{n}$, and this is done in most textbook presentations of the subject. However, the resulting expressions are quite messy, and most of the qualitative consequences of the theory come out in an approximation in which we include in the energies only terms independent of $\boldsymbol{v}_\mathrm{s}$ and $\boldsymbol{v}_\mathrm{n}$. For this reason I shall confine myself to this approximation (indicating the omitted terms by the notation "$o(v^2)$").

As orientation for the ensuing discussion, let's first remind ourselves of a few salient points concerning the hydrodynamics of a *normal* liquid (e.g. Putterman (1974), Section 1). For a normal liquid a state of local thermodynamic equilibrium is characterized by five variables, which we can choose as the pressure $P(\boldsymbol{r}, t)$, the temperature $T(\boldsymbol{r}, t)$ and the three components of the local fluid velocity $\boldsymbol{v}(\boldsymbol{r}, t)$. Other thermodynamic quantities such as the mass, energy and entropy densities are expressed in terms of $P(\boldsymbol{r}, t)$ and $T(\boldsymbol{r}, t)$ by using the first law of thermodynamics and the equation of state. To determine these five variables we have five equations each first-order in time, namely the scalar continuity equations for the mass and entropy densities[9] which define the corresponding currents $\boldsymbol{J}, \boldsymbol{J}_\mathrm{s}$:

$$\frac{\partial \rho(\boldsymbol{r}, t)}{\partial t} + \nabla \cdot \boldsymbol{J}(\boldsymbol{r}, t) = 0 \tag{3.4.1a}$$

$$\frac{\partial S(\boldsymbol{r}, t)}{\partial t} + \nabla \cdot \boldsymbol{J}_\mathrm{s}(\boldsymbol{r}, t) = 0 \tag{3.4.1b}$$

and the three components of Euler's equation for the mass current:

$$\frac{\partial \boldsymbol{J}(\boldsymbol{r}, t)}{\partial t} + \nabla P(\boldsymbol{r}, t) = o(v^2) \tag{3.4.2}$$

To complete our description we need explicit expressions for the mass and entropy current densities in terms of the fluid velocity $\boldsymbol{v}(\boldsymbol{r}, t)$ and the thermodynamic variables.

[9] Recall that in the hydrodynamic limit there is by definition no dissipation and hence no entropy production, so that the entropy density as well as the mass density must satisfy a continuity equation.

Both of these equations can be straightforwardly obtained without further assumptions by making a Galilean transformation to a frame moving with the velocity $\boldsymbol{v}(\boldsymbol{r},t)$:

$$\boldsymbol{J}(\boldsymbol{r},t) = \rho(\boldsymbol{r},t)\boldsymbol{v}(\boldsymbol{r},t) \tag{3.4.3a}$$

$$\boldsymbol{J}_s(\boldsymbol{r},t) = S(\boldsymbol{r},t)\boldsymbol{v}(\boldsymbol{r},t) \tag{3.4.3b}$$

Thus, given the values of P,T and $\boldsymbol{v}$ as functions of $\boldsymbol{r}$ at $t = 0$, Eqns. (3.4.1)–(3.4.2) determine their behavior at all subsequent times.

Turning now to the superfluid case, we note that both Eqns. (3.4.1) and, in the limit that we ignore the $o(v^2)$ terms, Eqn. (3.4.2) retain their validity, while Eqn. (3.4.3a) is replaced, according to the considerations of Section 3.3, by the expression

$$\boldsymbol{J}(\boldsymbol{r},t) = \rho_s(\boldsymbol{r},t)\boldsymbol{v}_s(\boldsymbol{r},t) + \rho_n(\boldsymbol{r},t)\boldsymbol{v}_n(\boldsymbol{r},t) \tag{3.4.4}$$

where for any given physical system the coefficients $\rho_{s,n}$ are uniquely fixed by P and T (generalized "equation of state"). Since we now have eight rather than five variables in the problem, we need three more equations to determine them (i.e. presumably one more vector equation). Also, we still need to know what is the correct replacement for Eqn. (3.4.3b) (since the concept of a single "fluid velocity" $\boldsymbol{v}(\boldsymbol{r},t)$ has in general no meaning within the two-fluid model).

I quote the solution and will subsequently justify it: The necessary extra vector equation is of the form

$$\frac{\partial \boldsymbol{v}_s(\boldsymbol{r},t)}{\partial t} = -\frac{1}{m}\boldsymbol{\nabla}\mu(\boldsymbol{r},t) + o(v^2) \tag{3.4.5}$$

and the appropriate replacement for (3.4.3b) is

$$\boldsymbol{J}_s(\boldsymbol{r},t) = S(\boldsymbol{r},t)\boldsymbol{v}_n(\boldsymbol{r},t) \tag{3.4.6}$$

In words: The scalar thermodynamic variable $\Omega(\boldsymbol{r},t)$ whose gradient drives the superfluid velocity (see Section 3.3) is (apart from a factor of m) nothing but the chemical potential $\mu \equiv (\partial E/\partial N)_S$, and the transport of entropy is associated exclusively with the normal component. Both of these statements are intuitively reasonable: The condensate is associated with a single one-particle quantum state, and therefore not only should transport no entropy (Eqn. 3.4.6) but should be insensitive to variations of the energy associated with input of heat or entropy (Eqn. 3.4.5).

It turns out that once we postulate that $\boldsymbol{v}_s$ is irrotational (so that its time derivative is the gradient of a scalar quantity) *both* the results (3.4.5) and (3.4.6) follow from considerations of thermodynamics and Galilean invariance alone, without the necessity for further microscopic arguments. This remarkable result is well discussed in Putterman (1974), Section 3. However, the derivation requires explicit consideration of the $o(v^2)$ terms in the hydrodynamics, and does not give much insight into the connection of (3.4.5) and (3.4.6) with the phenomenon of BEC. I shall therefore in the following attempt to give some less formal arguments for these equations.

Equation (3.4.5) actually follows from a more fundamental equation which gives the time variation of the phase of the condensate wave function (order parameter). Consider two points $\boldsymbol{r}, \boldsymbol{r}'$ in the system and denote the phase difference at time t, $\phi(\boldsymbol{r},t)-\phi(\boldsymbol{r}',t)$, by $\Delta\phi(t)$ and the difference in local chemical potential $\mu(\boldsymbol{r},t)-\mu(\boldsymbol{r}',t)$

by $\Delta\mu(t)$. Then the equation in question (which is often known as the second Josephson equation, see Chapter 5, Section 5.10) states that

$$\frac{\partial \Delta\phi(t)}{\partial t} = -\frac{\Delta\mu(t)}{\hbar} \tag{3.4.7}$$

Given (3.4.7), Eqn. (3.4.5) then follows by spatial differentiation and use of the definition (2.2.7).

A relatively rigorous derivation of Eqn. (3.4.7) can be given in certain limits, e.g. in the dilute and weakly interacting limit (Chapter 4, Section 4.3; cf. Equation 4.3.28) or for a Bose system confined to a two-dimensional Hilbert space (cf. Chapter 5, Section 3.10). If we are away from both these limits, as in the case of liquid helium, it is a little more problematic.[10] The derivation I shall now give is plausible rather than rigorous. Consider a system of N spinless Bose particles in a pure state Ψ_N and define a quantity which is a generalization of the single-particle density matrix defined in (2.1.2):

$$\begin{aligned}\rho_1(\boldsymbol{r}t\boldsymbol{r}'t') &\equiv \int \Psi_N^*(\boldsymbol{r}\boldsymbol{r}_2\boldsymbol{r}_3\cdots\boldsymbol{r}_N t)\Psi_N(\boldsymbol{r}'\boldsymbol{r}_2\boldsymbol{r}_3\cdots\boldsymbol{r}_N t')\mathrm{d}\boldsymbol{r}_2\mathrm{d}\boldsymbol{r}_3\cdots\mathrm{d}\boldsymbol{r}_N \\ &\equiv \langle\psi^\dagger(\boldsymbol{r}t)\psi(\boldsymbol{r}'t')\rangle\end{aligned} \tag{3.4.8}$$

The quantity (3.4.8), like (2.1.2), is Hermitian and can thus be diagonalized:

$$\rho_1(\boldsymbol{r}t:\boldsymbol{r}'t') = \sum_i \lambda_i\chi_i^*(\boldsymbol{r}t)\chi_i(\boldsymbol{r}'t') \tag{3.4.9}$$

Note that in the general case the functions χ_i defined by (3.4.9) need not coincide with those defined by (2.1.4). Unlike in the case of (2.1.4), there is no "natural" normalization of the eigenfunctions $\chi_i(\boldsymbol{r}t)$ and the latter must be chosen arbitrarily,

$$\int d\boldsymbol{r}\int_0^T dt|\chi_i(\boldsymbol{r}t)|^2 = 1 \tag{3.4.10}$$

It is convenient to choose T to be long compared to any microscopic time scales in the problem (cf below) but short compared to the time for the condensate fraction (if any) to change appreciably. We write the eigenfunctions χ_i in the form

$$\chi_i(\boldsymbol{r}t) \equiv |\chi_i(\boldsymbol{r}t)|e^{i\phi_i(\boldsymbol{r}t)} \tag{3.4.11}$$

and assume that after some time τ (the "microscopic timescale") the various ϕ_i's are out of phase by amounts $\gtrsim 2\pi$. Now, following our discussion in Chapter 2, Section 2.1, we assume that there is one and only one eigenvalue, say λ_0, which is of order N while the rest are of order unity; under the stated conditions λ_0 is, apart from a trivial factor of T^{-1}, independent of T and equal to the condensate number N_0 at $t = 0$ (and $\chi_0(\boldsymbol{r}t)$ can be identified with the corresponding quantity in Eqn. (2.2.5)). We then see that for $t \gg \tau$ the quantity $\rho_1(\boldsymbol{r}0:\boldsymbol{r}t)$ is given by an expression (reflecting "long-range order in time"!) of the form

$$\rho_1(\boldsymbol{r}0:\boldsymbol{r}t) = \lambda_0|\chi_0^*(\boldsymbol{r}0)\chi_0(\boldsymbol{r}t)|e^{i[\phi(\boldsymbol{r}t)-\phi(\boldsymbol{r}0)]} \tag{3.4.12}$$

[10]Though cf. e.g. Nozières and Pines (1989), Section 5.7.

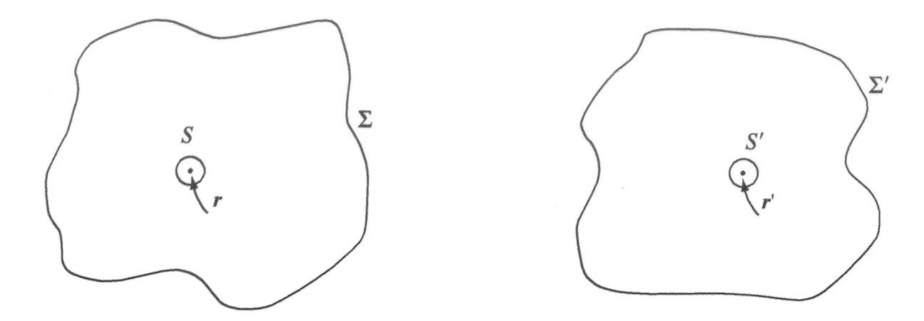

Fig. 3.3 The geometry used in the thought-experiment leading to Eqn. (3.4.7).

where we have dropped the subscript zero on the condensate phase $\phi(\boldsymbol{r}t)$. Thus,

$$\frac{\partial}{\partial t}\arg[\rho_1(\boldsymbol{r}0:\boldsymbol{r}t)] = \frac{\partial}{\partial t}\phi(\boldsymbol{r}t) \tag{3.4.13}$$

This result is general; note that the arbitrary normalization of the χ_i nowhere appears.

Let us now represent $\rho_1(\boldsymbol{r}0:\boldsymbol{r}t)$ in terms of the many-body wavefunction Ψ_N as in (3.4.8), and consider the following thought-experiment.[11] We consider a system of N particles described by a Hamiltonian $\hat{H}_0$, which may contain an external c-number potential as well as interactions between the particles. We choose two points of interest, say $\boldsymbol{r}$ and $\boldsymbol{r}'$, and surround each of them by a surface S (S') and a much larger surface Σ (Σ') (see Fig. 3.3). To avoid confusion we define the mean number of particles within Σ (Σ') as ν (ν'); let the values of these quantities in the ground state of $\hat{H}_0$ be ν_0 and ν_0' respectively.[12] Consider now the minimum expectation value of $\hat{H}_0$ subject to the constraint that the mean number of particles inside Σ is ν and that inside Σ', ν'; call this energy $E_0(\nu,\nu')$. We assume that for $\boldsymbol{r}$ and $\boldsymbol{r}'$ sufficiently different from one another $E_0(\nu,\nu') = E(\nu) + E'(\nu')$ (where the forms of the functions E and E' need not be identical, since e.g. the external potential may be different in the two regions). We can now define the coarse-grained chemical potential associated with the point $\boldsymbol{r}(\boldsymbol{r}')$ as the derivative of $E(\nu)$ with respect to ν (etc.):

$$\mu(\boldsymbol{r}) \equiv \frac{\partial E}{\partial \nu}, \quad \mu(\boldsymbol{r}') \equiv \frac{\partial E'}{\partial \nu'} \tag{3.4.14}$$

Imagine now that at a large negative time we apply some (additional) potential $U(\boldsymbol{r})$ in the region inside S, and similarly some potential $U'(\boldsymbol{r}')$ in the region of $\boldsymbol{r}'$; let the strength of the potential in each case be such that the induced value of ν in the ground state (of the new Hamiltonian) is exactly ν_0+1, and that of ν' exactly $\nu_0'+1$. Since in the original ground state of $\hat{H}_0$ the chemical potential is uniform in space, the condition for this is

$$\int^{\Sigma} U(\boldsymbol{z})\delta\rho(\boldsymbol{z})\,d\boldsymbol{z} - \int^{\Sigma'} U'(\boldsymbol{z}')\delta\rho(\boldsymbol{z}')\,d\boldsymbol{z}' = [\mu(\boldsymbol{r}) - \mu(\boldsymbol{r}')] \tag{3.4.15}$$

[11] For the sake of simplicity we assume that in the absence of external potentials the mean density is uniform, but this is not an essential simplification.

[12] Note that the subscript 0 has nothing to do with the condensate!

where $\delta\rho(\boldsymbol{z})$ is the extra density induced at point $\boldsymbol{z}$. We assume that the true ground state, satisfying (3.4.15), is well established (by unspecified weak interaction processes) by time $t=0$. Now, starting at $t=0$, we adiabatically turn off the potentials U, U' ("adiabatically" in this context means on a scale slow compared to the rearrangement processes taking place within each of Σ and Σ' but fast compared to the time scales of any transfer of particles between them). At the end of this process the wave function in the region of $\boldsymbol{r}$ will be approximately that corresponding to the constrained ground state of $\hat{H}_0$ with $\nu=\nu_0+1$; similarly in the region of $\boldsymbol{r}'$ it will be that with $\nu'=\nu_0'+1$. Thus, the many-body wave function $\Psi_N(\boldsymbol{r}\,\boldsymbol{r}_2\boldsymbol{r}_3\cdots\boldsymbol{r}_N\,t)$ satisfies the equation (where I omit the indices $\boldsymbol{r}_2\ldots\boldsymbol{r}_N$)

$$i\hbar\frac{\partial\Psi_N(\boldsymbol{r}t)}{\partial t}=\hat{H}_0\Psi_N(\boldsymbol{r}t)\cong E_{\nu+1}\Psi_N(\boldsymbol{r}t) \tag{3.4.16}$$

and similarly $i\hbar\,\partial\Psi_N(\boldsymbol{r}'t)/\partial t\cong E_{\nu'+1}\Psi_N(\boldsymbol{r}'t)$. But since $E_{\nu_0}\equiv E_{\nu_0'}$, this implies by (3.4.14) that

$$-\hbar\frac{\partial}{\partial t}\left(\arg[\Psi_N(\boldsymbol{r}t)]-\arg[\Psi_N(\boldsymbol{r}'t)]\right)=\mu(\boldsymbol{r})-\mu(\boldsymbol{r}') \tag{3.4.17}$$

But the argument of $\Psi_N(\boldsymbol{r}t)$ is, from the definition (3.4.8), identical to that of $\rho_1(\boldsymbol{r}0:\boldsymbol{r}t)$, so comparing (3.4.17) with (3.4.12) we finally obtain our desired result, namely Eqn. (3.4.7).

In the case when the many-body state is a mixture of different states i with weights p_i, we can repeat the above argument, but we must now specify that in setting up the perturbed states the p_i's are not changed; thus the derivatives in (3.4.12) are replaced by $\sum_i p_i\partial E_i/\partial\nu$ (etc.). But keeping the p_i's constant is equivalent to keeping the entropy constant, so this quantity is just $(\partial\overline{E}/\partial\nu)_S\equiv\mu(\boldsymbol{r})$, and Eqn. (3.4.7) again follows.

Equation (3.4.5) may be regarded as the key equation of two-fluid hydrodynamics. It has at least two immediate consequences:

1. Let us consider zero temperature and combine (3.4.5) with the generally valid (up to terms $o(v^2)$) hydrodynamic Eqn. '(3.4.2):

$$\frac{\partial\boldsymbol{v}_{\rm s}}{\partial t}=-\frac{1}{m}\boldsymbol{\nabla}\mu,\quad \frac{\partial\boldsymbol{J}}{\partial t}=-\boldsymbol{\nabla}P \tag{3.4.18}$$

   But at $T=0$ the Gibbs–Duhem relation $d\mu=mdP/\rho+SdT$ implies that (since by the third law of thermodynamics $S=0$: ρ is the *mass* density)

$$\boldsymbol{\nabla}P=\frac{\rho}{m}\boldsymbol{\nabla}\mu \tag{3.4.19}$$

   Thus, for a translationally invariant motion ($\rho=$ const. in space) Eqns. (3.4.18) imply (apart from a time-independent constant which we set equal to zero on symmetry grounds) $\boldsymbol{J}=\rho\boldsymbol{v}_{\rm s}$, i.e.[13]

$$\rho_{\rm s}(T=0)=\rho \tag{3.4.20}$$

[13] It is an interesting question whether one can prove (3.4.20) without explicit use of the two-fluid model. This question is discussed in Leggett (1998), with the conclusion that (3.4.20) indeed holds for any single-species system with unbroken translational and time-reversal invariance.

Experiments in (pure) He-II are certainly consistent with this conclusion. It is worth noting that for ^{3}He–^{4}He mixtures (3.4.20) fails experimentally; on the theoretical side, the above derivation fails because the Gibbs–Duhem relation now involves a third quantity, the "force" conjugate to the ^{3}He impurity concentration.

2. Consider two baths of liquid helium-4 connected by a narrow superleak (in practice, a tube filled with emery powder will provide this). In the normal phase neither mass nor heat can flow through the superleak; as a result, the differences in pressure, temperature and hence chemical potential between the two baths can be completely arbitrary. In the superfluid phase, however, the attainment of equilibrium ($\boldsymbol{v}_{\mathrm{s}} = 0$) or even of a steady state ($\partial \boldsymbol{v}_{\mathrm{s}}/\partial t = 0$) requires, by spatial integration of (3.4.5), that the chemical potentials $\mu_{1,2}$ of the two baths be equal. That is, since μ can be expressed as a function of P and T,

$$\mu(P_1, T_1) = \mu(P_2, T_2) \tag{3.4.21}$$

(It is not necessary that $P_1 = P_2$, since the forces exerted by the capillary walls prevent mass flow.) It thus follows that in the superfluid phase in equilibrium, a difference in pressure implies a difference in temperature and vice versa; if the difference ΔP, etc. are small then the Gibbs–Duhem relation implies the relation

$$\frac{\Delta P}{\Delta T} = -s \tag{3.4.22}$$

where s is the entropy density. If we violate (3.4.22), e.g. by applying heat suddenly to one bath, the system will try to remedy the situation by creating a superfluid mass flow through the superleak, which then tends to stabilize ΔP at the value required by (3.4.21). It is the "overshoot" in this process which produces the spectacular "fountain effect"; however, (3.4.22) can also be checked directly in other less dramatic experiments, see e.g. Donnelly (1967), pp. 22–28.

I now turn to the expression (3.4.6) for the entropy current $\boldsymbol{J}_{\mathbf{s}}$. A simple if not entirely rigorous argument for this expression runs as follows: From Galilean invariance, $\boldsymbol{J}_{\mathbf{s}}$ must have the form

$$\boldsymbol{J}_{\mathbf{s}} = A\boldsymbol{v}_{\mathrm{s}} + B\boldsymbol{v}_{\mathrm{n}} + o(v^2) \tag{3.4.23}$$

where

$$A(P,T) + B(P,T) = S(P,T) \tag{3.4.24}$$

Consider now a "superleak" geometry in which the pressure and temperature, and hence ρ_{s}, may be spatially varying subject to the constraint $\mu(P(\boldsymbol{r}), T(\boldsymbol{r})) = \text{constant}$; by definition, in such a geometry $\boldsymbol{v}_{\mathrm{n}}$ is zero (since the normal component is clamped by the walls) but $\boldsymbol{v}_{\mathrm{s}}$ need not be zero, even in (metastable) thermal equilibrium. However, in such equilibrium the density $\rho(\boldsymbol{r}t)$ cannot be a function of time, and hence by the continuity equation (3.4.1a) we have

$$\boldsymbol{\nabla} \cdot \boldsymbol{J} = \boldsymbol{\nabla} \cdot (\rho_{\mathrm{s}}(\boldsymbol{r})\boldsymbol{v}_{\mathrm{s}}(\boldsymbol{r})) = 0 \tag{3.4.25}$$

However, under these conditions the entropy density $S(r)$ is also time-independent, so from (3.4.16) and (3.4.23) we have, setting $\boldsymbol{v}_{\mathrm{n}} \equiv 0$.

$$\nabla \cdot \boldsymbol{J}_s(\boldsymbol{r}) \equiv \nabla \cdot (A(\boldsymbol{r})\boldsymbol{v}_s(\boldsymbol{r})) = 0 \tag{3.4.26}$$

where $A(\boldsymbol{r}) \equiv A(P(\boldsymbol{r}), T(\boldsymbol{r}))$. Since there are no restrictions on the spatial variation of $\rho_s(\boldsymbol{r})$, or on that of $\boldsymbol{v}_s(\boldsymbol{r})$ save for the irrotationality condition (2.2.8), we see that the combination of (3.4.25) and (3.4.26) can be fulfilled in the general case only if A and ρ_s are proportional, i.e. we must have

$$A(P, T) = k(\mu(P, T))\rho_s(P, T) \tag{3.4.27}$$

where $k(\mu)$ may at first sight be an arbitrary function for any given physical system. However, it is clear that at $T = 0$ the quantity $A(P, 0)$ must be zero for any P even though ρ_s is nonzero (no entropy can be carried at $T = 0$!); and since k is a function only of μ and not of P, T separately, it then follows barring pathology, that $k(\mu) \equiv 0$ for all P, T and thus, from (3.4.23) and (3.4.24), that (3.4.6) holds.

Collecting our various results, we have the following set of equations, valid up to terms of relative order v^2, for the two-fluid model:

$$\frac{\partial \rho}{\partial t} + \nabla \cdot \boldsymbol{J} = 0 \tag{3.4.28a}$$

$$\frac{\partial \boldsymbol{J}}{\partial t} + \nabla P = 0 \tag{3.4.28b}$$

$$\boldsymbol{J} = \rho_s \boldsymbol{v}_s + \rho_n \boldsymbol{v}_n \tag{3.4.28c}$$

$$\frac{\partial \boldsymbol{v}_s}{\partial t} + \frac{1}{m}\nabla \mu = 0 \tag{3.4.28d}$$

$$\frac{\partial S}{\partial t} + \nabla \cdot (S\boldsymbol{v}_n) = 0 \tag{3.4.28e}$$

In Eqn. (3.4.28c) ρ_s and ρ_n are taken to be known functions of T and P. One immediate consequence of Eqns. (3.4.28) is the possibility of a steady state with spatially uniform pressure P and chemical potential μ, and hence by the Gibbs–Duhem relation zero temperature gradient, and with zero mass flow, but with $\boldsymbol{v}_n = -(\rho_s/\rho_n)\boldsymbol{v}_s \neq 0$ and hence, by (3.4.28e), a finite entropy (heat) current. This represents a convective counterflow of the two "fluids" without mass flow, and explains the apparent divergence of the thermal conductivity in liquid ^{4}He below the λ-temperature.[14]

The general two-fluid Eqns. (3.4.28) should apply to any BEC (or pseudo-BEC) system with a scalar order parameter in the limit of sufficiently slow space and time variation, and they can be linearized and solved to predict various kinds of collective excitation, etc. However, in real-life ^{4}He at $T \lesssim T_\lambda$, there is an important simplifying factor: The thermal and mechanical properties to all intents and purposes decouple. Formally, the very small value of the thermal expansion coefficient (or more precisely of the quantity $(\partial P/\partial T)_v$) implies, by the appropriate Maxwell relation, that at constant temperature the entropy *per particle* (S/ρ) is independent of density; conversely, at constant entropy per particle the temperature is independent of density. Thus, we

[14]In real life the heat flux will be limited by boundary effects which tend to partially "clamp" $\boldsymbol{v}_n$, and this will give rise to an apparent "thermal conductivity" which is nonlinear and geometry-dependent.

can find a nontrivial solution to the linearized equations of motion (3.4.23) with the properties

$$T = \text{const.}, \quad \boldsymbol{v}_s(\boldsymbol{r}t) = \boldsymbol{v}_n(\boldsymbol{r}t) \tag{3.4.29}$$

and it may be verified, using the Gibbs–Duhem relation that Eqns. (3.4.28b) and (3.4.28d) are then mutually consistent. The speed c_1 of this mode is determined by the quantity $(\partial P/\partial \rho) \equiv c_1^2$ (which is in this approximation the same whether evaluated under "isentropic" or "isothermal" conditions); it is just in effect an ordinary sound wave, a pure density fluctuation in which the liquid moves as a whole without distinction between the normal and superfluid components. For historical reasons this mode is known in He-II as "first" sound; its velocity is only weakly temperature dependent, and varies from ~240 m/sec at s.v.p. to ~360 m/sec at the melting curve.

A second type of oscillation is possible, in which

$$P = \text{const.}, \quad \boldsymbol{v}_s(\boldsymbol{r}t) = \frac{-\rho_n}{\rho_s}\boldsymbol{v}_n(\boldsymbol{r}t) \tag{3.4.30}$$

According to (3.4.28a), (3.4.28b) and (3.4.28c), there is no mass current or density fluctuation in this mode; it is a "pure" temperature wave (accompanied by the necessary fluctuation of the chemical potential), with a velocity which follows from (3.4.28d and 3.4.28e) plus the relation (3.4.30) as[15]

$$\begin{aligned} c_2^2 &= S\frac{\rho_s}{\rho_n}\left(-\frac{\partial \mu}{\partial S}\right)_P \\ &= \frac{TS^2\rho_s}{C_1\rho_n} \end{aligned} \tag{3.4.31}$$

This is the celebrated "second sound" oscillation whose existence was predicted by Tisza and Landau and shortly thereafter verified by Peshkov; its velocity varies from a value comparable to c_1 at $T = 0$ to zero at T_λ.

It should be emphasized that the simple bifurcation of the collective oscillations into a pure pressure wave with no temperature fluctuation (first sound) and a pure temperature wave with no pressure fluctuation (second sound) is not a generic consequence of the two-fluid model but is specific to a system such as He-II where the quantity $(\partial P/\partial T)_V$ (or equivalently the difference between isentropic and isothermal compressibility) is negligible.[16] In the general case, there are still two modes, but each involves fluctuations of both pressure and temperature. For the explicit form of the relevant formulae, see, e.g. Putterman (1975), Section 7.

For completeness, I should note that in narrow superleaks, when $\boldsymbol{v}_n$ is completely clamped by the walls, Eqns. (3.4.28) tolerate a single type of solution, with $T = S =$ const., $\boldsymbol{v}_n = 0$. From Eqns. (3.4.28a), (3.4.28c) and (3.4.28d) we see that its velocity is

$$c_4^2 = \rho_s\left(\frac{\partial \mu}{\partial \rho}\right)_T \equiv \frac{\rho_s}{\rho}\left(\frac{\partial P}{\partial \rho}\right)_T \cong \frac{\rho_s}{\rho}c_1^2 \tag{3.4.32}$$

[15]We use the fact that from the Gibbs–Duhem relation $(\partial\mu/\partial T)_P = -S$.

[16]Except close to the λ-point, where the thermal expansion is large and the separation is no longer valid.

(where we used the Gibbs–Duhem relation). This oscillation is called "fourth sound"; like second sound, it is evidently peculiar to the superfluid phase. (The "missing" excitation, third sound, is a wave which propagates on a thin helium film, see, e.g. Putterman (1975), Section 41.)

3.5 Quantized vortices, phase slip and the Josephson effect

Fluid mechanics is one of the oldest branches of classical physics, and the study of vortices and vorticity goes back at least to the time of Descartes. A classical fluid in local thermodynamic equilibrium may be characterized by (e.g.) its density ρ and temperature T, and by the mean fluid velocity $\boldsymbol{v}$. The *vorticity* $\boldsymbol{\omega}$ is then defined by

$$\boldsymbol{\omega}(\boldsymbol{r}t) \equiv \boldsymbol{\nabla} \times \boldsymbol{v}(\boldsymbol{r}t) \tag{3.5.1}$$

Under certain circumstances it may turn out that the vorticity is zero everywhere except on a particular line, where it is singular in such a way that the line integral of the fluid velocity taken around a contour enclosing the line is some nonzero constant κ (see below):

$$\oint_C \boldsymbol{v}(\boldsymbol{r}t) \cdot d\boldsymbol{l} = \kappa \neq 0 \tag{3.5.2}$$

We then speak of a vortex line, if the line in question extends throughout the system, or a vortex ring if it closes in on itself; κ is called the circulation associated with the vortex. The study of the statics and dynamics of such vortex configurations is a major area of classical fluid mechanics.

As we have seen in Section 3.4, a BEC system in local thermodynamic equilibrium is characterized not by a single fluid velocity $\boldsymbol{v}(\boldsymbol{r}t)$, but by two independent velocities, the superfluid velocity $\boldsymbol{v}_\mathrm{s}(\boldsymbol{r}t)$ and the normal velocity $\boldsymbol{v}_\mathrm{n}(\boldsymbol{r}t)$. The normal velocity $\boldsymbol{v}_\mathrm{n}(\boldsymbol{r}t)$ behaves in a way which is qualitatively similar to the fluid velocity $\boldsymbol{v}(\boldsymbol{r}t)$ in a normal system; it can have a nonzero distributed vorticity in bulk ($\boldsymbol{\omega}(\boldsymbol{r}t) \neq 0$ over a region) and in principle it can sustain vortices with arbitrary κ. However, the superfluid velocity is constrained by Eqns. (2.2.8) and (2.2.9), meaning (a) that it cannot sustain bulk vorticity,[17] and (b) that the circulation κ of a vortex line or ring is *quantized* in units of h/m. When one speaks of "vortices" in He-II (or other BEC systems) it is the universal convention that one is thinking of these quantized vortices associated with the superfluid component.

A number of important theorems about vortices in a BEC system may be regarded as special cases of more general theorems which can be derived for a normal system in the ideal-fluid approximation, i.e. under the assumption that entropy production (dissipation) is negligible. It is therefore useful to review these results of classical fluid mechanics briefly; since the proofs may be found in standard textbooks (e.g. Landau and Lifshitz 1987, Acheson 1990), I will omit them and just quote the results.

First, from the fact that the divergence of the curl of any vector is zero, it follows (using Stokes' and Gauss' theorems) that a line of singularities around which the expression (3.5.2) is nonzero cannot terminate in the liquid; it must either close on itself

[17] That is, an $\boldsymbol{\omega}(\boldsymbol{r}t)$ which is a smooth function of position on a genuinely microscopic scale. After coarse-graining, the vorticity may look "continuous" on a macroscopic scale, see below.

(vortex ring) or terminate on the boundaries, and moreover the value of κ must be the same for any contour encircling it. Next, as to the time-dependence: In the ideal-fluid limit and in the absence of external forces such as gravity, the motion of a classical fluid in local thermodynamic equilibrium is completely determined (given a specification of the equation of state and the variables at $t = 0$) by the three equations (3.4.1a–c), namely the continuity equation, Euler's equation (Newton's second law applied to a fluid) and the equation of entropy conservation. In order these read (we now include the $o(v^2)$ terms)

$$\frac{\partial \rho}{\partial t} + \nabla \cdot (\rho \boldsymbol{v}) = 0 \tag{3.5.3a}$$

$$\frac{\partial \boldsymbol{v}}{\partial t} + (\boldsymbol{v} \cdot \nabla)\boldsymbol{v} (\equiv \frac{D\boldsymbol{v}}{Dt}) = -\frac{1}{\rho}\nabla P \tag{3.5.3b}$$

$$\frac{\partial S}{\partial t} + \nabla \cdot (\boldsymbol{v} S) = 0 \tag{3.5.3c}$$

where P is the pressure and S the entropy density.

If the entropy per particle (S/ρ) is initially constant in space, it follows from (3.5.3a) and (3.5.3c) that it will remain so throughout the motion. For such "isentropic" flows one can derive from (3.5.3b) the useful relation

$$\frac{\partial \boldsymbol{\omega}(\boldsymbol{r}, t)}{\partial t} = \nabla \times (\boldsymbol{v}(\boldsymbol{r}t) \times \boldsymbol{\omega}(\boldsymbol{r}t)) \tag{3.5.4}$$

From (3.5.4) it follows, in particular, that if we start off with a single line of singularities surrounded by a vorticity-free region (i.e. a vortex) then this situation must persist, and furthermore that the quantity κ defined by (3.5.2) must be constant in time (Kelvin's theorem), provided that the contour in (3.5.2) is always taken to encircle the vortex:

$$\frac{\partial \kappa}{\partial t} = 0 \tag{3.5.5}$$

It further follows (by a rather delicate argument, for which see, e.g. Acheson 1990, Section 5.3) that the rate of motion of the point of singularity, i.e. of the vortex line at a given point, is just the "background" fluid velocity at that point, i.e. the velocity averaged over a small cylindrically symmetric region centered on the vortex (the Helmholtz theorem).

The above can be summarized by saying that for a classical fluid within the approximation of ideal isentropic flow, vortices once created are *topologically stable* objects, with a unique and conserved value of the circulation κ; they move with the background fluid velocity, but are not created, destroyed or otherwise modified. Needless to say, these statements fail (as does Eqn. (3.5.4)) once we introduce the effects of dissipation.

Let us now discuss more specifically the flow pattern and the energetics of straight vortex lines and vortex rings. For a stationary straight vortex line along the z-axis, Eqn. (3.5.2) immediately implies that apart from a possible irrelevant circulation-free contribution, the velocity pattern must be

$$\begin{aligned} \boldsymbol{v}(\boldsymbol{r}) &= \frac{\kappa}{2\pi} \frac{(\hat{\boldsymbol{z}} \times \boldsymbol{r})}{|\hat{\boldsymbol{z}} \times \boldsymbol{r}|^2} \\ &\equiv \frac{\kappa}{2\pi} \frac{\hat{\boldsymbol{\theta}}}{r_\perp} \quad (r_\perp \equiv \sqrt{x^2 + y^2}) \end{aligned} \tag{3.5.6}$$

i.e. the velocity is tangential and proportional to $1/r$. Clearly if we take (3.5.6) seriously for all $\boldsymbol{r}$, the velocity is not defined on the z-axis, so we must postulate a "core" on which the density of fluid vanishes. A rough estimate of the size of this core (call it a) may be attained by balancing the energy of flow against the compressibility energy, see below: for the moment we just anticipate the fact that for real liquid helium, and circulations of the order of h/m, the quantity a is of the order of the interatomic spacing, a point by which the continuum description has already failed. At values of r large compared to a the flow does not result in appreciable compression, so we can treat the density ρ as equal to the "bulk" value, and the extra energy associated with the vortex is pure energy of flow and given by the simple expression

$$E_{\text{flow}} = \frac{1}{2}\rho \int \boldsymbol{v}^2(r) d\boldsymbol{r} \tag{3.5.7}$$

Substituting the form (3.5.6) and considering for definiteness a single vortex line threading the center of a cylinder of radius R, we find[18] for the flow energy per unit length

$$\mathcal{E} \equiv \frac{E_{\text{flow}}}{L} = \frac{1}{2}\rho \frac{\kappa^2}{2\pi} \ln\left(\frac{R}{a}\right) \tag{3.5.8}$$

It should be remembered that the "core size" a appearing in (3.5.8) is only an order of magnitude; moreover, in less symmetrical geometries R will be replaced by some quantity "of the order of" the cell dimensions. Since for any reasonably macroscopic geometry the argument of the logarithm is large (typically $\sim 10^8$) these considerations generally have only a small quantitative effect.

It is also straightforward to apply Eqn. (3.5.7) to obtain the "potential energy" and thus the "force" per unit length between a pair of parallel vortex lines a distance $r \gg a$ apart. I quote the results for equal circulations of either the same or opposite sign, which will be the case applicable to He-II: for "parallel" vortices (same sign of circulation) we have

$$\mathcal{E}(r) = \frac{1}{2} \frac{\rho}{2\pi} 2\kappa^2 \ln\left(\frac{R^2}{ar}\right) \tag{3.5.9}$$

For "antiparallel" (opposite-circulation) vortices the flow falls off fast enough at large distances that the integral (3.5.7) is independent of R (for $r \ll R$) and we find

$$\mathcal{E}(r) = \frac{1}{2} \frac{\rho}{2\pi} \kappa^2 \ln \frac{r}{a} \tag{3.5.10}$$

Equation (3.5.10) implies that antiparallel vortex lines attract one another (since this increases the mutual cancellation of their velocity fields and hence of the total flow

[18] In practice, in a classical fluid, viscous effects will push $\boldsymbol{v}$ to zero at the cylinder walls, but this at most adds a constant term to the logarithm.

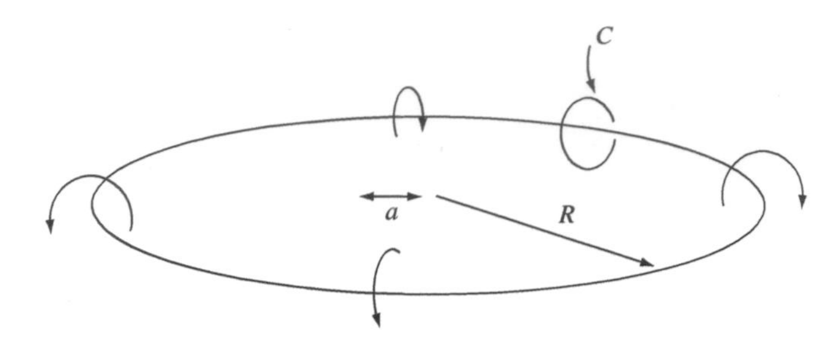

Fig. 3.4 The flow pattern in a vortex ring.

energy (3.5.7)), whereas (3.5.8) implies not only that parallel vortices repel one another but that, if one is constrained to have a given total circulation with no other constraints, it is energetically favorable to break up a single vortex line into two (and by extension of the argument to let the vorticity become diffuse). (It is a different question whether the kinetics allow this over the time-scale of interest.)

A vortex *ring* may be thought of as effectively a vortex line closed on itself (see Fig. 3.4). I shall consider here only circular rings of radius $R_v \gg a$; then the flow pattern is cylindrically symmetric around a line threading the center of the ring perpendicularly, and is schematically as shown in Fig. 3.4. For details see Putterman, 1974, Appendix V. At large distances ($r \gg R_v$) the fluid velocity falls off as r^{-2}, so the total flow energy (3.5.7) is finite and may be shown (see, e.g. Donnelly 1991, Section 1.6) to be given by the expression

$$\mathcal{E}(R_v) = \frac{1}{2}\kappa^2 \rho R_v \left[\ln\left(\frac{8R_v}{a}\right) - \alpha\right] \tag{3.5.11}$$

(where κ is the circulation around the contour shown in Fig. 3.4). The vortex ring also possesses a finite momentum[19] given by

$$\boldsymbol{P}(R_v) = \int \rho \boldsymbol{v}(\boldsymbol{r})\, \mathbf{d}\boldsymbol{r} = \pi\rho\kappa R^2 \hat{\boldsymbol{n}} \tag{3.5.12}$$

where $\hat{\boldsymbol{n}}$ is the direction perpendicular to the plane of the ring, such that the fluid velocity at the center is along $\hat{\boldsymbol{n}}$.

I now turn to the dynamics of (classical) vortex lines and rings. The general principle, as stated above, is very simple: the vortex moves exactly with the velocity of the "background" flow (i.e. that part of the flow not due to the vortex itself). However, the consequences of this principle are somewhat counterintuitive; for example, a pair of vortices of equal and parallel circulation κ will circle around one another at an angular velocity κ/r_{12}^2 where r_{12} is their separation, while a pair of equal and opposite circulation will move together with a linear velocity κ/r_{12} in a direction perpendicular to their separation. After that, it should come as no great surprise that a vortex

[19] A single straight vortex line in an infinite system is constrained by symmetry to have total momentum zero; however, a pair of lines can have a finite momentum, cf. below. For the complications associated with finite geometries, see Putterman 1974 Appendix V.

ring of circulation κ and radius R_v moves in a direction parallel to $\hat{\boldsymbol{n}}$ with a velocity $\boldsymbol{v} = \partial E/\partial \boldsymbol{P}$ given by (cf. Donnelly, loc. cit., for the value (~1) of α).

$$\boldsymbol{v}(R_v) = \frac{\kappa}{4\pi R}\left[\ln\left(\frac{8R_v}{a}\right) - (\alpha - 1)\right]\hat{\boldsymbol{n}} \tag{3.5.13}$$

Finally, we may consider the energy of a vortex in the presence of a current background flow $\boldsymbol{v}_\mathrm{b}$. For a straight vortex line in a semi-infinite slab of liquid with $\boldsymbol{v}_\mathrm{b}$ parallel to the boundary wall, it is straightforward to show by the method of images that the flow energy is proportional to const. $(-\rho v_\mathrm{b}\kappa z)$, where z is the distance from the wall; and it is only slightly more complicated to generalize this calculation to a slab of finite width w, provided that $w \gg a$. Thus, the force on the vortex core, which is the gradient of the above "potential" energy, has the form

$$\boldsymbol{F} = \rho \boldsymbol{v} \times \boldsymbol{\kappa} \tag{3.5.14}$$

Equation (3.5.14) is actually quite general, and is the expression for the celebrated *Magnus force*. For a vortex ring, the extra energy of flow in a constant background flow $\boldsymbol{v}_\mathrm{b}$ perpendicular[20] to the ring turns out to be, by Galilean invariance

$$\Delta\mathcal{E} = -\boldsymbol{v}_\mathrm{b} \cdot \boldsymbol{P}(+\mathrm{const.}) \tag{3.5.15}$$

where $\boldsymbol{P}$ is the rest-frame momentum, given by (3.5.12). This must be added to the original energy for $\boldsymbol{v}_\mathrm{b} = 0$, given by (3.5.11). As a result, in a finite perpendicular background flow $\boldsymbol{v}_\mathrm{b}$ (of a sign parallel to $\hat{\boldsymbol{n}}$) the total ring energy has a *maximum* at a "critical radius" $R_\mathrm{c}(\boldsymbol{v}_\mathrm{b})$ given by

$$\boldsymbol{v}(R_v = R_\mathrm{c}) = \boldsymbol{v}_\mathrm{b} \tag{3.5.16}$$

Rings of radius smaller than R_c will be unstable against collapse, while those larger than R_c will have a tendency (if the dynamics allow it) to expand indefinitely. Rings with $\hat{\boldsymbol{n}}$ antiparallel to $\boldsymbol{v}_\mathrm{b}$ will always tend to collapse.

After this brief review of some of the main properties of vortices in a classical liquid, let's turn to the system of actual interest, namely He-II. Let's first consider the case $T = 0$ (or, more realistically, temperatures so low that the normal component is negligible). In this limit almost all the relevant behavior can be obtained simply by replacing the fluid velocity $\boldsymbol{v}(\boldsymbol{r})$ in the above discussion by the superfluid velocity $\boldsymbol{v}_\mathrm{s}(\boldsymbol{r})$. The circulation is then quantized, according to Eqn. (2.2.4), in multiples of the fundamental unit of circulation κ_0:

$$\kappa = n\kappa_0, \quad \kappa_0 \equiv \frac{h}{m} \cong 10^{-7}\mathrm{m}^2\,\mathrm{sec}^{-1} \quad (n = 0, \pm 1, \pm 2, \ldots) \tag{3.5.17}$$

However, by the argument given above, in bulk He-II vortices with $|n| > 1$ are unstable against decay into two or more vortices each with $|n| = 1$, and are therefore not observed. Thus the magnitude of circulation of a vortex line or ring in bulk is uniquely h/m. Once given this, both the energetics and the dynamics are given by the classical formulae.

[20] Any components of $\boldsymbol{v}_\mathrm{b}$ in the plane of the ring have no effect.

At finite temperatures the situation is a bit more complicated because one now has a second velocity field, namely that of the normal velocity $\boldsymbol{v}_n(\boldsymbol{r})$. In particular, the question of the various dissipative forces acting on a quantized vortex line, and hence of its dynamics, is a complicated one and has not yet attained a universally agreed resolution. We shall not find it necessary to go into these questions in this book.

The presence of quantized vortex lines and rings in He-II may be inferred in a number of different ways. Historically, the observation which led to the postulation of vortex lines was that if a reasonably macroscopic (say radius ~1 cm) cylinder containing He-II is rotated, the equilibrium state has a surface which, when observed with the naked eye, appears to have the same meniscus as that of any ordinary fluid such as water under the same rotation. Application of Bernoulli's theorem then allows us to conclude that, at least on this coarse scale, the velocity pattern of the liquid is identical to that of a normal liquid, namely

$$\boldsymbol{v}(\boldsymbol{r}) = \boldsymbol{\omega} \times \boldsymbol{r} \tag{3.5.18}$$

where $\boldsymbol{\omega}$ is the angular velocity of the container. However, if we identify $\boldsymbol{v}$ with $\boldsymbol{v}_s$ the pattern (3.5.18) clearly violates the irrotationality condition (2.2.8). The obvious way out of the dilemma is to postulate that in equilibrium the liquid contains a constant density n of vortex lines, with circulation parallel to the rotation, given by

$$n = \frac{2\omega}{\kappa_0} \tag{3.5.19}$$

If we then average the superfluid velocity over a length scale large compared to the distance ($n^{-\frac{1}{2}}$) between vortices, we recover the "normal" pattern given by (3.5.18).

More recently it has been possible to exploit the fact that impurities in the liquid, in particular ions, tend to be attracted to the cores of vortex lines, to image the distribution of the lines in the rotating liquid directly (Yarmchuk et al. 1979). Electrons are drawn up the lines by an electric field and registered on a screen just above the liquid surface, so that their points of arrival on the screen reflect the positions of the vortex lines. In this way one can study not only the equilibrium configuration but also the kinetics of entry of the lines into the liquid after the rotation is started. Actually, it is only under rather special conditions that the true equilibrium configuration is easily attained; in general, if one quenches liquid He through T_λ, whether or not it is rotating, one risks generating a tangle of vortex lines and rings which is far from global equilibrium but may be quite persistent (see, e.g. Awschalom and Schwarz 1984).

Neither of the above techniques will allow us to observe vortex rings as such. However, their existence may be inferred from experiments on the mobility of ions in the (stationary) liquid under the influence of an external electric field (Rayfield and Reif 1964): all the evidence is consistent with the hypothesis that an ion tends either to attach itself to a pre-existing vortex ring, or to generate one from scratch, and that the subsequent dynamics is essentially that of the vortex ring.

A major role played in He-II by quantized vortex lines and, particularly, by vortex rings is in the decay, under appropriate conditions, of macroscopic superflow. We have seen (Chapter 2, Section 2.5) that the metastability of superflow in an annular (torroidal) geometry can be explained in terms of the topological conservation law for the winding number $n \equiv 1/2\pi \oint_E \boldsymbol{\nabla}\phi \cdot \mathbf{d}\boldsymbol{l}$. Suppose now that it is possible for a vortex

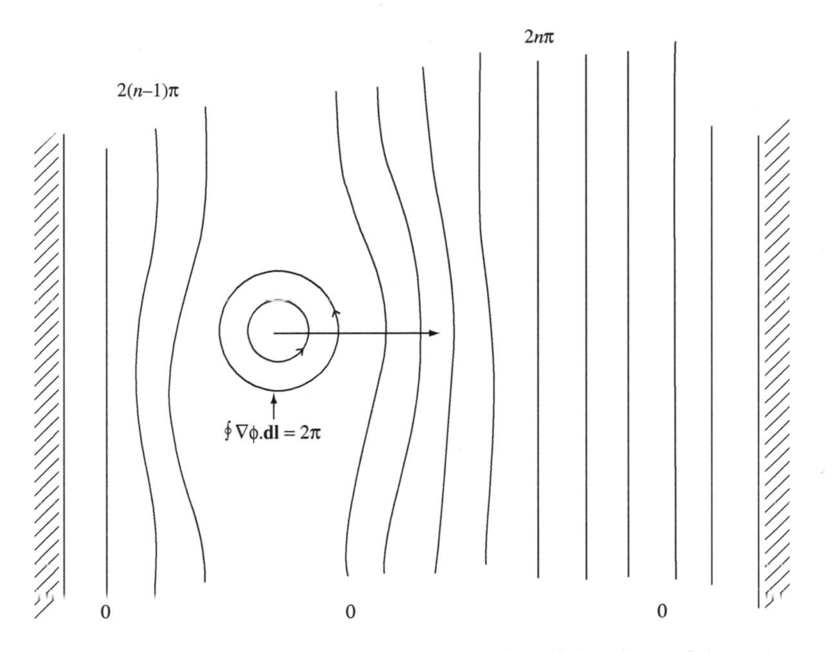

Fig. 3.5 Decay of superflow by vortex motion. The value of the phase of the order parameter at various points is indicated.

line to nucleate on one wall of the container and move across it (Fig. 3.5). We see that the effect is to reduce or to increase the winding number by 1, depending on the sense of the circulation of the line. Similarly, if we could nucleate a vortex *ring* in the middle of the tube and expand it outwards until it annihilates on the walls, the effect is again to change n by ± 1. The second type of process is believed to be responsible for the slow decay of circulating supercurrents in narrow channels close to T_λ. As to the first, it may be responsible for the critical velocities observed in wide channels (see, e.g. Putterman 1974, Section 42) and in addition can be used to give a schematic account of the Josephson effect in small orifices; I now discuss this question briefly (A more detailed discussion of the Josephson effect in general will be given in Chapter 5.)

Imagine that we have two bulk baths of liquid ^{4}He connected by a small orifice. Assume that we apply a chemical potential difference $\Delta\mu$ between the two baths; this could be done, for example, by raising one relative to the other in the earth's gravitational field. According to Eqn. (3.4.7), the effect is to give rise to a time-dependence of the phase difference $\Delta\phi$ of the condensate between the two baths:

$$\Delta\phi(t) = \frac{\Delta\mu \cdot t}{\hbar} \tag{3.5.20}$$

This will, of course, give rise to a flow through the superleak. However, once $\Delta\phi(t)$ exceeds π, the energy of the state can be lowered by changing (decreasing) $\Delta\phi$ by 2π. Such an event (already anticipated in Section 3.3) is called a *phase slip*. In a bulk geometry the energy barrier against a phase slip is usually sufficiently high that it can be neglected, but in a small orifice this may no longer be so; one obvious possibility (not the only one, as we shall see in Chapter 5) is for a vortex line to nucleate on one side of the orifice and pass across it. In the extreme case such a process may happen on every cycle; the system then shows an *ac* mass flow, with a characteristic frequency given by

$$\omega = \frac{\Delta\mu}{\hbar} \tag{3.5.21}$$

This is the AC Josephson effect, which will be discussed in more generality in the context of superconductors in Chapter 5. As we shall see there, it is possible to synchronize the effect with an external AC drive so as to produce a DC current.

3.6 The excitation spectrum of liquid He-II

At first sight, one might think that in order to calculate the excitation spectrum of superfluid He-II one would need detailed information about the ground state of the system, that is, a microscopic theory. Rather surprisingly, this turns out not to be so. In fact, it is possible to draw quite quantitative conclusions about the excitation spectrum without any but the most qualitative information about the nature of the ground state. I will first state these conclusions, and then review the arguments which lead to them.

According to our current theoretical understanding, the *only* low-energy (say $\epsilon \lesssim$ 5 K) elementary excitations of superfluid He-II are long-wavelength density fluctuations (phonons) with an energy–momentum relation given by

$$\epsilon(\boldsymbol{p}) = c_s p \equiv \hbar c_s k \tag{3.6.1}$$

where c_s is the speed of hydrodynamic sound, which is related to the bulk modulus $B \equiv \partial^2 E/\partial V^2$ by

$$c_s = \sqrt{\frac{B}{\rho}} \tag{3.6.2}$$

where ρ is the density. It should be emphasized that this conclusion *does not* follow trivially from Eqns. (3.4.1a) and (3.4.1b), since these *assume* hydrodynamic conditions. That the density fluctuations indeed have the spectrum (3.6.1) for small k is beautifully confirmed by inelastic neutron scattering experiments; that there are no other low-lying excitations (which, since by definition they do not induce any density fluctuation, would not show up in the neutron scattering data) is confirmed by the fact that the specific heat contributed by the phonons, namely

$$\frac{c_v}{k_B} = \frac{2\pi^2}{15}\left(\frac{k_B T}{\hbar c}\right)^3 \tag{3.6.3}$$

appears to exhaust the experimentally measured value of that quantity.

The simple dispersion relation (3.6.1) holds experimentally, to a good approximation, for $k \lesssim 0.6$ Å^{-1}. At larger values of k the spectrum observed in neutron scattering is still quite sharp, but has a dip, as shown in Fig. 3.6; for historical reasons the elementary excitations in the region of the minimum at $\sim$2 Å^{-1} are known as "rotons" (and the excitations near the maximum at $k \sim 1$ Å^{-1} are sometimes called "maxons"). We shall see that although some theoretical arguments can be given for

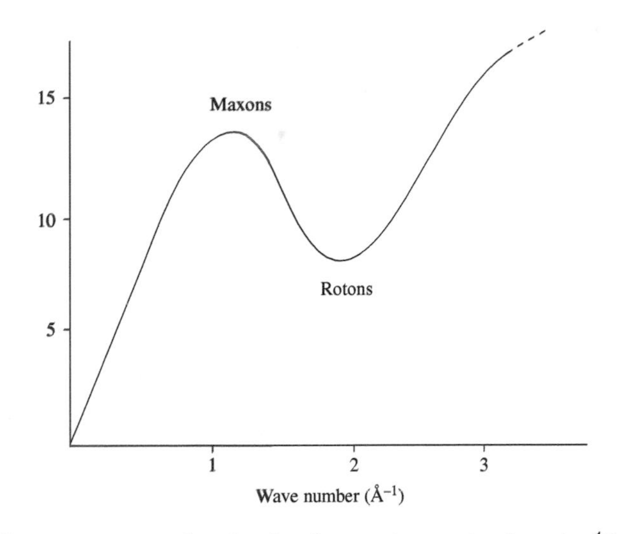

Fig. 3.6 The dispersion curve for density-fluctuation excitations in ^{4}He as observed by neutron scattering.

the form of the dispersion relation in the roton region, they are not generic to an arbitrary system of bosons but require some information on the general properties of the ground state wave function. Beyond about 2.5 Å^{-1} the energy spectrum as seen in neutron scattering is no longer well-defined, i.e. for given neutron momentum loss $\boldsymbol{p}$ the energy loss is quite diffuse; this is usually attributed to the fact that beyond this point a decay of a single excitation into two is compatible with the conservation of energy and momentum (as it is not for smaller values of $\boldsymbol{k}$).

Let us now turn to the theoretical justification for the statements made above, and in particular for the dispersion relation (3.6.1). We start from two well-known sum rules involving the matrix elements of the operator $\rho_k \equiv \sum_i e^{i\boldsymbol{k}\cdot\boldsymbol{r}_i}$ (see e.g. Nozières and Pines 1989, Section 2.4). If $|n\rangle$ denotes the nth excited state, with energy ϵ_n, and we write $\omega_{n0} \equiv (\epsilon_\text{n} - \epsilon_0)/\hbar$, then these two sum rules are, respectively, the "compressibility sum rule"

$$\sum_n \frac{|\langle n|\rho_k|0\rangle|^2}{\omega_{n0}} = \chi_k \tag{3.6.4}$$

(where χ_k is the static compressibility at wave vector k), and the "f-sum rule"

$$\sum_n |\langle n|\rho_k|0\rangle|^2 \omega_{n0} = \frac{N\hbar^2 k^2}{m} \tag{3.6.5}$$

We note for future reference that the static ground state correlation $\langle \rho_k \rho_{-k} \rangle_0 (k \neq 0)$ is given by a third "sum rule", namely

$$S(k) \equiv \langle \rho_k \rho_{-k} \rangle_0 = \sum_n |\langle n|\rho_k|0\rangle|^2 \tag{3.6.6}$$

Let us suppose (for the moment for no obviously good reason!) that the state $\rho_k|0\rangle$ is actually an exact energy eigenstate, so that the sums in (3.6.4)–(3.6.6) each reduce to a single term. Under this assumption we see that the energy spectrum is given by

$$\epsilon(k) = \sqrt{\frac{\hbar^2 k^2}{\chi_k}} = \frac{\hbar^2 k^2}{2mS(k)} \tag{3.6.7}$$

In the limit $\boldsymbol{k} \to 0$ the susceptibility χ_k reduces to the inverse of the uniform bulk modulus B divided by ρ, so we recover Eqn. (3.6.2). The second expression in (3.6.7) will be more useful in the roton regime, see below.

None of the above argument is in any way specific to the case of a Bose condensate or even a system of bosons; all the statements made would be equally true for a Fermi system such as normal liquid ^{3}He. However, we know that (3.6.1) is not true for such a system (in which the elementary excitations in fact do not possess a unique energy–momentum relation, even in the limit $k \to 0$), and this immediately shows that the assumption that the state $\rho_k|0\rangle$ is an exact energy eigenstate (rather than a superposition of eigenstates) cannot be true there. Why then should it be true for the case of He-II? Moreover, even supposing this to be so, why should the excitations described by (3.6.1) exhaust the spectrum of low-lying elementary excitations?

In his famous 1948 paper, Feynman gave a series of plausibility arguments for the thesis that the only low-lying excitations of a Bose system are simple density fluctuations, that is, that states of the form $\rho_k|0\rangle$ are (in the limit $k \to 0$) exact energy eigenstates and that there are no others. It is interesting to note that while the bosonic nature of the system is essential (cf. below) there is no requirement that it be Bose-condensed, and indeed part of Feynman's arguments and conclusions apply equally to *solid* ^{4}He (where they are perhaps less striking). It appears difficult to improve on these plausibility arguments at this level of generality; as we shall see (Section 7) a systematic perturbation-theoretic discussion does reproduce the result (3.6.1).

Once we know that the only low-lying excitations are phonons with the spectrum (3.6.1), a famous argument due to Landau gives us the normal density $\rho_n(T)$ in the low-temperature limit (where any phonon–phonon or phonon–roton interaction may be neglected). This argument will be given in detail in Chapter 5, Section 5.7 in the superconducting context, so here I just summarize the conclusion: In a frame moving with the superfluid velocity $\boldsymbol{v}_s$ the effective free energy which must be minimized in thermal equilibrium is given by $F + \boldsymbol{v}_s \cdot \boldsymbol{p}$, so that the distribution of excitations is $n_B(\tilde{E}_p)$ where $n_B(E) \equiv (\exp \beta E - 1)^{-1}$ is the Bose distribution and $\tilde{E}_p \equiv E_p + \boldsymbol{v}_s \cdot \boldsymbol{p} (\boldsymbol{p} \equiv \hbar \boldsymbol{k})$. Thus the total momentum as viewed from this frame, which is $-\rho_n(T)\boldsymbol{v}_s$, is given by the expression (taking $\boldsymbol{v}_s$ to be small and in the x-direction for definiteness)

$$\boldsymbol{P} = \sum_p \boldsymbol{p} n_B(E_p + \boldsymbol{v}_s \cdot \boldsymbol{p}) \cong \sum p_x^2 \frac{\partial n_p}{\partial E_p} \boldsymbol{v}_s \tag{3.6.8}$$

When the relevant $E(\boldsymbol{p})(\equiv \epsilon(\boldsymbol{p}))$ are given by the formula (3.6.1), formula (3.6.8) can be evaluated explicitly and yields

$$\rho_n(T) = aT^4, \quad a \equiv \frac{2\pi^2 k_B^4}{45\hbar^3 c^5} \tag{3.6.9}$$

This formula was verified in the original oscillating-disk experiments of Andronikashvili in 1946. It should be approximately valid provided that (a) $k_B T$ is well below the energies for which (3.6.1) begins to fail (i.e. crudely speaking, the roton energy (~8 K)), *and* (b) the total number of excitations (which is *not* equal to ρ_n!) is small compared to the total number of atoms in the system. Since the number of excitations is given from (3.6.8), by

$$n_{\text{exc}}(T) = bT^3, \quad b = 2\pi^2 k_B^4 / 45\hbar^3 c^3 \tag{3.6.10}$$

these criteria should both be fulfilled for temperatures $\lesssim 0.4$ K.

We finally turn briefly to the roton region of the spectrum. If we assume that the elementary excitation spectrum continues to be exhausted by single density fluctuations in this region, then Eqn. (3.6.7) tells us that excitation energy is $\hbar^2 k^2 / 2mS(k)$, where $S(k)$ is the static ground state density correlation

$$S(k) \equiv \frac{1}{N} \sum_{ij} \langle e^{i\boldsymbol{k}\cdot(\boldsymbol{r}_i - \boldsymbol{r}_j)} \rangle_0 \sim \int \langle \rho(0)\rho(\boldsymbol{r}) \rangle e^{i\boldsymbol{k}\cdot\boldsymbol{r}}\, d\boldsymbol{r} \tag{3.6.11}$$

The dip in the experimentally observed spectrum then indicates that $S(k)$ is quite sharply peaked in the region of 2 Å^{-1}, which in turn suggests that the real-space correlation function $\langle \rho(0)\rho(\boldsymbol{r}) \rangle$ is peaked in the region of $(2\pi/2)$ Å ~ 3.2 Å^{-1}. This is not surprising, since in the solid phase near melting pressure this function is sharply peaked in this region and one may expect the short-range order in the liquid phase to contain at least some vestiges of its form in the solid. In fact, the roton minimum may be viewed as a sort of incipient crystallization (cf. Nozières 2004). Microscopic theories of liquid ^{4}He (see next section) generally produce many-body ground state wave functions which to one extent or another indeed predict the above behavior of $S(k)$; however, it should be emphasized that the latter does not in itself actually have much to do with BEC (or even with Bose statistics – the form of $S(k)$ for liquid ^{3}He is not qualitatively different), and that while as we approach T_λ the roton spectrum appears to become so diffuse as to be no longer interpretable as an elementary excitation, this may be to some extent accidental; there is no reason to believe that $S(k)$ changes its form abruptly at T_λ.

The form of the elementary excitation spectrum shown in Fig. 3.6 has an important consequence for the mobility of foreign objects such as ions moving in He-II. Consider such an object, with a mass M large compared to that (m) of a He atom, moving in the liquid at $T = 0$ (so that there is no normal component). To lose momentum to the atoms of the liquid, it must create one or more elementary excitations; it is easy to show that if the creation of more than one excitation is allowed by energy and momentum conservation, then so must be that of a single excitation, so we just consider this latter possibility. Then it follows immediately from the need to conserve energy and momentum, in the limit $M \gg m$, that an excitation of momentum $\boldsymbol{p}$ can be excited only if the velocity v of the moving object exceeds $E(p)/p$. Thus, no excitations can be created if the ion (etc.) is moving with velocity less than a *critical velocity* given by

$$v_c = \min \left[\frac{\epsilon(p)}{p} \right] \tag{3.6.12}$$

Were the spectrum that of a free gas ($\epsilon(p) = p^2/2m$), the critical velocity would be zero; however, since at long wavelengths the He spectrum is linear in p, v_c is finite and is in fact given approximately by $\epsilon(p_0)/p_0$ where p_0 is the momentum corresponding to the roton minimum; the value of v_c is approximately 60 m/sec in He II at saturated vapor pressure. At any finite temperature there will still be some viscous drag on the ion even at velocities less than v_c, because of the effect of the normal component (i.e. the pre-existing excitations which can scatter off the ion); however, we expect the drag to increase dramatically and suddenly at v_c. This prediction is spectacularly confirmed in experiments on the mobility of ions in He-II. The details of the processes which give rise to drag for velocities above v_c are still not entirely understood, see e.g. Allum et al. 1976.

The critical velocity (3.6.12) was originally introduced by Landau in his discussion of the metastability of superflow, and is thus known as the Landau critical velocity. While the question of the general validity of the Landau argument in that context raises some delicate issues, the latter do not affect the very straightforward argument given above for the ion mobility problem, and there seems little doubt that its conclusions are correct. However, it should be mentioned that while a few experiments on ion mobility have "seen" the Landau critical velocity, in real life the situation is often complicated by the presence of pre-existing vortices, which affect the mobility below v_c: see e.g. Wilks (1967), Section 13.8.

3.7 Microscopic theories of He-II

The main emphasis of this book, as explained in the introduction, is on the qualitatively anomalous properties of superfluid systems and the way in which they are explained by the phenomenon of BEC; and in the case of liquid He-II, the considerations of the previous sections (and of Chapter 8) are entirely adequate for an explanation at this level. On the other hand, as our oldest neutral superfluid, He-II has been the subject of a vast amount of quantitative experimental study, and much theoretical effort over the last 60 years has gone into trying to understand these data; so it may be worthwhile to spend a few pages reviewing the principal types of microscopic theoretical approach which have been used. What follows is at best a lightning review of a complex and multi-faceted subject, and the reader is referred for details to the original literature and to several good monographs (Feenberg 1969, Griffin 1993, Nozières and Pines 1989). Some of the approaches to be discussed are likely to be much better approximations for the dilute alkali gases than they are for helium, and will be discussed in greater detail in the next chapter. I will start with the most naive (but exactly soluble) models, and move progressively in the direction of greater realism, ending with exact approaches which have to be implemented numerically.

3.7.1 The ideal Bose gas

Needless to say, no one takes the ideal Bose gas seriously as a realistic model for liquid ^{4}He, but it is interesting to recapitulate briefly which features of the behavior of that system it gets qualitatively right or wrong. An ideal gas of bosons with the mass and density (at SVP) of liquid ^{4}He undergoes a transition to the BEC phase at approximately 3.3 K, which is not too different from the actual λ-temperature of He, 2.17 K; however, the T_c of the ideal gas is proportional to $\rho^{2/3}$, so that the slope of the λ-line in the $P-T$ plane is predicted to be positive, in contradiction to experiment (see Fig. 3.1). The transition in the ideal gas is third-order in the Ehrenfest classification, so that in particular the specific heat has a cusp rather than the weak divergence which is observed experimentally. The BEC phase of such a gas shows one fundamental manifestation of "superfluidity," namely the Hess–Fairbank effect (nonclassical rotational inertia) (see Chapter 2, Section 2.5, point 3); however, it does *not* show the second fundamental ingredient, namely the phenomenon of persistent currents, since there is no energy barrier against the decay of a circulating state (ibid, point 6). Finally. the excitation spectrum at $T = 0$ is just that obtained by taking a single particle out of the condensate ($\boldsymbol{p} = 0$) and transforming it to a state of momentum $\boldsymbol{p}$, i.e. $\epsilon(\boldsymbol{p}) = p^2/2m$; this bears little resemblance to the observed spectrum of He-II (see Fig. 3.6).

3.7.2 The Gross–Pitaevskii ansatz

The ansatz associated with the names of Gross and Pitaevskii has a rather peculiar history: originally conceived as a phenomenological description of He-II analogous to the Ginzburg–Landau description of superconductivity (see Chapter 5, Section 5.7), it has really come into its own in the context of the more recently stabilized systems of dilute alkali gases, where its theoretical foundation is much firmer. I shall explain this application in detail in Chapter 4; for the moment I just note that for a hypothetical system of neutral atoms with a very weak *bare* interatomic potential $V(r)$ with Fourier transform V_k (*not* the case for alkali gases!), the Gross–Pitaevskii (GP) ansatz is equivalent to a well-defined approximation for the many-body wave function (density matrix) which is nothing but a special case of the standard Hartree approximation appropriate to a Bose system with condensation. In particular, for states (not necessarily equilibrium ones) with 100% condensation the ansatz takes the simple form

$$\Psi_N(\boldsymbol{r}_1\boldsymbol{r}_2\cdots\boldsymbol{r}_N : t) = \prod_{i=1}^{N} \chi_0(\boldsymbol{r}_i : t) \tag{3.7.1}$$

where $\chi_0(\boldsymbol{r} : t)$ is the wavefunction of the condensate. For a state of the form (3.7.1) it is clear, according to the definitions of Chapter 2, Section 2.2, that the order parameter is given by the simple expression

$$\Psi(\boldsymbol{r}t) = \sqrt{N}\chi_0(\boldsymbol{r} : t) \tag{3.7.2}$$

and moreover that provided the variation of $\chi_0(\boldsymbol{r})$ over the range of the weak interatomic potential $V(r)$ is small, the expectation value of the energy is given by the expression

$$\langle E\rangle = \int d\boldsymbol{r} \left\{ \frac{\hbar^2}{2m}|\boldsymbol{\nabla}\Psi(\boldsymbol{r}t)|^2 + U(\boldsymbol{r}t)|\Psi(\boldsymbol{r}t)|^2 + \frac{1}{2}V_0|\Psi(\boldsymbol{r}t)|^4 \right\} \tag{3.7.3}$$

where V_0 is the $\boldsymbol{k} = 0$ transform of $V(r)$; I have added for completeness a possible c-number external potential $U(\boldsymbol{r}t)$.

For any system even slightly more complicated than the above[21], we should expect the ansatz (3.7.1) to fail, even as regards the ground state. However, we can still define the order parameter $\Psi(\boldsymbol{r}t)$ by the general prescription (2.2.1)(with $\chi_o(\boldsymbol{r}t)$ defined as in Chapter 2, Section 2.1), and it is then possible to argue that at least for phenomena

[21]Including the dilute alkali gases, see Chapter 4.

involving only the condensate, the expression (3.7.3) is not a bad approximation to the energy. One can actually go further, and argue that if we are interested in (stable or metastable) thermal equilibrium at finite temperature, the bulk Helmholtz free energy may be expanded in a Taylor series in $|\Psi|^2$:

$$F = F_0(T) + \alpha(T)|\Psi|^2 + \frac{1}{2}\beta(T)|\Psi|^4 + \cdots \tag{3.7.4}$$

If Ψ is sufficiently slowly varying in space, the total free energy should be expressible in the form of a space integral of the free energy density $F\{\Psi(\boldsymbol{r}) : T\}$. Adding a gradient term for the reasons outlined in Chapter 2, Section 2.5, we find a phenomenological form for the "constrained" Helmholtz free energy $F\{\Psi(\boldsymbol{r})\}$ (defined below):

$$\mathcal{F}\{\Psi(\boldsymbol{r}) : T\} = \mathcal{F}_0(T) + \int \mathbf{d}\boldsymbol{r} \left\{ \alpha(T)|\Psi(\boldsymbol{r})|^2 + \frac{1}{2}\beta(T)|\Psi(\boldsymbol{r})|^4 + \gamma(T)|\boldsymbol{\nabla}\Psi(\boldsymbol{r})|^2 \right\} \tag{3.7.5}$$

Here the quantity (functional) $F\{\Psi(\boldsymbol{r} : T)\}$ has the following meaning: We consider the submanifold of the total Hilbert space of the system which satisfies the constraint that the order parameter, as defined by Eqn. (2.2.1), has the specified functional form $\Psi(\boldsymbol{r})$. We then define the partition functional $Z\{\Psi(r)\}$ as the trace of $e^{-\beta\hat{H}}$ taken over this submanifold, and define $\mathcal{F}\{\Psi(r)\}$ to be equal to $-\frac{1}{\beta}\ln Z\{\Psi(r)\}$. The usual (unconstrained) free energy of the system is then given by the expression

$$F(T) = -\frac{1}{\beta}\ln \int \mathcal{D}\Psi(\boldsymbol{r})e^{-\beta\mathcal{F}\{\Psi(\boldsymbol{r}):T\}} \tag{3.7.6}$$

where the $\mathcal{D}$ denotes a standard functional integration (see, e.g. Amit 1978, Chapter 2).

The general expression (3.7.5), contains so far unspecified values of the temperature-dependent coefficients α, β, γ. When we are sufficiently close to the transition temperature $T_{\mathrm{c}}(\equiv T_\lambda)$ to the BEC phase we we may reasonably assume (cf. Chapter 5, Section 5.7) the approximate temperature-dependences

$$\alpha(T) = \alpha_0(T - T_c), \quad \beta(T) = \text{const.}, \quad \gamma(T) = \text{const.} \tag{3.7.7}$$

Equation (3.7.5) thus forms a firm and very useful jumping off point for discussions of the critical behavior of liquid ^{4}He: see, for example, Amit (1978), Chapter 6.

The quantity $\mathcal{F}\{\Psi(\boldsymbol{r}) : T\}$ can of course be defined (by the above prescription) for arbitrary values of T (including those above T_{c}). However, in the general case it will contain higher-order terms, e.g. proportional to $|\Psi|^6$ or $|\Psi|^2(\boldsymbol{\nabla}|\Psi|)^2$, the only reason we could omit these near T_{c} is because of our confidence, on both theoretical and experimental (cf. Section 3.2) grounds that "typical" values of the order parameter are small in that region. Rather than introducing these terms and the accompanying multiplicity of unknown coefficients, it is more usual to treat the simple expression (3.7.5) (with or without the specific temperature-dependences (3.7.7)), as an approximate description of the equilibrium or near-equilibrium behavior of liquid ^{4}He at arbitrary temperatures below T_{c}. As such, it is most useful in analyzing those properties which depend on the macroscopic behavior, and particularly on the topology, of the condensate wave function. For example, it is possible, by minimizing the quantity $\mathcal{F}$ with respect to $\Psi(\boldsymbol{r})$ subject to the constraint of unit circulation, to discuss the structure of a vortex core; while the fluctuations around a metastable state which lead to "phase slip" may be analyzed with the help of the Gibbs distribution in (3.7.6) (cf. Section 3.5).

It is also possible to use the phenomenological description given by (3.7.5) to discuss time-dependent phenomena, at least schematically. To do this we allow the order parameter to be a function of time as well as position, but keep for $\mathcal{F}\{\Psi(\boldsymbol{r}t)\}$ the functional form (3.7.5). Since the condition for equilibrium is $\delta F/\delta\Psi = 0$ (cf. above), $\delta^2 F/\delta\Psi^2 > 0$, we expect that the time rate of change of Ψ should be proportional to $-\delta F/\delta\Psi$. The general belief is that, at least for phenomena occurring on a relatively macroscopic scale and for not too low temperatures, the dynamics is relaxational rather than oscillatory, i.e. if displaced from the equilibrium configuration the system will relax exponentially towards it. Such behavior can be generated by an equation of the form

$$\eta \frac{\partial \Psi(\boldsymbol{r}t)}{\partial t} = -\frac{\delta F\{\Psi(\boldsymbol{r}t)\}}{\delta \Psi(\boldsymbol{r}t)} \tag{3.7.8}$$

where η, which is some as yet undetermined real positive function of T, is a sort of viscosity coefficient. This "time-dependent Gross–Pitaevskii" (TDGP) equation may be used, for example, to discuss the generation of vortices in ^{4}He following a "pressure quench" through the λ-line. It should be emphasized that it is *not* identical to the "TDGP" equation routinely used for the study of confined alkali gases at low temperatures, since the coefficient which replaces η in the latter is pure imaginary (see Chapter 4, Section 4.3).

3.7.3 The Bogoliubov approximation

Like the GP ansatz, the Bogoliubov approximation really comes into its own in the context of a dilute Bose gas, and it will be discussed in detail in the next chapter. In the alkali-gas systems discussed in that chapter, the "bare" interatomic interaction, while short-ranged, is not weak, and a renormalization procedure which is not entirely trivial is necessary before we can apply the approximation. By contrast, in his original work Bogoliubov considered an interaction which is itself weak, and possesses a Fourier transform V_k which is positive in the limit $k \to 0$. In free space the theory is then a special case of that considered in Chapter 4, Section 4.4; omitting the details of the derivation, I quote the salient results: The ground state many-body wave function is approximated by the form[22]

$$\Psi_N = \text{const.}\left(a_0^\dagger a_0^\dagger - \sum_k c_k a_k^\dagger a_{-k}^\dagger\right)^{N/2} |\text{vac}\rangle \tag{3.7.9}$$

with c_k a set of real positive constants, which are determined by minimizing the energy. The elementary excitations are of the form

$$\chi_k = (u_k a_k^\dagger a_0 + v_k a_0^\dagger a_{-k})|\Psi_N\rangle \text{ with } u_k \equiv 1/\sqrt{1-c_k^2},$$
$$v_k \equiv c_k/\sqrt{1-c_k^2} \tag{3.7.10}$$

[22]In Bogoliubov's original paper (and much of the subsequent literature) the formula for Ψ_N is only implicit because of the "number-nonconserving" trick used: see Chapter 4, Section 4.4.

and have energies given by

$$E(k) = \sqrt{(\epsilon_k + nV_k)^2 - n^2V_k^2} \equiv \sqrt{\epsilon_k(\epsilon_k + 2nV_k)} \qquad \epsilon_k = \frac{\hbar^2 k^2}{2m} \tag{3.7.11}$$

where n is the particle density. If we assume that in the relevant regime V_k does not vary very much from its $k \to 0$ limit V_0, then Eqn. (3.7.11) defines a natural length scale $\xi \equiv \sqrt{\hbar^2/2mnV_0}$ at which ϵ_k and nV_k become approximately equal. For $k \gg \xi^{-1}$ we have $E(k) \cong \epsilon(k)$, so that the spectrum is essentially that of an ideal Bose gas; by contrast, for $k \ll \xi^{-1}$ we have

$$E(k) = \hbar c_s k \quad c_s \equiv \sqrt{\frac{nV_0}{m}} \tag{3.7.12}$$

The ground state energy turns out to be given by

$$E(N) = \frac{1}{2}\frac{N^2}{\Omega}V_0(1 + o(\zeta)) \quad \zeta \equiv \sqrt{n\left(\frac{mV_0}{4\pi\hbar^2}\right)^3} \tag{3.7.13}$$

so to lowest order in the small dimensionless parameter ζ the quantity c_s is just the speed of sound at $T = 0$, and Eqn. (3.7.12) just says that the elementary excitations are quantized sound waves (phonons) with a linear dispersion relation. This is of course, just the conclusion we obtained from general arguments, in the limit $k \to 0$, in Section 3.6. If, furthermore, we were to imagine that the expression (3.7.12) is valid under arbitrary conditions of density (etc.) then it is easy to see that a sharp decrease of V_k with k could give rise to a maximum and subsequent minimum in the excitation spectrum, analogous to what is seen in He-II.

Unfortunately this latter conclusion cannot be taken seriously. We need to enquire about the conditions of validity of the "primitive" Bogoliubov approximation which gives rise to the result (3.7.12). The first necessary condition is that the free-space two-particle scattering should be adequately described by the lowest-order Born approximation (if the condition is not fulfilled, we need to eliminate the bare interatomic potential in favor of a scattering length, as described in Chapter 4, Section 4.2). This condition is roughly equivalent (for not too pathological forms of potential) to the condition (where we abbreviate the quantity $2mV_0/\hbar^2$ by $\widetilde{V}_0$)

$$\widetilde{V}_0 R_0^2 \ll 1 \tag{3.7.14}$$

where R_0 is the range of the potential. The second condition turns out to be that the depletion of the condensate, i.e. the relative difference between N_0 and N caused by the mutual scattering of pairs $(\boldsymbol{k}, -\boldsymbol{k})$ out of the $\boldsymbol{k} = 0$ state (cf. Eqn. 3.7.9) is small compared to 1. We shall see in Chapter 4, Section 4.4 that the depletion is of order ζ, so from the definition (3.7.13) of the latter this requires

$$n\widetilde{V}_0^3 \ll 1 \tag{3.7.15}$$

Now, if we make the order-of-magnitude estimate $|\partial^2 V_k/\partial k^2| \sim V_0 R_0^2$, it is clear that a necessary (though not sufficient) condition for the excitation spectrum (3.7.11) to have a maximum is $nR_0^2\widetilde{V}_0 \gtrsim 1$; and it is immediately clear that such a condition is incompatible with the conjunction of (3.7.14) and (3.7.15). We conclude that the Bogoliubov approximation can never produce a "maxon–roton" energy spectrum in any parameter regime where it is formally valid. In any case, it is clear that neither of the conditions (3.7.14) or (3.7.15) is anywhere near satisfied for real-life liquid ^{4}He.

In the light of this it may be illuminating to discuss briefly an alternative approach (essentially that of Sunakawa et al. 1962) which also leads to the spectrum (3.7.11). Suppose we write down the second-quantized form of the Hamiltonian in terms of the Bose field operator $\hat{\psi}(r)$, which satisfies the commutation relations

$$[\hat{\psi}(\boldsymbol{r}), \hat{\psi}^{\dagger}(\boldsymbol{r}')] = \delta(\boldsymbol{r} - \boldsymbol{r}') \tag{3.7.16}$$

$$\hat{H} = \frac{\hbar^2}{2m} \int \boldsymbol{\nabla}\hat{\psi}^{\dagger}(\boldsymbol{r}) \cdot \boldsymbol{\nabla}\psi(\boldsymbol{r}) d\boldsymbol{r} + \frac{1}{2} \iint d\boldsymbol{r}\, d\boldsymbol{r}' V(\boldsymbol{r} - \boldsymbol{r}')\hat{\psi}^{\dagger}(\boldsymbol{r})\hat{\psi}^{\dagger}(\boldsymbol{r}')\hat{\psi}(\boldsymbol{r}')\hat{\psi}(\boldsymbol{r}) \tag{3.7.17}$$

We now attempt to express the field operator in terms of "density" and "phase" operators. This actually involves considerable technical complications similar to (but more severe than) those encountered in the attempt to define a relative phase operator $\Delta\hat{\varphi}$ (Appendix 2B). Blithely ignoring these complications, let us postulate the approximate expression

$$\hat{\psi}(\boldsymbol{r}) = \hat{\rho}^{1/2}(\boldsymbol{r}) e^{i\hat{\phi}(\boldsymbol{r})} \tag{3.7.18}$$

where to maintain the commutation relation (3.7.16) the operators $\hat{\rho}(\boldsymbol{r})$ and $\hat{\phi}(\boldsymbol{r}')$ should satisfy the (approximate) commutation relation

$$[\hat{\rho}(\boldsymbol{r}), \hat{\phi}(\boldsymbol{r}')] \equiv i\delta(\boldsymbol{r} - \boldsymbol{r}') \tag{3.7.19}$$

In terms of $\hat{\rho}$ and $\hat{\phi}$ the Hamiltonian (3.7.16) takes the form

$$\hat{H} = \frac{\hbar^2}{2m} d\boldsymbol{r} \left\{ \frac{1}{4}\hat{\rho}^{-1}(\boldsymbol{r})(\boldsymbol{\nabla}\hat{\rho}(\boldsymbol{r}))^2 + \hat{\rho}(\boldsymbol{r})(\boldsymbol{\nabla}\hat{\varphi}(\boldsymbol{r}))^2 \right\} + \frac{1}{2} \iint d\boldsymbol{r}\, d\boldsymbol{r}' V(\boldsymbol{r} - \boldsymbol{r}')\hat{\rho}(\boldsymbol{r})\hat{\rho}(\boldsymbol{r}') \tag{3.7.20}$$

Now let us write $\hat{\rho}(\boldsymbol{r}) \equiv n + \delta\hat{\rho}(\boldsymbol{r})$, where $\delta\hat{\rho}(\boldsymbol{r})$ obviously satisfied the commutation relation (3.7.19), and expand the Hamiltonian (3.7.20) in a Taylor series in $\delta\hat{\rho}(\boldsymbol{r})$. Since the ground state minimizes the expectation value of $\langle H \rangle$, the term linear in $\delta\hat{\rho}(\boldsymbol{r})$ drops out; neglecting terms of third and higher order in $\delta\hat{\rho}$, we find

$$\hat{H} = \text{const.} + \frac{\hbar^2}{2m} \int d\boldsymbol{r} \left\{ \frac{1}{4n}(\boldsymbol{\nabla}\delta\hat{\rho}(\boldsymbol{r}))^2 + n(\boldsymbol{\nabla}\hat{\phi}(\boldsymbol{r}))^2 \right\} + \frac{1}{2} \iint d\boldsymbol{r}\; d\boldsymbol{r}' V(\boldsymbol{r} - \boldsymbol{r}')\delta\hat{\rho}(\boldsymbol{r})\delta\hat{\rho}(\boldsymbol{r}') \tag{3.7.21}$$

or in Fourier-transformed form:

$$\hat{H} = \text{const.} + \frac{1}{2} \sum_k \left\{ \left(\frac{1}{2n} \cdot \epsilon_k + V_k \right) \hat{\rho}_k \hat{\rho}_{-k} + 2n\epsilon_k \hat{\varphi}_k \hat{\varphi}_{-k} \right\} \tag{3.7.22}$$

where $\hat{\phi}_k$, $\hat{\rho}_k$ satisfy the commutation relation

$$[\hat{\rho}_{\boldsymbol{k}}, \hat{\phi}_{-\boldsymbol{k}'}] = i\delta_{\boldsymbol{k},\boldsymbol{k}'} \tag{3.7.23}$$

Equations (3.7.22) and (3.7.23) simply describe a set of independent harmonic oscillators with the Bogoliubov spectrum (3.7.11)! While it is intriguing that this spectrum emerges from a calculation which assumes that the fluctuations in the total density $\hat{\rho}(\boldsymbol{r})$ are small (an assumption one might think is most likely to be valid in the regime of high density and strong interactions), the result (and the argument leading to it) should probably not be taken too seriously, if only because it gives unphysical results if we take $V(\boldsymbol{r})$ to be the real potential (including the hard core) between ^{4}He atoms. It is nevertheless qualitatively suggestive of how a phonon–roton type spectrum might emerge from a proper calculation: see below, and Subsection 3.5.

3.7.4 Analytical first-principles theories

In the literature there exists a bewildering variety of approaches, going back nearly 50 years which attempt to calculate the experimental properties of ^{4}He-II by various analytical approximation (or in some cases exact) techniques; here I shall do little more than steer the reader to some of the significant literature. One may try to distinguish three principal methods of attack:

1. General analyses using field-theoretic techniques. In such an approach one typically sums various infinite series of graphs which occur in the field-theoretic formulation of the problem, with the aim of establishing rigorous and generally applicable results; the obverse of the coin is that it is usually quite difficult to extract numbers applicable to a specific system such as ^{4}He at a particular temperature and pressure.[23] Typical examples of such calculations are the paper of Hugenholtz and Pines (1959), which established in full generality the absence of a gap in the energy spectrum of a Bose liquid, and that of Gavoret and Nozières (1964) which went further by showing (inter alia) that for such a system the $T = 0$, $k \to 0$ limit of the spectrum must be simple phonons with a speed of sound determined by the classical compressibility (cf. Section 3.6), and moreover that the superfluid fraction is unity in the limit $T \to 0$.[24]
2. Approaches based on a specific ansatz for the ground-state wave function (GSWF). Generally speaking, these approaches are variational in nature: one generates a particular kind of functional form for the GSWF, parametrized by various numbers (or functions) which one then varies; as usual, the "optimal" choice of parameters is usually taken to be that which gives the minimum value of the ground state energy. The archetypal form of such a variational GSWF is the Jastrow function, defined by

$$\Psi_J(\boldsymbol{r_1 r_2} \cdots \boldsymbol{r_N}) = \prod_{i<j} f(r_{ij}) \equiv \exp\left(\sum_{i<j} u(r_{ij})\right) \quad (r_{ij} = |\boldsymbol{r_i} - \boldsymbol{r_j}|) \tag{3.7.24}$$

[23] Thus, it might have been regarded as equally natural to discuss this work in Section 3.6 above.

[24] See their Equation (5.3.4). The proof requires the convergence of perturbation theory starting from the free Bose gas.

This kind of wave function[25] is intuitively plausible as a zeroth-order approximation for a system like ^{4}He possessing strong short-range repulsive ("hard-core") interactions, since by choosing $u(r_{ij})$ to be large and negative for small r_{ij} we automatically keep the particles out of one another's hard cores. More sophisticated treatments improve on the ansatz (3.7.24) by introducing explicit three- and higher-particle correlations. An extended account of this kind of approach may be found in Feenberg's (1969) book: for some more recent developments along these general lines, see, e.g. Lindenau et al. (2002).

3. Finally, there exists a third line of approach which is in a sense a hybrid of (1) and (2): One does not write down an explicit form of GSWF, but rather tries to use sum rules or other arguments to relate excited-state properties (in particular the excitation spectrum) to properties of the ground state which, rather than being calculated from first principles, we take from experiment. The prototype of this kind of approach is the work of Feynman in the early 1950s (see Section 3.6); this was considerably improved on by Feynman and Cohen, who obtained remarkably good agreement with the experimentally observed phonon–roton spectrum. Much subsequent work has focussed on trying to improve these techniques further; see for example Jackson and Feenberg (1962) or Boronat et al. (1995).

While analyses of type (1) of their nature cannot be implemented by numerical approximation, the spectacular advances in computer power over the last half-century have made it possible in some sense to sidestep the considerations involved in arguments of type (2) and to a lesser extent those of type (3); I discuss this question below in subsection 3.7.5. Nevertheless, the attempt to construct reasonably quantitative analytical approximations still has some point to it, inter alia since it provides a natural bridge between the results of pure numerical simulation and the qualitative considerations reviewed in earlier sections.

3.7.5 Numerical simulations

The spectacular advances in computer power which have occurred over the last two or three decades have meant that in many areas of condensed-matter physics approximate analytical calculations can be supplemented, or in some cases replaced, by purely numerical solutions of Schrödinger's equation or of an equivalent problem. Generally speaking, such methods tend to be very accurate for the calculation of the equilibrium properties of many-body systems, somewhat less so for near-equilibrium dynamical properties such as the excitation spectrum; for behavior very far from equilibrium (such as the behavior following a rapid quench through the λ-transition) their application is still in its infancy.

In the case of liquid helium, an excellent review of numerical simulations based on the path-integral technique has been written by Ceperley (1995); what follows is largely based on Sections II and III of this reference. I follow Ceperley's notation

[25]Which, contrary to an impression sometimes given in the literature, is *not* equivalent to the Bogoliubov approximation beyond first order in $u(r_{ij})$, see Chapter 4, section 4.4.

and denote by $\boldsymbol{R}$ the $3N$-dimensional position vector $\{\boldsymbol{r}_1, \boldsymbol{r}_2, \ldots, \boldsymbol{r}_N\}$; a general pure state of the system is then written $\Psi(\boldsymbol{R} : t)$. To answer the most general possible question concerning the behavior of the system, one might well have to solve the time-dependent Schrödinger equation for $\Psi(\boldsymbol{R} : t)$ subject to appropriate initial conditions – a formidable task indeed. Fortunately, if we confine ourselves for the moment to the calculation of equilibrium properties at some temperature $T \equiv (k_B\beta)^{-1}$, then it suffices to know a much more "coarse-grained" quantity, namely the (unnormalized) N-particle thermal density matrix $\hat{\rho}_N(\beta)$. It is convenient to write this quantity in the position representation:

$$\hat{\rho}_N(\beta) \equiv \rho(\boldsymbol{R}, \boldsymbol{R}' : \beta) \equiv \langle \boldsymbol{R}| e^{-\beta \hat{H}} |\boldsymbol{R}'\rangle \equiv \sum_i \phi_i^*(\boldsymbol{R})\phi_i(\boldsymbol{R}')\, e^{-\beta \mathrm{E}_i} \tag{3.7.25}$$

where ϕ_i are the eigenfunction of the N-particle Schrödinger equation and E_i the corresponding eigenvalues. This has the advantage that it will turn out (though it is not immediately obvious from the representation (3.7.25), that in this (position) representation *all* the elements of $\rho(\boldsymbol{R}, \boldsymbol{R}')$ (not just the diagonal ones!) are positive,[26] a feature which is of enormous advantage in the numerical simulations. Once $\rho(\boldsymbol{R}, \boldsymbol{R}' : \beta)$ is known, the expectation value of any operator $\hat{\theta}$ of interest can be calculated by the standard formula

$$\langle \hat{\theta} \rangle \equiv \frac{\mathrm{Tr}(\hat{\theta}\hat{\rho})}{\mathrm{Tr}(\hat{\rho})} = Z^{-1} \int d\boldsymbol{R}\, d\boldsymbol{R}' \rho(\boldsymbol{R}, \boldsymbol{R}' : \beta) \langle \boldsymbol{R}|\theta|\boldsymbol{R}'\rangle$$

$$Z \equiv \mathrm{Tr}\hat{\rho} \equiv \int \rho(\boldsymbol{R}, \boldsymbol{R} : \beta) d\boldsymbol{R} \tag{3.7.26}$$

So the problem reduces to the calculation of $\rho(\boldsymbol{R}, \boldsymbol{R}' : \beta)$. Although various numerical techniques can be used for this purpose, the most widely used is the "path integral Monte Carlo" (PIMC) method, which as its name implies is based on the original Feynman path-integral approach to quantum mechanics. This method not only has substantial technical advantages, but in many cases suggests a rather intuitive interpretation of the formulae in (pseudo-) classical terms.

As is well known, an exact path-integral representation of $\rho(\boldsymbol{R}, \boldsymbol{R}' : \beta)$ is given by the Feynman–Kac formula (see, e.g. Feynman and Hibbs 1965, Chapter 10)

$$\rho_{\mathrm{d}}(\boldsymbol{R}, \boldsymbol{R}' : \beta) = \int_{\boldsymbol{R}(0)=\boldsymbol{R}}^{\boldsymbol{R}(\beta)=\boldsymbol{R}'} \mathcal{D}\boldsymbol{R}(\tau) \exp\left(- \int_0^\beta d\tau \left\{ V(\boldsymbol{R}) + \frac{m}{2}\dot{\boldsymbol{R}}^2 \right\} \right) \tag{3.7.27}$$

Here $\dot{\boldsymbol{R}} \equiv \mathrm{d}\boldsymbol{R}/\mathrm{d}\tau$, and $\mathcal{D}\boldsymbol{R}$ indicates functional integration: the integral is over all paths running from the value $\boldsymbol{R}$ at "imaginary time" $\tau = 0$ to $\boldsymbol{R}'$ at $\tau = \beta$. (I ignore for the moment the question of indistinguishability, cf. below, and thus have put a subscript "d" for "distinguishable" on ρ.) Unfortunately the expression (3.7.27), while exact, is not susceptible as it stands to numerical simulation. As explained in detail by

[26]This is not true for the N-fermion case, and this circumstance leads to the notorious "sign problem" in path integral calculations for fermion systems (see, e.g. Kalos et al. 2005).

Ceperley (Section II.B), a good approximation to (3.7.27) under most circumstances is the *discrete* path integral

$$\rho_d(\boldsymbol{R_0}, \boldsymbol{R_M} : \beta) = \left(\frac{2\pi\hbar^2 m}{\tau}\right)^{-3NM/2} \int d\boldsymbol{R_1}\, d\boldsymbol{R_2} \cdots\, d\boldsymbol{R_{M-1}} \times \exp\left(-\sum_{m=1}^{M}\left\{\tau V(\boldsymbol{R_m}) + \frac{(\boldsymbol{R_{m-1}} - \boldsymbol{R_m})^2}{2\hbar^2\tau/m}\right\}\right) \tag{3.7.28}$$

where now $\tau \equiv \beta/M$. In words: the "imaginary time" axis is approximated by a set of M discrete points $k\tau$, $k = 1, 2, \ldots, M$, and the value of $\boldsymbol{R}$ is allowed to jump from $\boldsymbol{R_k}$ at $k\tau$ to $\boldsymbol{R_{k+1}}$ at $(k+1)\tau$. Thus the continuous "paths" in imaginary time of Eqn. (3.7.27) become a series of jumps between the values of $\boldsymbol{R}$ at the discrete times $k\tau$. In the approximation of neglect of relativistic effects, and for a system of distinguishable particles, Eqn. (3.7.28) is exact in the limit $M \to \infty$.

Equation (3.7.28) is actually identical to the expression for the thermal density matrix of a set of classical polymers with an unusual type of interaction; for a discussion of this analogy and the ways in which it can be exploited in visualizing the behavior of the quantum system see Ceperley 1995, Section II.D.

Equation (3.7.28) is not the correct expression for the density matrix of a set of indistinguishable bosons (or fermions); this may be explicitly checked by evaluating it in the special case of a noninteracting gas, for which the expression (3.7.28) is in fact exact and turns out to be nothing but the density matrix of a gas of *distinguishable* particles. However, the correction needed to apply the path-integral technique to a system of n indistinguishable bosons is very simple: One simply takes into account that the configurations $\hat{P}\boldsymbol{R}$ obtained by interchanging any number pairs of indices i, j in $\boldsymbol{R} \equiv \{\boldsymbol{r}_1, \boldsymbol{r}_2, \ldots, \boldsymbol{r_N}\}$ represent the same physical state as the original configuration $\boldsymbol{R}$, and therefore generalizes (3.7.28) (or (3.7.27)) to allow the final configuration $\boldsymbol{R}_{\mathrm{M}}$ to be not just $\boldsymbol{R}'$, but any permutation $\hat{P}\boldsymbol{R}'$. For an eventual correct normalization we should divide the resultant expression by the total number of permutations, namely $N!$ Equivalently, we can write for the true thermal density $\rho(\boldsymbol{R}, \boldsymbol{R}' : \beta)$ of N indistinguishable bosons

$$\rho(\boldsymbol{R}, \boldsymbol{R}' : \beta) = \frac{1}{N!}\sum_{P} \rho_{\mathrm{d}}(\boldsymbol{R}, \hat{P}\boldsymbol{R}' : \beta) \tag{3.7.29}$$

where the sum is over all possible permutations and ρ_{d} is given by the expression for distinguishable particles, Eqn. (3.7.28) (with $\boldsymbol{R_0} \to \boldsymbol{R}, \boldsymbol{R_M} \to \boldsymbol{R}'$)

Although in (3.7.28) and (3.7.29) the integral over time has been discretized, the integrals over $\boldsymbol{R}$ are still continuous. To evaluate (3.7.29) numerically one needs to discretize these too. However, it is easy to see that even the resulting purely arithmetical problem rapidly becomes intractable with increasing N and increasing demand for accuracy (i.e. increasing M). Thus, much of the technical effort associated with path-integral calculations has gone into devising ways of avoiding the complete evaluation by choosing some representative "sample" of paths whose multiplicity increases more slowly than that of the real problem; the most favored technique is then known as "path integral Monte Carlo" (PIMC) and I simply refer the reader to Section V of

Ceperley's review for a description of this and alternative methods (and for improvements on the expression (3.7.28) for finite M).

The PIMC method has been applied with great success to calculate the ground state energy (and hence the compressibility), transition temperature, specific heat, structure function, momentum distribution, and superfluid density[27] of liquid ^{4}He. The agreement with experiment is overall excellent, often being within the experimental error bars. It is interesting that this success has been achieved by the use of the "best" two-body potential for $V(\boldsymbol{R})$, without account of the three-body terms, thus confirming the usual assumption that the latter contribute at most a slowly varying background energy.

In principle, it is possible to use the (imaginary-time) path integral technique to calculate real-time quantities such as the (dynamical) density–density correlation function and hence the excitation spectrum, by doing an imaginary-time calculation and then an analytic continuation to the real axis. Unfortunately, when one has only a numerical approximation to the imaginary-time quantity the latter step is likely to induce appreciable error, and while such calculations have reproduced the phonon–roton spectrum qualitatively, the agreement for the correlation function is much less impressive than for the static properties: see Ceperley, Section III.I.

[27]Although the superfluid density is an equilibrium property (cf. Chapter 2), the generalization of the path-integral method to calculate it is not entirely trivial: see Ceperley, 1995, Sections 3.E–F.

4 The Bose alkali gases

A new chapter in the history of BEC and Cooper pairing was opened up by the extraordinary success of atomic physicists in the 1980s and 1990s in using laser and other techniques to cool dilute systems of monatomic (mostly alkali) gases into the nano K temperature range;[1] this culminated in the attainment of BEC in such systems in the summer of 1995, and more recently has permitted investigation not only of Cooper pairing but of the "crossover" between this phenomenon and BEC proper. While the present chapter is explicitly devoted to BEC and thus to the case when the atoms in question are bosonic, the discussion of the "atomic-physics" aspects of the problem in the first two sections is of course equally applicable to the fermion systems which I shall consider in Chapter 8, Section 8.4.

For the discussion of the experimental properties of liquid ^{4}He in the last chapter, it was sufficient to regard the ^{4}He atom as a fixed entity, ignoring its detailed internal structure and describing the interatomic interaction by a fixed central potential $V(r)$ whose origins we did not need to investigate in detail. Moreover, the only property of the physical vessel in which the helium was contained which is of much relevance is its topology. By contrast, in the case of the dilute atomic gases an understanding of existing and current experiments requires at least a modicum of familiarity both with the atomic physics of these systems and with the principal mechanisms by which they are trapped, so I start with a discussion of these.

4.1 The atoms: structure, trapping, and diagnostics

At the time of writing, the atomic species in which the attainment of BEC in the dilute-gas phase has been reported are ^{87}Rb, ^{23}Na, ^{7}Li, ^{85}Rb, ^{1}H, ^{41}K, ^{133}Cs, ^{174}Yb, ^{52}Cr and ^{4}He in the metastable ^2S state (^{4}He*). In addition, cooling into the nano K temperature range has been reported for ^{40}K and ^{6}Li, Fermi species in which one can look for Cooper pairing. With the exception of ^{1}H, ^{4}He, Yb and Cr, these are all alkali atoms and are very similar both in structure and with respect to their trapping and diagnostic behavior, so I devote most of this section to them and comment briefly on the four exceptions at the end.

[1] For some history of this remarkable effort, see the 1997 Nobel lectures.

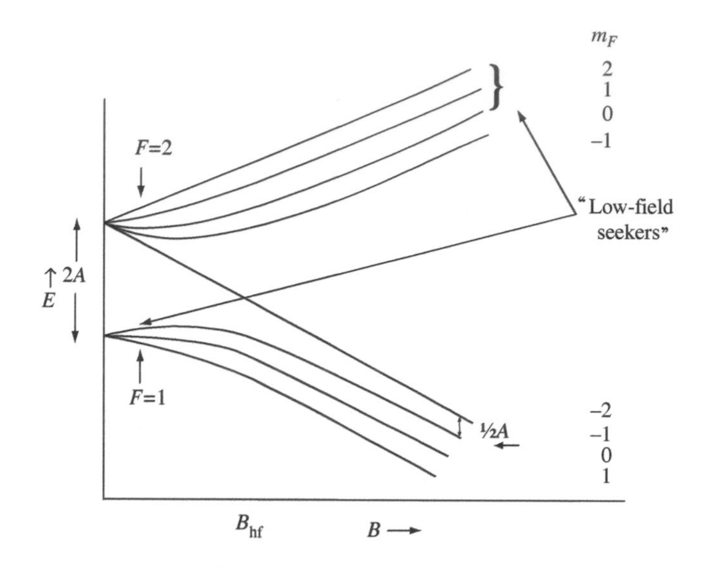

Fig. 4.1 The energies of the different hyper-Zeeman states as a function of magnetic field.

4.1.1 Structure[2]

By definition, a (neutral) alkali atom has a single valence electron outside one or more closed shells, and the electronic ground state, omitting the closed shells, is therefore the doublet $(ns)^2S_{1/2}$, where n ranges from 2 for Li to 6 for Cs. The only electronic excited states with which we need to be concerned in the present context are those to which electric dipole radiation overwhelmingly couples the ground state, namely the $(np)^2P_{1/2}$ and $^2P_{3/2}$ manifolds: the excitation energy is typically in the range 1.4–2 eV, corresponding to an optical wavelength in the visible regime (6000–9000 Å). The nuclear spin I is nonzero for all the alkali species listed above, so that in zero magnetic field the ground state is split by the hyperfine interaction into two multiplets with total atomic spin F equal to $I \pm \frac{1}{2}$, the splitting (denoted $2A$) being of the general order of 10^{-5} eV, or in temperature units ~0.1 K. It is important to note that the sign of the splitting is that of the nuclear g-factor, so that in the usual case of positive g the multiplet with the larger value of F lies higher; however, in ^{40}K the g-factor is negative, so that the ground state multiplet actually has $F = I + \frac{1}{2}(= \frac{9}{2})$.

In the presence of a nonzero external magnetic field B there is a competition between the hyperfine interaction and the electron Zeeman energy (the nuclear Zeeman energy is a much smaller effect and is usually negligible). As a result, the energy levels as a function of B have the generic structure shown in Fig. 4.1 (which is plotted for $I = \frac{3}{2}$ and positive[3] g, with the "crossover" field $B_{\text{hf}} \equiv A/|\mu_{\text{B}}|$ being of the order of a few hundred G). Existing experiments have usually operated either in the "low-field" regime $B \ll B_{\text{hf}}$ or in the "Paschen–Back" regime $B \gg B_{\text{hf}}$. In the former case the good quantum numbers are the total atomic spin F and its projections m_F, and if

[2]For an account of the relevant atomic physics, see, e.g. Woodgate, 1970, Chapter 9.

[3]The negative-g case is essentially obtained by turning the figure upside down.

we take the zero of energy to lie midway between the two multiplets the energies are given approximately by the formula

$$E = \pm \left\{ A + \frac{2}{2I+1} |\mu_{\mathrm{B}}| m_F B \right\} \tag{4.1.1}$$

so that in this approximation the slopes of (e.g.) the $F = 2, m_F = 1$ and $F = 1, m_F = -1$ states of ^{87}Rb are identical. In the literature these two states are often denoted as "maximally stretched". In the Paschen–Back limit the good quantum numbers are the projections m_J and m_I on the field axis of the electron and nuclear spins respectively (hence also $m_F \equiv m_I + m_J$), but it is conventional to continue to specify the levels by the indices F and m_F, with the former indicating the F-value from which the state has evolved as B is increased from zero. The energy in this regime is given by the approximate formula

$$E = \pm |\mu_{\mathrm{B}}| \left(B + \frac{2}{2I+1} m_F B_{\mathrm{hf}} \right) \tag{4.1.2}$$

so that the splitting within a given (electronic) multiplet is approximately constant and equal to $2/(2I+1)$ of the zero-field hyperfine splitting. Formulae (4.1.1) and (4.1.2) are valid for any value of I, not just for the case $I = \frac{3}{2}$ illustrated in the figure; for the exact (Breit-Rabi) formula valid for all B, see Woodgate (1970), Eqn. (9.80).

As we shall see below, the different hyperfine states can also be split by laser irradiation, the splitting being proportional to the power of the laser and dependent on its frequency and polarization.

4.1.2 Trapping techniques

In view of the dependence of the various hyperfine atomic levels on the external magnetic field shown in Fig. 4.1, an atom in a given hyperfine state placed in a spatially varying magnetic field will have a spatially dependent potential energy. (In analyzing this situation, it is necessary to bear in mind that in the relevant temperature regime the atomic velocities are so slow that a moving atom will adiabatically adjust its spin direction to that of the local field, so that the hyperfine index is conserved; thus we can write $V_i(r) = E_i(|\boldsymbol{B}(r)|)$ where i denotes the hyperfine state in question and $E_i(B)$ is the curve shown in Fig. 4.1 for that state.) Because it is possible to generate a minimum of a dc magnetic field in free space but not a maximum, this permits magnetic trapping of some but not all hyperfine states, namely the subset (known in the literature as "low-field-seekers") which in the relevant region of field values have a positive value of $dE(B)/dB$. Thus, for example, in the $I = \frac{3}{2}$ case illustrated in Fig. 4.1, the states $F = 2, m_F = 2, 1, 0, -1$ can be trapped for any value of the minimum field[4] while the state $F = 1, m_F = -1$ can be trapped provided this minimum is small enough. I will not go into the details of the construction of the various kinds of magnetic trap used in existing experiments (on this see, e.g. Ketterle et al. 1999), but will note that the resulting potentials are usually of anisotropic harmonic-oscillator form. Experimental papers tend to specify the trap parameters in terms of the three associated harmonic

[4]Though the $m_F = -1$ state may be trapped away from the position of the minimum.

frequencies ν_x, ν_y, ν_z for the relevant hyperfine state, whose values typically lie in the range between a few Hz and a few kHz.

While magnetic trapping is the technique used in the earliest experiments reporting the observation of BEC, it is somewhat inconvenient, both because not all levels can be trapped (and a low-field-seeking state may sometimes relax, by collisions, into a "high-field-seeking" and thus untrappable one) and, more seriously, because it has become common to use the magnetic field to tune the interaction (for example near a Feshbach resonance, see Section 4.2), and one usually does not wish the latter to depend on position in the trap. For this and other reasons, an increasing proportion of current experiments on BEC and related phenomena use an alternative technique, namely laser trapping. I now explain the essentials of this technique.

Imagine for a moment that we were dealing with an atom of nuclear spin zero, so that the electronic ground state is a simple doublet ($^2S_{1/2}$). As noted above, the electric dipole interaction with the laser field couples this state overwhelmingly to the $(np)^2P_{1/2}$ and $^2P_{3/2}$ fine-structure manifolds; coupling to $(n'p)(n' \neq n)$, and higher multipole couplings, are of negligible significance. For the moment let us neglect the fine-structure splitting, i.e. treat the six ^2P states as degenerate. Imagine now that we turn on a single laser beam, with a frequency comparable to the (gross-structure) ns–np splitting but detuned from the resonance by an amount large compared to the linewidth of the ^2P states. Then, irrespective of the laser polarization, there will be a nonzero matrix element of the electric dipole interaction between our initial (^2S) state and the ^2P manifold; and a straightforward application of second-order quantum-mechanical perturbation theory then leads to the result that when averaged over times long compared to the period ($\sim 10^{-15}$ sec) of the laser radiation the energy of the atom is shifted by an amount (see Eqn. (4.6.3) for details)

$$\Delta E = \text{const.} I/\Delta \tag{4.1.3}$$

where I is the laser intensity and $\Delta \equiv \hbar\omega_{\text{laser}} - (E_p - E_s)$ is the detuning. Thus, when the laser is focused so that the intensity I varies in space, the atom moves in an effective "potential" proportional to $I(\boldsymbol{r})$, which is repulsive for $\Delta > 0$ ("blue detuning") and attractive for $\Delta < 0$ ("red detuning"). This is the basic principle which is used in laser trapping of cold atoms. A particular advantage of this trapping technique is that the timescale for turning on the "potential" (4.1.3) is essentially the ^2P lifetime, which is of the order of 10^{-8} secs and hence tiny compared to the typical timescales of the atoms themselves (see below). Another advantage is that under suitable conditions the potential is independent of the hyperfine index (though see Section 4.6).

However, this is only the beginning of the story. With a single beam, the length scale over which the intensity can be made to vary significantly is generally of the order of a few microns. However, by using two counter propagating laser beams of the same frequency and polarization one can create a one-dimensional periodic "potential" with a lattice constant equal to $\lambda_{\text{laser}}/2$; and by using three pairs of such beams, propagating in mutually orthogonal directions, one can produce a three-dimensional "optical lattice". Further, by using slightly different frequencies for members of a pair, one can make this lattice time-dependent. Such lattice-like configurations are now extensively used, in particular for simulations of various systems of interest in condensed-matter physics; see below.

A further interesting twist is provided by the fact that all the relevant alkalis have a nonzero nuclear spin, so that not only the fine structure (in the ^2P manifold) but the hyperfine structure becomes relevant. It is important in this context to bear in mind that in a typical alkali the zero-field hyperfine splitting $2A$ is large compared to the ^2P linewidth Γ (as is shown by the fact that the different hyperfine transitions are clearly resolved in the optical spectrum). Without going into details,[5] I state the result that by working with detunings large compared to Γ but small compared to A, and/or adjusting the laser polarization appropriately, it is possible to make the energy shift (4.1.3) substantial for one particular hyperfine state and close to zero for the rest, i.e. to provide a "potential" which is sensitive to the hyperfine index. In particular, it is possible to construct optical lattices in which the minima for one hyperfine species are displaced relative to those for another.

4.1.3 Imaging

Almost without exception, the raw data obtained in experiments on the alkali gases reduces in the last resort to a measurement of either the total density or of that of one or more hyperfine species as a function of position and possibly time. Although it is possible to use phase-contrast techniques, the commonest method is simple absorption: one simply shines a laser beam of specified frequency and polarization on the gas and detects the fraction of the power transmitted. Because of the circumstance mentioned in the last subsection, by adjusting the frequency and polarization appropriately one can pick out a single hyperfine species. The spatial resolution of this technique is typically of the order of a few microns; since it works better at lower densities, it is common practice to remove the trap and let the gas expand before taking the image.

4.1.4 Nonalkali cold gases

The first nonalkali to be cooled unto the BEC phase was hydrogen. Although this system is in many ways similar to a typical alkali, there are several important differences:

1. There is of course no $1p$ state, so that electromagnetic dipole radiation couples the ground state predominately to the $2p$ state. It is also possible, by using a Raman laser, to couple it to the metastable (lifetime $\sim$0.15 sec) $2s$ state. In either case the frequency of the radiation involved is of the order of the Rydberg, i.e. an order of magnitude higher than for the alkalis.
2. Because of the light mass of the H atom, the temperatures at which BEC can be attained are considerably higher than for the alkalis ($\sim$50 μK versus a few hundred nK).
3. While the singlet ground state of the H_2 molecule is strongly bound (dissociation energy 4.7 eV) and has many excited vibrational levels, the $^3\sum$ configuration (which is realized if all the electron spins are strongly polarized by an external field) does not tolerate a bound state at all (contrary to the case of the alkalis). Moreover, the s-wave scattering length is very small ($\sim$0.6 Å).

A second non-alkali which has been cooled into the BEC phase is ^{4}He in the metastable $(1s2s)^3S$ state (^{4}He*). This system is unique among the BEC gases in

[5]See e.g. Corwin et al. (1999).

that since the ^{4}He nucleus has zero spin, the ground state in a high magnetic field has no hyperfine degree of freedom.

Finally, Yb is distinguished by having a ground state with electronic spin zero, and Cr by its large (6 μ_B) electronic magnetic moment, which has the consequence that the long-range anisotropic dipole-dipole interactions may be nonnegligible compared to the short-range repulsion (see next section).

4.2 s-wave scattering and effective interaction

The interaction energy $V(r)$ between any two neutral atoms, whether the same or different, is in general a complicated function of their relative separation r. However, for the analysis of the behavior in the dilute-gas phase it turns out that it is usually the long-distance behavior of $V(r)$ which is important.[6] In the limit $r \to \infty$ the dominant term is actually the r^{-3} interaction between the electron spins, but the coefficient of this term is so small that it is usually neglected (as noted in Section 4.1, it may be non-negligible for Cr). The leading term in this approximation is then due to the van der Waals interaction, and is of the form $-C_6/r^6$, when the coefficient C_6, expressed in atomic units (bohrs and hartrees) ranges from ~1200 for a pair of Li atoms to ~6300 for two Cs atoms. We can use C_6 to define a characteristic "van der Waals length" $r_0 \equiv (2m_r C_6/\hbar^2)^{1/4}$ (where m_r is the reduced mass, $m/2$ for identical atoms); the physical significance of this length is that it is the "typical" extent of the last bound state in the potential, and as we shall see below sets the scale of the s-wave scattering length. For identical atoms the quantity r_0 is approximately equal to $6 \cdot 5(AC_6)^{1/4} a_0$ (A = atomic mass, a_0 = Bohr radius) and thus lies in the range 30–100 Å for all the alkalis.

A circumstance which greatly simplifies the consideration of the alkali gases, be they Bose or Fermi, in the quantum degenerate regime is that under the currently attainable experimental conditions the interparticle spacing is very large compared to r_0. Indeed, as we have seen, r_0 is specifically of the order of 50–100 Å, while the interparticle spacing is never less than 1000 Å and is often considerably greater. Since the criterion for degeneracy is, crudely speaking, that the thermal de Broglie wavelength $\lambda_T \equiv 2\pi/k_T$ is comparable to or greater than the interparticle spacing, this means that in the degenerate regime all states with appreciable probability of occupation have wave vectors k satisfying the condition $kr_0 \ll 1$; this then clearly implies the same condition if k is interpreted as the wave vector of relative motion of two atoms. Now, for two atoms in a state of nonzero relative angular momentum ℓ, the probability of approaching within a distance $r \ll k^{-1}$ is proportional to $(kr)^{2\ell}$; thus, in the degenerate regime the probability of two atoms with $\ell \neq 0$ approaching within a distance where the interaction is appreciable is negligible. Consequently, it is usually an excellent approximation, when considering the interaction between atoms in the degenerate regime, to restrict oneself to the case when the relative state is s-wave ($\ell = 0$), and unless otherwise stated I will do so throughout this chapter. It should be noted that in this approximation two identical fermions in the same hyperfine state cannot scatter at all, a circumstance which has severe consequences for attempts to

[6] The $^3\sum$ state of H is an exception; here short-distance effects are important and in fact there is no bound state at all.

cool such a system into the degenerate regime (since, generally speaking, effective cooling relies on the existence of collisions).

4.2.1 The *s*-wave scattering length

Consider two atoms (which need not be identical) in a relative *s*-state such that in the center-of-mass frame the energy at infinity is zero.[7] In the region where $V(r) = 0$ (i.e. crudely speaking, for $r \gg r_0$) the relative (three-dimensional) wave function satisfies the equation

$$\frac{d}{dr} r^2 \frac{d\psi(r)}{dr} = 0 \tag{4.2.1}$$

and thus has the general solution (up to normalization)

$$\psi(r) = 1 - a_s/r \tag{4.2.2}$$

where a_s is a quantity with the dimensions of a length, whose value in any particular case is determined by the details of the potential $V(r)$. The quantity a_s is known as the (*s*-wave) scattering length and is a fundamental input to theories of a dilute gas of the atoms in question; in general it is a function not only of the two atomic (and isotopic) species involved but of the hyperfine states they occupy. It can be measured by a number of different techniques, of which the most precise is probably photoassociation: see e.g. Tiesinga et al. 1996. Rather general considerations (cf. e.g. Gribakin and Flambaum 1993, or Pethick and Smith 2002, Chapter 5) lead to the conclusion that, statistically speaking, a_s is likely to be of order of magnitude r_0 and positive values are three times as likely as negative ones; the values of a_s measured to date for the alkalis roughly conform to this prediction (see Pethick and Smith 2002, Section 3.5.1). It is particularly interesting to study the behavior of a_s for values of the parameters describing $V(r)$ close to those corresponding to the onset of a bound state. Consider for example a three-dimensional potential well of depth V_0 and radius a. As is well known, such a potential will tolerate a bound state if and only if V_0 satisfies the condition

$$V_0 > \pi^2\hbar^2/2m_r a^2 \equiv V_c \tag{4.2.3}$$

(where m_r is the reduced mass of the two-particle system). Imagine that we start from a value of V_0 substantially greater than V_c (but still not too close to the value ($4V_c$) necessary for the onset of a second bound state). Then, generally speaking, the scattering length a_s will be of order of a (its sign and exact value depends on the value of V_0). Now imagine that we gradually decrease V_0 towards the critical value V_c defined by Eqn. (4.2.3). The value of a_s becomes positive and tends to infinity as V_0 approached V_c from above. In fact, it is easy to show[8] that in this limit a_s is equal to the "radius" a_b of the bound state, which is defined by the expression $\psi(r)_b = \text{const}.r^{-1}e^{-r/a_b}$

[7] Or more precisely the de Broglie wavelength is long compared to any length scale in which we are interested (cf. below).

[8] The simplest argument is that the bound state and the $E = 0$ scattering state must be mutually orthogonal, and that in this limit the overlap integral has negligible contribution from $r < a_s$.

for $r > a$, i.e.

$$a_b \equiv (2m_r|E|/\hbar^2)^{-1/2} = a_s \tag{4.2.4}$$

where $-|E|$ is the energy of the bound state. In this case, a_s^{-1} is linear in $(V_0 - V_c)$ and thus the binding energy is proportional to $(V_0 - V_c)^2$. For V_0 just less than V_c (when there is no bound state) the scattering amplitude a_s is negative and its magnitude is proportional to $(V_c - V_0)^{-1}$, with the same constant as for $V_0 > V_c$. This behavior is rather generic; if we define (in a "nonpathological" way) some control parameter δ such that positive (negative) values of δ compared to the absence (presence) of a bound state,[9] then we can say quite generally that for $|\delta| \to 0$ the behavior of a_s, is

$$a_s(\delta) = -c/\delta \tag{4.2.5}$$

where c is a positive constant. Correspondingly, for $\delta < 0$ (and $|\delta| \to 0$), the energy of the bound state is given by

$$-E = \hbar^2/2m_r a_s^2(\delta) = \text{const.}\delta^2 \tag{4.2.6}$$

Actually, if we are given a single potential curve $V(r)$, it is not that easy in practice to tune the parameter close to the point $\delta = 0$ when a new bound state appears. However, if we have two different potential curves, e.g. corresponding to different pairs of hyperfine levels for the two atoms, it is often possible to "tune" them relative to one another, e.g. by a magnetic field, and thereby to achieve at least qualitatively the same effect. Such a situation is known as a Feshbach resonance: I defer explicit consideration of it until after exploring the generic implications of a particular value of a_s.

4.2.2 The effective interaction

A very fundamental result in the theory of dilute gases is that for two atoms of reduced mass m_r, the existence of a scattering length a_s implies, under appropriate conditions and in a sense to be specified below, an effective interaction equal to $(2\pi\hbar^2 a_s/m_r)\delta(\boldsymbol{r})$. The case most commonly considered is that of two (chemically and isotopically) identical atoms of mass M, for which $m_r = M/2$, so that the result is commonly stated in the form

$$V_{\text{eff}}(\boldsymbol{r}) = (4\pi\hbar^2 a_s/M)\delta(\boldsymbol{r}) \tag{4.2.7}$$

Equation (4.2.7) needs to be interpreted carefully. A more exact statement would be that in a situation where, at distances $\gg a_s$, the relative wave function tends to a constant value[10] which I schematically denote $\Psi(0)$, the expectation value of the interatomic potential energy is given by the expression

$$\langle E_{\text{int}} \rangle = (4\pi\hbar^2 a_s/M)|\Psi(0)|^2 \tag{4.2.8}$$

which is the result we would get by evaluating the expectation value of (4.2.7) on the *unperturbed* wave function $\Psi(0)$. Note that it would *not* be correct to try to go to higher than first order in V_{eff}.

[9]Thus in our example δ would be some constant times $V_c - V_0$.

[10]For technical reasons it is actually safer to specify that $\Psi(0)$ is the constant multiplying the 1 in Eqn. (4.2.2), which should hold for all $r > r_0$.

Equation (4.2.7), or better (4.2.8), can be derived by any number of methods, of which to my mind the most intuitive (though not perhaps the most rigorous) is the following: Let us first consider the problem of a single particle moving in a spherically symmetric potential $V(r)$ characterized by a zero-energy s-wave scattering length a_s, and enclose it in a hard-walled container of radius $R \gg a_s$, so that the wave function must vanish for $r = R$. First consider a state with nonzero angular momentum ($\ell \neq 0$). *Provided* that we are in the "ultracold" region where all relevant k-values satisfy the condition $kr \ll 1$, both the quantity $\Psi(0)$ as defined above (in footnote 10) and the interaction energy are zero, so (4.2.8) is trivially satisfied. Now consider an s-wave state with small $k(ka_s \ll 1)$. In the absence of the potential (so that $a_s \equiv 0$) the normalized wave function must be of the form,

$$\Psi_n(r) = \frac{1}{\sqrt{4\pi}}\sqrt{\frac{2}{R}}\frac{\sin k_n r}{r}, \qquad k_n = n\pi/R \tag{4.2.9}$$

In the presence of the potential, the solution for $r > r_0$ is of the form

$$\Psi_n(r) = \frac{1}{\sqrt{4\pi}}\sqrt{\frac{2}{R}}\frac{\sin k'_n(r - a_s)}{r}, \qquad k'_n = n\pi/(R - a_s) \tag{4.2.10}$$

which reduces to (4.2.2) for $k'_n r \ll 1$. (Here I neglected the small difference in the normalization constant introduced by the potential, which is an effect of relative order $a_s/R \ll 1$.) In each case the energy is simply $\hbar^2k^2/2m$, so the difference in energy ΔE introduced by the presence of the potential is, to lowest order in a_s/R,

$$\Delta E = (\hbar^2/2M)(2n^2\pi^2 a_s/R^3) \tag{4.2.11}$$

On the other hand, the quantity $|\Psi_n(0)|^2$ (obtained by evaluating $\Psi_n(r)$ at a distance r such that $a_s \ll r \ll k^{-1}$) is equal to $k_n^2/2\pi R = n^2\pi/2R^3$. Consequently, irrespective of the values of n and of R, we obtain (4.2.8) except for the replacement of the factor 4 by 2.

It is now obvious that we can reproduce the argument word for word for the case of two particles interacting by a central potential $V(r)$, provided only that we are content to employ the artificial device of restricting the relative coordinate r to lie within a fictitious "hard-walled sphere" of radius R. The only difference is that M is replaced by the reduced mass m_r; for the case of identical atoms ($m_r = M/2$) this just supplies the necessary extra factor of 2 in (4.2.8).

An alternative derivation, which does not require us to put any artificial restriction on the relative motion and thus may be more appropriate in the many-particle case, is given in Leggett (2001), Section IV.C. In any event, the upshot of those considerations is that in a system of identical "ultracold" atoms of mass m occupying the same hyperfine state (and thus characterized by a unique value of a_s) the interaction energy is given to a good approximation by the expression

$$E_{\text{int}} = \frac{1}{2}\left(4\pi\frac{\hbar^2}{M}a_s\right)\sum_{ij}|\Psi(r_{ij} \xrightarrow{\sim} 0)|^2 \tag{4.2.12}$$

when the notation "$r_{ij} \xrightarrow{\sim} 0$" is introduced, following Leggett (2001), to indicate that the relative coordinate r_{ij} should be taken to be large compared to a_s but small compared to any other characteristic lengths in the problem such as the interparticle spacing.

4.2.3 Feshbach resonances[11]

Let us consider a pair of atoms of which at least one has available to it more than one hyperfine state, and let us suppose for the sake of simplicity that the interaction Hamiltonian may be treated to a first approximation as diagonal in the hyperfine indices. We can then define a given "channel" as corresponding to a particular pair of hyperfine indices, and the interaction potential will then in general be different for different channels.[12] Generally speaking, the (z-component of) magnetic moment will also be different for the different channels, and thus the relative "zero" of energy ΔE (i.e. the energy distance, in the two channels, of the two atoms at infinite distance apart) may be adjusted by tuning the magnetic field:

$$\Delta E = \Delta\mu \cdot B$$

where $\Delta\mu$ is the difference in the magnetic moment.

Consider now two atoms approaching one another in a given channel (pair of hyperfine states) which we label the "open" channel. It may happen that the energy difference $\Delta E \equiv E_{\text{closed}} - E_{\text{open}}$ between this channel and a particular other channel, which we denote as the "closed" channel, is close to the binding energy $|E_0|$ of a particular bound state in the closed channel, so that this bound state is almost degenerate with the $E = 0$ scattering state in the open channel (see Fig. 4.2). Suppose now that we take into account the small part of the interaction, thus far neglected, which does not conserve the channel index; then, however small those terms, we would expect that in the limit that the "detuning parameter" $\delta \equiv \Delta\mu \cdot B - |E_0|$ tends to zero the two channels would be strongly hybridized. This is the effect known as Feshbach resonance.

For the two-particle problem it is straightforward in principle to solve the coupled equations and obtain a solution which is exact to the extent that we may legitimately neglect all the nonresonant closed-channel states (including the continuum); see Appendix 4A. In this limit we can define a "critical" value $\delta_c > 0$ of the detuning, which is a function of the details of all the potential functions involved, including the hybridizing term; as explained in Appendix 4A, δ_c can be extracted directly from the experimental data. The physical significance of the quantity δ_c is that for $|\delta| \gg \delta_c$ the two channels may be regarded as decoupled to a first approximation, while for $|\delta| \lesssim \delta_c$ the closed-channel state has in some sense no real existence; in particular, if one solves the Schrödinger equation for $\delta < 0$ and $|\delta| \lesssim \delta_c$ and seeks a bound-state solution, one finds that such a solution indeed exists, but that the weight is largely in the open channel and its characteristic radius is proportional to $|\delta|^{-1}$ and thus, in the limit specified, very much greater than that of the original resonant closed-channel state. In fact, for $|\delta| \ll \delta_c$ the Feshbach-resonance problem looks

[11] Feshbach resonances will not be relevant in the rest of this chapter, but will be essential in the considerations of Chapter 8, Section 8.4.

[12] For example, in the "Paschen–Back" regime $B \gg B_{\text{hf}}$ the different channels can be approximately classified by the value (1 or 0) of the total electronic spin and by the z-component of the nuclear spin; in this case the potential is to a first approximation independent of the nuclear spin.

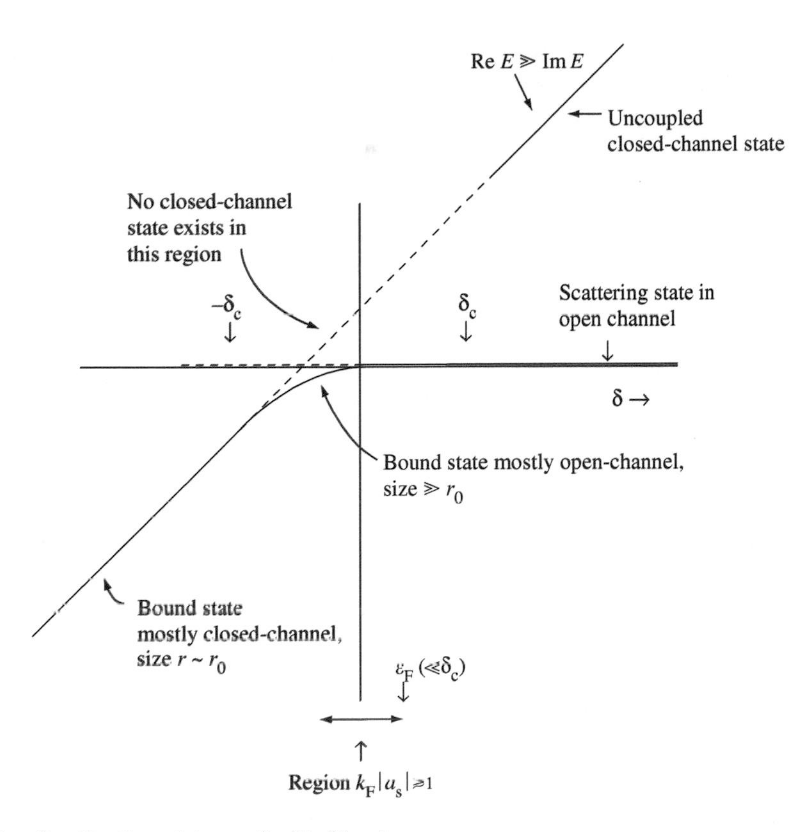

Fig. 4.2 Qualitative picture of a Feshbach resonance.

exactly like the simple "potential-resonance" problem described above, with the role of the original closed-channel state being simply in effect to provide a "knob" to tune the effective open-channel potential. For $|\delta| \sim \delta_c(\delta < 0)$ the weight in the bound state shifts over into the closed channel and the situation is complicated; for $|\delta| \gg |\delta_c$, as noted, the channels are decoupled to a first approximation (but the resonance can still contribute a term to the open-channel scattering length which may be large compared to the "background" value which would obtain in the absence of the interchannel coupling).

4.3 The Gross–Pitaevskii equation: some simple applications

The Gross–Pitaevskii (GP) equation is nothing more nor less than the Hartree equation applied to a large number of condensed bosons. It was suggested many years ago, as a description of the Bose-condensed (superfluid) phase of liquid ^{4}He, but in view of the strong interactions in that system can claim there at best a qualitative validity. By contrast, in the case of the dilute alkali gases, which are far more weakly interacting, the GP equation is for many purposes an excellent quantitative description, and it has been very widely applied in the interpretation of experiments. In this section I will introduce it for the spinless case[13] and discuss a few of its immediate consequences. Consider a system of N identical bosons moving in a given external potential $V(\boldsymbol{r}t)$, and suppose that it satisfies the so-called "diluteness condition", namely

$$\zeta \equiv (na_s^3)^{1/2} \ll 1 \tag{4.3.1}$$

where n is a typical value of the density. It will further be assumed that the s-wave scattering length, while possibly large compared to the range r_0, of the interatomic potential, is small compared to any length characterizing the *external* potential $V_{\text{ext}}(\boldsymbol{r})$ (such as for example the zero-point length $a_{zp} \equiv (\hbar/M\omega_0)^{1/2}$ in the case of an oscillator potential with force constant $M\omega_0^2$).

The very simplest ansatz we can make for the many-body wave function Ψ_N is of the Hartree form, i.e. we assume that all the bosons occupy the same (possibly time-dependent) state:

$$\Psi_N(\boldsymbol{r}_1\boldsymbol{r}_2\ldots\boldsymbol{r}_N : t) = \Pi_{i=1}^N \chi_0(\boldsymbol{r}_i, t) \tag{4.3.2}$$

where $\chi_0(\boldsymbol{r}, t)$ is some normalized single-particle wave function which need not necessarily be an eigenfunction of the single-particle Schrödinger equation. We will call $\chi_0(\boldsymbol{r}, t)$ the "condensate wave function": note that in contrast to the "order parameter" to be introduced below it is independent of the total particle number N.

Assuming that the form of Ψ_N is indeed (4.3.2), what should be the equation obeyed by the condensate wave function $\chi_0(\boldsymbol{r}t)$? In the absence of any interatomic interactions, it is clearly simply the familiar one-particle Schrödinger equation

$$i\hbar\frac{\partial\chi_0}{\partial t} = -\frac{\hbar^2}{2M}\nabla^2\chi_0(rt) + V_{\text{ext}}(rt)\chi_0(rt) \tag{4.3.3}$$

Now we have seen in the last section that the effect of interatomic interactions, in the low-density limit, is to contribute to the total energy a term of the form (4.2.12). Since for two particles in *the same* single-particle state we have

$$|\Psi(r_{ij} \rightsquigarrow 0)|^2 = |\chi(\boldsymbol{r}_i)|^2|\chi(\boldsymbol{r}_j)|^2 \tag{4.3.4}$$

it follows that (4.2.12) is equivalent to the existence of an extra single-particle potential

$$V_{\text{eff}}(\boldsymbol{r}_i, t) = \frac{4\pi a_s\hbar^2}{M}\sum_{j\neq i}|\chi_j(\boldsymbol{r}_i, t)|^2 \cong \frac{4\pi N a_s\hbar^2}{M}|\chi_0(\boldsymbol{r}_i, t)|^2 \tag{4.3.5}$$

where the factor of $\frac{1}{2}$ is omitted to avoid undercounting terms, and we have approximated $N-1$ by N. Adding (4.3.5) to the external potential $V(r)$ in (4.3.3), we obtain the equation

$$i\hbar\frac{\partial\chi_0(\boldsymbol{r}t)}{\partial t} = -\frac{\hbar^2}{2M}\nabla^2\chi_0(\boldsymbol{r}t) + V(\boldsymbol{r}t)\chi_0(\boldsymbol{r}t) + \frac{4\pi N a_s}{M}\hbar^2|\chi_0(\boldsymbol{r}t)|^2\chi_0(\boldsymbol{r}t) \tag{4.3.6}$$

The explicit factor of N in the last term is a bit awkward, and it is conventional to remove it by introducing[14] the "order parameter" $\Psi(\boldsymbol{r}t)$ of the system by the

[13] By "spinless case" I mean, here and below, that we deal with a situation where only a single hyperfine state of the relevant atoms occurs, so that the hyperfine index need not be written out explicitly; all quantities such as a_s then refer to this state.

[14] $\Psi(\boldsymbol{r}t)$ should not be confused with the many-body wave function $\Psi_N(\{\boldsymbol{r}\}, t)$, which I will not subsequently mention explicitly.

definition (cf. 2.2.1)

$$\Psi(\boldsymbol{r}t) \equiv \sqrt{N}\chi_0(\boldsymbol{r}t) \tag{4.3.7}$$

Equation (4.3.6) then reads

$$i\hbar\frac{\partial\Psi(\boldsymbol{r}t)}{\partial t} = -\frac{\hbar^2}{2M}\nabla^2\Psi(\boldsymbol{r}t) + V_{\text{ext}}(\boldsymbol{r}t)\Psi(\boldsymbol{r}t) + \frac{4\pi a_{\text{s}}\hbar^2}{M}|\Psi(\boldsymbol{r}t)|^2\Psi(\boldsymbol{r}t) \tag{4.3.8}$$

It is the form (4.3.8) which is conventionally known as the (time-dependent) Gross–Pitaevskii equation (or, sometimes, the "nonlinear Schrödinger equation"). Note that the order parameter $\Psi(\boldsymbol{r}t)$ is normalized to the total particle number N, that is,

$$\int |\Psi(\boldsymbol{r}t)|^2 d\boldsymbol{r} = N \tag{4.3.9}$$

Before proceeding to discuss the consequences of (4.3.8), let us make a number of digressions. First, why should we in fact assume that the many-body wave function is in any sense "likely" to be of the form (4.3.2)? A partial answer can be given from the following consideration (cf. Chapter 2, Section 2.3). Suppose that we have two (spinless) bosons in *different* (mutually orthogonal and normalized) single-particle states, let us say $\varphi(\boldsymbol{r})$ and $\chi(\boldsymbol{r})$. Then the identity of the particles requires the two-particle wave function $\Psi(\boldsymbol{r}_{ij})$ has the correctly symmetrized and normalized form

$$\Psi(\boldsymbol{r}_{ij}) = 2^{-1/2}(\varphi(\boldsymbol{r}_i)\chi(\boldsymbol{r}_j) + \chi(\boldsymbol{r}_i)\varphi(\boldsymbol{r}_j)) \tag{4.3.10}$$

and (4.3.4) is replaced by

$$|\Psi(r_{ij} \rightharpoonup 0)|^2 = 2|\varphi(\boldsymbol{r}_i)|^2|\chi(\boldsymbol{r}_i)|^2 \tag{4.3.11}$$

Thus, the interaction energy is in effect multiplied by a factor of 2. (The extra term is of course nothing but the "Fock" term, which in the case of bosons enters with a + sign relative to the Hartree one.) Provided that the s-wave scattering length a_{s} is positive (as it has been in most though not all experiments on the dilute alkali gases to date) this means that the system has, as it were, a strong energy incentive to maintain the maximum possible degree of BEC, as implied by the ansatz (4.3.2).

Let us next note that the apparent simplicity of the derivation just given of the GP equation conceals a subtle point. For definiteness, let us anticipate the result that in a large box the form of $\chi_0(\boldsymbol{r})$ corresponding to the many-body ground state is constant ($\boldsymbol{k} = 0$) over most of the box. Then the ansatz (4.3.2) says in effect that all atoms are uniquely in the $\boldsymbol{k} = 0$ state (with corrections that vanish in the thermodynamic limit) and hence have zero relative momentum. Yet, according to the considerations of the last section, the last term in (4.3.8) owes its very existence to the fact that, at distances $\lesssim a_{\text{s}}$, the relative wave function is *not* a constant! It is tempting to think that the problem is simply one of notation, and that it should be straightforward to correct the ansatz (4.3.2) to take into account the short-range correlations discussed in the last section. However, it turns out that the matter is more subtle than this: as discussed in detail in Leggett (2003)[15] and more

[15]In the context of a hard-sphere model this issue is actually resolved, at least implicitly, in the classic paper of Lee and Yang (1957).

briefly in the next section, the ansatz (4.3.2) does not actually correspond to any well-defined "approximation" for the many-body wave function, and the results of applying the GP equation should actually be interpreted as the result of expanding the predictions of the more exact Bogoliubov theory to lowest nontrivial order in ζ (Eqn. 4.3.1).

Finally, I note that the definition of the order parameter given above, Eqn. (4.3.7), is a simple special case of the more general prescription given in Chapter 2, Section 2.2; in the present case, by construction (see Eqn. 4.3.2) the condensate number N_0 is equal to N. In this special case, as in the more general one discussed in Chapter 2, I believe that the alternative definition of $\Psi(\boldsymbol{r}t)$ (the one the reader is likely to encounter most frequently in the literature) as the expectation value of the Bose field operator $\hat{\psi}(\boldsymbol{r}t)$ has no advantages and is liable, if not treated with care, to lead to unphysical conclusions.

I now turn to the consequences of the GP equation (4.3.8). Let us first set $V_{\text{ext}}(\boldsymbol{r}t) = V_{\text{ext}}(\boldsymbol{r})$ and consider Eqn. (4.3.2) as an ansatz for the many-body ground state: then the condensate wave function $\chi_0(\boldsymbol{r}t)$ should be time-dependent only through a phase factor which we write as $\exp(-i\mu t/\hbar)$ (where the quantity μ is for the moment unidentified), and substitution of this form into (4.3.8) yields the time-independent GP equation (where $\Psi(\boldsymbol{r})$ is now the time-independent factor)

$$-\frac{\hbar^2}{2m}\nabla^2\Psi(\boldsymbol{r}) + V_{\text{ext}}(\boldsymbol{r})\Psi(\boldsymbol{r}) + \frac{4\pi\hbar^2 a_s}{M}|\Psi(\boldsymbol{r})|^2\Psi(\boldsymbol{r}) = \mu\Psi(\boldsymbol{r}) \tag{4.3.12}$$

To identify the physical significance of the quantity μ, I note that the form of $\Psi(r)$ must be that which minimizes the expression for the energy which follows from (4.2.12) and (4.3.2), namely

$$E = \frac{\hbar^2}{2M}\int |\boldsymbol{\nabla}\Psi(r)|^2 dr + \int V_{\text{ext}}(\boldsymbol{r})|\Psi(\boldsymbol{r})|^2 d\boldsymbol{r} + \frac{1}{2}\cdot\frac{4\pi\hbar^2 a_s}{M}\int |\Psi(\boldsymbol{r})|^4 d\boldsymbol{r} \tag{4.3.13}$$

subject to the condition of conservation of total particle number, that is, to the constraint

$$\int |\Psi(\boldsymbol{r})|^2 d\boldsymbol{r} = N \tag{4.3.14}$$

If we handle the constraint in the standard way by introducing a Lagrange multiplier, and vary (4.3.13) with respect to the form of $\Psi(\boldsymbol{r})$, we obtain (4.3.12) with μ equal to the multiplier in question. Thus, μ is just the Lagrange multiplier associated with the conservation of particle number, i.e. it is the standard chemical potential, as the notation chosen for it implies:[16]

$$\mu = \frac{\partial E(N)}{\partial N} \tag{4.3.15}$$

The actual value of μ is fixed, in any given situation, by the condition that the solution of (4.3.12) (which is uniquely determined by μ) should satisfy (4.3.14).

[16]Note that this means that the time-dependence of the many-body wave function is $\exp -iN\mu t/\hbar$ rather than $\exp -iE_N t/\hbar$. This does not matter, since for fixed particle number N the zero of E_N is arbitrary, and for small variations of N the two expressions give the same results. An alternative derivation of (4.3.15) is to express (4.3.13) explicitly in terms of N and χ_0, differentiate with respect to N and compare with (4.3.12).

The time-independent GP equation (4.3.12) has many important applications. We see that it is characterized, at any given point in space, by a characteristic length $\xi(\boldsymbol{r})$ defined by

$$\xi(\boldsymbol{r}) \equiv (8\pi n(\boldsymbol{r}) a_{\mathrm{s}})^{-1/2} \tag{4.3.16}$$

(where $n(\boldsymbol{r}) \equiv |\Psi(\boldsymbol{r})|^2$ is the density of atoms at the point in question). In the "dilute-gas" limit $na_{\mathrm{s}}^3 \ll 1$ the quantity ξ is large compared both to the interparticle spacing $n^{-1/3}$ and to the (much smaller) s-wave scattering length a_{s}. By studying specific cases such as that of a box (where $V_{\mathrm{ext}}(\boldsymbol{r}) \equiv 0$ but Ψ must tend to zero on the walls) one sees that ξ is a measure of the distance over which the order parameter, when forced to vary away from its equilibrium value $[\bar{n}(\boldsymbol{r})]^{1/2}$, recovers to this value; the standard name for ξ is therefore the "healing length". If ξ is small enough compared to other characteristic lengths in the problem, then the kinetic-energy term in (4.3.12) may be neglected to a first approximation and the equation reduces to the "Thomas–Fermi" equation

$$V_{\mathrm{ext}}(\boldsymbol{r}) + n(\boldsymbol{r}) U_0 = \mu, \quad U_0 \equiv 4\pi\hbar^2 a_{\mathrm{s}}/M \tag{4.3.17}$$

which simply states that the total effective potential, external plus interatomic, in which a given atom moves is constant.

A good example of this behavior is that of a typical alkali gas ($a_{\mathrm{s}} \sim 50$ Å) in a typical (magnetic or laser) harmonic trap, with a zero-point length $a_{zp} \equiv (\hbar/M\omega_0)^{1/2} \sim 1\ \mu$m. Recall (Chapter 1, Section 1.2) that in such a trap the critical temperature T_{c} is of order $N^{1/3}$ times $\hbar\omega_0/k_{\mathrm{B}}$, and thus in the normal state just above T_{c} the width of the thermal distribution is greater than the groundstate width a_{zp} by a factor of order $N^{1/6}$ (typically $\sim$10). Were we able to neglect the effect of interactions, then at temperatures well below T_{c} most of the atoms would be in the noninteracting harmonic-oscillator ground state and the width of the distribution would have shrunk by a factor $\sim$10, corresponding to a density increase $\sim$1000, a dramatic effect indeed. In the presence of (repulsive) interactions, however, such a large increase of density is highly unfavorable energetically. In fact, suppose that we are at zero temperature and we imagine that the typical extent of the distribution is R. Then from Eqn. (4.3.13) we see that the expectation values (per atom) of the kinetic energy T, external potential energy V_{ext} and atomic interaction energy $\bar{U}$ have the following orders of magnitude:

$$\bar{T} \sim \hbar^2/MR^2 \sim \hbar\omega_0 (a_{zp}/R)^2 \tag{4.3.18a}$$

$$\bar{V}_{\mathrm{ext}} \sim M\omega_0^2 R^2 \sim \hbar\omega_0 (R/a_{zp})^2 \tag{4.3.18b}$$

$$\bar{U} \sim \bar{n} U_0 \sim N U_0 R^{-3} \sim \hbar\omega_0 (N a_{\mathrm{s}}/a_{zp})(a_{zp}/R)^3 \tag{4.3.18c}$$

The qualitative behavior depends on the value of the dimensionless ratio $\lambda \equiv N a_{\mathrm{s}}/a_{zp}$. If λ is small compared to 1, then the interatomic interaction is just a small perturbation on the noninteracting-gas behavior. In the opposite limit $\lambda \gg 1$ (the situation which has obtained in most of the experiments conducted to date on the alkali gases) the behavior is determined primarily by the competition of the external and interatomic potential energies, with the kinetic energy constituting only a small correction (so that we are in the Thomas–Fermi limit discussed above). It is easy to see that the order of magnitude of the extent of the cloud is

$$R \sim \lambda^{1/5} a_{zp} \tag{4.3.19}$$

where the factor $\lambda^{1/5}$ is typically of the order of 5–10. A more quantitative calculation starting from Eqn. (4.3.17) shows that the distribution has in the Thomas–Fermi limit a simple parabolic shape:

$$n(r) = n_0(1 - r^2/R_0^2) \tag{4.3.20}$$

with an extent R_0 given (Baym and Pethick 1996) by

$$R_0 = (15Na_s/a_{zp})^{1/5} a_{zp} \tag{4.3.21}$$

in agreement with the estimate (4.3.19). Thus the actual width of the low temperature atomic cloud, while large compared to a_{zp}, is still considerably smaller than the width of the thermal cloud above T_c. As a consequence, the onset of BEC in a harmonically trapped BEC is signaled by the approximate of a sharp "spike" in the density profile rising above the diffuse thermal background (see the pictures in Andrews et al. (1996)). Under typical conditions the scale of this spike is only slightly below what could (under other circumstances!) be detected with the naked eye, so that this phenomenon is probably the most dramatic direct manifestation of BEC known to date.

Let us now turn to the time-dependent GP (TDGP) equation, Eqn. (4.3.8). It should be emphasized that this equation only makes sense to the extent that the many-body wave function not only has the form of the ansatz (4.3.2) at $t = 0$, but retains this form throughout the period of motion of interest.[17] Given this assumption, let us explore its consequences. First, let us separate $\Psi(\boldsymbol{r}t)$ into amplitude and phase:

$$\Psi(\boldsymbol{r}t) \equiv (n(\boldsymbol{r}t))^{1/2} \exp i\varphi(\boldsymbol{r}t) (n, \varphi \text{ real}) \tag{4.3.22}$$

and define as in Chapter 2, Section 2.2 the superfluid velocity $\boldsymbol{v}_s(\boldsymbol{r}t)$ by

$$\boldsymbol{v}_s(\boldsymbol{r}t) \equiv \frac{\hbar}{m} \boldsymbol{\nabla}\varphi(\boldsymbol{r}t) \tag{4.3.23}$$

The physical significance of the quantity $\boldsymbol{v}_s(\boldsymbol{r}t)$ was discussed in the cited section. Substituting (4.3.22) and (4.3.23) into the TDGP equation and equating real and imaginary parts, we find

$$\frac{\partial \rho}{\partial t}(\boldsymbol{r}t) = -\text{div}(\rho(\boldsymbol{r}t)\boldsymbol{v}_s(\boldsymbol{r}t)) \tag{4.3.24}$$

$$m\frac{\partial \boldsymbol{v}_s}{\partial t} = -\boldsymbol{\nabla}\left\{V_{\text{ext}}(\boldsymbol{r}t) + n(\boldsymbol{r}t)U_0 + \tfrac{1}{2}mv_s^2(\boldsymbol{r}t) + \Phi_Q(\boldsymbol{r}t)\right\} \tag{4.3.25}$$

where $\Phi_Q(rt)$ is the so-called "quantum pressure" term defined by

$$\Phi_Q(rt) \equiv -\frac{\hbar^2}{2m}\frac{1}{[\rho(\boldsymbol{r}t)]^{1/2}}\boldsymbol{\nabla}^2\rho^{1/2}(\boldsymbol{r}t) \tag{4.3.26}$$

[17]Note in particular that if particles are being scattered out of the condensate, then Eqn. (4.3.6), even if true, is no longer equivalent to (4.3.8).

Equation (4.3.24) is just the equation of continuity. As to Eqn. (4.3.25), in the limit of slow spatial variation we can neglect the quantum pressure term (4.3.26) and define a space and time-dependent chemical potential $\mu(\boldsymbol{r}t)$ by the formula

$$\mu(\boldsymbol{r}t) \equiv V_{\text{ext}}(\boldsymbol{r}t) + n(\boldsymbol{r}t)U_0 \qquad (4.3.27)$$

Then, using the irrotationality of $\boldsymbol{v}_{\text{s}}(\boldsymbol{r}t)$, we can rewrite Eqn. (4.3.25) in the form

$$m\frac{D\boldsymbol{v}_{\text{s}}}{Dt}(\boldsymbol{r}t) = -\boldsymbol{\nabla}\mu(\boldsymbol{r}t) \qquad (4.3.28)$$

where D/Dt denotes the so-called hydrodynamic derivative $(\partial/\partial t + \boldsymbol{v}\cdot\boldsymbol{\nabla})$. Thus we can interpret the TDGP equation in this limit as simply equivalent to the equations for a classical ideal fluid (compare the discussion of Eqn. 3.5.3(b) of Chapter 3). In the limit of small amplitudes ($\boldsymbol{v}_{\text{s}} \to 0$) Eqns. (4.3.24) and (4.3.27) evidently have solutions corresponding to sound waves with velocity c, given by

$$c_S^2 \equiv m^{-1}(\partial\mu/\partial n) = n_0U_0/m \qquad (4.3.29)$$

Let us now discuss the nature of the small oscillations more generally. Assume that the ground state of the many body system has the form of the ansatz (4.3.2), with

$$\chi_0(\boldsymbol{r}t) = \exp(-\mathrm{i}\mu t/\hbar)\chi_0(\boldsymbol{r}) \qquad (4.3.30)$$

so that the equilibrium order parameter has the form

$$\Psi_0(\boldsymbol{r}t) = \exp(-\mathrm{i}\mu t/\hbar)\Psi_0(\boldsymbol{r}) \qquad (4.3.31)$$

where $\Psi_0(\boldsymbol{r})$ solves the time-independent GP equation (4.3.12) with $V_{\text{ext}}(\boldsymbol{r}t) \equiv V_{\text{ext}}(\boldsymbol{r})$.[18] We now consider a solution of the TDGP equation of the form

$$\Psi(\boldsymbol{r}t) \equiv \Psi_0(\boldsymbol{r}t) + \delta\Psi(\boldsymbol{r}t) \qquad (4.3.32)$$

where $\delta\Psi(\boldsymbol{r}t)$ is assumed small. Inserting (4.3.32) into (4.3.8) and keeping only terms linear in $\delta\Psi$ and its complex conjugate, we obtain

$$\begin{aligned} i\hbar\frac{\partial}{\partial t}\delta\Psi(\boldsymbol{r}t) &= -\frac{\hbar^2}{2M}\nabla^2\delta\Psi(\boldsymbol{r}t) + V_{\text{ext}}(\boldsymbol{r})\delta\Psi(\boldsymbol{r}t) \\ &\quad + U_0 n(\boldsymbol{r})(2\delta\Psi(\boldsymbol{r}t) + \exp -2\mathrm{i}\mu t/\hbar\delta\Psi^*(\mathrm{rt})) \end{aligned} \qquad (4.3.33)$$

with $n_0(\boldsymbol{r}) \equiv |\Psi_0(\boldsymbol{r})|^2$.

It is convenient to seek a solution in the form

$$\delta\Psi(rt) = \exp(-i\mu t/\hbar)[u(r)\exp -i\omega t + v^*(r)\exp i\omega t] \qquad (4.3.34)$$

Then, if we introduce the shorthand

$$\hat{H}_0 \equiv -(\hbar^2/2M)\nabla^2 + V_{\text{ext}}(\boldsymbol{r}) \qquad (4.3.35)$$

[18]For simplicity I assume below that $\Psi_0(r)$ can be chosen real (as is the case provided the Hamiltonian contains no time-reversal violating terms such as the effects of rotation). In the more general case the $n_0(r)$ in the last term (only) of (4.3.36) should be replaced by $\Psi_0^2(r)$ in (4.3.36a) and by $(\Psi_0^*(r))^2$ in (4.3.36b).

the quantities $u(\boldsymbol{r})$ and $v(\boldsymbol{r})$ must satisfy the *Bogoliubov-de Gennes equations*[18]

$$\hbar\omega u(r) = \left\{\hat{H}_0 - \mu + 2U_0 n_0(\boldsymbol{r})\right\} u(\boldsymbol{r}) + U_0 n_0(\boldsymbol{r}) v(\boldsymbol{r}) \tag{4.3.36a}$$

$$-\hbar\omega v(\boldsymbol{r}) = \left\{\hat{H}_0 - \mu + 2U_0 n_0(\boldsymbol{r})\right\} v(\boldsymbol{r}) + U_0 n_0(\boldsymbol{r}) u(\boldsymbol{r}) \tag{4.3.36b}$$

In the special case of a constant $V_{\text{ext}}(\boldsymbol{r})$ and periodic boundary conditions, we have $n_0(\boldsymbol{r}) = \text{const.} \equiv \text{n}_0$ and $\mu = n_0 U_0$; we may then take both $u(\boldsymbol{r})$ and $v(\boldsymbol{r})$ to be proportional to exp $i\boldsymbol{k}\cdot\boldsymbol{r}$, and substitution in (4.3.36a) and (4.3.36b) then gives the celebrated Bogoliubov spectrum

$$\hbar\omega(k) = \sqrt{\varepsilon_k(\varepsilon_k + 2n_0U_0)}; \quad \varepsilon_k \equiv \hbar^2 h^2/2M \tag{4.3.37}$$

or equivalently in terms of the sound velocity c_s (Eqn. (4.3.29))

$$\omega(k) = (c_s^2 k^2 + \hbar^2 k^4/4M^2)^{1/2} \tag{4.3.38}$$

The spectrum is thus sound-wave-like for $k \ll \hbar/Mc_s(\sim \xi^{-1})$, in agreement with our general result (4.3.29), and free-particle-like for $k \gg \xi^{-1}$. Note however that in both cases, in fact for all k, the density variation in the excitation is of the simple form

$$\delta\rho(\boldsymbol{r}t) = \text{const.} \cos(\boldsymbol{k}\cdot\boldsymbol{r} - \omega_{\text{k}}\text{t}) \tag{4.3.39}$$

To conclude this section, let us note that it is obviously possible to extend the GP approximation to cases where the condensate number N_0, though macroscopic, is not equal to the total particle number N (the most obvious example would be the equilibrium behavior at nonzero temperature). One then has to consider many-body wave functions which have the "Hartree–Fock" rather than the simple "Hartree" form: apart from normalization

$$\Psi_N(\boldsymbol{r}_1\boldsymbol{r}_2\ldots\boldsymbol{r}_N : t) = \mathcal{S}\Pi_{i=1}^{N_0}\chi_0(\boldsymbol{r}_i,t)\cdot\Pi_{j=N_0+1}^{N_0+n_\mu}\chi_\mu(\boldsymbol{r}_j:t)\Pi_{k=N_0+n_\mu+1}^{N_0+n_\mu+n_\nu}\chi_\nu(r_k:t)\ldots \tag{4.3.40}$$

where $\mathcal{S}$ denotes the operation of symmetrization, and where the states $\chi_\mu(\boldsymbol{r},t)$, $\chi_\nu(\boldsymbol{r},t)\ldots$ are mutually orthogonal and orthogonal[19] to the condensate wave function $\chi_0(\boldsymbol{r}t)$, and the occupation numbers $n_\mu, n_\nu\ldots$ (which I will assume, with N_0, are not themselves a function of time) are of order 1 rather than N. With the ansatz (4.3.40), both $\chi_0(\boldsymbol{r}t)$ and the other χ_μ satisfy a nonlinear Schrödinger equation of a form similar to (4.3.6), with however the following important caveat: If we define the condensate density $n_0(\boldsymbol{r}t)$ and the density of uncondensed particles[20] $n_{\text{T}}(\boldsymbol{r}t)$ by the formulae

$$n_0(\boldsymbol{r}t) \equiv N_0|\chi_0(\boldsymbol{r}t)|^2 \tag{4.3.41}$$

$$n_{\text{T}}(\boldsymbol{r}t) \equiv \sum_{j\neq 0} n_j|\chi_j(\boldsymbol{r}t)|^2 \tag{4.3.42}$$

then according to the considerations given above, the effective nonlinear potential which enters the equation for $\chi_0(\boldsymbol{r}t)$ is $U_0(n_0(\boldsymbol{r}t) + 2n_{\text{T}}(\boldsymbol{r}t))$, while (neglecting terms

[19]Because the "potentials" seen by the condensate and the noncondensate particles are different, the solutions of the relevant nonlinear Schrödinger equation (see below) do not automatically satisfy this condition. For a treatment of this difficulty, see Huse and Siggia (1982).

[20]I use the notation n_{T} (for "thermal") for convenience, although the considerations advanced here are not restricted to the case of thermal equilibrium.

of relative order N^{-1}) that which enters the equation for any of the $\chi_j (j \neq 0)$ is $2U_0(n_T(\boldsymbol{r}t)+n_T(\boldsymbol{r}t))$. When this is taken into account, the resultant equations appear to give a good description of the finite-temperature equilibrium behavior and, with appropriate account taken of collisions by introducing phenomenological relaxation times, of the collective excitations at finite temperature: see for example Pitaevskii and Stringari (2003), Chapter 13.

4.4 The Bogoliubov approximation

Let us imagine that we take seriously the difficulty raised in the last section concerning the internal inconsistency of the GP description, and try to rewrite that description so as to take into account the two-particle correlations which give rise to the effective interaction (4.2.7). For simplicity I confine myself to the problem of the ground state in the homogenous case, that is, the case of N atoms moving in a volume Ω with $V_{\text{ext}}(\boldsymbol{r}) \equiv 0$ and periodic boundary conditions, so that the ansatz (4.3.2) simply reduces to a constant and the total interaction energy (4.3.12) is just $\frac{1}{2}NnU_0$, where $n \equiv N/\Omega$ and I assume $U_0 > 0$. Since the relative wave function of two atoms in the limit $r \gg r_0$ has the form (4.2.2), the natural choice for the many-body ground state would be of the Jastrow form, i.e. up to normalization

$$\Psi_N(\boldsymbol{r}_1\boldsymbol{r}_2 \ldots \boldsymbol{r}_N) = \prod_{ij} f(r_{ij}) \tag{4.4.1}$$

when $f(r_{ij})$ has the form of the two-particle $E = 0$ relative wave function, and in particular has for $r \gg r_0$ the form (up to normalization)

$$f(r_{ij}) = 1 - a_s/|\boldsymbol{r}_{ij}| \tag{4.4.2}$$

However, the ansatz (4.4.1), with the choice (4.4.2), immediately leads to two difficulties. In the first place, while it is certainly possible to normalize the wave function (4.4.1), if we calculate the probability of any given atom being in the zero-momentum state we will find it to be negligible in the thermodynamic limit,[21] so that the representation in the form (4.3.2) with χ_0 = constant is nowhere near a good approximation. This could perhaps be regarded as a "cosmetic" difficulty which need not necessarily lead us to challenge the validity of the ansatz (4.4.1). However, a closely related and more serious difficulty arises from the "Fock" term in the energy associated with Eqn. (4.4.1): Consider the contribution to the many-body wave function of the mutual scattering of two particles, say 1 and 2, namely the factor $1 - a_s/|r_{12}|$. Unless a_s is zero, this indicates that there is a nonzero probability for atom 1 to be in a state of nonzero momentum $\boldsymbol{k}$ and atom 2 to have simultaneously momentum $-\boldsymbol{k}$. We do not, of course, need to introduce an extra "Fock" term to take this into account, since by definition (4.2.2) is the exact solution of the two-particle problem and the associated energy is rigorously given by the formula (4.2.8) (no Fock term!). The crunch, however, comes in the interactions of (say) atom 1 with a third atom, 3. Atom 1, once scattered by atom 2 into a state with $\boldsymbol{k} \neq 0$, is no longer (except by improbable coincidence) in the same single-particle state as atom 3, and thus its interactions with the latter

[21]Each value of j subtracts from the 1 a term of order $\Omega^{-1/3}$, so the total subtraction is $O(N\Omega^{-1/3}) \sim N^{2/3}$ in the thermodynamic limit.

should indeed contain the Fock term. Since, as remarked above, the probability of 1 remaining in the original $\boldsymbol{k} = 0$ state is negligible, the net upshot is that the expectation value of the energy in the (normalized) state described by the ansatz (4.4.1) and (4.4.2) is approximately *twice* the "naive" GP energy $\frac{1}{2}NnU_0$, and that for $U_0 > 0$ this state is highly disadvantageous energetically.

It is clear that a possible solution to this difficulty is to maintain the Jastrow ansatz (4.4.1), but to modify the form of $f(r)$ so that the correction to 1 falls off faster than $1/r$ at large differences. Let us estimate the optimum choice of the distance R beyond which the behavior (4.4.2) is thus modified. Let $N_R \equiv nR^3$ be the (order of magnitude of) the number of atoms within R of a given atom. The extra kinetic energy needed to drive the term $a_s/|\boldsymbol{r}_{ij}|$ to zero for $|r_{ij}| \sim R$ will be $\sim\hbar^2 a_s^2/mR^4$ for each neighbor j, and hence the total extra kinetic energy per atom will be $\sim N_R\hbar^2 a_s^2/mR^4 \sim na_s^2\hbar^2/mR$. The Fock potential energy will be of the order of the mean-field energy nU_0 times the probability of the atom not being in the $\boldsymbol{k} = 0$ state, which has a contribution $\sim(a_s/R)^2$ from each neighbor and hence a total value of $N_R(a_s/R)^2$; thus the total extra energy per atom beyond the "naive GP" expression $\frac{1}{2}nU$, has the form (using $U \sim \hbar^2 a_s/m$)

$$E(R) \sim na_s^2\hbar^2/mR + n^2 a_s^3\hbar^2 R/m \tag{4.4.3}$$

Minimizing this expression with respect to R, we find (not surprisingly with hindsight) that R is of the order of the "healing length" $\xi \equiv (8\pi n a_s)^{-1/2}$ introduced in the last section, and that the extra energy is of order $(\hbar^2/m)n^{3/2}a_s^{5/2}$, i.e. of order $(na_s^3)^{1/2}$ relative to the GP result. Note also that the total "depletion", i.e. the probability of not being in the $\boldsymbol{k} = 0$ state, is now finite and of order $N_\xi(a_s/\xi)^2 \sim (na_s^3)^{1/2}$.

The Bogoliubov approximation is simply a quantitative implementation of the above qualitative arguments: in effect, it starts from the ansatz (4.4.1)[22] and minimizes the total energy so as to find the specific form of the function $f(r)$. For the free-space case it can be carried out quantitatively to the end *provided* that the "diluteness parameter" $\zeta \equiv (na_s^3)^{1/2}$ is assumed small compared to unity (a condition satisfied by the alkali gases in most though not all experiments to date). Most textbook presentations (e.g. Huang 1987, Pethick and Smith 2002) follow Bogoliubov's original presentation, relaxing the condition of total particle number conservation and treating the condensate creation operator as a classical variable. It is also possible, and arguably more illuminating, to derive the standard results by writing down an explicit ansatz for the N-body wave function and optimizing the coefficients. This is done for example in Leggett (2001), Section VIII,[23] and I will simply quote the salient results, using for convenience second-quantized notation: The ground state of the N-body system is taken to be of the form (which is equivalent to the Jastrow form (4.4.1) to the extent that three-body correlations may be neglected)

$$\Psi_N = \text{const.}(a_0^+ a_0^+ - \sum_{q\neq 0} c_q a_q^+ a_{-q}^+)^{N/2} \tag{4.4.4}$$

[22]Or more precisely from an ansatz (the "first-quantized" version of (4.4.4) below) which is equivalent to (4.4.1) provided that we can neglect three-body correlations.

[23]Note the corrections in the cited erratum.

where the quantity c_q is positive and of magnitude less than unity. Thus, intuitively, the ground state contains pairs of atoms which spend most of their time in a relative state of zero momentum, but sometimes scatter into states $\boldsymbol{q}$ and $-\boldsymbol{q}$. Minimization with respect to the coefficients c_q gives an explicit expression for the latter (Eqn. (8.17) of Leggett 2001) in terms of the excitation energies (see below); it is not worth writing out this somewhat messy expression here, and it is sufficient to note that c_q tends to 1 in the limit $q \to 0$ and to nU_0/ε_q for $q \to \infty$ (but qr_0 still $\ll 1$). The elementary excitations of wave vector q are, as might be guessed from the form of (4.4.4), a linear combination of a state in which a particle originally in the condensate is scattered into state $\boldsymbol{q}$, and a state in which a particle originally in state $-\boldsymbol{q}$ is scattered into the condensate: explicitly.

$$\Psi_q = (u_q a_q^+ a_0 + v_q a_0^+ a_{-q}|\Psi_N\rangle \tag{4.4.5}$$

where $v_q/u_q = c_q$ and (for normalization) $u_q^2 - v_q^2 = 1$. The energy of the excitation (4.4.5) has the Bogoliubov form

$$E_q = (\varepsilon_q(\varepsilon_q + 2nU_0))^{1/2} \quad (\varepsilon_q \equiv \hbar^2 q^2/2m) \tag{4.4.6}$$

and thus coincides exactly with $\hbar\omega(q)$, where $\omega(q)$ (Eqn. 4.3.38) is the frequency of a small oscillation of the order parameter calculated from the TDGP equation; needless to say, this is not an accident. In the limit of small and large q the quantity E_q has a simple form:

$$E_q = \hbar c_s q, \qquad q\xi \ll 1 \tag{4.4.7}$$

$$E_q = \hbar^2 q^2/2m + nU_0, \quad q\xi \gg 1 \tag{4.4.8}$$

where the nU_0 in (4.4.8) is the Fock energy.

One may use the above results to calculate various properties of the system such as the depletion K, the ground state energy E_0 relative to the GP value $\frac{1}{2}NnU_0$, and the two-particle correlation function $F(r) \equiv \langle\rho(0)\rho(r)\rangle$ (i.e. the quantity $\sum_{ij}\overline{|\Psi(r_{ij})|^2}$ where the bar indicates an average for given i and j over the other $N-2$ particles). In agreement with the qualitative results obtained above, we find

$$K = \frac{8}{3\sqrt{\pi}} \quad \zeta = \frac{8}{3\sqrt{\pi}}(na_s^3)^{1/2} \tag{4.4.9}$$

$$E_0 = \frac{64}{15\sqrt{\pi}} N(nU_0)(na_s^3)^{1/2} \quad (\sim na_s^3)^{1/2} E_{\rm GP}) \tag{4.4.10}$$

$F(r)$ is given by a rather complicated expression (see Eqns. (43)–(48) of Lee et al. 1957), which however is proportional to $(1 - a_s/r)^2$ for $r \ll \xi$ and to $1 + o(1/r^4)$ for $r \gg \xi$.

In view of Eqn. (4.2.12), it is tempting to think that the role of the correlations introduced by the Bogoliubov approximation is to reduce the value of $|\Psi(r_{ij} \xrightarrow{\sim} 0)|^2$, i.e. of the probability of finding two atoms close to one another, relative to its value in the two-particle problem and thereby to reduce the energy (4.2.12). However, Eqns. (4.4.10) and the above behavior of $F(r)$ show that in reality precisely the opposite happens: the value of the probability of close approach is *increased* relative to the two-particle result, and the total energy is *larger* than that calculated from (4.3.13). This latter fact shows right away that the GP energy cannot be the expectation value of the Hamiltonian on any consistent ansatz for the N-body ground state wave function (since if it were, this ansatz would be better than the Bogoliubov one!) In fact, the correct way to look at the GP "approximation" is that it is the limit of the Bogoliubov approximation in which terms of relative order $(na_s^3)^{1/2}$ can be neglected. (These points are at least implicit in the 50-year-old work of Lee et al. (1957), and are discussed at greater length in Leggett (2003).)

Once one has to deal with spatially inhomogeneous situations (e.g. those involving a nontrivial external potential V_{ext}) the formulation of the Bogoliubov approximation becomes a little more complicated: while it is still possible to write an ansatz for the N-body ground state which is a generalization of (4.4.4), the relation between the coefficients appearing in it and the properties of the elementary excitation becomes quite complicated: see Fetter (1972). In this case, to obtain the latter, it is actually simpler to use as a starting-point the "semiclassical" Bogoliubov–de Gennes equations (4.3.35) which are obtained by solving the TDGP equation for small deviations from equilibrium, construct explicitly a Lagrangian and go through the standard canonical quantization procedure. Since all the modes considered in Section 4.3 are of harmonic-oscillator type, it is not surprising that the result of such a calculation is to give an excitation spectrum $E_i = \hbar\omega_i$, where the ω_i are the various independent solutions of Eqn. (4.3.35). The free-space problem which I have discussed in detail is just a special case. Note that nowhere in this procedure do we have to invoke nonconservation of the total particle number.

4.5 Coherence and interference in dilute alkali Bose gases

Perhaps the most spectacular manifestation of BEC in the Bose alkali gases is the dramatic appearance, below T_c, of a "spike" in the density distribution, with a width which is only a fraction of the width of the thermal distribution above T_c (see Andrews et al. 1996). Nevertheless, a determined skeptic could argue that since the width of the peak is still a factor ~5–10 times greater than that of the oscillator zero-point wave function, all that appearance of this peak really shows is that a finite fraction of the atoms has occupied the lowest few hundred states of the trap out of the $\sim 10^6$ states occupied in the thermal distribution. So the question arises: how do we know that a *single* state is being macroscopically occupied, i.e. that the BEC is "simple" rather than "fragmented"?

To see how we might go about answering this question, let us go back to the definition of BEC in terms of the single-particle density matrix given, for a set of spinless particles, in Chapter 1:

$$\rho_1(\boldsymbol{r}\boldsymbol{r}' : t) \equiv \langle \psi^\dagger(\boldsymbol{r})\psi(\boldsymbol{r}')\rangle = \sum_i n_i(t)\chi_i^*(\boldsymbol{r}t)\chi_i(\boldsymbol{r}'t) \tag{4.5.1}$$

Recall that we are said to have "simple" BEC at time t if a *single* state i has associated with it an eigenvalue $N_0 \sim N$, while all the other n_i's are of order 1. Consider now values of $\boldsymbol{r}$ and $\boldsymbol{r}'$ which, while still such that the density at both points is appreciable, are far apart on the scale of the thermal-component correlation length λ_T (to be defined below). Then the different states $i \neq 0$ which are not macroscopically occupied will have relatively random phases and will tend to interfere destructively, while the condensate contribution will still, in general, be finite, and we find

$$\rho_1(\boldsymbol{r}\boldsymbol{r}' : t) \cong N_0 \chi_0^*(\boldsymbol{r}t)\chi_0(\boldsymbol{r}'t), \quad |\boldsymbol{r}-\boldsymbol{r}'| \gg \lambda_T \tag{4.5.2}$$

Indeed, as we saw in Chapter 2, Section 2.2, Eqn. (4.5.2) has in effect been proposed, under the name of "off-diagonal long-range order", as an alternative definition of (simple) BEC (Yang, 1962). If more than one state i were macroscopically occupied, then the form of $\rho_1(\boldsymbol{r}\boldsymbol{r}' : t)$ would be more complicated and in general it would not reduce to a single product of functions of $\boldsymbol{r}$ and $\boldsymbol{r}'$.

We can confirm this general conclusion by analyzing the specific case of a non-interacting Bose gas in thermal equilibrium at some temperature T below its BEC transition temperature, so that the thermal distribution $n_i (i \neq 0)$ is given by $(\exp \beta\epsilon_i - 1)^{-1} (\beta \equiv 1/k_B T)$. For a gas in free space the result is

$$\rho_1(\boldsymbol{r}\boldsymbol{r}') = N_0(T) + \frac{\Omega}{2\pi\lambda_T^2 |\boldsymbol{r}-\boldsymbol{r}'|} F(|\boldsymbol{r}-\boldsymbol{r}'|/\lambda_T) \tag{4.5.3}$$

where[24] $\lambda_T \equiv (h^2/2\pi k_B T)^{1/2}$ and the function $F(\alpha)$ is defined by

$$F(\alpha) \equiv \int_0^\infty \frac{\chi d\chi \sin \alpha\chi}{\exp \chi^2 - 1} \tag{4.5.4}$$

and tends to a constant for $\alpha \to \infty$: thus, the second term in (4.5.3) falls off as $|\boldsymbol{r}-\boldsymbol{r}'|^{-1}$. For a gas in a one-dimensional harmonic trap of frequency ω_0 (a case of some practical interest, as we shall see), if we take $\boldsymbol{r} \equiv (x, y, z)$ and $\boldsymbol{r}' \equiv (x, y, z')$ where the trapping potential is a function of z,

$$\rho_1(z, z') = N_0 \exp -(z^2 + z'^2)/z_0^2 + \ldots \tag{4.5.5}$$

where "..." indicates a (calculable) function which falls off fast as a function of $|z-z'|$; for $T > T_c$ ($N_0 = 0$) the fall-off is exponential (as in the free particle case). For an interacting gas in a one-dimensional harmonic trap, the condensate contribution to $\rho_1(\boldsymbol{r}\boldsymbol{r}')$ is replaced by $\Psi_0^*(\boldsymbol{r})\Psi_0(\boldsymbol{r}')$ where Ψ_0 is the ground state order parameter; the contribution of the thermal cloud cannot in general be evaluated exactly, but provided we are at temperatures high enough that most of the excitations are in the "free-particle" limit of the Bogoliubov spectrum, (roughly speaking, this requires $k_B T \gg n_{\max} U_0$ when $n_{\max}$ is the maximum condensate density) then it should not be very different from the second term in (4.5.5).

Rather than measuring $\rho_1(\boldsymbol{r}\boldsymbol{r}' : t)$ directly (see below) one may prefer to measure its Fourier transform, which is the single-particle momentum distribution $n(\boldsymbol{p})$. This has been done for a sample of trapped ^{23}Na gas by Stenger et al. (1999) by a technique which is conceptually similar to that used to measure $n(\boldsymbol{p})$ in liquid ^{4}He (see Chapter 3, Section 3.2); the difference is that rather than using neutrons one employs two Raman lasers to scatter the atoms, see Section 4.6 below. However, the principle is the same: if the momentum distribution in the initial state is strongly peaked around zero momentum, that should show up as a strong dependence, for given $\boldsymbol{\Delta k}$ (the difference in $\boldsymbol{k}$-vector of the two laser beams) on their frequency difference $\Delta\omega$

[24]Beware of different factors of $(2\pi)^{1/2}$ in the definitions of λ_T in the literature.

(which is an independently adjustable parameter).[25] This is indeed what is seen in the experiments (see Fig. 1 of Stenger et al. 1999), and from the observed width of the peak these authors were able to infer that the spread in momentum was considerably smaller than that of the noninteracting ground state in the trap, and consistent with the value of order $\hbar/R$ which is predicted from the uncertainty principle for a spread $\sim R$ in coordinate space, which is calculated as in Section 4.3 above.

If one wishes to measure $\rho_1(\boldsymbol{r}\boldsymbol{r}' : t)$ itself rather than its Fourier transform, a possible way to do it is to propagate atoms from the points $\boldsymbol{r}$ and $\boldsymbol{r}'$ to some common point and measure the probability density at that point (a sort of generalized Young's slits experiment, except that the initial position of the atoms is at the analog of the slits rather than at a "source"). Experiments of this type have been done by Hagley et al. (1999), by giving the condensate successive "kicks" with laser pulses, and by Bloch et al. (2000) by tipping magnetically trapped atoms into an untrapped state and letting them fall under gravity. As the latter experiment is particularly easy to visualize and moreover was extended to temperatures $\gtrsim T_c$, I will discuss it in some detail.

In the experiment one starts with a set of ^{87}Rb atoms in the (trapped) $F = 1, m_F = -1$ state in a magnetic trap whose "tight" axis is in the z- (vertical) direction with the minimum at $z = 0$. Because the Zeeman splitting $\Delta\varepsilon$ between this state and the untrapped $F = 1, m_F = 0$ state is a function of the local magnetic field and hence of z, a pulse of rf radiation at frequency ω selectively tips into the latter state only those atoms which are close to the point at which $\Delta\epsilon(z) = \hbar\omega$. Once in the $m_F = 0$ state, the atoms scarcely feel the magnetic field[26] and simply fall under gravity. Suppose now that we consider a single atom which is initially in a definite Schrödinger state in the trap, with finite amplitude $\psi(z)(\psi(z'))$ to be at the point $z(z')$; we assume that any time-dependence of the phase of ψ is common to the two points and so can be ignored. Imagine that our rf field actually has two components, with frequencies ω and ω' such that $\Delta\varepsilon(z) = \omega$ and $\Delta\varepsilon(z') = \omega'$. After the rf has been on for some time, the probability amplitude to have $m_F = 0$ and be at a point Z in the gravitational field will be a superposition of the amplitudes which have been propagated from z and z' respectively, that is, it will be of the form

$$\psi(Z,t) = \psi_z(Z,t) + \psi_{z'}(Z,t) \tag{4.5.6}$$

Let us assume for simplicity that the effects of the rf pulses and of the subsequent propagation in the gravitational field on the amplitudes (as distinct from the phases) of ψ_z and $\psi_{z'}$ are the same (this should be approximately true for $Z \gg |z - z'|$), so that

$$\Psi_z(Z,t) = \alpha\psi(z) \exp i\varphi_z(Z,t), \quad \psi_{z'}(Z,t) = \alpha\psi(z') \exp i\varphi_{z'}(Z,t) \tag{4.5.7}$$

when α is some constant factor. Then we see that the probability density to find the atom at point A at time t is given by an expression of the form

$$P(Z,t) = \text{const.}(1 + V \cos \Delta\varphi(Z,t)) \tag{4.5.8}$$

[25] For a good discussion of this "Bragg spectroscopy" technique, see Pitaevskii and Stringari (2003), Section 12.9.

[26] Although the absolute value of the field is not stated, it is presumably much less than the value of B_{hf} for ^{87}Rb, so that one is on the "flat" part of the $m_F = 0$ curve in Fig. 4.1.

when $\Delta\varphi(Z,t) \equiv \varphi_z(Z,t) - \varphi_{z'}(Z,t)$ and the "visibility" V is given, for this simple case, by the expression[27]

$$V = \frac{2\mathrm{Re}\,\psi^*(z)\psi(z')}{|\psi(z)|^2 + |\psi(z')|^2} \tag{4.5.9}$$

Before proceeding to the many-body case, let us work out an explicit formula for $\Delta\varphi(Z, t)$, confining ourselves to the limit $Z \gg |z - z'|$. To do this we can use a semiclassical argument: To arrive at the point Z at time t, the two wave packets must have left at approximately the same time, namely $t - (2gZ)^{1/2} \equiv t_c$. Since they have (conserved) energies $-mgz$ and $-mgz'$ respectively, the phase difference accumulated during their acceleration by the gravitational field will be approximately $(2Z/g)^{1/2}(z - z')$. On the other hand, they will actually have been created in the $m_F = 0$ state with a phase difference $(\omega - \omega')(t_c - t_0)$ (where t_0 is a constant which depends on the details of the switching-on of the field, etc.); this may be verified by following the motion of the "spin" as it is rotated from "up" ($m_F = -1$) to "down" ($m_F = 0$). Consequently the total phase difference will be, up to a constant,

$$\Delta\varphi(Z,\, t) = mg(2Z/g)^{1/2}(z - z') + (\omega - \omega')t(+\text{const.}) \tag{4.5.10}$$

or, since $\omega - \omega' = -mg(z - z')$,

$$\Delta\varphi(Z,t) = mg(z - z')\left\{(2Z/g)^{1/2} - t\right\}(+\text{const.}) \tag{4.5.11}$$

Note that the locus of constant phase follows the Newtonian trajectory $Z = \frac{1}{2}gt^2$, as we expect in the semiclassical limit.

It is now straightforward to apply those results to the many-body problem. We just take account of the fact that in the initial state we had n_i atoms in the single-particle-like state $\chi_i(r)$ (and implicitly assume that once the atoms are tipped out of the trap we can neglect their interactions). Then we recover Eqn. (4.5.8), with the visibility V now given by the expression[28]

$$V = \frac{2\mathrm{Re}\,\sum_i n_i\chi_i(z)\chi_i(z')}{\sum_i n_i(|\chi_i(r)|^2 + |\chi_i(z')|^2)} \equiv \frac{2\mathrm{Re}\,\rho_1(z,z')}{\rho(z) + \rho(z')} \tag{4.5.12}$$

(where $\rho(z) \equiv \rho_1(z,z)$ is the total density at point z). (In the case of a three-dimensional system, the ρ's should be interpreted as averaged over the x- and y-dimension.) Thus, in the case where V is of order 1, we expect to see (literally, e.g. by absorption spectroscopy) a set of "drops" whose motion in time exactly follows the Newtonian acceleration under gravity. This is exactly what we in fact see (cf. Fig. 1(a) of Bloch et al. 2000), and inspection of the dependence of the magnitude of V on z and z' then gives us direct information about the behavior of $\rho_1(z, z')$. As expected, it is found that for $T > T_c$ $\rho_1(z, z')$ falls off exponentially in $|z - z'|$ over a length of the order of the thermal correlation length, while for $T < T_c$ it tends to

[27]Here I implicitly assume that $\psi(z)$ and $\psi(z')$ are relatively real (true in the absence of time-reversal-violating effects). Failure of this condition would involve the addition of an offset to $\Delta\varphi$ which is in general a function of z and z'.

[28]Again I assume for simplicity that $\rho_1(\boldsymbol{r}\boldsymbol{r}')$ is real.

a constant value provided z and z' are much less than the dimensions of the trapped cloud.[29]

An experiment somewhat similar in spirit to that of Bloch et al. had been done somewhat earlier by Anderson and Kasevich (1998). Instead of starting from a single harmonic trap, they constructed a one-dimensional vertical optical lattice (see below, Section 4.6) and allowed the atoms to leak out of the various wells by tunneling and thereafter fall in the gravitational field. If we consider two wells at mean height z and z', respectively, the phase difference accumulated in free fall is the same as in the Bloch et al. experiment, and the initial phase difference is, up to a constant, $mg(z - z')/\hbar$; thus the formulae for $P(Z,t)$ and $\Delta\varphi(Z, t)$ are identical to (4.5.8) and (4.5.10), respectively. Actually we have several wells, so the probability density is proportioned to the expression

$$\left|\sum_n \exp\left\{in\delta((2Z/g)^{1/2} - t)\right\}\right|^2 \tag{4.5.13}$$

(where $\delta = mg\lambda/2$ is the gravitational energy difference between neighboring wells, with λ the laser wavelength). Thus the density of the falling cloud is more strongly peaked in the "droplets" than in the Bloch et al. 2000 experiment, but the latter still fall with the classical gravitational acceleration. Note that in this experiment the clouds in the different wells are essentially uncoupled, so that their relative phase is a function of time, whereas in the experiment of Bloch et al. the phases at z and z' are strongly coupled and the t-dependent phase difference is a consequence of the out-coupling ($m_F = -1 \Rightarrow m_F = 0$) process.

Probably the most famous experiment done to date on interference in alkali Bose gases is the historically earliest, namely that of the MIT group (Andrews et al. 1997). As we shall see, the interpretation of this experiment involves some conceptual issues which go beyond those relevant to the experiments so far discussed. The authors start with a trap which is split into two by a laser-induced barrier so high that the single-atom tunneling time between the two wells is greater than the age of the universe. They then condense clouds of ^{87}Rb atoms independently in the two wells and allow them to come to thermal equilibrium. At this point there seems no doubt that the correct quantum-mechanical wave function of the system is, schematically, of the form (up to symmetrization)

$$\Psi_N = \Pi_{i=1}^{N_\mathrm{L}} \chi_\mathrm{L}^0(r_i) \Pi_{j=1}^{N_\mathrm{R}} \chi_\mathrm{R}^0(r_j) \quad (N_\mathrm{L} + N_\mathrm{R} = N) \tag{4.5.14}$$

i.e. we have some number N_L of atoms in the ground state $\chi_\mathrm{L}(r)$ of the left-hand trap, and some number N_R in the ground state χ_R of the right-hand trap. Note that this description is, formally, quite different from the wave function

$$\Psi_N = \Pi_{i=1}^{N} 2^{-1/2}(\chi_\mathrm{L}^0(r_i) + (\exp i\Delta\varphi)\chi_\mathrm{R}^0(r_i)) \tag{4.5.15}$$

which describes all N atoms condensed into a linear superposition of the left and right ground states with some definite relative phase $\Delta\varphi$. In the literature (4.5.15) is often

[29]In view of the considerations to be advanced below in the context of the MIT interference experiment, it is worth mentioning that for one value of $\Delta z \equiv |z - z'|$ the authors checked explicitly that the behavior was reproducible (i.e. the offset in $\Delta\varphi(Z,t)$ did not vary at random from shot to shot).

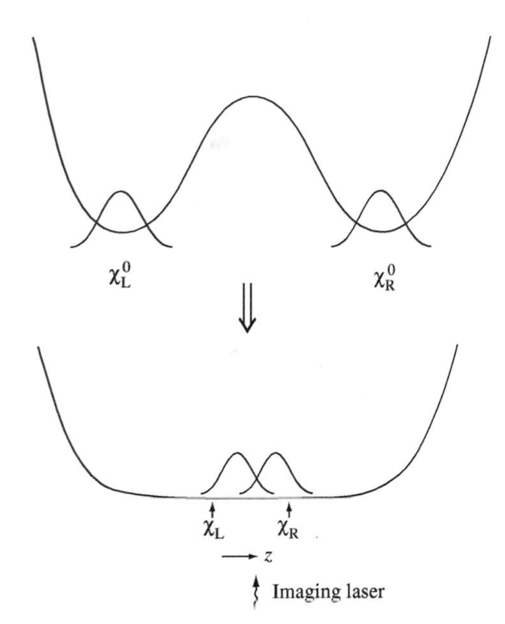

Fig. 4.3 The MIT interference experiment (schematic; in the actual experiment the outer walls are also taken down).

called a coherent or GP state, while (4.5.14) is sometimes called a (relative) Fock state (cf. Chapter 2, Section 2.3).

At some initial time the MIT group now take down the laser barrier,[30] allow the two clouds to expand into the region between them where they can overlap, and detect the (appropriately integrated) density of atoms as a function of the "interesting" coordinate, z (see Fig. 4.3). What should one predict that they see? If for a moment we imagine that the starting description is (4.5.15) rather than (4.5.14), with a known value of $\Delta\varphi$ (e.g. zero) then the situation seems to be straightforward, at least to the extent that by the time the clouds overlap the density is so low that one can neglect the nonlinear terms in the GP equation (a condition which is almost certainly met in the real-life experiments): According to the TDGP equation, each of the wave functions χ_L^0 and χ_R^0 will evolve into an appropriate time-dependent function $\chi_L(\boldsymbol{r}t)$ and $\chi_R(\boldsymbol{r}t)$ respectively, which by the time of overlap will effectively be described by a simple Schrödinger equation (the nonlinear terms being negligible). Since each cloud will have acquired some (time-dependent) velocity $v(t) \equiv \hbar k(t)/m$ which is, by symmetry, opposite for L and R, the functions $\chi_{L,R}(\boldsymbol{r}t)$ will be schematically of the form

$$\chi_{L,R}(\boldsymbol{r}t) \sim G_{L,R}(\boldsymbol{r}t)\exp[\pm ik(t)z] \tag{4.5.16}$$

where the envelope functions $G_{L,R}(\boldsymbol{r}t)$ are smoothly varying in space. The total condensate wave function $\chi(\boldsymbol{r}t)$ will then be a simple superposition of χ_L and χ_R with the appropriate initial relative phase $\Delta\varphi$, i.e. the N-body wave function will be of

[30]They actually release the whole trap, but this feature is inessential in the present context.

the form

$$\Psi_N = \Pi_{i=1}^N \chi(\boldsymbol{r}_i, t) \tag{4.5.17}$$

$$\chi(\boldsymbol{rt}) \equiv (\chi_{\mathrm{L}}(\boldsymbol{rt}) + \exp i\Delta\varphi \cdot \chi_{\mathrm{R}}(\boldsymbol{rt})) \tag{4.5.18}$$

Substituting the form (4.5.16), we see that the density of particles in the overlap region is schematically of the form

$$\begin{aligned}\rho(zt) &= \frac{N}{2}\left\{|G_{\mathrm{L}}(\boldsymbol{rt})|^2 + |G_{\mathrm{R}}(\boldsymbol{rt})|^2 + 2\mathrm{Re}\left[G_{\mathrm{L}}^*(\boldsymbol{rt})G_{\mathrm{R}}(\boldsymbol{rt})\exp i[\{2k(t)+\Delta\varphi\}\right]\right. \\ &\equiv f(\boldsymbol{rt}) + g(\boldsymbol{rt})\cos(2\pi z/\lambda(t) + \Delta\varphi)\end{aligned} \tag{4.5.19}$$

when the functions $f(\boldsymbol{rt})$ and $g(\boldsymbol{rt})$ are slowly varying, $\lambda(t) \equiv \pi/k(t)$ and we have arbitrarily fixed the origin of z so as to absorb any extra phase factor arising from the (calculable) relative phase of φ_{L} and φ_{R}. In other words, we predict that if we image our system at time t, we will see *interference fringes* with a definite spacing $\lambda(t)$ and a definite "offset" $\Delta\varphi$ which is just the relative phase appearing in the original wave function (4.5.15).[31]

But what if the starting description of our system is (as it almost certainly was in the MIT experiments) not the coherent (GP) state (4.5.15) but the Fock state (4.5.14)? We can then recalculate the single-particle density $\rho(zt)$ and find simply

$$\rho(zt) = N_{\mathrm{R}}|G_{\mathrm{R}}(zt)|^2 + N_{\mathrm{L}}|G_{\mathrm{L}}(zt)|^2 \tag{4.5.20}$$

with G_{L} and G_{R} smooth functions as above. The expression (4.5.20) clearly contains no terms corresponding to interference fringes, and it is therefore at first sight tempting to conclude that the experimental images will not show any.

Such a conclusion, however, would be quite wrong: the actual experimental images show spectacular interference fringes (see, e.g. Fig. 2 of Andrews et al. 1997) which correspond precisely to Eqn. (4.5.19), with the given value of $\lambda(t)$ but with an offset $\Delta\varphi$ that appears to vary at random from one experimental run to the next. It should be emphasized that this result is not in conflict with the result (4.5.20), which is formally correct: the reason is that when we calculate the "density" $\rho(zt)$, what we are actually calculating is the expectation value of a quantum-mechanical operator, and that expectation value has to be defined *with respect to the correct quantum-mechanical ensemble.* In this case the correct definition of the relevant "ensemble" is not the collection of atoms on a single run, but the *collection of different runs.* And indeed, if we superpose the images taken on different runs, i.e. average over runs, then because $\Delta\varphi$ is random from run to run the result is indeed just (4.5.20)!

The crucial experimental observation – that a definite fringe pattern is seen on each run, but with a random offset – can be interpreted theoretically in any number of different ways, and it has by now generated a considerable literature. One possibility is to observe that, as already noted in Chapter 2, Section 2.3, the Fock state (4.5.14) can be interpreted as a quantum superposition of coherent states (4.5.15) with all possible

[31] In practice, even if we know $\Delta\varphi$ (e.g. because the Josephson coupling is appreciable and we know that we start from the ground state) it may be impossible to calculate the absolute position of the fringes because small and uncontrollable fluctuations, e.g. in the shape of the potential may affect the relative phase of φ_{L} and φ_{R}.

values of $\Delta\varphi$, and to regard the measurement of the density distribution as effectively a "measurement" of $\Delta\varphi$: then by the standard axioms of quantum measurement theory we are guaranteed to get a definite value of this quantity. This approach (or rather a variant of it, see below) seems particularly natural in the context of the experiments in spin space which I shall discuss later. An alternative point of view is to regard the MIT experiment as a sort of macroscopic analog of the well-known Hanbury-Brown–Twiss experiment in quantum optics. To explain this, let us imagine for a moment that we have only two particles, and that we start with a "Fock" state, which in this case has the simple symmetrized form

$$\psi(\boldsymbol{r}_1\boldsymbol{r}_2) = 2^{-1/2}(\chi^0_{\mathrm{L}}(\boldsymbol{r}_1)\chi^0_{\mathrm{R}}(\boldsymbol{r}_2) + \chi^0_{\mathrm{R}}(\boldsymbol{r}_1)\chi^0_{\mathrm{L}}(\boldsymbol{r}_2)) \tag{4.5.21}$$

After the expansion process we will have similarly

$$\psi(\boldsymbol{r}_1\boldsymbol{r}_2 : t) = 2^{-1/2}(\chi_{\mathrm{L}}(\boldsymbol{r}_1 t)\chi_{\mathrm{R}}(\boldsymbol{r}_2 t) + \chi_{\mathrm{R}}(\boldsymbol{r}_1 t)\chi_{\mathrm{L}}(\boldsymbol{r}_2 t)) \tag{4.5.22}$$

Evaluation of the single-particle density on the wave function (4.5.22) yields the result (4.5.20), with $N_{\mathrm{L}} = N_{\mathrm{R}} = 1$, i.e. no interference pattern. However, if we evaluate the *two-particle* correlation function (density matrix) $\rho_2(\boldsymbol{r}\boldsymbol{r}' : t) \equiv |\Psi(\boldsymbol{r}, \boldsymbol{r}' : t)|^2$, we find

$$\rho_2(\boldsymbol{r}\boldsymbol{r}' : t) = f(\boldsymbol{r}\boldsymbol{r}'t) + \mathrm{Re}\,\rho_{\mathrm{L}}(\boldsymbol{r}\boldsymbol{r}'t)\rho_{\mathrm{R}}(\boldsymbol{r}'\boldsymbol{r}t) \tag{4.5.23}$$

where the function f is smooth, and where for this simple case the quantities $\rho_{\mathrm{L,R}}(\boldsymbol{r}\boldsymbol{r}'t)$ are given by

$$\rho_{\mathrm{L}}(\boldsymbol{r}\boldsymbol{r}' : t) \equiv \chi^*_{\mathrm{L}}(\boldsymbol{r}t)\chi_{\mathrm{L}}(\boldsymbol{r}'t) \tag{4.5.24}$$

and similarly for ρ_{R}. With $\chi_{\mathrm{L,R}}(\boldsymbol{r}t)$ given by an expression of the form (4.5.6), it is clear that $\rho_2(\boldsymbol{r}\boldsymbol{r}'t)$ shows oscillations as a function of the *relative* coordinate $z - z'$:

$$\rho_2(zz' : t) = f_2(zz' : t) + g_2(zz' : t)\cos\{2k(t)(z - z')\} \tag{4.5.25}$$

where f_2 and g_2 are smooth functions of both the COM and the relative variable. This is, in slightly different language, the original Hanbury-Brown–Twiss effect: light (particles) from two independent sources (in this case the L and R wells) cannot show interference effects in the one-particle distribution (intensity) but can and do show them in the two-particle correlations. Note that in obtaining this result the symmetrized nature of the wave function was essential. An intuitive way of understanding the effect is to note that if the particles are identical, there is no way of telling whether the one arriving at $\boldsymbol{r}$ came from source L and that arriving at $\boldsymbol{r}'$ from R, or vice versa: see Fig. 4.4.

Now let us apply a similar analysis to the many-body case. For the moment we will assume that the initial state is a generalized "Fock" state, i.e. that corresponds to a definite number of particles $N_{\mathrm{L}}, N_{\mathrm{R}}$ in each well, but will not necessarily assume that these are Bose-condensed. While it is of course entirely possible to do this in the "first-quantized" notation used above, use of the second-quantized formalism permits a very concise derivation:[32]

$$\rho_2(z, z' : t) \equiv \langle\rho(zt)\rho(z't)\rangle \equiv \langle\psi^\dagger(zt)\psi(zt)\psi^\dagger(z't)\psi(z't)\rangle \tag{4.5.26}$$

[32]For simplicity, I assume $N_{\mathrm{L}}, N_{\mathrm{R}}$ are much greater than 1 and thus neglect subtleties related to the proper ordering of the operators $\psi, \psi^\dagger$.

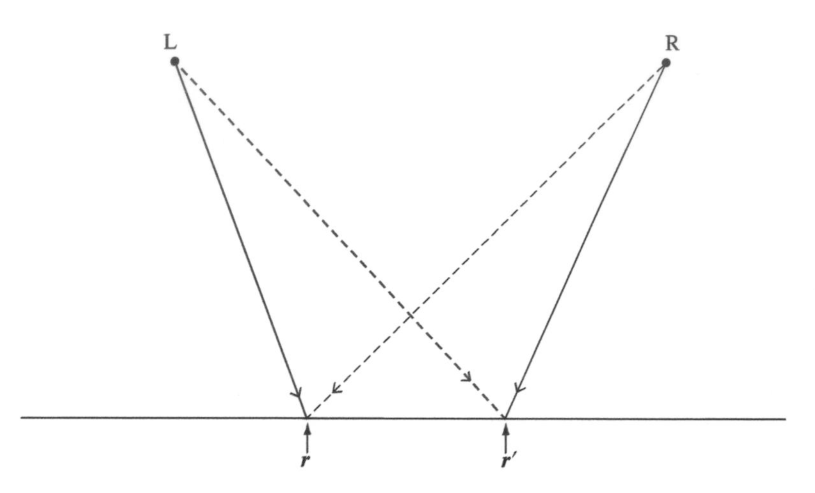

Fig. 4.4 The Hanbury-Brown–Twiss effect (interference between two mutually incoherent sources).

Writing the single-particle amplitude $\psi(zt)$ as the sum of L and R contributions,

$$\psi(zt) = \psi_{\rm L}(zt) + \psi_{\rm R}(zt)$$

and using the fact that, for any state of the "generalized Fock" type, operators which change $N_{\rm L} - N_{\rm R}$ automatically vanish, while operators of the form $\langle \hat{\Omega}_{\rm L} \hat{\Omega}_{\rm R} \rangle$ can be factorized as $\langle \hat{\Omega}_{\rm L} \rangle \langle \hat{\Omega}_{\rm R} \rangle$, we find that only six of the 16 terms survive, and can be collected to give (to order N^{-1})

$$\rho_2(z, z' : t) = \rho(zt)\rho(z't) + 2{\rm Re}\, \rho_{\rm L}(zz' : t)\rho_{\rm R}(z'z : t) \tag{4.5.27}$$

Equation (4.5.27) is valid for any generalized Fock state. If now we suppose that the original state has 100% BEC in each well separately, i.e. is given by (4.5.14), then evidently $\rho_{\rm L}(zz't)$ is simply given by

$$\begin{aligned} \rho_{\rm L}(zz't) &= N_{\rm L}\chi_{\rm L}^*(zt)\chi_{\rm L}(z't) \\ &\equiv N_{\rm L}\Phi(zz' : t) \exp ik(t)(z - z') \end{aligned} \tag{4.5.28}$$

and similarly for $\rho_{\rm R}(zz' : t)$ (with $k \to -k$). Substituting these forms into (4.5.27), we find a result of the general form (4.5.25) (but with the functions f and g now of course of order N^2).

What this result, (4.5.25) coupled to the valid statement (4.5.20), means is intuitively the following: We have no information on where a particular atom will arrive. However, once we know that one atom arrived at point z, we know from (4.5.25) that we have a much enhanced probability of finding a second atom at a point $z + n\lambda$ where n is integral (and correspondingly a much reduced probability of finding it at $z + (n + \frac{1}{2})\lambda$). This is exactly a description of the experimental result. A very elegant way of visualizing the situation, by imagining the atoms to arrive sequentially in real time, is given by Castin and Dalibard (1997).

One advantage of the "Hanburg-Brown–Twiss" approach is that it makes it clear that the necessary and sufficient condition to obtain an interference pattern on each shot is not BEC itself, but rather that the quantities $\rho_L(zz' : t)$ and $\rho_R(zz' : t)$ should be non-negligible (and possess interesting, i.e. oscillating, structure) over a range of $|z - z'|$ large compared to λ. While BEC and the consequent long-range order is a sufficient condition for this property to obtain, it is not obviously a necessary one; however, in practice it seems difficult to construct an experimentally realistic counterexample.

4.5.1 Interference in spin space

While experiments involving the coherence of different hyperfine species in principle form a nice complement to the real-space experiments, in practice the complications associated with spatial inhomogeneity, etc., make it more difficult to isolate the essential point. I will therefore discuss this type of experiment rather briefly and schematically, without going into all the details that are necessary in practice to interpret the raw data. (We already previewed some of this material in Chapter 2, Section 2.5.)

First, let us review the basic experimental set-up. We need an atomic species in which the electronic ground state contains (at least) two different hyperfine levels (e.g. the $F = 2, m_F = 2$ and $F = 1$, $m_F = -1$ states of ^{87}Rb) such that a pair of Raman lasers can drive transitions between them without introducing appreciable dissipation. We also need an imaging technique (e.g. optical absorption) which can resolve the density of these two species. We start with all atoms in one hyperfine state, say A. Then by a so-called "$\pi/2$" Raman laser pulse we drive the atoms into a linear superposition of states A and B, allow it to evolve for a given time t, then apply a second $\pi/2$ pulse and finally measure the density distribution (or failing that the total number) of A and B atoms separately. In the language of optics this is a "Ramsey-fringe" experiment.

It is actually rather more convenient to analyze the experiment in the language of the analogous NMR experiment, which would be a so-called "free precession" one. We can think of the A and B hyperfine states as the states of a fictitious spin $-\frac{1}{2}$ particle corresponding to projection $S_z = \pm\frac{1}{2}$ respectively; the role of the dc magnetic field is then played by the hyperfine splitting of the A and B states. In NMR language we then start from (say) the $S_z = -\frac{1}{2}$ state, apply an initial ($t = 0$) $\pi/2rf$ pulse which flips the spins parallel to the x-axis, allow a period t for the precession, and finally apply a second $\pi/2$ of pulse and measure S_z. If the two $\pi/2$ pulses are similarly polarized and are synchronized to a clock running at the nominal hyperfine splitting ("Larmor") frequency, then it is clear that in the ideal case one will measure all the atoms to have $S_z = +\frac{1}{2}$ (total $\langle S_z\rangle = \frac{1}{2}N$); any deviation from this result will be a measure of (deterministic or random) extra fluctuating "magnetic fields" acting in the system. In a real NMR experiment done on (e.g.) an ordinary liquid where there is no question of BEC, the effect of such fluctuating fields is to cause $\langle S_z\rangle$ to decay exponentially over the length t of the free-precession period, with a time constant which is the so-called transverse relaxation time T_2 (or T_2^*, if deterministic spatial inhomogeneity is involved): see e.g. Slichter (1992), Section 2.7.

Let us now analyze this experiment for a degenerate alkali Bose gas such as ^{87}Rb. I will assume that, as has been the case in most existing experiments, the initial ($S_z = -\frac{1}{2}$) state of the atoms is close to 100% Bose condensed, i.e. that all atoms are in the same orbital state $\chi_0(\boldsymbol{r})$ (in practice, the GP ground state in the relevant trap) as well as the same spin (hyperfine) state. Then, immediately following the first $\pi/2$ pulse (assumed short on the time scale of the orbital motion) the wave function will be, in an obvious notation.

$$\Psi_N(t = 0+) = \Pi_{i=1}^N \chi_0(r_i) \cdot 2^{-1/2}(|\uparrow\rangle_i + |\downarrow\rangle_i)$$

Now let us consider the period of free precession. I will make the initial assumption that throughout this whole period we maintain 100% BEC, i.e. that the orbital wave function is uniquely $\chi_0(\boldsymbol{r}t)$. If that is the case, then (trivially) the complete many-body wave function must be of the generic form

$$\Psi_N(t) = \Pi_{i=1}^N \chi_0(\boldsymbol{r}t)(\alpha(\boldsymbol{r}t)|\uparrow\rangle_i + \beta(\boldsymbol{r}t)|\downarrow\rangle_i)$$

where $|\alpha(\boldsymbol{r}t)|^2 + |\beta(\boldsymbol{r}t)|^2 \equiv 1$. Now in real life, the dynamics (which is in general nonlinear) may be sufficiently sensitive to (e.g.) small unknown variations in the trap potential that we cannot reliably calculate the functions $\alpha(\boldsymbol{r}t)$ and $\beta(\boldsymbol{r}t)$; nevertheless, this does not change the fact that all atoms in a given region must have *the same* values of α and β. In particular, if α and β are slowly varying and we average over a distance small compared to their characteristic scale of variation and containing N atoms, then the value of $\boldsymbol{S}^2$ we obtain must be $\frac{1}{4}N^2$ (compare the discussion in Chapter 2, Section 2.6). Note that this is very different from the situation in a normal system, in which the different orbital states $\chi_i(\boldsymbol{r}t)$ may be associated with quite different values $\alpha_i(\boldsymbol{r}t)$ and $\beta_i(\boldsymbol{r}t)$ of α and β.

What do we now expect to see in our final measurement of $\langle S_z\rangle$ following the second $\pi/2$ pulse? For simplicity let me assume (somewhat unrealistically) (a) that the motion conserves the total S_z (which is zero with fluctuations $\sim\sqrt{N}$) i.e. that any fluctuating fields are only in the z-direction, and (b) that such fields, while possibly unknown and even random functions of time from shot to shot, are constant in space over the volume of the system, so that the precession they induce (i.e. the relative phase of α and β) is a function only of time. Under these conditions the *magnitude* of the transverse (xy-plane) component of the total spin is always equal (to order N^{-1}) to $\frac{1}{2}N$, and the uncertainty about the random fields translates into uncertainty about its *direction* in the plane (recall that $\langle S_x\rangle = N \cdot \text{Re}\ \alpha^*\beta, \langle S_y\rangle = N \cdot \text{Im}\ \alpha^*\beta$). Thus, when we perform our second $\pi/2$ pulse and subsequently measure $\langle S_z\rangle$, we will in all cases find a result which is "of order" $N/2$; however, in the case when the "uncompensated" field, while possibly nonzero, is constant from run to run, then the final value of $\langle S_z\rangle$, while not in general equal to $N/2$, is reproducible, while if the field is random then the values or $\langle S_z\rangle$ will be randomly distributed between $+N/2$ and $-N/2$ and will average to zero. Unfortunately, the original experimental paper (Hall et al. 1998) reports only the average value of $\langle S_z\rangle$ as a function of t and the rms scatter, so it is not entirely clear how far the latter situation is being approached at long times.

While this "spin-space interference" experiment is at first sight closely analogous to the MIT real-space one, with S_z the analog of $N_\text{L} - N_\text{R}$ and the angle θ of $\boldsymbol{S}_\perp$ the analog of $\Delta\varphi$, there is actually an important difference: in this case the state of the system on which θ is effectively "measured" is a *mixture* of different θ-values, not a superposition[33] and correspondingly the fluctuation (before the second $\pi/2$ pulse) in S_z is of order $N^{-1/2}$ rather than rigorously zero as in a true "Fock" state. It would in principle be possible to do the exact analog of the MIT experiment, e.g. by conditioning the sample on an appropriate value of "S_z" ($= N_A - N_B$) as measured say by differential absorption imaging; but to my knowledge no such experiments have been done.

It is an interesting question whether the effect just discussed could be used to estimate the degree of BEC as a function of time. In the case explicitly treated, where the fluctuating field is constant in space over the volume of the system, it clearly cannot, because a normal system which starts with the same initial condition ($S_z = -\frac{1}{2}$) will behave in exactly the same way (the dependence on the orbital and spin degrees of freedom factors, and the former is irrelevant). However, consider a case where the fields depend on spatial position as well as on time, but in such a way that we can be reasonably sure that the characteristic scale of variation of α and β is large compared to the resolution of our measurement of $\langle S_z(\boldsymbol{r}t)\rangle$, while the (average) number of atoms N in that latter region is still $\gg 1$. Then, as argued above, if the system is 100% Bose-condensed the expectation value of $\boldsymbol{S}_\perp^2$ will automatically be $(N/2)^2$, irrespective of how complicated the evolution of the condensate wave function may have been. On the other hand, if the system is normal then the particles in different orbital states i, even though they all started in the same spin state ($S_z = -\frac{1}{2}$ before the first $\pi/2$ pulse) will have experienced different histories and thus accumulated different phases (angles) θ_i in the xy-plane. Provided that the difference between the various θ_i's is $\gtrsim 2\pi$, they will interfere destructively and the expectation value of $\langle S_\perp^2\rangle$ will be much less than $N/2$; consequently, on *all* runs the final measurement of S_z will yield a value $\ll N/2$. It is plausible that this situation can be reached while still meeting the stated condition on α and β, but this may depend on the details of the trap and the interactions, and at the time of writing the question is not completely resolved.

4.6 Optical lattices

A single linearly polarized laser gives rise to an electric field

$$\boldsymbol{E}(\boldsymbol{r}t) = \boldsymbol{E}_0 \cos(\boldsymbol{k}\cdot\boldsymbol{r} - \omega t) \tag{4.6.1}$$

and hence to an effective energy for a single atom which after averaging over the (very fast) optical-frequency oscillation has the form

$$V(\boldsymbol{r}) = \frac{1}{2}\alpha(\omega)\overline{(\boldsymbol{E}(\boldsymbol{r}t))^2} \tag{4.6.2}$$

when $\alpha(\omega)$ is the atomic polarizability. If ω is close to the frequency ω_0 of a particular atomic transition (and well separated from others) the expression (4.6.2) can be written in the form (cf. e.g. Leggett 2001, Section II.B.1)

$$V(\boldsymbol{r}) = (I(\boldsymbol{r})/I_0)\Gamma^2/\Delta \tag{4.6.3}$$

[33]This feature is not removed by treating the fluctuating "magnetic fields" as themselves quantum mechanical.

where Γ is the linewidth of the transition in question, $\Delta \equiv \omega - \omega_0$ is the "detuning", $I(r)(\propto \overline{(E(rt)^2)}$ is the laser intensity at point $\boldsymbol{r}$ and I_0 is the saturation intensity, typically $\sim$100 W/m^2. Consider now the effect of two laser beams, with wave vectors $\boldsymbol{k}_1$ and $\boldsymbol{k}_2$ and frequencies ω_1, ω_2, (where $|\omega_1 - \omega_2| \ll \omega_1, \omega_2$); for definiteness let us assume that they have identical polarizations and intensities I. Using a simple trigonometric identity, we can write the total electric field in the form

$$\boldsymbol{E}(\boldsymbol{r}t) = 2\boldsymbol{E}_0 \cos\left\{\frac{(\boldsymbol{k}_1 + \boldsymbol{k}_2)\cdot \boldsymbol{r}}{2} - \frac{(\omega_1 + \omega_2)t}{2}\right\} \times \cos\left(\frac{\boldsymbol{k}_1 - \boldsymbol{k}_2}{2}\cdot \boldsymbol{r} - \frac{(\omega_1 - \omega_2)t}{2}\right) \tag{4.6.4}$$

Provided that $|\omega_1 - \omega_2| \ll \Gamma$ (so that $\alpha(\omega)$ and Δ may be evaluated at (say) the frequency $(\omega_1 + \omega_2)/2$), the potential $V(\boldsymbol{r}t)$ arising from (4.6.4) after averaging over the "fast" oscillation at $(\omega_1 + \omega_2)/2$ is

$$V(\boldsymbol{r}t) = (2I/I_0)(\Gamma^2/\Delta)(1 + \cos(\Delta\boldsymbol{k}\cdot \boldsymbol{r} - \Delta\omega \cdot t)) \tag{4.6.5}$$

where $\Delta\boldsymbol{k} \equiv \boldsymbol{k}_1 - \boldsymbol{k}_2, \Delta\omega \equiv \omega_1 - \omega_2$. The result (4.6.5) is the basis of the "Bragg spectroscopy" technique, in which two lasers of slightly different $\boldsymbol{k}$ and ω are used to transfer quanta of momentum $\hbar\boldsymbol{k}$ and energy $\hbar\omega$ to the system, thereby playing much the same role as neutrons do with respect to liquid ^{4}He: see, e.g. Pitaevskii and Stringari (2003), Section 12.9.

For our present purposes it is convenient to concentrate on the case where $\boldsymbol{k}_1 = -\boldsymbol{k}_2$, so that $\Delta\omega = 0$ (two counterpropagating laser beams). Then we see that the potential (4.6.5) is time-independent and periodic with a wavelength $\lambda/2$ which is half that of the laser (thus typically a few hundred nm). It is clear that by employing two or three pairs of lasers (with slightly different frequencies for each pair, so that the cross-terms average to zero) one can create potentials which are periodic in two or three dimensions respectively; for example, with three pairs of equal intensity and (almost) equal frequencies we get a potential of the form (up to an addition constant).

$$V(\boldsymbol{r}) = \tfrac{1}{2}V_0(\cos 2kx + \cos 2ky + \cos 2kz) \quad (k \equiv \omega/c) \tag{4.6.6}$$

All sorts of variations on this basic theme are possible; for example, by using more lasers and tuning them sufficiently finely one can simultaneously create different potentials for different hyperfine atomic species, and even make their relative displacement time-dependent (cf. Mandel et al. 2003). While such possibilities may well have important applications, e.g. in quantum computing, for our purposes here it will generally be enough to consider either simple three-dimensional potentials of the form (4.6.6), or the one-dimensional versions formed by keeping only the first term. It is important to bear in mind that in most real-life experiments a potential of the form (4.6.6) is superposed on the much more slowly varying background potential of a magnetic or other trap (this is necessary simply to avoid losing atoms). It is conventional in the literature to specify the "depth" of the periodic potential (the constant V_0 in (4.6.6)) in units of the "recoil energy" defined by

$$E_{\mathrm{R}} \equiv \hbar^2 k_{\mathrm{R}}^2/2M \tag{4.6.7}$$

when M is the mass of the atomic species in question, and $k_{\mathrm{R}} \equiv \omega_0/c$.

Consider now a single atom moving in the potential (4.6.6). The qualitative nature of the motion will depend on the ratio of the depth V_0 of the wells relative to the zero-point energy which we would calculate on the assumption that the well is harmonic with the curvature which the actual well has near its minimum. Since from (4.6.6) the (one-dimensional) harmonic frequency $(V''/M)^{1/2}$ is $(2k^2V_0/M)^{1/2}$, the ratio $V_0/(\frac{1}{2}\hbar\omega)$ is just $(V_0/E_{\mathrm{R}})^{1/2}$. Thus, when $V_0 \ll E_{\mathrm{R}}$ the periodic potential is a small perturbation on the free motion of the atom; by contrast, in the limit $V_0 \gg E_{\mathrm{R}}$ the lowest energy eigenstates are to a first approximation Wannier states $\psi_0(\boldsymbol{r} - \boldsymbol{R}_i)$ localized in the individual walls, with a spread a_{zp} which is of order $(\hbar/M\omega_0)^{1/2}/(\lambda_{\text{laser}}/2) = \pi^{-1}(E_{\mathrm{R}}/V_0)^{1/4}$ relative to the lattice spacing. Because of the exact degeneracy of the states in the different wells, we would expect the true eigenstates to be "tight-binding" Bloch states of the form (do not confuse $\boldsymbol{q}$ with the laser wave vector $\boldsymbol{k}$!)

$$\psi_q(\boldsymbol{r}) = \sum_i \exp i\boldsymbol{q} \cdot \boldsymbol{R}_i \psi_0(\boldsymbol{r} - \boldsymbol{R}_i) \tag{4.6.8}$$

(with i labeling the various sites (well minima)) with a dispersion relation

$$E(\boldsymbol{q}) = -2t(\cos q_x + \cos q_y + \cos q_z) \tag{4.6.9}$$

and thus a bandwidth $12t$, where t is the tunneling matrix element between neighboring wells, which in this "extreme tight-binding" limit is given by the WKB expression

$$t = \text{const.}\omega_0 \exp(-\alpha) \tag{4.6.10}$$

$$\alpha \equiv \int_0^{\pi/h} \sqrt{2M(V(x) - V(0))}dx/\hbar = (8V_0/E_{\mathrm{R}})^{1/2} \tag{4.6.11}$$

The expression (4.6.10) is actually not a bad approximation for values of V_0/E_{R} as small as 1.5. Let us now consider the mutual interaction of two atoms in the potential (4.6.6). I shall assume that the s-wave scattering length is small compared not only to the laser wavelength but to the zero-point length a_{zp}; this has almost always been satisfied in experiments to date. I will also assume that $V_0/E_{\mathrm{R}} \gtrsim 1$. Then two atoms situated in Wannier states in different wells will be described by wave functions whose overlap is exponentially small, and thus will effectively experience no interaction. On the other hand, two atoms in the same well will experience a nonzero interaction. Provided that the condition $a_{\mathrm{s}} \ll a_{zp}$ is well satisfied, this will not be sufficient to mix in higher Wannier states appreciably, and the interaction energy U may be calculated simply by evaluating the expression (4.2.8) over a two-body wave function which is just the product of the Wannier ground states, i.e. $\Psi(\boldsymbol{r}_1, \boldsymbol{r}_2) = \psi_0(\boldsymbol{r}_1 = \boldsymbol{R}_i)\psi_0(\boldsymbol{r}_2 - \boldsymbol{R}_i)$. Clearly the result is of the form

$$\begin{aligned} U &= \frac{4\pi\hbar^2}{M} a_{\mathrm{s}} \int |\psi(\boldsymbol{r})|^4 d\boldsymbol{r} \\ &= (2\pi)^{-3/2} \frac{4\pi\hbar^2}{M} a_{\mathrm{s}} a_{zp}^{-3} \end{aligned} \tag{4.6.12}$$

which under the condition $a_s \ll a_{zp}$ is small compared to $\hbar\omega_0 \equiv \hbar^2/Ma_{zp}^2$, as stated. Note that U scales with the quantity V_0/E_R only as $(V_0/E_R)^{3/4}$, in contrast to the exponential dependence on this quantity of the single-particle tunneling matrix element t, Eqn. (4.6.10).

In the light of the above results, we see that to the extent that we can neglect occupation of higher Wannier states (and any harmonic background) the Hamiltonian for N *bosonic* atoms in the optical lattice consists of just two terms: the single-particle tunneling between neighboring walls, with matrix element t, and the interaction between atoms in the same well, given by U. In second-quantized notation we have

$$\hat{H} = -\frac{t}{2}\sum_{i,j=n.n.}(a_i^+ a_j + h.c.) + \frac{U}{2}\sum_i n_i(n_i - 1) \qquad (4.6.13)$$

where i, j label different wells and "n.n." stands for "nearest neighbor". When supplemented by a specification of the total particle number N, or alternatively of the chemical potential μ, the Hamiltonian (4.6.13) defines the so-called *Bose-Hubbard model* - a model which has been the subject of appreciable research[34] in a condensed-matter context, see especially Fisher et al. (1989) and Sheshadri et al. (1993). The great advantage of using optical lattices to implement this model is that by varying the laser intensity we can adjust the ratio U/t to any desired value, and moreover do so virtually instantaneously[35] on the relevant time scales, which are of order $U/\hbar, t/\hbar$.

If we start by considering the case when N is exactly equal to the number of wells, and imagine for the moment that the tight-binding description implicit in (4.6.13) remains valid for arbitrarily large as well as arbitrarily small values of t/U, we can anticipate the behavior in these two limits: For very large values of t/U the ground state of the system will be a state in which we have (almost) 100% BEC in the lowest ($\boldsymbol{k} = 0$) Bloch state, and the interaction term will then scatter two condensate particles weakly into Bloch states ($\boldsymbol{k}, -\boldsymbol{k}$), in analogy to what happens in free space. On the other hand, in the limit $t/U \to 0$ it is intuitively obvious that it will be advantageous to allow each well to be occupied by exactly one atom; by analogy with the corresponding state of electrons in solids, this state is usually called the "Mott-insulator" state. In this state all eigenvalues of the single-particle density matrix are 1, so there is obviously no BEC. It is then plausible that there is some critical value U_c of U in units[36] of t at which we lose the property of BEC. Actually, there has been a considerable amount of theoretical work on this model over the last 15 years (see especially Fisher et al. (1989) and Sheshadri et al. (1993)), and the consensus is that the "superfluid-to-Mott-insulator" transition should occur at a U_c of $\sim$35 and should be second order (something which is not totally obvious a priori).

The above analysis is for the special case when the "filling fraction" f – that is the ratio of the total number of atoms to the total number of wells, is exactly 1 (a similar analysis can be carried out for any integer value). If f is nonintegral and we assume

[34] Though much less than the original Hubbard model, which describes fermions and has very recently also been implemented in optical lattices (see Kohl et al. 2005).

[35] Actually, one usually wants to adjust the intensity and a time scale slow compared to ω_0^{-1} so as to avoid possible excitation of higher Wannier states.

[36] Beware of different definitions of t in the literature.

the system is uniform, then it is rather obvious we must always have a finite degree of BEC. It might therefore be thought that it would be almost impossible to observe the MI phase in any real-life experiment, particularly since in the presence of the magnetic background trap the "number of wells" is not really well-defined. However, what the theoretical work actually finds is that if we plot the phase diagram in a plane whose axes correspond to the *mean* value of f and to U_2, a finite region is occupied by a MI phase in which phase separation has occurred (so that different spatial regions have different integral values of f).

Given the theoretical interest of this transition, it is very tempting to simulate it in an optical lattice. But this raises the question: how do we tell whether the system is in the BEC ("superfluid") or Mott-insulating state? It is clear that a simple in situ Bragg diffraction experiment, which simply measures the nonzero Fourier components of the single-particle density distribution $\langle\rho(\boldsymbol{r})\rangle$, will not distinguish between those states: one needs a handle on the *momentum* distribution. The trick is to remove the optical lattice (and the magnetic background), allow the cloud to expand freely and then to do a straight absorption imaging of its density (*not* Bragg diffraction!). If the ramp-down of the potential is sufficiently fast, we thereby effectively obtain a snapshot of the velocity (or momentum) or wave vector $\boldsymbol{k}$ distribution of the original state of the gas in the optical lattice. If this is characterized by a finite degree of BEC, then we expect a delta-function peak in the distribution at $\boldsymbol{k} = 0$, with subsidiary peaks at the reciprocal lattice vectors $\boldsymbol{G}$, the relative weight in those being determined by the ratio $a_{zp}/\lambda_{\text{laser}}$. Thus, the signature of BEC should be a set of regularly spaced peaks in the absorption images, at the origin and at positions[37] $\boldsymbol{r} = \hbar\boldsymbol{G}t/M$ where t is the time allowed for expansion before imaging: by contrast, the Mott-insulator state (or indeed any state in which in effect all states in the first Bloch band are roughly equally occupied) should give rise to a fuzzy image with no marked peaks.

Experiments along the above lines in a fully 3D optical lattice have been done by Greiner et al. (2002a) (experiments which are similar in general concept but in a 2D lattice had been done somewhat earlier by Orzel et al. 2001). Figure 4.5 reproduced from their paper, shows the absorption images obtained after a "fast" ramp-down ($\ll 1$ msec) and a 15 msec expansion for different values of V_0/E_{R} in the original state: one sees that as V_0/E_{R} is increased, initially the $\boldsymbol{k} \neq 0$ maxima grows, reflecting the decreased value of a_{zp}/λ, but at higher values of V_0/E_{R} both they and the central peak become blurred, and eventually disappear for $V_0 \sim 14E_{\text{R}}$. This corresponds to a U/t value of approximately 36, which is in good agreement with the theoretically predicted value of 35 for the superfluid-MI transition.

A very interesting further feature of these experiments is the dependence of the final pattern observed on the ramp-down time, that is, the time over which the optical depth of the optical lattice potential is decreased from its original (large) value to a value low enough that the BEC phase is the equilibrium one. Suppose we start from a value of U well above U_{c}, so that the original state is the MI one. If then we ramp down very fast, say over 0.1 msec, then the above results are consistent with the assumption that we are simply taking a "snapshot" of the momentum distribution in this phase. However, as we increase the ramp-down time t, we find that the peaks

[37]In practice one of course only observes the components of $\boldsymbol{r}$ perpendicular to the imaging beam.

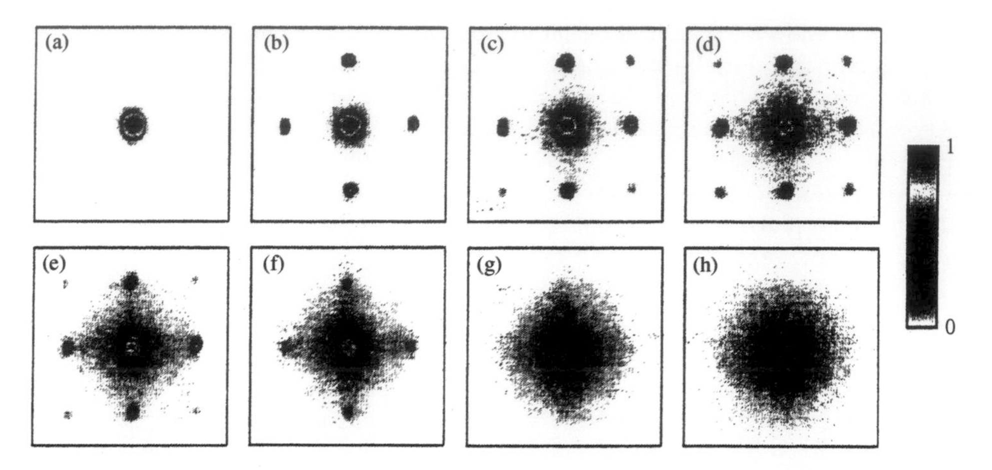

Fig. 4.5 Absorption images of multiple matter wave interference patterns. These were obtained after suddenly releasing the atoms from an optical lattice potential with different potential depths V_0 after a time of flight of 15 ms. Values of V_0 were: **a**, 0 E_r; **b**, 3 E_r; **c**, 7 E_r; **d**, 10 E_r; **e**, 13 E_r; **f**, 14 E_r; **g**, 16 E_r; and **h**, 20 E_r. (Reprinted with permission from Macmillan Publishers Ltd from M. Greiner et al., published in *Nature* vol. 415, p.39, Copyright (2002).)

in the absorption images are rapidly restored (see Fig. 3 of Greiner et al. 2002a), so that by the time $\tau = 4$ msec we recover essentially the pattern we would have got by starting well below U_c. It is very significant that the time τ is comparable to $\hbar/t$, the characteristic time taken to tunnel (in the final state) between neighboring wells, and much shorter than the time which a single atom would need to explore the whole array. It is equally interesting that when the authors create what they call a "phase-incoherent" state, by applying a magnetic field gradient during the ramp-up time of the potential, the time to recover the characteristic BEC absorption-image position is much larger, in fact more than 40 msec. While the difference is probably related to the fact that in the MI case one stays in the many-body ground state, whereas in the case of the "phase-incoherent" state one is probably creating rather highly excited states within the first Brillouin zone, the detailed kinetics of these processes is not to my knowledge well understood at present.

I should note that the account above does not cover all the interesting aspects of the experimental data obtained by the München group (see also Greiner et al. 2002b) but only those most directly related to BEC. Note that I say "BEC" not "superfluidity" because it is only the former, not the latter, which is demonstrated in these experiments. Other experiments in optical lattices, to be discussed in the next section, relate more directly to the superfluid properties.

4.7 Signatures of superfluidity in the BEC alkali gases

As we saw in Chapter 3, in the case of liquid ^{4}He the phenomenon of superfluidity is very easy to detect, while the occurrence of the BEC which is believed to underlie it has proved extremely difficult to see in any direct way. In the alkali gases the situation is reversed: the occurrence of BEC manifests itself at the level of direct inspection, notably in the spectacular peak which arises below T_c in the density of a harmonically trapped gas, while the phenomenon of superfluidity has proved more elusive. Of course, it is to some extent a matter of taste which of the consequences of BEC one chooses to class under the heading of "superfluidity" (e.g. does the Josephson effect count?), but I believe most people would include nonclassical rotational inertia, persistent currents, vortices and reduced friction on an impurity moving in the system, and it is those which I will address in this section.[38]

4.7.1 Rotational properties and vortices

Perhaps the most fundamental defining property of a "superfluid" is the phenomenon of nonclassical rotational inertia (Hess–Fairbank effect): when the "container" rotates at a sufficiently small angular velocity ω, the angular momentum L of the system is less than $I_{\text{cl}}\omega$ where $I_{\text{cl}} \equiv \int \rho(\boldsymbol{r}) r_\perp^2 \, d\boldsymbol{r} \equiv M\langle R^2 \rangle$ is the classical moment of inertia. It is relatively easy to verify this property in a dense liquid such as ^{4}He, e.g. (as in the original Hess–Fairbank experiment) by measuring the angular momentum transferred to and from the container, but this is less feasible in the case of the alkali gases where the "container" is a magnetic or laser trap. So it is a nontrivial question how to detect the value of the angular momentum and/or the presence of vortices (as we shall see, these two questions are not quite the same).

A very elegant technique to detect the angular momentum of a harmonically trapped gas makes use of the precession of certain collective modes. Imagine a cloud of atoms (not necessarily Bose-condensed) trapped in a potential which is spherically symmetric around the z-axis. If the cloud is at rest and we are in the hydrodynamic limit, then one of the normal modes is a "quadrupole" ("$m = 2$") deformation, in which an elongation along (say) the x-axis is accompanied by a compression along the z-axis. Evidently, by symmetry this mode must be degenerate with a similar one in which the elongation is along the y-axis (and the compression still along z); thus, if we elongate the system along an arbitrary axis in the xy-plane and release it, it will continue to oscillate along this same axis. Now suppose the cloud is rotating with angular momentum L. In this case, as is shown by Zambelli and Stringari (1998), the normal modes become linear combinations of x-and y-deformations with a $\pi/2$ phase difference, i.e. they are "circularly polarized" ("+" and "−"), with a frequency splitting

$$\Delta\omega \equiv \omega_+ - \omega_- = 2\langle L_z \rangle / I_{\text{cl}} \tag{4.7.1}$$

Thus an initial linear deformation is a linear combination of normal modes, and the plane of the deformation will now precess as a function of time at a rate which is $\frac{1}{4}$ of the $\Delta\omega$ given by (4.7.1). Note that this result, while it assumes hydrodynamic conditions, is independent of the presence or not of BEC. If the trap is slightly anisotropic in the xy-plane (as it is for some of the experiments, see below) the result (4.7.1) should still work provided that $\Delta\omega$ is large compared to the difference in $|\omega_x - \omega_y|$ of the original linear harmonic frequencies.

This technique was used by Chevy et al. (2000) to measure the value of $\langle L_z \rangle$ in a cloud of trapped ^{87}Rb gas as a function of the angular velocity Ω of the trap around

[38] It might well be argued that the effect of BEC on the collective excitations of the system should be included. However, I have deliberately chosen not to discuss this topic in this book.

the z-axis (but cf. below). They started with a cylindrically symmetric potential with radial frequency $\omega_\perp$, and provided the rotation of the "container" by using an appropriately focused laser beam to provide an extra potential which up to a symmetric time-independent term is of the form

$$V(\boldsymbol{r}t) = \epsilon \cdot \tfrac{1}{2} M\omega_\perp^2 (\boldsymbol{r} \cdot \hat{\boldsymbol{n}}(t))^2 \tag{4.7.2}$$

where the unit vector $\hat{\boldsymbol{n}}(t)$ rotates in the xy-plane at angular frequency Ω. This "stirring" phase lasted about a second, and after stopping it they excited the quadrupole mode and looked for its precession rate, if any. It should be noted that this is not the exact analog of the Hess–Fairbank experiment in helium, in the sense that the angular momentum $\langle L_z \rangle$ is not actually measured while the rotation of the "container" is taking place but only after is has stopped; however, it is plausible in the extreme that had it been possible to measure $\langle L_z \rangle$ during the rotation itself, the value obtained would have been the same.[39]

The results found by Chevy et al. are clear-cut: For values of Ω less than about 116 Hz the plane of quadrupole mode does not precess, indicating that $\langle L_z \rangle = 0$. For frequencies somewhat higher than this, a precession is found at a rate consistent with a value of $\langle L_z \rangle$ equal to $N\hbar$. Since this is exactly the value of $\langle L_z \rangle$ which would be present by a state with a single vortex line centered at the origin, this result strongly supports the hypothesis that for frequencies $\gtrsim$118 Hz the rotation process nucleates such a single vortex line. For frequencies higher than about 125 Hz the behavior is more complicated: it rises to a maximum at about 132 Hz and then falls to zero at around 145 Hz, and there is evidence for the onset of turbulent behavior. While not all details are understood, this pattern is not entirely surprising, since one is approaching the value of the original harmonic frequency (170–210 Hz). At any rate the experimental results seem entirely consistent with the hypothesis that the phenomenon of NCRI is indeed occurring in this system, and that the critical velocity for introduction of a single vortex line is $\sim$117 Hz. This value is about 50% greater than predicted by calculations done on the GP approximation, so it may be a "superheating" critical frequency rather than the true thermodynamic equilibrium one.

If one wishes to "see" vortices themselves rather than the angular momentum associated with them, the most straightforward method is simply absorption imaging: if this is done along the direction of the vortex line axis, the vortex core should show up as a region of reduced absorption. In this way images have been obtained not only of single vortices, but of large arrays (obtained by relatively fast rotation of the trap): they occur, as theoretically predicted, in a regular pattern, see e.g. Abo-Shaeer et al. (2001).

The absorption imaging technique is of course sensitive only to the density of the cloud as a function of position, not to its velocity. A nice way to verify that the latter does indeed have the circulation around the vortex core given by (3.5.2) with κ taking the quantized value h/m is to use an interferometric technique which is a generalization of that of the MIT interference experiment discussed in Section 4.4

[39]With the value of $\epsilon(= 0.1)$ used in the experiment, the condition for applicability of (4.7.1) is not satisfied during the rotation process. However, in principle there seems no reason why one should not use smaller values of ϵ so that is it satisfied.

(Inouye et al., 2001). As we saw in that section, the position of the interference fringes reflects the phase difference $\Delta\varphi$ (possibly indeterminate before the actual "measurement"!) of the condensates in the two wells. If neither well initially contains a vortex, then the phase within each well separately should be constant as a function of position, and the interference fringes should be (approximately)[40] straight lines in the xy-plane. Suppose now we introduce a single vortex line parallel to the y-axis in the L well, with core at (x_0, z_0). Then the condensate phase in this well will have the form

$$\varphi(x, z) = \theta(+\text{const.})$$
$$\theta \equiv \tan^{-1}\{(x - x_0)/(z - z_0)\} \tag{4.7.3}$$

and in particular, for large $|x|$ will be approximately constant, but different by π for positive and negative x. Hence, when we allow the condensates to expand and interfere and take absorption images, the pattern at large positive x should be displaced by half a period relative to that at negative x – there should be a "dislocation" at $x = x_0$. This effect is clearly seen in the data of Inouye et al. (2001), see their Fig. 4.3.

A rather different kind of phase interference technique was used for detection in the first experiment to produce vortices in a BEC alkali gas (Matthews et al. 1999a). These authors used two different hyperfine states of $^{87}\text{Rb}, F = 1, m_F = -1$ ("$|1\rangle$") and $F = 2, m_F = +1$ ("$|2\rangle$"), driving transitions between them with a Raman laser; their imaging technique was sensitive only to atoms in the $|2\rangle$ state. They produced a vortex in the $|2\rangle$ component in a region where the density of $|1\rangle$ is nonzero (the method by which this was done is itself of interest and will be discussed below) and, to detect it, shone on a uniform Raman pulse of length π, so as to convert the $|2\rangle$ component completely into $|1\rangle$ and vice versa, and imaged the system (that is, the density of $|2\rangle$) as a function of position both at the final time and half-way through the pulse. To analyze this problem, let us imagine for the sake of simplicity that at a given point $\boldsymbol{r}$ the density of $|1\rangle$ and $|2\rangle$ in the initial state is the same, so that the initial condensate wave function has the form[41]

$$\psi(\boldsymbol{r}) = 2^{-1/2}(|1\rangle + \exp i\theta(\boldsymbol{r})|2\rangle) \tag{4.7.4}$$

Since the effect of a $\pi/2$ pulse is that $|1\rangle \to 2^{-1/2}(|1\rangle + |2\rangle)$ and $|2\rangle \to 2^{1/2}(|1\rangle - |2\rangle)$, we find that half-way through the pulse the state is

$$\psi(r) = 2^{-1}\{[1 + \exp i\theta(r)]|1\rangle + [1 - \exp i\theta(r)]|2\rangle\} \tag{4.7.5}$$

so the density of the $|2\rangle$ component, which is what we image, is $\sin^2(\theta/2)$. Hence, if the initial state corresponds to a vortex ($\theta = \varphi + \delta$, where φ is the angle around the axis of the vortex), we expect an (absolute) image intensity proportional to $\sin^2 \frac{1}{2}(\varphi + \delta)$. A similar and only slightly more complicated analysis applied to the general case when the initial densities of $|1\rangle$ and $|2\rangle$ are unequal shows that quite generally the

[40]They will not actually be quite straight because of uninteresting effects associated with the expansion process.

[41]It might well be asked whether it is reasonable to take the initial state to be "GP" rather than "Fock" type (i.e. rather than saying we have N_1 atoms in $|1\rangle$ and N_2 in $|2\rangle$ without phase coherence, cf. Section 4.4.). However even if we do so, we find that just as in the MIT experiment the measurement "selects" a definite *overall* relative phase.

"normalized density difference" (see Matthews et al. (1999a), footnote 14) should be simply equal to $\cos\varphi$ and hence, for an initial vector state, to $\cos(\varphi+\delta)$; this is exactly what is seen in the experiments, see Fig. 4.2(a) of Matthews et al. (1999a).

4.7.2 Persistent currents

To the best of my knowledge, no experiments have been done up to now on a single-species alkali gas in an annular geometry, and thus it has not been possible to verify that just as in liquid ^{4}He a circulating current in such a geometry would be to one degree or another metastable. On the other hand, experiments such as those of Chevy et al. (2000) discussed in (4.7.1) have verified that the single-vortex state created by stirring the trap persist for times of the order of a minute after the stirring has been switched off, when they are manifestly not the thermodynamic equilibrium state (Madison et al. 2000); to this extent one could say that the circulating currents which constitute the vortex are 'persistent". The degree of metastability is much less than one would expect in an annular geometry, because the vortex core can move out of the trap altogether – a process for which there is no analog in the annular geometry, where there is of course no question of a "core". The detailed kinetics of the process by which the vortices migrate in this way is still an ongoing topic of research.

The one experiment in which something like an annular geometry was realized was that of Matthews et al. (1999a), since then the $|2\rangle$ component, once formed, tends to form a ring surrounding the $|1\rangle$ component (this happens simply because the effective self-interaction of $|2\rangle$ is more repulsive than that of $|1\rangle$ and it therefore prefers to occupy a region of the trap corresponding to lower density). Actually, in the context of persistent currents this experiment has a special significance, since it in effect shows that a two-component BEC does *not* have the same kind of topological stability as a single-component one such as ^{4}He! This was already briefly mentioned in Chapter 2, Section 2.5, and I now flesh out the argument somewhat (cf. Matthews et al. 1999b, Ho 1982). Consider a system with two hyperfine states which can be represented formally as the $S_z = \pm\frac{1}{2}$ states $|\uparrow\rangle$ and $|\downarrow\rangle$ of a spin $-\frac{1}{2}$ particle, and suppose that in this representation the Hamiltonian is invariant under simultaneous rotation of all the spins (that is, it is invariant under the group $SU(2)$). Suppose that we have somehow created a simple vortex in the $|\uparrow\rangle$ state, with the $|\downarrow\rangle$ state being initially unoccupied. That is, the wave function has the form

$$\psi(\boldsymbol{r}) = e^{i\varphi}|\uparrow\rangle \quad (|\psi(\boldsymbol{r})|^2 = 1) \tag{4.7.6}$$

where φ is the angle of $\boldsymbol{r}$ with respect to the vortex core (taken along the z-axis). Then it is easily verified that by a θ-dependent rotation of the "spin", around an axis in the xy-plane whose angle with (say) the x-axis is θ, we can continuously convert (4.7.6) into the *vortex-free* state

$$\psi(r) = |\downarrow\rangle \quad (|\psi(\boldsymbol{r})|^2 = 1) \tag{4.7.7}$$

and the condition $|\psi(\boldsymbol{r})|^2 = 1$ will hold not just in the initial and final states but everywhere along the intermediate trajectory. Since in the case of a simple complex scalar order parameter such as that of ^{4}He it was precisely the necessity for $|\psi(\boldsymbol{r})|^2$ to deviate from 1 which provided the free energy barrier against relaxation, we see that at least in the ideal case of exact $SU(2)$ invariance, the two-component system should have no such "topological" barrier against the decay of circulating current. While in the original experiment (Matthews et al. 1999a) the eventual decay of the vortex was probably not due to this process, it is precisely the inverse of this which was used in creating it starting from a uniform (vortex-free) state of component $|1\rangle$ ("$|\downarrow\rangle$" in the above notation); and while no experiment in an annular geometry has been done, further experiments by the same group (Matthews et al. 1999b) saw essentially this "untwisting" effect in a simply connected geometry. It should also be noted that in a more recent experiment (Leanhardt et al. 2002) a vortex was created in a single hyperfine state by a technique which is conceptually closely similar to the above one: in this case the relevant inversion of the "spin" was effected by adiabatically changing the magnetic field so that its direction, and hence that of the spin, follows the θ-dependent rotation from up to down specified above. The difference is that in this case the "spin" is not the fictitious spin associated with any two-level system, but the real angular momentum of a particular atomic hyperfine state; this is the m_F-value of the state and is in general integral (1 or 2 in the actual experiment) rather than $\frac{1}{2}$, and as a result, according to arguments originally given by Ho (1982), the final circulation of the vortex thus created is not 1 but $2m_F$, i.e. 2 or 4.

4.7.3 Reduced friction on impurities

As we saw in Chapter 3, according to a famous argument of Landau an object moving in a gas or liquid whose excitation spectrum is $\epsilon(p)$ can create excitations only if the velocity exceeds a critical velocity v_c given by

$$v_c \equiv \min_p \epsilon(p)/p \tag{4.7.8}$$

For velocities less than this "Landau critical velocity" the object can lose energy and momentum only by colliding with existing excitations, and if the latter are few enough then the mobility will show a sharp drop when the velocity reaches v_c. An ordinary gas or liquid has a zero value of the Landau critical velocity, while for liquid ^{4}He the latter is believed to be determined by the roton minimum and to have the large value of $\sim$60 m/sec. A dilute Bose alkali gas should have no roton minimum in its spectrum, which should be of the sound-wave form $\epsilon(p) = c_s p$ for $p \lesssim \hbar/\xi$ (see Section 4.4), switching over to the free-particle form $\epsilon(p) = p^2/2m$ for $p \gtrsim \hbar/\xi$. Thus we expect that for such a gas v_c should be simply the sound velocity c_s.

In liquid ^{4}He, the "objects" which have been used to demonstrate the reduced friction for $v < v_c$ in the superfluid phase are physical impurities, usually ions, whose dimensions are comparable to the healing length of the liquid. In the alkali gases, for various reasons, experiments of this type are difficult and have not so far been done, but a conceptually similar experiment has been done (Chikkatur et al. 2000) by using a Raman laser to transfer some fraction of the atoms into a different hyperfine state, in which they do not experience the magnetic force felt by the original species and fall under gravity; by tracking their progress as a function of time one can infer their mobility and thus the friction on them. In those experiments a fairly sharp transition between a low- and a high-friction regime was seen, at a velocity which was approximately equal to the local speed of sound. In this experiment the dimensions of the "impurity" were obviously much less than the healing length of the condensate. In a related series of experiments, a density "hole" was created in the gas by shining on it a well-focused blue-detuned laser, and this hole was then moved around. The dimensions of the hole were large compared to the healing length. Evidently in this case one cannot directly measure the frictional force acting on the hole, but we can infer it as follows (Onofrio et al. 2000): the pressure difference across the hole should be equal to $n(r)\Delta\mu \cong gn(r)\Delta n$, where Δn is the difference in density immediately ahead of and behind the hole, which can be measured by an appropriate imaging technique. The measurements showed that Δn is close to zero until the velocity of the probe reaches a value equal to about 0.1 of the speed of sound, after which it increases sharply. The heating of the sample by the "stirring" laser beam is also very weak at low velocities, though the critical velocity for the onset of substantial heating appears to be somewhat higher than $0.1c_s$ (Raman et al. 1999).

While the macroscopic nature of the "impurity", and the resulting substantial direction of the flow field of the condensate, are expected to decrease the critical velocity (measured with respect to the condensate at infinity) by some factor, it is difficult to explain a factor of 10 this way, and it seems likely that the process which initiates friction is connected with the creation of vortices; such effects are well-established for ^{4}He flowing past relatively macroscopic obstructions. At any rate, all the above data seem consistent with the idea that at sufficiently low velocity the condensate exerts no friction.

4.7.4 "Superflow" in an optical lattice

I would like to mention briefly a series of experiments (Burger et al. 2001, Cataliotti et al. 2001) which, while their detailed interpretation is still somewhat controversial, show at least that the flow of the condensate in an optical lattice is qualitatively different from that of an uncondensed gas (or of the thermal cloud in the BEC phase[42]). Those experiments were conducted in a one-dimensional optical lattice superimposed on a harmonic magnetic trap; an initial "kick" was given to the system by changing the zero of the trap, and the subsequent motion was examined by optical imaging of the density (after expansion). If this experiment is done in the absence of the optical lattice, one obviously expects the cloud, whether condensed, thermal or both, to excite simple harmonic motion at the trap frequency $\omega_0 \equiv (V''/m)^{1/2}$, and indeed this is found. If a fairly weak optical lattice ($V_0/E_R \lesssim 1.5$) is superimposed on the magnetic background, and the "kick" is small, so that the subsequent velocity is low, then the motion of the condensate is still harmonic, with a reduced frequency which is consistent with the replacement of m by the effective mass m^* in the periodic lattice. However, above T_c the behavior is quite different: the cloud executes no harmonic motion, but remains at its original equilibrium position. This is also true of the thermal cloud in the BEC phase, so that the two components may end up well separated. If the strength of the kick is small, the motion of the condensate also eventually gets locked; the data are consistent with this happening when the velocity reaches a critical velocity approximately equivalent to the Landau value.

One way of looking at this experiment (Cataliotti et al. 2001) is to regard the barriers between neighboring wells as equivalent to Josephson junctions: the different

[42]Since we are dealing with a very dilute gas, it is unnecessary in the present context to distinguish between the "thermal (noncondensed) cloud" and the "normal component"; cf. Section 4.4.

behavior of the condensate and the thermal cloud at low velocities then reflects the fact that the former can tunnel coherently by the usual Josephson mechanism, while the latter sees the junction as essentially a "normal" one and tunnels only in the normal coherent (probabilistic) way. As mentioned, some aspects of the interpretation of these experiments are currently controversial (cf. Wu and Niu 2002, Cataliotti et al. 2002).

4.7.5 The Josephson effect

Finally, I would like to mention briefly a very elegant set of experiments on the Josephson effect and phenomena closely related to it, which appeared when manuscript of this book was at a late stage of preparation. In the first round of this experiment (Albiez et al. 2005) a Josephson junction was realized for a Bose-condensed trapped gas of 87-Rb atoms, and the authors were able to observe both the Josephson effect itself and the phenomenon of "non-linear self-trapping" (see Chapter 6, Section 6.4). In a second round (Gati et al. 2006), they expanded the trap and used the interference between the two bulk condensates to measure the thermal fluctuations of the phase difference across the junction. This latter experiment is yet another nice example of how the techniques peculiar to atomic physics can be used to measure relatively easily quantities which, while of equal theoretical interest in the context of traditional condensed matter systems, are difficult or impossible to measure in those systems.

All in all, one can say that while the evidence for "superfluidity" in the alkali gases is less clear-cut and unambiguous than in He-II, most of the major expected manifestations of the phenomenon have been verified to occur.

Appendix

4A Feshbach resonances

As described in the text, a "Feshbach resonance" is said to occur when the energy of the zero-energy two-body scattering state in one channel (the "open" channel), that is for a specified pair of hyperfine indices, crosses that of a bound state in a different channel (the "closed" channel); close to the resonance considerable hybridization of the two channels occurs, and it is the purpose of this appendix to investigate this behavior, and in particular its consequences for the open-channel scattering length. However, for orientation I start by briefly discussing the behavior of the scattering length for the simple "single-channel" problem close to the onset of a bound state.

Consider then two atoms whose hyperfine indices we regard as fixed throughout, interacting via a central potential $V(r)$ which is characterized by some parameter δ which we imagine we can adjust ($V(r) \equiv V_\delta(r)$). We choose the sign convention so that increasing δ corresponds to increasing (less attractive) $V(r)$. We imagine that at some value of δ, which by convention we take equal to zero, the potential tolerates an s-wave bound state with exactly zero energy. thus, confining ourselves to s-wave states ($\psi(\boldsymbol{r}) \equiv \psi(r)$) and introducing as usual the reduced radial wave function $\chi(r) \equiv r\psi(r)$, we find that the general form of the time-independent Schrödinger equation is

$$-\frac{\hbar^2}{2m_{\mathrm{r}}}\frac{d^2}{dr^2}\chi_\delta(r) + V_\delta(r)\chi_\delta(r) = E\ \chi_\delta(r) \tag{4.A.1}$$

where m_r is the reduced mass and E will in general be a function of δ and of the boundary condition imposed at infinity. In particular, denoting the bound state wave function by $\chi_0(r)$, we have for $\delta \to 0$

$$-\frac{\hbar^2}{2m_\mathrm{r}}\frac{d^2}{dr^2}\chi_0(r) + V_{\mathrm{cr}}(r)\chi_0(r) = 0 \tag{4.A.2}$$

where $V_{\mathrm{cr}}(r) \equiv V_{\delta=0}(r)$ is the "critical" form of the potential. Also, defining for general δ $\chi(r)$ to be the form of $\chi_\delta(r)$ for the zero-energy scattering state (the state such that $\psi(r) \to$ const. as $r \to \infty$), we have

$$-\frac{\hbar^2}{2m_\mathrm{r}}\frac{d^2\chi}{dr^2} + V(r)\,\chi(r) = 0 \quad (V(r) \equiv V_\delta(r)) \tag{4.A.3}$$

Now let us multiply (4.A.2) by $\chi(r)$ and (4.A.3) by $\chi_0(r)$, and integrate the difference of the expressions so obtained up to some cutoff r_c which is large compared to the range r_0 of $V(r)$ but is otherwise for the moment unspecified. Using Green's theorem and the fact that both $\chi_0(r)$ and $\chi(r)$ are constrained to vanish in the limit $r \to 0$, we obtain the result

$$[\chi_0'\chi - \chi'\chi_0]_{r_\mathrm{c}} = -\frac{2m_\mathrm{r}}{\hbar^2}\int_0^\infty \delta V(r)\chi_0\chi\,dr \tag{4.A.4}$$

where $\delta V(r) \equiv V(r) - V_{\mathrm{cr}}(r)$, and we have used the fact that $\delta V(r)$ vanishes for $r > r_0$ to extend the upper limit of the integral to infinity. While formula (4.A.4) is of course valid for arbitrary normalization of χ and χ_0, it is convenient to choose the latter so that for $r \gg r_0$ we have $\chi_0 = 1$, $\chi = 1 - r/a_\mathrm{s}$; then in the RHS of (4.A.4) we may set $\chi(r) = \chi_0(r)$ (any corrections being of relative order a_s^{-1}, which as we shall see vanishes on resonance), and the LHS is just a_s^{-1}. Thus we find the fundamental result

$$a_\mathrm{s}^{-1} = -\frac{2m_\mathrm{r}}{\hbar^2}\int_0^\infty \delta V(r)\,|\chi_0(r)|^2\,dr \sim -\mathrm{const.}\delta \tag{4.A.5}$$

Thus, when the potential is just too weak to tolerate the bound state, the s-wave scattering length is negative and its magnitude is proportional to δ^{-1}, while when the state is just bound it is again proportional to δ^{-1} but positive.

A similar analysis can be applied to the bound state itself, provided we are sufficiently close to the resonance ($\delta \to 0^-$). We know that for $r > r_0$ the form of the bound-state wave function $\chi_\mathrm{b}(r)$ must be $\exp(-r/a_\mathrm{b})$, where $a_\mathrm{b}^{-1} \equiv (2m|E|/\hbar^2)^{1/2}$. Since $E \to 0$ for $\delta \to 0^-$, we may assume $a_\mathrm{b} \gg r_0$, and hence for $r_0 < r \lesssim r_\mathrm{c}$ can approximate $\chi_\mathrm{b}(r)$ by $1 - r/a_\mathrm{b}$. The RHS of the equation corresponding to (4.A.4) now contains (cf. 4.A.1) an extra term $(2mE/\hbar^2)\int_0^{r_\mathrm{c}}\chi_0\chi dr$, but this is evidently of order a_s^{-2} and hence is a higher-order correction. Dropping this correction, we find by an argument analogous to the one used above that for $\delta \to 0^-$,

$$a_\mathrm{b}^{-1} = -\frac{2m_\mathrm{r}}{\hbar^2}\int_0^\infty \delta V(r)\,|\chi_0(r)|^2\,dr \equiv a_\mathrm{s}^{-1}\,(>0) \tag{4.A.6}$$

Thus in this limit the bound-state radius a_b is exactly equal[43] to the zero-energy s-wave scattering length a_s, and both diverge as δ^{-1} for $\delta \to 0^-$.

[43]It may be verified that this condition is necessary to ensure mutual orthogonality of the two states $\chi_\mathrm{b}(r)$ and $\chi(r)$.

We now turn to the case of actual interest, a Feshbach resonance. Suppose that the two hyperfine states involved in the open channel are a and b, and those involved in the closed channel c and d; it is irrelevant, in the context of the two-body problem, whether two or more of a, b, c, d happen to be the same.[44] Let $E_i(B)$ be the single-atom energies as a function of B (cf. Fig. 4.1) and suppose that the energy of the closed-channel bound state is $-\epsilon_0$. Then, if we define an unrenormalized control parameter (or "detuning parameter") $\tilde{\delta}(B)$

$$\tilde{\delta}(B) \equiv E_c(B) + E_d(B) - E_a(B) - E_b(B) - \epsilon_0 \tag{4.A.7}$$

we see that in the absence of interchannel coupling the Feshbach resonance occurs at $\tilde{\delta} = 0$. Moreover, to the extent that we are either in the low-field ($B \ll B_{\text{hf}}$) or high-field ($B \gg B_{\text{hf}}$) limit, so that the dependence of the $E_i(B)$ on B is linear, we can write

$$\tilde{\delta}(B) = \text{const.}(B - B_o) \tag{4.A.8}$$

where B_o is the value of field at the resonance. (Eqn. (4.A.7) is obviously valid quite generally close enough to the resonance.) Since in most cases of current interest states a and b are in the lower hyperfine manifold and at least one of states c and d in the upper one, the constant is generally positive.

In writing the time-independent Schrödinger equation (TISE) for our problem we shall schematically denote the closed channel (i.e. the relevant pair of atomic states) by $|\uparrow\rangle$ and the open channel by $|\downarrow\rangle$, and take the zero of energy to be that of the open channel at infinite separation, i.e. $E_a(B) + E_b(B)$. Let $V(r)$ be the potential in the open channel and $V_c(r)$ that in the closed one. The "hybridizing" term in the Hamiltonian has matrix elements $\langle\uparrow|\hat{H}|\downarrow\rangle = \langle\downarrow|\hat{H}|\uparrow\rangle \equiv gf(\mathbf{r})$, where we take $f(r)$ to be a dimensionless function whose magnitude is of order 1 for small r; note that the range of $f(r)$ (which comes predominantly from exchange effects; see, e.g. Pethick and Smith (2002), Section 5.4.1) is likely to be quite small even on the scale of the closed-channel bound state. With this notation the general form of the wave function (assumed to be s-state) is

$$\psi(r) = r^{-1}\{\chi(r)|\downarrow\rangle + \chi_c(r)|\uparrow\rangle\} \tag{4.A.9}$$

and the two-channel Schrödinger equation has the form

$$\left(-\frac{\hbar^2}{2m_r}\frac{d^2}{dr^2} + V(r) - E\right)\chi(r) + gf(r)\chi_c(r) = 0 \tag{4.A.10a}$$

$$\left(-\frac{\hbar^2}{2m_r}\frac{d^2}{dr^2} + V_c(r) - E + \epsilon_0 + \tilde{\delta}\right)\chi_c(r) + gf(r)\chi(r) = 0 \tag{4.A.10b}$$

Equations (4.A.10) are exact within the two-channel approximation.

We now make a further important approximation, namely that any coupling to states of the closed channel other than the single resonant one, including the continuum, is negligible. Let the *normalized* (reduced radial) wave function of this resonant state, which must be real for an s-state, be denoted $\phi_0(r)$ and its energy, as above, $-\epsilon_0$.

[44]Though of course the possibilities ($c = a$, $d = b$) and ($c = b$, $d = a$) are excluded.

Then by expanding $\chi_c(r)$ in the eigenfunctions $\phi_i(r)$ of the closed-channel Hamiltonian and keeping only the term in ϕ_0, we find from (4.A.10b)

$$\chi_c(r) = \frac{g\phi_0(r)}{E-\tilde{\delta}} \int_0^\infty \phi_0^*(r')f(r')\chi(r')dr' \tag{4.A.11}$$

Substituting (4.A.11) into (4.A.10a), we find an integrodifferential equation for the open-channel amplitude $\chi(r)$:

$$\left(-\frac{\hbar^2}{2m_r}\frac{d^2}{dr^2} + V(r) - E\right)\chi(r) + \frac{g^2}{E-\tilde{\delta}}\int_0^\infty K(rr')\chi(r')dr' = 0 \tag{4.A.12}$$

where the (real) kernel $K(rr')$ has the form

$$K(rr') \equiv \phi_0(r)\phi_0(r')f(r)f(r') \equiv K(r'r) \tag{4.A.13}$$

Note that K has the dimensions of inverse length. The solution $\chi(r)$ of (4.A.12) must satisfy the boundary condition $\chi(r) \to 0$ for $r \to 0$.

Consider now the zero-energy scattering state in the open channel in the *absence* of any interchannel coupling ($g = 0$); call this $\chi_0(r)$. This function satisfies the simple equation

$$\left(-\frac{\hbar^2}{2m_r}\frac{d^2}{dr^2} + V(r)\right)\chi_0(r) = 0 \tag{4.A.14}$$

with the boundary condition $\chi_0(r) \to 0$ as $r \to 0$; for $r \to \infty$ it has the form const.$(1 - r/a_{bg})$ where a_{bg} is, by definition, the "background" scattering length (which may have either sign). In the following I will exclude the possibility that a_{bg} is zero, a circumstance which would require a formally different derivation.

We now proceed exactly as in the single-channel case, by multiplying (4.A.12) by χ_0 and (4.*A*.14) by χ, subtracting and using Green's theorem; however, it is convenient to keep for the moment the term which is explicitly proportional to E. The result is (for arbitrary value of the upper cutoff r_c)

$$\begin{aligned}\chi_0(r_c)\chi'(r_c) - \chi_0'(r_c)\chi(r_c) + \frac{2m_r E}{\hbar^2}\int_0^{r_c}\chi_0(r)\chi(r)dr \\ = \frac{2m_r g^2/\hbar^2}{E-\tilde{\delta}}\int_0^{r_c} dr \int_0^{r_c} dr'\; \chi_0(r)K(rr')\chi(r')\end{aligned} \tag{4.A.15}$$

Equation (4.A.15) is exact in the limit of a single resonant state, and is of course valid independently of the normalization of χ and of χ_0; it is however convenient to choose the latter so that as $r \to \infty$ $\chi_0(r) \to 1 - r/a_{bg}$ (cf. above) and $\chi(r) \to 1 - r/a_s(\tilde{\delta})$, where $a_s(\tilde{\delta})$ is the actual open-channel scattering length in the presence of the interchannel coupling.

Consider first the $E = 0$ scattering state in the open channel. In this case, choosing $r_c \gg r_0$, we immediately obtain from (4.A.15) the result

$$a_s^{-1}(\tilde{\delta}) - a_{bg}^{-1} = \frac{2m_r g^2/\hbar^2}{\tilde{\delta}}\int_0^\infty dr \int_0^\infty dr'\; \chi_0(r)K(rr')\chi(r') \tag{4.A.16}$$

where $\chi(r)$ (but nothing else on the RHS) is implicitly a function of $\tilde{\delta}$. It is convenient to define a quantity κ by the implicit equation

$$\kappa \equiv -\frac{2m_r a_{bg} g^2}{\hbar^2} \int_0^\infty dr \int_0^\infty dr' \chi_0(r) K(rr') \chi_{\tilde{\delta}=\kappa}(r') \tag{4.A.17}$$

so that a_s diverges at the point $\tilde{\delta} = \kappa$; thus, κ is essentially the shift of the position of the Feshbach resonance due to the interchannel coupling. It is a crucial assumption in what follows that κ is small compared to ϵ_0; since we will see below that κ can be obtained directly from experiment, the consistency of this assumption can be checked in any given case.

We now come to a crucial point: From the point of view of the integrodifferential equation (4.A.16), the point $\tilde{\delta} = \kappa$ (as distinct from $\tilde{\delta} = 0$) is not a singular point. Hence, at least provided we do not have to deal with values of the detuning comparable to κ (cf below), we can take $\chi(r)$ in (4.A.16) to have the value it has for $\tilde{\delta} = \kappa$, and hence write this equation in the simple form

$$a_s^{-1}(\tilde{\delta}) - a_{bg}^{-1} = -\kappa/(\tilde{\delta} a_{bg}) \tag{4.A.18}$$

If now we introduce the detuning δ from the "actual" resonance (i.e. the point at which $a_s^{-1} \to 0$) by

$$\delta \equiv \tilde{\delta} - \kappa \tag{4.A.19}$$

then Eqn. (4.A.18) yields the result

$$a_s(\delta) = a_{bg}\left(1 + \frac{\kappa}{\delta}\right) \tag{4.A.20}$$

It is plausible (though not completely obvious[45]) that the integral in (4.A.17) is positive, so that the sign of κ is opposite to that of a_{bg}. In that case we can rewrite (4.A.20) in the form

$$a_s(\delta) = a_{bg} - \text{const}.\delta^{-1} \tag{4.A.21}$$

where the constant is positive and proportional to a_{bg}^2. Thus, we expect that for small positive detuning ($|\kappa| > \delta > 0$) the scattering length is negative; this seems to be the case in the resonances investigated experimentally so far. In any case we see that $|\kappa|$ can be read off directly from experiment, as the value of δ at which the contribution to a_s from the Feshbach resonance becomes comparable to the background value a_{bg}. In the literature it is conventional to write (4.A.20) in the form $a_s(\delta) = (1 + (\Delta B/(B - B_0)))a_{bg}$ where $\Delta B \equiv \kappa(\partial\tilde{\delta}/\partial B)^{-1}$ and B_0 is the experimental value of the field at resonance.

Now let's turn to the bound state. We start from Eqn. (4.A.15), and again make the crucial assumption that the integral on the RHS may be approximated by its value for $\tilde{\delta} = \kappa$ ($\delta = 0$), so that the RHS is simply $-\kappa/(E - \tilde{\delta})$. The term linear in E on the LHS turns out to contribute an effect of order δ_c/ϵ_0 in the "interesting" region (where δ_c is defined below), so, anticipating that this ratio will be small, we drop it.

[45]It follows from assumption (b) defined below.

We consider the case $E < 0$ and define the (positive) quantity

$$a_{\mathrm{b}}(E) \equiv \frac{\hbar}{(2m_{\mathrm{r}}|E|)^{1/2}} \tag{4.A.22}$$

so that for sufficiently large r (where $V(r)$ and $K(rr')$ both vanish) the quantity $\chi(r)$ which solves (4.A.12) is proportional to $\exp(-r/a_{\mathrm{b}}(E))$; note that we should *not* a priori identify $a_{\mathrm{b}}(E)$ with the open-channel zero-energy scattering length a_{s}. We will assume that $E \ll \epsilon_0$ and hence $a_{\mathrm{b}}(E) \gg r_0$, and choose a cutoff r_{c} in (4.A.15) such that $r_0 \ll r_{\mathrm{c}} \ll a_{\mathrm{b}}(E)$, so that we can approximate $\chi(r)$ by $1 - r/a_{\mathrm{b}}(E)$. Then from (4.A.15) and the definitions of κ and δ we obtain

$$\frac{a_{\mathrm{bg}}}{a_{\mathrm{b}}(E)} - 1 = \frac{-\kappa}{\delta + \kappa - E} \tag{4.A.23}$$

Provided that both δ and E are small compared to κ, this yields

$$a_{\mathrm{b}}(E) = \frac{\kappa a_{\mathrm{bg}}}{\delta - E} \equiv a_{\mathrm{b}}(\delta) \tag{4.A.24}$$

Comparing (4.A.24) with (4.A.21), we see that if (and *only* if) E is small compared to δ as well as to κ, $a_{\mathrm{b}}(\delta)$ is indeed equal to the open-channel scattering length $a_{\mathrm{s}}(\delta)$.

To solve explicitly for $a_{\mathrm{b}}(\delta)$ in the general case, we introduce a "characteristic" value δ_{c} of the detuning by the formula

$$\delta_{\mathrm{c}} \equiv \frac{\kappa^2}{\hbar^2/m_{\mathrm{r}}a_{\mathrm{bg}}^2} \quad (\sim a_{\mathrm{bg}}^4) \tag{4.A.25}$$

and find from (4.A.24) and (4.A.22) that for $E < 0$ its magnitude $|E|$ satisfies the quadratic equation

$$(|E| + \delta)^2 - 2\delta_{\mathrm{c}}|E| = 0 \tag{4.A.26}$$

Provided that the quantity κa_{bg} is negative (so that a_{s} has the opposite sign to δ for $\delta \to 0$, see Eqn. (4.A.20)), Eqn. (4.A.23) implies that the only acceptable solutions satisfy $\delta < E < 0$, which immediately excludes positive δ and moreover forces us to choose the solution

$$|E| = \delta_{\mathrm{c}} - \delta - \sqrt{\delta_{\mathrm{c}}^2 - 2\delta\delta_{\mathrm{c}}} \quad (\delta < 0) \tag{4.A.27}$$

Thus, for $|\delta| \gg \delta_{\mathrm{c}}$ E is simply equal to δ, while for $|\delta| \ll \delta_{\mathrm{c}}$ it is approximately equal to $-\delta^2/2\delta_{\mathrm{c}}$.

To explore the physical significance of the quantity δ_{c}, we first consider the region $\delta > 0$, where the closed-channel state is metastable, and use simple golden-rule perturbation theory to calculate the energy width $\Gamma(\delta)$ of the latter due to its decay into the open channel. Taking into account the normalization of the functions $\chi_0(r)$ and $\chi(r)$, and assuming that their short-range forms do not differ substantially, we find

$$\Gamma(\delta) \sim \left(\frac{2m_{\mathrm{r}}}{\hbar^2}\right)^{1/2} |\kappa a_{\mathrm{bg}}|\delta^{1/2} \tag{4.A.28}$$

(where we neglect constants of order 1). Then we see from (4.A.25) that the value of δ below which $\Gamma(\delta) > \delta$ (and thus perturbation theory no longer makes sense) is precisely of order δ_{c}.

Next, let us consider the bound state which occurs for $\delta < 0$ and calculate the relative weights, in it, of the closed and open channels. According to (4.A.11), the *relative* weight (probability) λ of the closed channel is given by the expression

$$\lambda = \left(\frac{g}{E - \tilde{\delta}}\right)^2 \left| \int \phi_0(r') f(r') \chi_n(r') \, dr' \right|^2 \tag{4.A.29}$$

where $\chi_n(r')$ is the *normalized* open-channel bound-state component (i.e. such that $\int_0^\infty |\chi_n(r')|^2 dr' = 1$). Comparing (4.A.29) with the definition (4.A.17) of κ, assuming (cf above) that the short-range form of $\chi_n(r)$ does not differ much from that of $\chi_0(r)$ and bearing in mind the normalizations adopted in (4.A.17), we find

$$\lambda = (E - \tilde{\delta})^{-2} \frac{\hbar^2}{2m_r a_{bg}} \kappa a_b^{-1}(E) \tag{4.A.30}$$

Now for $\delta \lesssim \delta_c (\ll \kappa)$ the factor $E - \tilde{\delta}$ is slowly varying and can be set equal to κ (cf above). Thus,

$$\lambda = \frac{\hbar^2}{2m_r a_{bg}} (\kappa a_b(E))^{-1} \tag{4.A.31}$$

Finally, using (4.A.24), (4.A.22) and (4.A.26), we have for the relative weight of the closed channel

$$\lambda = \left(\frac{|E|}{2\delta_c}\right)^{1/2} \tag{4.A.32}$$

Comparing with (4.A.27), we see that $-\delta_c$ marks, approximately, the value of δ for which (as we increase δ) the weight of the bound state shifts from predominantly in the closed channel to predominantly in the open one. For $|\delta| \gg \delta_c$ the open-channel weight is $\sim (\delta_c/\delta)^{1/2}$, while for $\delta \ll \delta_c$ the closed-channel weight is $\sim \delta/\delta_c$. It is important to note that even at $|\delta| = \delta_c$ (and thus, a fortiori, over the whole region $\delta < 0$, $|\delta| < \delta_c$) the radius a_b of the open-channel component is much larger (by a factor $\sim \kappa/\delta_c$, see (4.A.24)) than the background scattering length a_{bg}. (Thus, the "characteristic field detuning" $\Delta B_c \equiv B(\delta_c) - B_0$ is typically small compared to the usually defined "width" ΔB of the Feshbach resonance.) In the region $\delta \ll \delta_c$ (which is usually the regime in which many-body effects are substantial) the behavior of the open-channel scattering length, and of the bound state, is exactly the same as in the simple "potential resonance" case discussed at the beginning of this appendix; the only role of the closed channel is to provide an effective coupling in the open channel.

The above analysis rests on two implicit assumptions: (a) that the short-range ($r \lesssim r_0$) behavior of the open-channel amplitude (i.e. the form of its dependence on r) is essentially independent of both the energy (over a scale $\sim\delta_c$) and of the presence or absence of coupling to the closed channel, and (b) that the quantity δ_c will turn out to be small compared to κ. With regard to (a), it is extremely plausible since the r-dependence of the (uncoupled) open-channel amplitude in the region $r \lesssim r_0$ is determined by the competition of kinetic and (negative) potential energies, each of which is individually large compared to g (even though their sum may be $\lesssim g$). As to (b), it is not guaranteed a priori for any particular resonance, but can be checked explicitly from the experimental data. In fact, from (4.A.20) and the definition (4.A.25) we have

$$\delta_c = \frac{m_r}{\hbar^2} \lim_{\delta \to 0} \left(\frac{da_s^{-1}}{d\delta} \right)^{-2} \tag{4.A.33}$$

For example, I calculate that for the 224 G Feshbach resonance of ^{40}K (one of the two used widely in the context of the "BEC-BCS crossover" problem, see Chapter 8, Section 8.4) the ΔB_c corresponding to δ_c is approximately 2.5 G; thus, since ΔB is 9.7 G, the conditions of validity of the above discussion are reasonably well satisfied. By contrast, for the other widely used fermion Feshbach resonance, the 822 G resonance of ^{6}Li, the ΔB_c calculated from (4.A.33) turns out to much exceed ΔB, which is itself anomalously large ($\sim$100 G). In such a case, while the parameter δ_c itself no longer has much physical meaning, the formulae derived above are still valid, in the limit $\delta \to 0$, and in particular it is still possible to treat the Feshbach resonance in this limit as equivalent to a simple potential resonance.

5
Classical superconductivity

In this book I use the term "classical" to refer to the large class of superconductors, most of them though not all discovered before 1975, whose behavior is generally believed to be well understood in terms of the 1957 theory of Bardeen, Cooper and Schrieffer (hereafter BCS). The superconducting metals in this class, which may be either crystalline or amorphous, show a number of common features: (a) the transition temperature to the superconducting state, T_c, is a small fraction[1] (typically 10^{-3}–10^{-4}) of the Fermi degeneracy temperature T_F (b) the normal state is well described, at temperatures of the order of room temperature and below, by the standard "textbook" theory based on the ideas of Sommerfeld, Bloch and Landau (see Section 5.1 below); (c) superconductivity does not occur near other kinds of phase transition; (d) the symmetry of the Cooper pairs is "conventional", i.e. the same as that of the crystal lattice;[2] and (e) the principal mechanism of Cooper pairing is (universally believed to be) the exchange of virtual phonons. In addition, most though not all of the crystalline materials in this class are more or less three-dimensional in structure (i.e. not filamentary or layered), and their electromagnetic behavior is almost always "Type I" (though alloys of this class often show type-II behavior). I will refer to superconductors which lack at least one of the features (a)–(e) as "exotic"; in some cases, such as some of the organics, the dichotomy may be somewhat ambiguous, but this need not concern us unduly. Exotic superconductivity is the subject of Chapter 7 and Chapter 8, section 1.

5.1 The normal state

In view of feature (a) above ($T_c \ll T_F$) one might anticipate that to build a viable theory of classical superconductivity one would first need an adequate description of the normal state. As mentioned under (b), for temperatures of the order of room temperature or less (in fact, in most cases, for all temperature up to melting) the normal state of the classical superconductors appears to be well described by the standard "textbook" theory of metals, and so I briefly review the barest essentials of the latter at this point; for a much more detailed account see, e.g. Ashcroft and Mermin (1976), Harrison (1970) or Ziman (1964). It is convenient to introduce three

[1] In absolute terms, with the exception of a few materials ($BaKBiO_3$, MgB_2, etc.), discovered after 1988, and the recently discovered metallic hydrides T_c is always below 25 K.

[2] Or in the case of amorphous materials, "s-wave".

successive levels of description, of increasing sophistication, which we may associate respectively with the names of Sommerfeld, Bloch, and Landau and Silin.[3]

The very simplest model of the normal state of a metal, which is actually adequate to a surprising degree as a qualitative basis for the description of the superconducting state, is that originally due to Sommerfeld. In this model one completely ignores both the presence of the crystalline lattice and the Coulomb interaction between the electrons, and treats the latter as N independent noninteracting fermions of spin 1/2 moving freely in a volume Ω (with periodic boundary conditions). Then the single-particle eigenstates are simply plane waves of the form

$$\psi_k(\boldsymbol{r}) = \frac{1}{\sqrt{\Omega}} e^{i\boldsymbol{k}\cdot\boldsymbol{r}} \tag{5.1.1}$$

with energy

$$\epsilon_k \equiv \frac{\hbar^2 k^2}{2m} \tag{5.1.2}$$

and the number of electrons in state $\boldsymbol{k}$ and with spin projection $\sigma(=\pm 1/2)$ is given, in thermal equilibrium at temperature T, by the standard Fermi distribution, which in the absence of a magnetic field is independent of σ and given by

$$n(\boldsymbol{k}, \sigma) = \frac{1}{e^{(\epsilon_k - \mu)/(k_B T)} + 1} \tag{5.1.3}$$

where μ is the chemical potential. At zero temperature the Fermi distribution (5.1.3) reduces to the step function $\Theta(\mu - \epsilon_k)$; the lowest-energy plane-wave states $\boldsymbol{k}$ are each filled by two electrons (one for each spin) up to the point at which $\epsilon_k = \mu$, and by equating the total number of electrons to N we find that the Fermi wave vector k_F, i.e. the maximum value of k which is occupied (or equivalently the Fermi momentum $p_F \equiv \hbar k_F$) is given by

$$k_F = (3\pi^2 n)^{1/3} \quad \left(n \equiv \frac{N}{\Omega}\right) \tag{5.1.4}$$

Correspondingly, the Fermi energy $\epsilon_F (\equiv k_B T_F \equiv \hbar^2 k^2/2m)$ is

$$\epsilon_F = \frac{\hbar^2}{2m}(3\pi^2 N/\Omega)^{2/3} \tag{5.1.5}$$

so that typical values of the Fermi temperature T_F are of order 10^4–10^5 K, much higher than the melting temperature of the metal in question. At temperatures $T \ll T_F$, i.e. for all temperatures at which the metal remains solid, the Fermi distribution (5.1.3) differs appreciably from its zero-temperature limit only in an energy shell of width $\sim k_B T$ around the Fermi energy. Consequently, it is only states within this shell, i.e. close to the Fermi surface (the locus in k-space corresponding to $\hbar^2 k^2/2m = \epsilon_F$),

[3]For the next few paragraphs I will assume that the metal in question is a regular crystalline solid; I return at the end of this section to the case of an amorphous metal. I will also neglect, in the ensuing discussion, the complications which arise from the fact that the conduction electrons are not really different from the electrons which we treat as part of the ionic "cores"; these complications are best handled by the pseudopotential method, see e.g. Harrison (1970), Chapter II, Section 5, and while considerably complicating the notation do not change the picture qualitatively.

that contribute to the reaction of the system to weak perturbations, with the states with $\epsilon \ll \epsilon_\mathrm{F} - k_\mathrm{B}T$ (the "Fermi sea") remaining inert; this fundamental feature persists, as we shall see, in more sophisticated models. For this reason a critical role is played by the density of (single-particle) states (DoS) per unit energy[4] at the Fermi surface, which I shall denote by $dn/d\epsilon$; in the Sommerfeld model, since $\epsilon_\mathrm{F} \sim n^{2/3}$, this turns out to be simply equal to $3n/2\epsilon_\mathrm{F}$. A number of basic properties of the system can be expressed, in the limit $T \ll T_\mathrm{F}$, in terms of $dn/d\epsilon$; e.g. the static (Pauli) spin susceptibility is given by

$$\chi_\mathrm{s} = \mu_\mathrm{B}^2 \frac{dn}{d\epsilon} \tag{5.1.6}$$

(where μ_B is the Bohr magneton) and the electronic specific heat is

$$c_V = \frac{\pi^2}{3} k_\mathrm{B}^2 T \frac{dn}{d\epsilon} \tag{5.1.7}$$

While in the "pure" version of the Sommerfeld model there is no mechanism of scattering of the electrons and the electrical conductivity is therefore infinite, it is conventional to add "by hand" a dilute concentration of static impurities which give rise to a mean free path, for electrons close to the Fermi surface, of l and hence a relaxation time $\tau \equiv l/v_\mathrm{F}$, where $v_\mathrm{F} \equiv p_\mathrm{F}/m \equiv \hbar k_\mathrm{F}/m$ is the velocity of these electrons. The standard result for the dc electrical conductivity σ is then the Drude expression

$$\sigma = \frac{ne^2\tau}{m} \quad (\equiv \epsilon_0 \omega_\mathrm{p}^2 \tau) \tag{5.1.8}$$

but it is worth noting that this can be equivalently expressed in terms of quantities referring only to the states near the Fermi surface:

$$\sigma = \frac{1}{3} e^2 v_\mathrm{F}^2 \frac{dn}{d\epsilon} \tau \tag{5.1.9}$$

Thus the conduction can be attributed either, as in (5.1.8), to the motion of the whole Fermi sea or, as in (5.1.9), to the perturbation only of states near the Fermi surface. As we shall see, a similar state of affairs obtains in the superconducting state.

Equation (5.1.8) is actually the DC limit of a more general expression in Sommerfeld (or Drude) theory for the complex frequency-dependent conductivity, namely

$$\sigma(\omega) = \frac{(ne^2/m)\tau}{1 + i\omega\tau} \equiv \frac{\epsilon_0 \omega_\mathrm{p}^2 \tau}{1 + i\omega\tau} \tag{5.1.10}$$

If we combine (5.1.10) with Maxwell's equations, we find that for $\omega \gg \tau^{-1}$ electromagnetic radiation is absorbed only at the plasma frequency ω_p given by

$$\omega_\mathrm{p}^2 = \frac{ne^2}{m\epsilon_0} \tag{5.1.11}$$

which is typically in the visible or UV frequency regime. ω_p is also the frequency of oscillations of the charge density.

[4]I define $dn/d\epsilon$ as the DoS for *both* spins. In the superconductivity literature it is conventional to introduce the DoS for a single spin, which is one-half of $dn/d\epsilon$ and is conventionally denoted $N(0)$; this notation is also used below.

The next level of sophistication is the Bloch theory, in which one continues to ignore the inter-conduction-electron Coulomb interaction but takes into account the periodic potential exerted on the conduction electrons by the ionic lattice, regarded as static. As is well known, in the presence of such a periodic potential the single-particle eigenstates no longer have the plane-wave form (5.1.1), but are "Bloch waves," that is, functions of the form

$$\psi_{\boldsymbol{k}n}(\boldsymbol{r}) = u_{\boldsymbol{k}n}(\boldsymbol{r})e^{i\boldsymbol{k}\cdot\boldsymbol{r}} \tag{5.1.12}$$

where the functions $u_{\boldsymbol{k}n}(\boldsymbol{r})$ are periodic with the lattice periodicity, and the wave vector $\boldsymbol{k}$ now runs only over the first Brillouin zone (see, e.g. Ashcroft and Mermin (1976) Chapter 8), so that the number of allowed values of $\boldsymbol{k}$ is equal to the number of unit cells in the crystal. Different values of n in (5.1.12) label different energy bands, within each of which one has an energy spectrum $\epsilon_n(\boldsymbol{k})$; note that in general $\epsilon_n(\boldsymbol{k})$ depends on the direction as well as the magnitude of $\boldsymbol{k}$. Unless the number of electrons per unit cell is even (in which case the crystalline solid in question may be a "band insulator"), the Fermi surface, that is the locus in $\boldsymbol{k}$-space corresponding to $\epsilon_n(\boldsymbol{k}) = \epsilon_{\mathrm{F}}$, must intersect at least one energy band; for simplicity I will assume in what follows that it is only one, and suppress the corresponding band index n. Then the qualitative considerations regarding the behavior at low temperatures ($T \ll T_{\mathrm{F}}$) are very similar to those in the Sommerfeld model; in particular, the response to weak perturbations is determined entirely by the properties of the states in a shell of width $\sim k_{\mathrm{B}}T$ around the Fermi surface. One can again define the total DoS $dn/d\epsilon$, and when expressed in terms of this quantity the static spin susceptibility χ and specific heat remain given by Eqns. (5.1.6) and (5.1.7) respectively; however, the expression (5.1.9) for the electrical conductivity must now be replaced by an appropriate average over the Fermi surface (which there is no point in writing out explicitly here). Overall, in the case of metals (as distinct from that of band insulators) the low-temperature, low-frequency properties in the Bloch model do not differ qualitatively from those in the Sommerfeld model.[5]

It is worth noting for future reference, however, that one quantity which is qualitatively modified in the Bloch picture is the AC conductivity $\sigma(\omega)$. This can now have contributions from "interband" transitions even when $\boldsymbol{q} \to 0$, and as a result the real part of $\sigma(\omega)$ can be nonzero even for $\omega \gg \tau^{-1}$, provided that ω is comparable to the band gap. As a result, the electromagnetic properties may be considerably more complicated than in the Sommerfeld model. While this consideration is not usually believed to be particularly significant for the classical superconductors, I mention it here because it may be relevant to the exotic superconductivity to be discussed in Chapter 7.

The third and most sophisticated level of description of a "textbook" normal metal is that given by the Landau–Silin theory, which is a generalization of the original Landau theory of a neutral, translationally invariant Fermi system. In this theory one incorporates, in a phenomenological way, the interactions between the conduction electrons (including the effective interactions induced by exchange of virtual

[5] With the exception of a few properties such as the Hall and thermoelectric coefficients which are sensitive to the details of the energy spectrum $\epsilon_n(\underline{k})$ and may even change sign in the Bloch picture; see, e.g. Ashcroft and Mermin (1976), Chapters 3 and 13.

lattice vibrations, see Section 2 below). The result is that the low-lying states of the many-body system must now be expressed not in terms of the excitation of real electrons, but rather in terms of "quasiparticles"; a quasiparticle may be thought of intuitively as a real electron accompanied by a "dressing cloud" of other electrons and of phonons. Quasiparticles, like real electrons, obey Fermi statistics, and thus we can define for them a Fermi surface (which need not be identical to that of the original electrons, but must enclose the same number of states) and a "single-quasiparticle" density of states, etc.; a formal justification of this idea is given in Appendix 5A. In the Landau–Silin theory most of the effects of the electron-electron and electron–phonon interactions are absorbed by (a) a modification of the single-electron energy spectrum $\epsilon_n(\boldsymbol{k})$, and (b) a set of "molecular fields" (a generalization of that occurring in the Weiss theory of magnetism) which are generated by various "polarizations" of the system: see Appendix 5A for details. However, it is still necessary to take out and handle explicitly the long-range effects of the Coulomb potential, and in addition there are always "residual" interactions which can lead to real quasiparticle–quasiparticle scattering processes; among them a specially important role is played, in the present context, by the special class of terms which can lead to the formation of Cooper pairs, see below.

The general belief is that in the problem of classical superconductivity in a crystalline metal the modifications to the simple Sommerfeld picture introduced by the considerations of Bloch and of Landau and Silin, while changing some of the details of the behavior,[6] do not radically modify the qualitative picture. For this reason, in this chapter I shall follow the approach which has become more or less traditional in this problem, in which one starts from a model of the normal phase which is essentially that of Sommerfeld plus the more or less phenomenological account of the effects of the electron–electron and electron–phonon interactions which will be described in the next section.

All of the above discussion refers to the case where the superconductor in question has a crystalline structure. When we turn to the important class of classical superconductors which are amorphous metals (disordered alloys), the considerations are a bit different. It is clear that in this case there is no Bloch theorem, and indeed the nature of the single-electron energy eigenstates is qualitatively different from that of the Sommerfeld model; they are not characterized by a wave vector $\boldsymbol{k}$, and indeed in some circumstances may even be spatially localized.[7] However, we can still define a Fermi *energy* (though not a "Fermi surface"!) and a single-particle DoS $dn/d\epsilon$, and the Pauli spin susceptibility and electronic specific heat are still given by Eqns. (5.1.6) and (5.1.7) respectively. Moreover, it is still possible to formulate a "Landau–Silin" theory (see Appendix 5A). However, it is clear that because of the very different nature of the single-particle eigenstates the electromagnetic behavior in the normal phase is likely to be rather different from that of a crystalline system, and we shall see below that this difference extends into the superconducting phase.

[6]For example, the Landau–Silin molecular fields can modify the temperature-dependence of quantities such as the spin susceptibility, cf. Leggett 1965.

[7]In that case the "normal" state may not correspond to a metal, cf. the end of Section 5.8.

5.2 The effective electron–electron interaction

A real metal is composed of a large number $N(\sim 10^{23})$ of nuclei of a certain type,[8] plus a number ZN of electrons. (where Z is the atomic number). In the nonrelativistic approximation the Hamiltonian consists of (a) electron kinetic energy (b) nuclear kinetic energy (c) the standard (unscreened) Coulomb interaction

$$\frac{1}{2}\sum_{ij} q_i q_j / 4\pi\epsilon_0 |\boldsymbol{r}_i - \boldsymbol{r}_j|$$

where i, j run over all particles (nuclei and electrons) and $q_i \equiv e \equiv -|e|$ for electrons, $+Z|e|$ for nuclei. Clearly, the solution of the many-body Schrödinger equation for this system is not analytically tractable, nor even amenable to numerical simulation in the foreseeable future, so one needs to make some simplifying approximations.

A first step is to group the electrons into a set (the "valence" electrons) which are attached to the individual nuclei, thereby forming ions, and the rest (the "conduction" electrons), and thereafter to consider the ions to be simple structureless charged particles. Since the "conduction" electrons are actually indistinguishable from the "valence" ones, even this step is not rigorously justified, but it is usually believed that the corrections can be handled by the method of pseudopotentials (see, e.g. Harrison (1970), Chapter II, Section 5) and do not affect the relevant results qualitatively, so I will not discuss them here. So our Hamiltonian now consists of (a) kinetic energy of the conduction electrons (b) kinetic energy of the ions (c) the long-range Coulomb interaction between conduction electrons, between ions and between the conduction electrons and the ions. Strictly speaking, the last term (c) should itself be screened by the dielectric response of the ionic cores; in addition, the "pseudopotential" corrections in general produce extra short-range (and in general nonlocal) conduction-electron–ion potentials. In the following I will neglect these complications and thus assume the Coulomb interaction to have the simple unscreened form specified above, where now $q_i \equiv e \equiv -|e|$ for the conduction electrons and $q_i \equiv +Z_{\text{ion}}|e|$ for the ions. The resulting problem is only slightly less intractable than the original one.

As the next step towards an approximate solution, let us imagine for the moment that the ions are replaced by a uniform static positive charge background whose only function is to keep the system as a whole electrically neutral (this is sometimes called the "jellium" model). Then we are left with the kinetic energy of the conduction electrons and the bare long-range Coulomb repulsion between them; note that the Fourier-transformed form of the latter is

$$V_c(q) = +\frac{e^2}{\epsilon_0 q^2} \tag{5.2.1}$$

which is strongly divergent at small q. To handle this problem we introduce the standard random-phase approximation (RPA); that is, we calculate the response of the electronic charge density to a given potential $V(\boldsymbol{r}t)$ by treating the electrons as noninteracting, then take $V(\boldsymbol{r}t)$ to be the sum of the original external perturbation plus

[8] For simplicity I assume an electrically neutral elemental metal with a single isotopic species.

the potential generated self-consistently by the calculated charge response. Formally, we write

$$V_{\mathrm{RPA}}(\boldsymbol{r}t) = V_{\mathrm{ext}}(\boldsymbol{r}t) + V_{\mathrm{ind}}(\boldsymbol{r}t) \tag{5.2.2}$$

$$\nabla^2 V_{\mathrm{ind}}(\boldsymbol{r}t) = -\frac{\rho_{\mathrm{ind}}(\boldsymbol{r}t)}{\epsilon_0} \tag{5.2.3}$$

$$\rho_{\mathrm{ind}}(\boldsymbol{r}t) = -\hat{\chi}_0 V_{\mathrm{RPA}}(\boldsymbol{r}t) \tag{5.2.4}$$

where $\hat{\chi}_0$ is a shorthand for an integral operator which need not be written out explicitly.[9] (It is easier to paraphrase Eqn. (5.2.4) in words: $\rho_{\mathrm{ind}}(\boldsymbol{r}t)$ is the charge response induced, *in a noninteracting gas*, by the potential $V_{\mathrm{RPA}}(\boldsymbol{r}t)$.) Poisson's equation, (5.2.3), is simply a restatement of (5.2.1). On Fourier-transforming Eqns. (5.2.1–5.2.4) and solving for $V_{\mathrm{RPA}}(\boldsymbol{q}\omega)$ in terms of $V_{\mathrm{ext}}(\boldsymbol{q}\omega)$ we find

$$V_{\mathrm{RPA}}(\boldsymbol{q}\omega) = \frac{V_{\mathrm{ext}}(\boldsymbol{q}\omega)}{1 + e^2\chi_0(q\omega)/\epsilon_0 q^2} \tag{5.2.5}$$

where $\chi_0(q\omega)$ is the (number) susceptibility of the noninteracting gas. For example, if V_{ext} is the potential of a static impurity of charge Z_{imp} ($V_{\mathrm{ext}}(\boldsymbol{q}0) = Z_{\mathrm{imp}}|e|/\epsilon_0 q^2$), then $V_{\mathrm{RPA}}(\boldsymbol{q}0)$) is given by an expression which in the limit $q \to 0$ reduces to

$$V_{\mathrm{RPA}}(q) = \frac{Z_{\mathrm{imp}}\kappa_0/|e|}{1 + q^2/q_{\mathrm{TF}}^2} \tag{5.2.6}$$

where $\kappa_0(\equiv 1/\chi_0(0,0)) = (dn/d\epsilon)^{-1}$ is the static bulk modulus of the noninteracting Fermi gas, and the Fermi–Thomas wave vector q_{TF} is given by the expression

$$q_{\mathrm{TF}} \equiv \sqrt{\frac{e^2}{\epsilon_0\kappa_0}} \tag{5.2.7}$$

At large distances the coordinate-space potential falls off as $r^{-1}e^{-q_{\mathrm{TF}}r}$, so that for $r \gg q_{\mathrm{TF}}^{-1}$ the effect of the impurity is completely screened out.

The punch-line, now, is that there is no reason why we should not apply Eqns. (5.2.6)–(5.2.7) to the screening of the Coulomb interaction energy $V_{\mathrm{eff}}(q)$ between two given electrons by the other background electrons simply by setting $V_{\mathrm{eff}}(q) = eV_{\mathrm{RPA}}(q)$ and, in (5.2.7), $Z_{\mathrm{imp}} = 1$! The result is that

$$V_{\mathrm{eff}}(q) = \frac{\kappa_0}{1 + q^2/q_{\mathrm{TF}}^2} \tag{5.2.8}$$

Thus, the effect of screening is to make the effective Coulomb interaction *very much less repulsive* at long wavelengths (or equivalently at large distances). The above argument implicitly assumes that $|\boldsymbol{r}_i - \boldsymbol{r}_j|$ is independent of time; if it is not (as will certainly be the case in general for two electrons moving in a real metal) then the effect will be to replace $\chi_0(q,0)$ by the full frequency-dependent susceptibility $\chi_0(q\omega)$. However, it is straightforward to show (e.g. from inspection of the exact (Lindhard) formula for $\chi_0(q\omega)$, see e.g. Ziman (1964), Chapter 5), that the deviation of $\chi_0(q\omega)$ from its static limit is small provided $\omega \ll v_{\mathrm{F}}q$ where $v_{\mathrm{F}} = p_{\mathrm{F}}/m$ is the Fermi velocity; as we shall see below, this condition is well satisfied for most of the processes important to

[9]The $-$ sign is introduced so that $\chi_0(q,0)$ comes out positive.

Cooper-pair formation in classical superconductors, so that we may safely use (5.2.8) in this context, at least when q is not too large.[10]

We must now try to take into account the effect of the conduction-electron–ion interaction and the ionic kinetic energy, both of which are ignored in the "jellium" approximation. Before going into the details, let us note a very generic result concerning the interaction of physical systems of any kind with harmonic oscillators: If two systems S_1 and S_2 interact in the same way with a harmonic oscillator of natural frequency ω_0, then the result is to produce an effective attraction between S_1 and S_2 at frequencies $\omega < \omega_0$ (and a repulsion for $\omega > \omega_0$). It is worth sketching the derivation of this result in both classical and quantum mechanical language.

First, a classical argument: Imagine that the (one-dimensional) coordinates of S_1 and S_2 are x_1 and x_2 respectively, and that of the oscillator X. Furthermore, let us assume, for simplicity only, that the interaction between the systems and the oscillator is of the simple bilinear form

$$V(t) = -g(t)\ (x_1 X + x_2 X) \tag{5.2.9}$$

(so that the force exerted on the oscillator by system S_i is independent of X). It is convenient to suppose that $g(t)$ is switched from zero to a constant value g at (say) time zero. Suppose now that in the absence of the coupling (5.2.9) the oscillator is at rest (cf. below) and the trajectories of the systems are $x_i(t)$ Then, the force exerted on the oscillator for $t > 0$ is $F(t) \equiv g(x_1(t) + x_2(t))$, and it is straightforward to show that the resultant motion is

$$X(t) = \int_0^t K(t-t')F(t')dt' \tag{5.2.10}$$

where the linear response function $K(t)$ is given by

$$K(t) \equiv \frac{1}{m\omega_0} \sin \omega_0 t \tag{5.2.11}$$

with m the oscillator mass and ω_0 its harmonic frequency. As a result, the force $F_1(t)$ on S_1 at time t, which according to (5.2.9) is equal to $gX(t)$, is given by the expression

$$F_1(t) = g^2 \int_0^t K(t-t')(x_1(t') + x_2(t'))dt' \tag{5.2.12}$$

The term in $x_1(t')$ on the RHS of (5.2.12) corresponds to an induced self-interaction of system 1 and will be ignored in the present context. The term in $x_2(t')$ may be seen to be equivalent to a term in the Hamiltonian of the form

$$H_{12} = -g^2 \int_0^t dt' K(t-t')x_1(t')x_2(t') \tag{5.2.13}$$

i.e. it corresponds to a *retarded* interaction $-g^2K(t-t')$ between S_1 and S_2. Now the Fourier transform $K(\omega)$ of $K(t)$ is easily shown to be $[m(\omega_0^2 - \omega^2)]^{-1}$. and thus we

[10]The characteristic scale of variation of $\chi_0(q0)$ is the Fermi wave vector q_F, which under realistic conditions is of the same order as q_{TF}.

find a Fourier-transformed effective interaction between S_1 and S_2 of the form

$$V_{\text{eff}}^{12}(\omega) = \frac{g^2}{m}\frac{1}{\omega^2 - \omega_0^2} \tag{5.2.14}$$

This is attractive for $\omega < \omega_0$ and repulsive for $\omega > \omega_0$ (and vanishes for $\omega \to \infty$). Because the *differential* response of a harmonic oscillator to an extra force is independent of its pre-existing motion, the result (5.2.14) is actually not restricted to the assumption that the oscillator was at rest in the absence of the coupling (5.2.9).

Since the harmonic oscillator has the (unique) property that its linear response to an external force is identical in classical and quantum mechanics, one would guess that a result similar to (5.2.14) would hold also in quantum mechanics (QM). In this case one would expect intuitively that the frequency ω should correspond to $\hbar^{-1}$ times the energy $\Delta\epsilon$ transferred between S_1 and S_2. However, this concept is unambiguous only for "real" transitions, in which the energy lost by S_1 is equal to that gained by S_2. To clarify this point we consider the standard QM formula for the effective matrix element for transitions between two states i, j mediated by virtual transitions into a state n (cf. e.g. Landau and Lifshitz (1977), Eqn. (38.12))

$$V_{ij}^{\text{eff}} = -\sum_n{}' V_{in}V_{nj}\left\{\frac{1}{\epsilon_i - \epsilon_n} + \frac{1}{\epsilon_j - \epsilon_n}\right\} \tag{5.2.15}$$

where to simplify the notation I have assumed that the matrix elements are real. In our case the relevant intermediate state n corresponds to S_1 having lost energy $\Delta\epsilon_1$ (with no change in S_2) and the oscillator being excited; if, as in (5.2.9), the coupling is linear in x then the only relevant excited state is the first one, so that $\epsilon_i - \epsilon_n$ is equal to $\Delta\epsilon_1 + \hbar\omega$. Similarly, the final state j is one in which the oscillator is de-excited back to the ground state and S_2 has gained energy $\Delta\epsilon_2$, so that $\epsilon_j - \epsilon_n = \hbar\omega - \Delta\epsilon_2$. Thus, if the original interaction is of the form (5.2.9), expression (5.2.15) becomes

$$V_{\text{eff}} = g^2\langle i_1|\hat{x}_i|f_1\rangle\langle i_2|\hat{x}_2|f_2\rangle|X_{10}|^2\left\{\frac{1}{-\Delta\epsilon_1 - \hbar\omega_0} + \frac{1}{\Delta\epsilon_2 - \hbar\omega_0}\right\} \tag{5.2.16}$$

where i_i, f_i denote the initial and final states of S_i.

In general this does not look like the classical expression (5.2.14). However, if we suppose that $|\Delta\epsilon_1 - \Delta\epsilon_2| \ll \hbar\omega_0$ (even though $\Delta\epsilon_1$ and $\Delta\epsilon_2$ themselves may be quite comparable to $\hbar\omega_0$), and thus set $\Delta\epsilon_1 \cong \Delta\epsilon_2 \equiv \Delta\epsilon$, and moreover take the into account that the oscillator matrix element X_{10} is just $(\hbar/2m\omega_0)^{1/2}$, then we precisely recover an expression corresponding to (5.2.14), with the expected identification of $\Delta\epsilon$ with $\hbar\omega$. In both the classical and the quantum cases, it is clear that the qualitative results are unchanged if x_1 and x_2 are replaced by arbitrary operators on S_1 and S_2 respectively, (but the coupling is still linear in X).

After these preliminaries we now return to the system of actual interest, namely the electrons and ions in a metal. The simplest generalization of the jellium model which gives a reasonable description is simply to make it two-component, that is to allow the ionic background to vibrate. One must then generalize the RPA treatment to take into account not only the ion kinetic energy but *all* components (electron–electron, electron–ion, and ion–ion) of the Coulomb interaction. The calculation is straightforward but somewhat lengthy, and is relegated to Appendix 5B; here I just quote the results for the effective electron–electron interaction in the approximation $q \ll k_F$, $\omega \ll c_s k_F$ (where c_s is the speed of sound), $c_s \ll v_F$, namely

$$V_{\text{eff}}(q\omega) = \frac{\kappa_0}{1 + q^2/q_{\text{TF}}^2} \left\{ 1 + \frac{\omega_{\text{ph}}^2(q)}{\omega^2 - \omega_{\text{ph}}^2(q)} \right\} \tag{5.2.17}$$

Here $\kappa_0 = (dn/d\epsilon)^{-1}$ is the static bulk modulus of the neutral Fermi gas, and $\omega_{\text{ph}}(q)$ is the frequency of longitudinal sound waves (phonons), which in this model is just $(\Omega_p/q_{\text{TF}})q \equiv c_s q$ ($\Omega_p \equiv (Z_{\text{ion}}^2(m/M)\omega_p^2)^{1/2}$ is the ionic plasma frequency). The first term in the curly brackets is just the effect of the direct (but suitably screened) interaction between the conduction electrons which we already calculated (see Eqn. 5.2.8), while the second expresses the effect of virtual polarization of the ionic lattice; as expected from the general discussion above, it is attractive for frequencies less than that of the relevant phonon and repulsive above it. The interaction (5.2.17) is known as the Bardeen–Pines interaction.

Equation (5.2.17) should not be taken too seriously; even within the "double-jellium" model it is not exact for $q \gtrsim k_F$, and moreover it will (for any q) have corrections from "nonjellium" effects (e.g. the short-range part of the conduction-electron–ion interaction). While most of these corrections do not affect the general structure of the effective interaction (crudely speaking, the main effect is to change the numerator (only) in the second term), one point is worth noting: Eqn. (5.2.17) has the slightly pathological feature that the effective interaction at $\omega = 0$ is exactly zero. This pathology is removed in a more general model, and in fact one finds that in such a model $V_{\text{eff}}(q, 0)$ may be either attractive or repulsive.

The derivation of Eqn. (5.2.17) given in Appendix 5B is semiclassical, in the sense that the motion of the ions is not explicitly quantized. It is possible to give a fully quantum-mechanical discussion, and one then finds that *provided* one considers only processes which are "non-resonant", i.e. in which the difference in the energy ($\Delta\epsilon_1$) lost by electron 1 and that gained ($\Delta\epsilon_2$) by electron 2 is small on the scale of $\omega_{\text{ph}}(q)$, then Eqn. (5.2.17) is justified provided that ω is identified with $\hbar^{-1}$ times the (approximate) energy transferred from 1 to 2. When this condition is not satisfied, the effective interaction (5.2.17) really has no meaning, and one must resort to the more first-principles calculational scheme devised by Eliashberg (see e.g. Scalapino 1969).

Very fortunately, it turns out that in most of the classical superconductors (the large class usually called "weak-coupling") not only $|\Delta\epsilon_1 - \Delta\epsilon_2|$ but the quantities $\Delta\epsilon_1$ and $\Delta\epsilon_2$ themselves are small compared to $\hbar\omega_{\text{ph}}(q)$ (i.e. $\omega \ll \omega_{\text{ph}}(q)$) for the processes most important for formation of the superconducting state. As a result, we can not only use the concept of an effective "retarded" interaction with the general structure of (5.2.17) but reasonably approximate it by its $\omega = 0$ limit. Moreover, any dependence on q does not usually seem to give rise to qualitatively important features, so that it seems reasonable to approximate $V_{\text{eff}}(q0)$ to be independent of q. The resulting interaction, when transformed back into coordinate space, has the simple "contact" form

$$V_{\text{eff}}(rt) = -V_0\,\delta(\boldsymbol{r}) \tag{5.2.18}$$

where the – sign is introduced for subsequent convenience. If we take the time-independent form (5.2.18) literally, it will turn out that we run into various problems with divergent integrals. This is essentially because arbitrarily large energy transfers are allowed. A standard way of avoiding this difficulty (cf. below) is to impose a cutoff not on the energy transfer as such but on the *absolute* energies, *relative to the Fermi energy*, of the electron states which are allowed to be coupled by the interaction (5.2.18). Thus, if we now change our notation so as to denote the energy of the plane-wave state *relative to the Fermi energy* by ϵ_k, so that in the jellium model

$$\epsilon_k \equiv \frac{\hbar^2 k^2}{2m} - \epsilon_{\mathrm{F}} \tag{5.2.19}$$

then the Fourier-transformed form of (5.2.18) is modified to

$$V_{kk'} \equiv \begin{cases} -V_0 & |\epsilon_k|, |\epsilon_{k'}| \text{ both} < \epsilon_{\mathrm{c}} \\ 0 & \text{otherwise} \end{cases} \tag{5.2.20}$$

In accordance with the general physical considerations involved in obtaining the phonon-induced attraction, the cutoff energy ϵ_{c} is usually taken to be of the order of the Debye energy (the maximum phonon energy), which is typically of the order of room temperature and hence much larger than T_{c}. In what follows I will usually use the interaction (5.2.18) (which is often called the "BCS contact potential").

Lest the reader meeting the considerations of this section for the first time should find them implausibly "hand-waving," I should emphasize that a much more complete and satisfactory justification of the use of the effective interaction (5.2.18) in the context of the calculation of most (though not quite all[11]) properties of the classical superconductor exists in the literature (one important aspect of the argument will be presented in Section 5.6 below.) The above discussion is intended primarily as motivational: I have simply tried to make plausible the fact that in some (though not all) metals there is an effective electron–electron interaction which is attractive for states close to the Fermi surface. It is, of course, this attraction which is, following the work of BCS, universally believed to be the reason for the formation of the superconducting state.

5.3 The Cooper instability

Before plunging into the full details of BCS theory, it may be useful to reproduce a very simple model calculation, originally due to Cooper, which historically played an important role in the genesis of the full theory. It seems not always to be appreciated how useful this "toy" model and simple generalizations of it can be, in particular in giving one a physical feel for which kinds of effects are likely to inhibit (or not) the formation of the superconducting state. In this Section I will present the model in its simplest (original) form, and in Section 5.9 will sketch some useful generalizations of it (see also Chapter 7, Appendix 7B).

Consider then two Fermi particles, constrained to be in a spin singlet state with center of mass at rest, so that the two-particle orbital wave function has the generic form

$$\psi_{\mathrm{orb}}(\boldsymbol{r}_1, \boldsymbol{r}_2) \equiv \psi_{\mathrm{orb}}(\boldsymbol{r}_1 - \boldsymbol{r}_2) = \sum_k c_k \, e^{i\boldsymbol{k}\cdot(\boldsymbol{r}_1 - \boldsymbol{r}_2)} \tag{5.3.1}$$

[11]In particular, a realistic calculation of tunneling spectra requires one to go back to the first-principles Eliashberg formalism: see McMillan and Rowell (1969).

subject to the conditions (of which the first is necessary to guarantee the overall Fermi antisymmetry)

$$c_k \equiv c_{-k}, \quad \sum_k |c_k|^2 = 1 \tag{5.3.2}$$

With Cooper, let us attempt to take into account the effect of the other $N-2$ particles by excluding possible values of $|\boldsymbol{k}|$ less than the Fermi momentum k_F. Then, if we define as above the single-particle kinetic energy $\epsilon_k \equiv \hbar^2/2m(k^2 - k_F^2)$ relative to the Fermi energy ϵ_F and moreover define the energy eigenvalue E to be relative to $2\epsilon_F$, the time-independent Schrödinger equation for the c_k takes the form

$$c_k = -\frac{1}{2\epsilon_k - E} \sum_{k'} V_{kk'} c_{k'} \tag{5.3.3}$$

where $V_{kk'}$ is the matrix element of the interelectron potential for scattering from states $(\boldsymbol{k}, -\boldsymbol{k})$ to $(\boldsymbol{k}', -\boldsymbol{k}')$. We take the latter to have the simple BCS form (5.2.18) with $\epsilon_c \ll \epsilon_F$, and seek a solution where c_k is independent of the direction of $\boldsymbol{k}$ (i.e. an s-state) and thus a function only of $\epsilon_k \equiv \epsilon$. Thus Eqn. (5.3.3) takes the form

$$c(\epsilon) = \frac{V_0}{2\epsilon - E} \int_0^{\epsilon_c} d\epsilon' \rho(\epsilon') c(\epsilon') \tag{5.3.4}$$

where $\rho(\epsilon) \equiv \sum_k \delta(\epsilon - \epsilon_k)$ is the (single-spin) density of states. We are interested in the possibility of a solution with $E < 0$. Now, were ϵ the *absolute* kinetic energy, the form of $\rho(\epsilon)$ in three space dimensions would be const. $\epsilon^{1/2}$, and Eqn. (5.3.4) would have no bound-state $(E < 0)$ solution for small enough V_0. However, in the present (Cooper) problem $\rho(\epsilon)$ tends to the constant $N(0)$ as $\epsilon \to 0$, and in fact may be reasonably approximated by this value for all $\epsilon < \epsilon_c$ (recall that in the BCS model for $V_{kk'}, \epsilon_c \ll \epsilon_F$). Thus (5.3.4) becomes

$$c(\epsilon) = \frac{N(0)V_0}{2\epsilon - E} \int_0^{\epsilon_c} d\epsilon' c(\epsilon') \tag{5.3.5}$$

This equation may be solved for the eigenvalue E by integration of both sides over ϵ, and cancellation of the integral, with the result (assuming $E < 0$)

$$1 = N(0)V_0 \int_0^{\epsilon_c} \frac{d\epsilon}{2\epsilon - E} = \frac{N(0)}{2} V_0 \ln\left(\frac{2\epsilon_c}{|E|} + 1\right) \tag{5.3.6}$$

i.e.

$$E = -2\epsilon_c \frac{1}{e^{2/N(0)V_0} - 1} \cong -2\epsilon_c e^{-2/N(0)V_0} \tag{5.3.7}$$

where the second approximate equality applies in the limit $V_0 \to 0$.

We thus find that however small the effective interelectron attraction V_0, a bound state (i.e. one with energy less than that of two free particles at the Fermi surface) always exists in the Cooper problem, although its binding energy $|E|$ tends to zero exponentially as $V_0 \to 0$. Note that an essential ingredient in this result is the logarithmic divergence of the integral on the RHS of Eqn. (5.3.6) in the limit $E \to 0$, which in turn is a result of the fact that the density of states available for scattering of the two particles remains constant in the limit that the energy (relative to the Fermi energy cutoff) tends to zero. It follows that any effect (such as the Zeeman effect of a magnetic field) which tends to destroy the availability of these low-energy scattering states will tend to suppress the bound state, cf. Section 5.9 below.

Let us now examine the wave function corresponding to the bound state as a function of the relative coordinate $r \equiv |\boldsymbol{r}_1 - \boldsymbol{r}_2|$. Taking into account that the angular dependence is that of an s-state, we find from (5.3.3) that up to normalization the wave function is given by the expression

$$\psi(r) = \frac{1}{r}\frac{\partial}{\partial r}\int_{k_F}^{k_c} \frac{\cos kr}{2\epsilon_k + |E|} dk \tag{5.3.8}$$

where k_c is the wave vector corresponding to ϵ_c. The general structure of $\psi(r)$ is a sum of terms of the form $(\cos k_F r, \sin k_F r)$ multiplied by functions of r which fall off as (at most) r^{-1} for small r but as (at least) r^{-2} for large r, with the crossover between the two types of behavior occurring at $r \sim \hbar v_F/|E| \sim (\hbar v_F/\epsilon_c)\exp[2/N(0)V_0] \equiv \xi_c$. Thus the state is bound[12] and of "radius" ξ_c, in the sense that the total probability of finding the particles beyond a distance $r \gg \xi_c$ apart tends to zero as r^{-1}; however, for $r \lesssim \xi_c$ the relative wave function is practically indistinguishable from that of two free particles with energy close to the Fermi energy in a relative s-state (which would have $\psi(r) \sim \sin k_F r/(k_F r)$).

It is tempting to try to generalize the Cooper calculation to finite temperature. The most natural way to do so would seem to be to note that the pair can only occupy the pair of states $\boldsymbol{k}\uparrow, -\boldsymbol{k}\downarrow$ if both of the latter are previously unoccupied by any of the other $N-2$ electrons, and that at finite temperature $T \equiv 1/k_B\beta$ the probability of this is $(1 + \exp -\beta\epsilon_k)^{-2}$ (where ϵ_k can have either sign). Thus, it is natural to replace the density of states used in (5.3.5), namely $\rho(\epsilon) = N(0)\theta(\epsilon)$, by the expression $N(0)(1 + e^{-\beta\epsilon})^{-2}$, and correspondingly Eqn. (5.3.6) is replaced by

$$1 = N(0)V_0 \int_{-\infty}^{\infty} \frac{d\epsilon}{(2\epsilon - E)(1 + \exp -\beta\epsilon)^2} \tag{5.3.9}$$

The singularity at $\epsilon = 0$ is now removed, and in fact it is clear (ignoring some messy details) that the effect is qualitatively similar to replacing the lower limit in (5.2.6) by $k_B T$. Thus the equation should have no negative-energy solution when it is no longer possible to satisfy a condition of the approximate form

$$\int_{k_B T}^{\epsilon_c} \frac{d\epsilon}{2\epsilon} > 1/N(0)V_0 \tag{5.3.10}$$

i.e. above a "critical temperature" T_c given by the order-of-magnitude estimate

$$T_c \sim (\epsilon_c/k_B)\exp -2/N(0)V_0 \sim |E|/k_B \tag{5.3.11}$$

We will see in the next two sections that the three fundamental properties of the "Cooper pair" state just described – the exponentially small binding energy and critical temperature, and the exponentially large radius – are reflected, albeit in a slightly different guise, in the *many-body* bound state which occurs in the full BCS theory.

[12]However, contrary to some statements in the literature (including Cooper's original Letter) the mean-square radius $\langle r^2 \rangle \equiv \int r^2|\psi(r)|^2 d^3r / \int |\psi(r)|^2 d^3r$ diverges. This is a consequence of the sharp cutoff at $\epsilon = 0$ (and $\epsilon = \epsilon_c$).

5.4 BCS theory at $T = 0$

In this section I shall outline the essentials of the BCS theory of the ground state of a superconducting metal; the generalization to finite temperature is given in Section 5.5. Details of the derivation of the formulae quoted below are relegated to Appendix 5C. The discussion given here differs somewhat from that of the original BCS paper, but the final outcome is the same.

Let us consider a fixed, even number N of fermions (in our case electrons) of spin $\frac{1}{2}$ moving in free space and subject to a weak interaction whose precise form we do not for the moment specify. It is convenient to subtract from the Hamiltonian the fixed quantity μN, which is equivalent to measuring the single-particle energies from μ; to make contact with a realistic situation in which the metal in question can exchange electrons with leads, etc., it will eventually be necessary to identify μ with the chemical potential $\partial E/\partial N$, but we will verify below that in the context of classical superconductivity[13] the latter may be consistently approximated by its $T = 0$ normal-state value, namely the Fermi energy ϵ_{F}. Our Hamiltonian then consists of a kinetic energy

$$\hat{T} \equiv \sum_{k,\sigma} \epsilon_k \hat{n}_{k\sigma} \tag{5.4.1}$$

(where $\hat{n}_{k\sigma}$ is the operator of the number of particles in plane-wave state $\boldsymbol{k}$ with spin σ), and a potential energy

$$\hat{V} \equiv \frac{1}{2} \sum_{ij} V(\hat{\boldsymbol{r}}_i - \hat{\boldsymbol{r}}_j) \tag{5.4.2}$$

We start by defining a class of many-body states which I shall call "generalized BCS states", among which we hope to find a good approximation to the true ground state of the system. A generalized BCS state is by definition a state of the form[14]

$$\Psi_N = \mathcal{N} \cdot \mathcal{A} \cdot \phi(\boldsymbol{r_1 r_2} \sigma_1 \sigma_2) \phi(\boldsymbol{r_3 r_4} \sigma_3 \sigma_4) \ldots \phi(\boldsymbol{r_{N-1} r_N} \sigma_{N-1} \sigma_N) \tag{5.4.3}$$

where $\mathcal{N}$ (like $\mathcal{N}'$, $\mathcal{N}''$ below) is a normalization factor and $\mathcal{A}$ antisymmetrizes with respect to the simultaneous exchange of coordinates and spins of any pair of electrons i, j; because of the presence of this operator, we may as well choose $\phi(\boldsymbol{r}_1\boldsymbol{r}_2\sigma_1\sigma_2)$ from the outset to be antisymmetric under the exchange $\boldsymbol{r_1} \rightleftharpoons \boldsymbol{r_2}$, $\sigma_1 \rightleftharpoons \sigma_2$. In second-quantized notation (see Appendix 2A) Ψ_N takes the more compact form

$$\Psi_N = \mathcal{N}' \left\{ \sum_{\alpha\beta} \iint d\boldsymbol{r}\, d\boldsymbol{r}' \phi(\boldsymbol{r}, \boldsymbol{r}' : \alpha, \beta) \psi_\alpha^\dagger(\boldsymbol{r})\, \psi_\beta^\dagger(\boldsymbol{r}') \right\}^{N/2} |\mathrm{vac}\rangle \tag{5.4.4}$$

where $|\mathrm{vac}\rangle$ indicates the vacuum state. Our task now is to find the form of the function $\phi(\boldsymbol{r}, \boldsymbol{r}', \alpha, \beta)$ which minimizes the sum of expressions (5.4.1) and (5.4.2). This formulation of the problem is very general, and applies not only to classic superconductivity but also to liquid ^{3}He (Chapter 6), the Fermi alkali gases (Chapter 8), and perhaps

[13]But not more generally, cf. Chapter 8, Section 8.4.

[14]In the most general case ϕ may also be a function of time t (cf. Section 5.7) but I omit this dependence since in the present section and all the next we are considering only the equilibrium state.

even cuprate superconductivity (Chapter 7). As we will see below, however, the most "physical" quantity is not the function $\phi(\boldsymbol{r}, \boldsymbol{r}' : \alpha, \beta)$ itself but a related quantity $F(\boldsymbol{r}, \boldsymbol{r}' : \alpha, \beta)$ which is essentially the eigenfunction of the two-particle density matrix associated with a macroscopic eigenvalue.

In the case of present interest, that of a superconducting metal, we will immediately restrict our search for the ground state to a sub-class of the class of functions (5.4.3), namely that defined by the conditions: (a) $\phi(\boldsymbol{r_1 r_2}\sigma_1\sigma_2)$ is a function only of the relative coordinate $\boldsymbol{r_1} - \boldsymbol{r_2}$ (i.e. corresponds to center-of-mass momentum zero); (b) the spin structure is a singlet. In other words we assume, in an obvious notation

$$\phi(\boldsymbol{r_1 r_2}\sigma_1\sigma_2) = 2^{-1/2}(\uparrow_1\downarrow_2 - \downarrow_1\uparrow_2)\phi(\boldsymbol{r_1} - \boldsymbol{r_2}) \tag{5.4.5}$$

Then Ψ_N can be written in the form

$$\Psi_N = \mathcal{N}'' \left(\sum_{\boldsymbol{k}} c_{\boldsymbol{k}} a^\dagger_{\boldsymbol{k}\uparrow} a^\dagger_{-\boldsymbol{k}\downarrow} \right)^{N/2} |\text{vac}\rangle \tag{5.4.6}$$

where the Fourier components $c_{\boldsymbol{k}}$ of $\phi(\boldsymbol{r_1} - \boldsymbol{r_2})$ satisfy the condition $c_{\boldsymbol{k}} \equiv c_{-\boldsymbol{k}}$ but are otherwise arbitrary. For convenience, however, we shall choose the normalization of the $c_{\boldsymbol{k}}$'s to satisfy the condition $\sum_k |c_k|^2/(1 + |c_k|^2) = N/2$ (c.f. Appendix 5C). Note that the *normal* ground state of the metal in the absence of interactions is a special case of the form (5.4.6), with $c_{\boldsymbol{k}} = \theta(k_{\mathrm{F}} - k)$.

The next step in the argument is to express the kinetic and potential energies in terms of the coefficients $c_{\boldsymbol{k}}$. Actually, it turns out to be more useful to work in terms of the quantity

$$F_{\boldsymbol{k}} \equiv \frac{c_{\boldsymbol{k}}}{1 + |c_{\boldsymbol{k}}|^2} \tag{5.4.7}$$

As shown in Appendix 5C, $F_{\boldsymbol{k}}$ is actually the value of the matrix element of the operator $a_{-k\downarrow}a_{k\uparrow}$ between the ground states of the $(N-2)$-particle and N-particle systems:

$$F_{\boldsymbol{k}} = \langle N - 2 | a_{-\boldsymbol{k}\downarrow} a_{\boldsymbol{k}\uparrow} | N \rangle \tag{5.4.8}$$

In terms of $F_{\boldsymbol{k}}$ the deviation of the expectation value of the single-particle operator $\hat{n}_{\boldsymbol{k}\sigma}$ from its normal-state value $\theta(k_{\mathrm{F}} - k)$ is given, from Eqn. (5.C.15) of Appendix 5C and Eqn. (5.4.7), by

$$\delta\langle \hat{n}_{\boldsymbol{k}\sigma}\rangle = \frac{1}{2}\left\{1 - \sqrt{1 - 4|F_{\boldsymbol{k}}|^2}\right\} \operatorname{sgn}[\epsilon_k] \tag{5.4.9}$$

and we can insert this expression in (5.4.1) to obtain the expectation value of the kinetic energy (cf. below). With regard to the potential energy $\hat{V}$ a delicate point arises. The general expression for $\langle \hat{V} \rangle$ in any many-body state is

$$\langle \hat{V} \rangle = \frac{1}{2} \sum_{ijmn} V_{ijmn} \langle a_i^\dagger a_j^\dagger a_m a_n \rangle \tag{5.4.10}$$

where i, j, m, n denote possible values of the plane-wave vector $\boldsymbol{k}$ and spin σ characterizing the single-particle states, and the matrix element V_{ijmn} is given explicitly by

$$V_{ijmn} = \Omega^{-2} \int dr \int dr' V(\boldsymbol{r} - \boldsymbol{r}') e^{i(\boldsymbol{k}_i - \boldsymbol{k}_n)\cdot \boldsymbol{r}} e^{i(\boldsymbol{k}_j - \boldsymbol{k}_m)\cdot \boldsymbol{r}'} \tag{5.4.11}$$

For a many-body wave function of the generic form (5.4.3) there are three types of terms $\langle a_i^\dagger a_j^\dagger a_m a_n \rangle$ which are nonzero:

1. $i = n$, $j = m$
   These are the Hartree terms: since from (5.2.18) the relevant matrix element V_{ijmn} is simply a constant $V(0)$ which is independent of i and j, the corresponding contribution to $\langle V \rangle$ is

$$\langle V \rangle_{\text{Hartree}} = \frac{1}{2} V(0) \langle \hat{N}^2 \rangle \equiv \frac{1}{2} V(0) N^2 \tag{5.4.12}$$

   where the second equality follows because (5.4.6) is by construction an eigenstate of $\hat{N}$. Thus this term is completely insensitive to the values of the individual F_k and can be ignored in the optimization process.

The second class of terms which are nonzero in (5.4.10) are

2. $i = m$, $j = n$
   These are the Fock terms: the corresponding contribution to $\langle V \rangle$ has the form

$$\langle V \rangle_{\text{Fock}} = \frac{1}{2} \sum_{ij} V_{ij,ji} \langle \hat{n}_i \hat{n}_j \rangle = \frac{1}{2} \sum_{ij} V_{ij,ji} \langle \hat{n}_i \rangle \langle \hat{n}_j \rangle$$

$$V_{ij,ji} = \Omega^{-1} \int dr \, e^{i(\boldsymbol{k}_i - \boldsymbol{k}_j) \cdot \boldsymbol{r}} V(r) \tag{5.4.13}$$

   where the second equality in (5.4.13) is valid, for a wave function of the generic form (5.4.6), to order N^{-1}. Inserting the value of $\langle \hat{n}_i \rangle$ from (5.4.9), we see that in general $\langle V \rangle_{\text{Fock}}$ is sensitive to the detailed form of the function $F_{\boldsymbol{k}}$, and therefore in principle needs to be taken into account in the optimization process. I will for the moment ignore, apparently arbitrarily, this term, and return in section 8 to the justification for doing so.[15]

The third class of terms which is nonzero in (5.4.10) are the so-called "pairing" terms:

3. $i = -j$, $m = -n$ (i.e. $\boldsymbol{k_i} = -\boldsymbol{k_j}$, etc.):

$$\langle V \rangle_{\text{pairing}} \equiv \sum_{\boldsymbol{k}\boldsymbol{k}'} V_{\boldsymbol{k}\boldsymbol{k}'} \left\langle a_{\boldsymbol{k}\uparrow}^{+} a_{-\boldsymbol{k}\downarrow}^{+} a_{-\boldsymbol{k}'\downarrow} a_{\boldsymbol{k}'\uparrow} \right\rangle \tag{5.4.14}$$

   where

$$V_{\boldsymbol{k}\boldsymbol{k}'} \equiv \int e^{i(\boldsymbol{k}-\boldsymbol{k}')\cdot\boldsymbol{r}} \, V(\boldsymbol{r}) \, d\boldsymbol{r} \quad (\equiv V(\boldsymbol{k} - \boldsymbol{k}')) \tag{5.4.15}$$

   As shown in Appendix 5C, these have a very simple form when expressed in terms of the functions F_k:

$$\langle V \rangle_{\text{pairing}} = \sum_{\boldsymbol{k}\boldsymbol{k}'} V_{\boldsymbol{k}\boldsymbol{k}'} F_{\boldsymbol{k}} F_{\boldsymbol{k}'}^* \tag{5.4.16}$$

[15] In the original BCS procedure, one starts from a model Hamiltonian which omits the Hartree and Fock terms.

Note that these are (the only) terms which would occur for scattering of two particles with CoM momentum zero.

Putting together the results (5.4.9) and (5.4.15), we find that for any state of the form (5.4.6) the expectation value of the Hamiltonian, up to a constant and the neglected Fock term, by the expression

$$\langle H \rangle = \langle T \rangle + \langle V \rangle_{\text{pair}}$$
$$= \sum_{\boldsymbol{k}} |\epsilon_k|(1 - \sqrt{1 - 4|F_{\boldsymbol{k}}|^2}) + \sum_{\boldsymbol{kk'}} V_{\boldsymbol{kk'}} F_{\boldsymbol{k}} F^*_{\boldsymbol{k'}} \tag{5.4.17}$$

Note that to the extent that we can expand the kinetic energy to lowest nontrivial order in $|F_{\boldsymbol{k}}|^2$, (5.4.17) becomes $\langle H \rangle = \sum_k 2|\epsilon_k||F_k|^2 + \sum_{kk'} V_{kk'} F_k F^*_{k'}$ and is identical to the expression for the total energy in the simple two- particle problem, with $F_{\boldsymbol{k}}$ interpreted as the Fourier component of the relative wave function and $|\epsilon_k| \equiv \epsilon_k = \hbar^2 k^2/2m$. This is, of course, not an accident, and we will explore this point further in Chapter 8, Section 8.4.

Equation (5.4.17) is valid for any generalized BCS wavefunction (5.4.3) subject to (5.4.5). We now optimize the choice of the function F_k by setting the derivative of (5.4.17) with respect to F^*_k equal to zero. This gives for $F_{\boldsymbol{k}}$ the self-consistent equation

$$\frac{2|\epsilon_k|F_{\boldsymbol{k}}}{\sqrt{1 - 4|F_{\boldsymbol{k}}|^2}} + \sum_{k'} V_{\boldsymbol{kk'}} F_{\boldsymbol{k'}} = 0 \tag{5.4.18}$$

Equation (5.4.18) is actually nothing but the celebrated BCS gap equation in disguise. We can make this explicit by introducing the notation

$$E_{\boldsymbol{k}} \equiv \frac{|\epsilon_k|}{\sqrt{1 - 4|F_{\boldsymbol{k}}|^2_{\text{eq}}}} \tag{5.4.19}$$

$$\Delta_k \equiv \left(\frac{F_k}{|F_k|}\right)_{\text{eq}} (E_k^2 - \epsilon_k^2)^{1/2} \tag{5.4.20}$$

so that

$$(F_k)_{\text{eq}} \equiv \Delta_k / 2E_k \tag{5.4.21}$$

where "eq" refers to the equilibrium value of F_k, i.e. that value which solves (5.4.18). Then Eqn. (5.4.18) becomes

$$\Delta_k = -\sum_{k'} V_{kk'} \frac{\Delta_{k'}}{2E_{k'}}$$
$$\left(E_k \equiv \sqrt{\epsilon_k^2 + |\Delta_k|^2}\right) \tag{5.4.22}$$

which is the standard form of the BCS gap equation at $T = 0$. (Recall that $V_{kk'}$ is defined as the matrix element $\langle k \uparrow, -k \downarrow |V| k' \uparrow, -k' \downarrow \rangle$.) I reemphasize that, subject to the neglect of the Fock term, the only assumption necessary to obtain (5.4.22) is that the (approximate) ground state indeed lies within the subclass (5.4.5) of generalized BCS functions (5.4.3).

Equation (5.4.22) always possess the trivial solution $\Delta_k \equiv 0$; according to (5.4.21) and (5.4.9) this corresponds to $F_k = 0$, $\langle n_k \rangle = \theta(-\epsilon_k)$, i.e. to the normal ground state (which as we have noted is a special case of (5.4.6) and thus of (5.4.3)). The existence and nature of one or more nontrivial solutions depends, of course, on the specific form of $V_{\boldsymbol{kk}'}$. At this point it is convenient to follow the original BCS paper and choose $V_{kk'}$ to have the simple form (5.2.18), that is,

$$V_{\boldsymbol{kk}'} \equiv \begin{cases} -V_0 & |\epsilon_k|, |\epsilon_{k'}| < \epsilon_c \\ 0 & \text{otherwise} \end{cases} \tag{5.4.23}$$

We will explore in Section 5.8 why this apparently arbitrary choice gives such good agreement with experiment for most of the classic superconductors.

With the choice (5.4.23) the RHS of (5.4.22) is independent of $\boldsymbol{k}$, so we can set Δ_k within the shell $|\epsilon_k| < \epsilon_c$ equal to a constant Δ:

$$\Delta_k = \begin{cases} \Delta, & |\epsilon_k| < \epsilon_c \\ 0, & |\epsilon_k| > \epsilon_c \end{cases} \tag{5.4.24}$$

where the magnitude of Δ is given by the implicit equation

$$1 = \frac{1}{2} V_0 \sum_k (\epsilon_k^2 + |\Delta|^2)^{-1/2} \tag{5.4.25}$$

where the sum goes over the states within the shell $|\epsilon_k| < \epsilon_c$. The phase of Δ, which according to (5.4.21) is the (uniform) phase of F_k, is not a physically meaningful quantity, any more than is the overall phase of the relative wave function in the two-particle problem. For $V_0 < 0$ (repulsive interaction) Eqn. (5.4.22) clearly has no solution, and the ground state within the ansatz (5.4.6) is the normal one. The interesting case is $V_0 > 0$, $N(0)V_0 \ll 1$ (weak attractive interaction). In this case Eqn. (5.4.25) yields the result

$$|\Delta| \cong 2\epsilon_c \exp[-1/N(0)V_0] \quad (\ll \epsilon_c) \tag{5.4.26}$$

Note that the exponent in (5.4.26) is half the value occurring in the expression for $|E|$ in the Cooper problem.[16]

With this solution both the value of F_k, and the deviation $\delta\langle n_{k\sigma}\rangle$ of the single-particle occupation number $\langle n_{k\sigma}\rangle$ from its normal-state value $\theta(-\epsilon_k)$, are substantial only for $|\epsilon_k| \lesssim |\Delta|$; in fact, from (5.4.21) and (5.4.9) we have (cf. Fig. 5.1)

$$F_k = \frac{\Delta}{2(\epsilon_k^2 + |\Delta|^2)^{1/2}} \sim |\epsilon_k|^{-1} \tag{5.4.27}$$

$$\delta\langle n_{k\sigma}\rangle = \frac{1}{2}\left(1 - \frac{\epsilon_k}{(\epsilon_k^2 + |\Delta|^2)}\right) - \theta(-\epsilon_k) \sim |\epsilon_k|^{-2} \tag{5.4.28}$$

where the asymptotic behavior for $|\epsilon| \gg |\Delta|$ is indicated. Figure 5.1 shows that the effect of pairing on the single-particle distribution $\langle n_{k\sigma}\rangle$ is qualitatively similar to that of a finite temperature of order Δ/k_B; the sharp discontinuity at $k = k_F$ is removed but the "average" Fermi surface is unchanged.

[16]This is because the sum over $\boldsymbol{k}$ now runs over negative as well as positive ϵ_k.

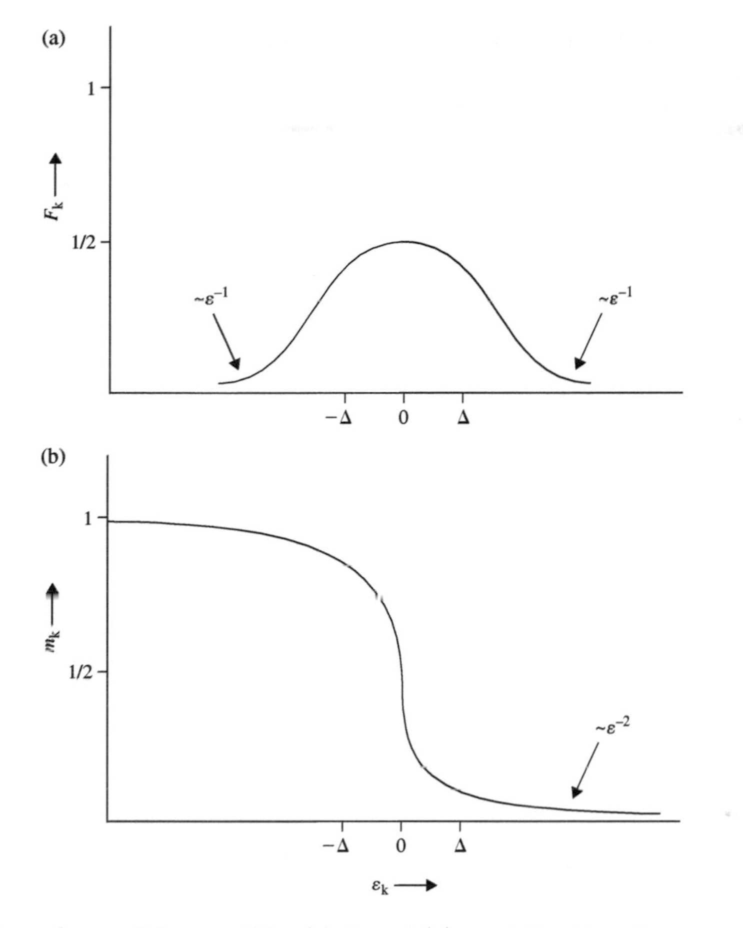

Fig. 5.1 Dependence of the quantities (a) F_k and (b) n_k at $T = 0$ on the energy ϵ_k relative to the Fermi energy.

Using (5.4.26)–(5.4.27) we can calculate the value of the terms in (5.4.17) in the limit $\epsilon_c \gg |\Delta|$:[17]

$$\langle T \rangle = N(0)\, \Delta^2 \left(\ln \left(\frac{2\epsilon_c}{\Delta} \right) - \frac{1}{2} \right) \tag{5.4.29}$$

$$\langle V \rangle_{\text{pair}} = -V_0 N^2(0) \ln^2 \left(\frac{2\epsilon_c}{\Delta} \right) \tag{5.4.30}$$

It should be noted that these results remain valid even if $|\Delta|$ does not have the equilibrium value specified by (5.4.26), provided F_k still has the form (5.4.27). Differentiation of $E(\Delta) \equiv \langle T \rangle(\Delta) + \langle V \rangle_{\text{pair}}(\Delta)$ with respect to Δ gives back the result

[17]The exact result replaces the ln in (5.4.29)–(5.4.30) by $\sinh^{-1}(\epsilon_c/\Delta)$ and the condensation energy (5.4.31) by $N(0)[\epsilon_c^2 - \epsilon_c(\epsilon_c^2 + \Delta^2)^{1/2}] = -\frac{1}{2}N(0)\Delta^2 + 0(\Delta/\epsilon_c)^2$.

(5.4.26), and a condensation energy relative to the normal groundstate energy of

$$E_{\text{cond}} = -\frac{1}{2}\, N(0)\, \Delta^2 \tag{5.4.31}$$

Note that relative to the energy of the filled Fermi sea E_{cond} is only a fraction of order $(\Delta/\epsilon_{\text{F}})^2$, which for classical superconductors is typically $\sim 10^{-7}$–10^{-8}. Consequently the relative shift in chemical potential is also only of this order, and can usually be safely neglected.

Let us study the structure of the Fourier transform[18] $F(\boldsymbol{r})$ of F_k. As noted in Appendix 5C, this quantity is proportional to the orbital part of the eigenfunction of the two-particle density matrix associated with the single macroscopic eigenvalue, and its norm (i.e. the integral of $|F(\boldsymbol{r})|^2$) just gives the value N_0 of this eigenvalue; thus, we may regard $F(\boldsymbol{r})$ (or more precisely the normalized quantity $N_0^{-1/2}F(\boldsymbol{r})$) as the "wave function of the condensate" and N_0 as the "number of Cooper pairs". The calculation of N_0 itself is straightforward: from (5.4.21), in the approximation of a constant density of states $N(0)$ and $|\Delta| \ll \epsilon_{\text{c}}$

$$N_0 \equiv \Omega \int d\boldsymbol{r} |F(\boldsymbol{r})|^2 = \Omega \sum_k |F_k|^2 = \Omega\Delta^2 \sum_k (4E_k^2)^{-1} = \frac{\pi}{4}\Delta\; N(0)\Omega \tag{5.4.32}$$

Since in the free-electron (Sommerfeld) approximation $N(0)$ is equal to $(3\epsilon_{\text{F}})^{-1}$ times the total density N/Ω of electrons, the "condensate fraction" N_0/N is of order $\Delta/\epsilon_{\text{F}} \sim 10^{-3}$–$10^{-4}$.

We now consider the general form of the "condensate wavefunction"

$$F(\boldsymbol{r}) = \Omega^{-1}\Delta \sum_k (2E_k)^{-1} e^{i\boldsymbol{k}\cdot\boldsymbol{r}} \tag{5.4.33}$$

where the sum goes over the shell $|\epsilon_k| < \epsilon_{\text{c}}$. This expression is isotropic ($F(\boldsymbol{r}) \equiv F(r)$, $r \equiv |\boldsymbol{r}|$), and in the approximation of a constant density of states may be written

$$F(r) = \Delta N(0) k_{\text{F}}^{-2} r^{-1} \frac{\partial}{\partial r} \left\{ \cos k_{\text{F}} r f\left(\frac{r}{\xi'}\right) \right\}, \quad \xi' \equiv \frac{\hbar v_{\text{F}}}{|\Delta|} \tag{5.4.34}$$

$$f(z) \equiv \int_{-\epsilon_{\text{c}}/\Delta}^{\epsilon_{\text{c}}/\Delta} \frac{\cos zx\, dx}{\sqrt{1+x^2}} \tag{5.4.35}$$

(where the coefficient of a term proportional to $\sin k_{\text{F}} r$ vanishes because of the "particle–hole" symmetry, i.e. the fact that the integrand is an even function of ϵ).

As in the Cooper problem, the existence of a sharp cutoff in F_k (in this case at $|\epsilon_k| = \epsilon_{\text{c}}$ only) leads to a power-law behavior of $F(r)$ at long distances. However, unlike in that problem, it is now possible to argue that this behavior is a pathological consequence of the artificial choice (5.2.18) of the pairing potential and will be removed by a more physically reasonable choice. This justifies the extension of the limits of integration in (5.4.35) to $\pm\infty$, whereupon we find that $f(z)$ is just the Bessel function $K_0(z)$; for $z \ll 1$ it diverges as $\ln z^{-1}$, but this singularity is removed when we go back to a finite cutoff, and is of no great interest to us. More interesting is the

[18] It should be emphasized that we are assuming throughout this section (and the next) that the center of mass of the pairs is at rest: $\boldsymbol{r}$ here denotes the *relative* coordinate.

behavior for $z \gtrsim 1$, where we find $f(z) \sim \exp(-\sqrt{2}z)$. Thus, anticipating the fact that the quantity ξ' will turn out to be $\gg k_F^{-1}$, we find that the approximate form of the pair (condensate) wave function is, for $r \gg k_F^{-1}$, v_F/ϵ_c,

$$F(r) \cong \Delta N(0) \frac{\sin k_F r}{k_F r} e^{-\sqrt{2}r/\xi'} \tag{5.4.36}$$

We see that the "radius" of a pair is of order ξ'. (For historical reasons, in the literature it is more common to use the quantity $\xi \equiv \pi^{-1}\xi' \equiv \hbar v_F/\pi\Delta$, which is called the "Pippard coherence length" (cf. the end of Section 5.7).) Using the fact that $N(0) \times k_F^{-2}$ is of order $(\hbar v_F)^{-1}$, we find that the integral of $|F(r)|^2$ is indeed of the order (5.4.31). For $r \lesssim \xi'$ the pair wavefunction is approximately proportional to that of two particles at the Fermi energy moving freely in a relative s-wave state. In a typical classical superconductor ξ' is of order 1 μm, much larger than the interelectron spacing (~1 Å).

To conclude this section, let us note an alternative particle-conserving ansatz for the many-body wave function of a Cooper-paired Fermi system which for the simple case discussed in this section (a pure system with s-wave pairing in the limit $\Delta/\epsilon_F \to 0$) is completely equivalent to (5.4.6), but which when generalized to more complicated situations, such as those discussed in Section 5.9 of this chapter, Chapter 6 or Chapter 8, Section 8.4, may be subtly different. We consider the limit $\Delta/\epsilon_F \to 0$ and thus set the chemical potential equal to the normal-state Fermi energy. The idea is to create our pairs not on the vacuum as in Eqn. (5.4.6), but rather out of the normal-state Fermi gas, which we denote as $|\text{FS}\rangle$: formally,

$$|\text{FS}\rangle = \prod_{\substack{\boldsymbol{k}\sigma \\ (k<k_F)}} a_{\boldsymbol{k}\sigma}^{+}|\text{vac}\rangle \equiv \left(\sum_{\substack{\boldsymbol{k} \\ k<k_F}} a_{\boldsymbol{k}\uparrow}^{+} a_{-\boldsymbol{k}\downarrow}^{+} \right)^{N/2} |\text{vac}\rangle \tag{5.4.37}$$

The ansatz is, apart from normalization and the caveat below,

$$\Psi_N = \left(\sum_{k>k_F} c_{\boldsymbol{k}} a_{\boldsymbol{k}\uparrow}^{+} a_{-\boldsymbol{k}\downarrow}^{+} \right)^{N_+} \left(\sum_{k<k_F} d_{\boldsymbol{k}} a_{-\boldsymbol{k}\downarrow} a_{\boldsymbol{k}\uparrow} \right)^{N_-} |FS\rangle \equiv \Psi_N(N_+) \tag{5.4.38}$$

where we set (for this case, $\mu = \epsilon_F$)

$$N_+ = N_- = \sum_{k>k_F} \frac{|c_{\boldsymbol{k}}|^2}{1 + |c_{\boldsymbol{k}}|^2} \tag{5.4.39}$$

and where the $c_{\boldsymbol{k}}$ are the same as in (5.4.6), while the $d_{\boldsymbol{k}}$ are the algebraic inverses of those $c_{\boldsymbol{k}}$s:

$$d_{\boldsymbol{k}} = c_{\boldsymbol{k}}^{-1} \tag{5.4.40}$$

Actually, Eqn. (5.4.38) as it stands does not give results identical to the treatment given above, for a subtle reason: since it conserves the number of "particles" (N_+) and "holes" (N_-) separately, the expectation value $\langle a_{\boldsymbol{k}\uparrow}^{+} a_{-\boldsymbol{k}\downarrow}^{+} a_{-\boldsymbol{k}'\downarrow} a_{\boldsymbol{k}'\uparrow} \rangle$ vanishes whenever $\boldsymbol{k}$ and $\boldsymbol{k}'$ are on opposite sides of the Fermi surface, in contrast to the standard result (cf. Eqn. (5.4.16), where there is no restriction on the sums). This problem however is easily remedied by replacing (5.4.38) by

$$\Psi_N = \sum_{N_+} c(N_+)\Psi_N(N_+) \tag{5.4.41}$$

where the coefficients $c(N_+)$ are strongly peaked around the value of the N_+ given by (5.4.39). Then for the case presently considered (5.4.41) is effectively equivalent to (5.4.6) , and thus gives identical expectation values for all physical quantities, in particular for the $F_{\boldsymbol{k}}$. Note that the quantity $\langle N_+\rangle$ is given, for the BCS model presented in this section, by

$$\langle N_+\rangle = \langle N_-\rangle = \frac{1}{2}\sum_{k>k_\mathrm{F}}\left(1-\frac{\varepsilon_k}{E_k}\right) = N(0)\Delta\Omega \tag{5.4.42}$$

Comparing this result with (5.4.32), we see that up to a factor of $4/\pi$ N_+ is equal to the "number of Cooper pairs" as we have defined it above. Note also that, from (5.4.40) the phase of $d_{\boldsymbol{k}}$ is opposite to that of $c_{\boldsymbol{k}}$; this remark is of course trivial for the s-wave case of present interest, where the $c_{\boldsymbol{k}}$ can always be taken real, but has some significance in the case, realized in superfluid ^{3}He, of pairing with nonzero angular momentum, see Appendix 6A.

5.5 Excited states and finite-temperature BCS theory

The ground state of the Hamiltonian within the manifold of generalized BCS states (5.4.6) is given by the particular choice of the $c_{\boldsymbol{k}}$ (or equivalently of the $F_{\boldsymbol{k}}$) specified by Eqns. (5.4.21) and (5.4.20), with $\Delta_{\boldsymbol{k}}$ satisfying the self-consistent equation (5.4.22). Let us start our consideration of the elementary excitations by imagining that we stay within the manifold (5.4.6) but allow a small number ($\ll N$) of the $c_{\boldsymbol{k}}$'s to differ from their ground state values. Since it is easy to convince oneself that under these conditions the contributions from the different $\boldsymbol{k}$'s are additive, we may as well focus on a single pair of states $(\boldsymbol{k}\uparrow, -\boldsymbol{k}\downarrow)$.

According to Appendix 5C, if we suppress the dependence on the other $\boldsymbol{k}$'s (i.e. the Ψ'_n) the state of the pair in question in the ground state (the "ground pair" (GP) state) can be written schematically

$$\Psi_{\mathrm{GP}} = (1+|c_k|^2)^{-1/2}(|00\rangle_k + c_k|11\rangle_k) \equiv u_k|00\rangle_k + v_k|11\rangle_k \tag{5.5.1}$$

Then the quantity $\langle a_{-k\downarrow}a_{k\uparrow}\rangle$ (or more strictly $\langle N-2|a_{k\downarrow}a_{-k\uparrow}|N\rangle$) is equal to $F_k \equiv c_k/(1+|c_k|^2)$, and the expectation value of the kinetic energy relative to that in the normal ground state is $2|\epsilon_k||c_k|^2/(1+|c_k|^2) = |\epsilon_k|(1-\sqrt{1-4|F_k|^2}) = |\epsilon_k|(1-|\epsilon_k|/E_k)$. Now, within the two-dimensional manifold spanned by $|00\rangle_k$ and $|11\rangle_k$ there exists exactly one other energy eigenfunction, which by a general theorem must be orthogonal to Ψ_{GP}: denoting this as the "excited pair" (EP) state, we see that its form is (up to an arbitrary overall phase factor)

$$\Psi_{\mathrm{EP}} = (1+|c_k|^2)^{-1/2}(c_k^*|00\rangle_k - |11\rangle_k) \equiv v_k^*|00\rangle_k - u_k^*|11\rangle_k \tag{5.5.2}$$

It is clear that the value of $\langle a_{k\downarrow}a_{-k\uparrow}\rangle$ in this state is just $-F_k$ (note no extra phase factor!), while the value of the kinetic energy is $2|\epsilon_k|/(1+|c_k|^2) = |\epsilon_k|(1+\sqrt{1-4|F_k|^2})$.

Consequently, the total energy *relative to the "normal state" value* $|\epsilon_k|$ is just opposite that in the ground state. Furthermore, the difference between the "excited-pair" and "ground-pair" energies is given from (5.4.17) by the expression[19]

$$E_{\text{EP}} - E_{\text{GP}} = 2|\epsilon_k|\sqrt{1 - 4|F_k|^2} + 2F_k \cdot 2\sum_{k'} V_{kk'}F^*_{k'} \tag{5.5.3}$$

which by Eqns. (5.4.19) and (5.4.22) becomes

$$E_{\text{EP}} - E_{\text{GP}} = 2\left(\frac{|\epsilon_k|^2}{E_k} + \frac{|\Delta|^2}{E_k}\right) \equiv 2E_k \tag{5.5.4}$$

Thus, we have for the "absolute" energies[20] of the GP and EP states

$$E_{\text{GP}} = |\epsilon_k| - E_k \quad E_{\text{EP}} = |\epsilon_k| + E_k \tag{5.5.5}$$

We now consider a different class of elementary excitations, which falls outside the manifold (5.4.6). Suppose we construct a pair wave function out of $N - 2$ particles, leaving two over. These two must then go into different single-particle plane-wave states $\boldsymbol{k}'\sigma'$ and $\boldsymbol{k}''\sigma''$, and the pair states $\boldsymbol{k}'\sigma'$, $-\boldsymbol{k}' - \sigma'$ and $\boldsymbol{k}''\sigma''$, $-\boldsymbol{k}'' - \sigma''$ must then be omitted from the "pair" part of the wave function, something which we denote schematically by a prime on the sum over $\boldsymbol{k}$. Thus the relevant ansatz for the many body wave function is

$$\Psi_N = \left(\sum_{\boldsymbol{k}}{}' c_{\boldsymbol{k}} a^\dagger_{\boldsymbol{k}\uparrow} a^\dagger_{-\boldsymbol{k}\downarrow}\right)^{N/2-1} a^\dagger_{\boldsymbol{k}'\sigma'} a^\dagger_{\boldsymbol{k}''\sigma''}|\text{vac}\rangle \tag{5.5.6}$$

It is clear that the kinetic energy (relative to the normal groundstate) associated with the "broken pair" states $\boldsymbol{k}'$ and $\boldsymbol{k}''$ is $|\epsilon_{k'}|$ and $|\epsilon_{k''}|$ respectively, and since they have no "partners" they cannot scatter into another pair state and hence make no contribution to the "pairing" terms in the potential energy.

Thus, considering the four-dimensional "occupation" space associated with the pair of plane-wave states $(\boldsymbol{k}\uparrow, -\boldsymbol{k}\downarrow)$, we find the following table of eigenstates and energies

$$\Psi_{\text{GP}} = u_k|00\rangle + v_k|11\rangle \quad E_{\text{GP}} = |\epsilon_k| - E_k \tag{5.5.7a}$$

$$\Psi_{\text{BP}} = |10\rangle, |01\rangle \quad E_{\text{BP}} = |\epsilon_k| \tag{5.5.7b}$$

$$\Psi_{\text{EP}} = v^*_k|00\rangle - u^*_k|11\rangle \quad E_{\text{EP}} = |\epsilon_k| + E_k \tag{5.5.7c}$$

I note that in the literature (with the usual relaxation of number conservation) it is conventional to describe these states in the language of "Bogoliubov quasiparticles"; the relevant quasiparticle creation operators are

$$\alpha^\dagger_{\boldsymbol{k}\uparrow} = u^*_k a^\dagger_{\boldsymbol{k}\uparrow} + v^*_{\boldsymbol{k}} a_{-\boldsymbol{k}\downarrow} \tag{5.5.8a}$$

$$\alpha^\dagger_{-\boldsymbol{k}\downarrow} = u^*_k a^\dagger_{-\boldsymbol{k}\downarrow} + v^*_k a_{\boldsymbol{k}\uparrow} \tag{5.5.8b}$$

and thus satisfy the standard fermion anticommutation relations. It is easy to verify that the operator $\alpha^\dagger_{k\uparrow}$, when acting on the GP state, creates the broken-pair

[19]Strictly speaking, we should omit the term $\boldsymbol{k}' = \boldsymbol{k}$ in the sum. However, the error introduced by not doing so is of order N^{-1}. In writing (5.5.3) we have used the reality of F_k up to a k-independent phase.

[20]i.e. the value of $\langle \hat{H} - \mu\hat{N}\rangle$ relative to that in the normal ground state.

state $|10\rangle$, and the operator $\alpha^{\dagger}_{k\downarrow}$ similarly creates (up to an irrelevant phase factor) the broken-pair state $|01\rangle$. The combination $\alpha^{\dagger}_{k\uparrow}\alpha^{\dagger}_{-k\downarrow}$ creates the excited-pair state, which can thus if we wish be regarded as consisting of two Bogoliubov quasiparticles each with excitation energy E_k. However, I believe that this point of view tends to obscure the fact that in some intuitive sense the excited-pair states, unlike the broken-pair ones, are still "part of" the condensate, and in particular contribute (negatively) to the quantity $\langle a^{\dagger}_{k\uparrow}a^{\dagger}_{-k\downarrow}\rangle$.

It is important to appreciate that provided we define E_k (and thus Δ_k) purely as parametrizations of the coefficients c_k (or equivalently F_k) occurring in the ansatz (5.4.6) for a generalized BCS state, that is, implicitly by

$$E_k \equiv (\epsilon_k^2 + |\Delta_k|^2)^{1/2}, \quad F_k \equiv \frac{\Delta_k}{2E_k} \tag{5.5.9}$$

(or explicitly by $E_k \equiv \epsilon_k/\sqrt{1-4|F_k|^2}$, $\Delta_k \equiv (F_k/|F_k|)(E_k^2 - \epsilon_k^2)^{1/2}$), then the results (5.5.3)–(5.5.5) are valid *irrespectively* of whether or not Δ_k satisfies the gap equation (5.4.22), i.e. whether the choice of the c_k in (5.4.6) minimizes the energy. However, under these more general conditions the energies E_{GP} and E_{EP} should be interpreted as expectation values of the energy rather than true eigenvalues. (E_{BP} of course remains an eigenvalue.)

With the above remark in mind we now consider the finite-temperature behavior in BCS theory. The technically correct description of the many-body system at $T \neq 0$ is by a density matrix $\hat{\rho}_N$; we expect the many-body states Ψ_N which contribute substantially to $\hat{\rho}_N$ to have the typical form

$$\Psi_N \sim \left(\sum_{\substack{\boldsymbol{k} \\ \boldsymbol{k}\neq\boldsymbol{k}',\boldsymbol{k}''\ldots}} c_{\boldsymbol{k}} a^{\dagger}_{\boldsymbol{k}\uparrow} a^{\dagger}_{-\boldsymbol{k}\downarrow} \right)^{N_{\text{p}}} a^{\dagger}_{\boldsymbol{k}'\sigma'} a^{\dagger}_{\boldsymbol{k}''\sigma''} \ldots |\text{vac}\rangle \tag{5.5.10}$$

where the number of single-particle operators $a^{\dagger}_{k'\sigma'}\ldots$ is $(N - 2N_{\text{p}})$. In the BCS particle-nonconserving representation (see Appendix 5C) the description simplifies: the many-body density matrix $\hat{\rho}$ can be written as a product of density matrices $\hat{\rho}_k$ referring only to the 4×4 "occupation space" spanned by the four states $|00\rangle$, $|01\rangle$, $|11\rangle$ and $|10\rangle$. Equivalently, for any given choice of parameters $c_{\boldsymbol{k}}$ and hence of $\Delta_{\boldsymbol{k}}$, etc., the occupation space is spanned by four functions of the form (5.5.7). Let us now assume for the moment that the choice of the Δ_k is indeed optimal, so that the states (5.5.7) are true energy eigenstates, and assign to them standard thermal weights. Since the four states are distinguishable, the respective probabilities P_{GP}, etc., are given by ($\beta \equiv 1/k_{\text{B}}T$)

$$P_{\text{GP}} : P_{\text{BP}} : P_{\text{EP}} = 1 : e^{-\beta E_k} : e^{-2\beta E_k} \tag{5.5.11}$$

$$P_{\text{GP}} + 2P_{\text{BP}} + P_{\text{EP}} = 1 \tag{5.5.12}$$

Rather than writing out the solutions for the P's explicitly, it is convenient to write down the expressions for the thermal averages of the single-particle occupation number n_k and the quantity $c_k/(1+|c_k|^2)$; from now on we will simply denote this latter average F_k, in the hope that this will not lead to confusion. Since the expectation values of $n_k - \frac{1}{2}$ and of $c_k/(1+|c_k|^2)$ are both opposite for the EP and GP states, we find

$$\delta\langle n_{k\sigma}\rangle \equiv \langle n_{k\sigma}\rangle - \langle n_{k\sigma}\rangle_0 = \frac{1}{2}\left(1 - \frac{|\epsilon_k|}{E_k}\tanh\frac{\beta E_k}{2}\right)\mathrm{sgn}[\epsilon_k] \tag{5.5.13}$$

$$F_k = \frac{\Delta_k}{2E_k}\tanh\frac{\beta E_k}{2}, E_k \equiv (E_k^2 + |\Delta_k(T)|^2)^{1/2} \equiv E_k(T) \tag{5.5.14}$$

(where $\langle n_k\rangle_0 \equiv \theta(k_\mathrm{F} - k)$ is the distribution in the normal ground state) We note that in the normal phase ($\Delta_k \equiv 0$, $E_k \equiv |\epsilon_k|$) F_k is zero and $\langle n_k\rangle - \langle n_k\rangle_0$ has the standard Fermi form $(\exp\beta|\epsilon_k| + 1)^{-1}$ sgn ϵ_k.

It remains to find the optimal choice of the parameters Δ_k. To do this at $T \neq 0$ we must minimize the Helmholtz free energy, that is the sum of the kinetic energy $\sum_k 2\epsilon_k(\langle n_k\rangle - \langle n_k\rangle_0)$, the potential energy (approximated, as at $T = 0$, by the pairing term $\sum_{kk'} V_{kk'} F_k F_{k'}^*$) and the term $-TS$, where the entropy S is a sum of contributions S_k given by

$$S_k \equiv -\sum_i P_i^{(k)} \ln P_i^{(k)} = -k_\mathrm{B}\{\beta E_k \tanh\tfrac{1}{2}\beta E_k - 2\ln(2\cosh\tfrac{1}{2}\beta E_k)\} \tag{5.5.15}$$

Relegating the rather messy algebra to Appendix 5D, we simply quote the result that minimization of the Helmholtz free energy with respect to Δ_k^* leads to the equation (which of course is derivable in many other ways, cf. Appendix 5D)

$$\Delta_k = -\sum_k V_{kk'} \frac{\Delta_{k'}}{2E_{k'}} \tanh\frac{\beta E_{k'}}{2} \tag{5.5.16}$$

Equation (5.5.16) is the finite-temperature BCS gap equation. Its most important property is that, irrespective of the detailed form of $V_{kk'}$, there is always a critical temperature T_c above which it has no solution. To find T_c we neglect terms of order higher than linear in Δ_k, i.e. set $E_k = |\epsilon_k|$. In the case of the BCS contact potential (5.2.20) this yields an implicit equation for T_c, namely

$$[N(0)V_0]^{-1} = \int_0^{\epsilon_c} \frac{\tanh(\beta\epsilon/2)}{\epsilon} d\epsilon = \ln(1.14\beta_\mathrm{c}\epsilon_\mathrm{c}) \quad \left(\beta_\mathrm{c} \equiv \frac{1}{k_\mathrm{B}T_\mathrm{c}}\right) \tag{5.5.17}$$

where in the second equality is assumed $k_\mathrm{B}T_\mathrm{c} \ll \epsilon_\mathrm{c}$. That is,

$$T_\mathrm{c} = 1.14\frac{\epsilon_\mathrm{c}}{k_\mathrm{B}} e^{-1/N(0)V_0} \tag{5.5.18}$$

Comparing (5.5.18) with the result (5.4.26) for the zero-temperature gap $\Delta(0)$, we find the celebrated relation

$$\Delta(0) = 1.76 k_\mathrm{B} T_\mathrm{c} \tag{5.5.19}$$

It is actually easy to see that in the BCS contact model (with $T_\mathrm{c} \ll \epsilon_\mathrm{c}$) the gap at any temperature below T_c can be expressed in the scaled form

$$\Delta(T) = k_\mathrm{B}T_\mathrm{c} f(T/T_\mathrm{c}) \tag{5.5.20}$$

where the universal function $f(x)$ is determined by the implicit equation

$$\int_0^\infty \left\{\frac{\tanh[(2x)^{-1}(z^z + f^2(x))^{1/2}]}{(z^2 + f^2(x))^{1/2}} - \frac{\tanh z/2}{z}\right\} dz = 0 \tag{5.5.21}$$

Although the function $f(x)$ must be obtained numerically, it is reasonably well approximated by the analytic expression $1.76(1-x^4)^{1/2}$.

To conclude this section, let us discuss the behavior of the finite-temperature pair wave function $F(r)$ as a function of the relative coordinate r. Actually the definition of $F(r)$ needs a little care: In general, the different many-body states Ψ_N occurring in (5.5.10), while they have similar values of Δ_k, and thus of the *magnitudes* of the coefficients c_k, will have fluctuations in the sign of c_k (corresponding to the fact that the relevant pair state is sometimes GP and sometimes EP); in addition, for some Ψ_N (those in which the pair state in question is BP) c_k will be zero. However, since according to (5.5.11) the behavior of different pair states $\boldsymbol{k}$, $\boldsymbol{k}'$ is uncorrelated even when $\boldsymbol{k}$ and $\boldsymbol{k}'$ are close, the sum over $\boldsymbol{k}$ for any given Ψ_N (and hence a fortiori for the statistical ensemble) may be safely approximated by assigning to the c_k, or equivalently to the F_k, their thermal average values (5.5.14). Thus,

$$F(r) = \sum_k \frac{\Delta_k}{2E_k} \tanh \frac{\beta E_k}{2} e^{i\boldsymbol{k}\cdot\boldsymbol{r}} \tag{5.5.22}$$

where Δ_k and E_k are the appropriate functions of T. Thus in the BCS contact model the effect of finite temperature is to replace the quantity $(\epsilon^2 + |\Delta(0)|^2)^{-1/2}$ occurring in (5.4.33) by $(\epsilon^2 + |\Delta(T)|^2)^{-1/2} \tanh \beta/2(\epsilon^2 + |\Delta(T)|^2)^{1/2}$ (as well as changing the overall magnitude). The net result is that while the Δ $(\equiv \Delta(0))$ appearing in (5.4.34) is replaced by $\Delta(T)$, the "pair radius" $\xi'(T)$ which replaces ξ' in that equation is *not* very different from its $T = 0$ value (5.4.34) for any $T < T_\mathrm{c}$, and in particular does not diverge in the limit $T \to T_\mathrm{c}$. The condensate fraction N_0/N, which is given by substituting the finite-temperature value (5.5.14) of F_k in (5.4.32), is temperature-dependent and in particular near T_c is of order $|\Delta(T)|^2/(T_\mathrm{c}\epsilon_\mathrm{F})$. (not of order $\Delta(T)/\epsilon_\mathrm{F}$). In trying to visualize the behavior of the condensate near T_c, it is crucial to remember that the pair radius is still of order $\hbar v_\mathrm{F}/|\Delta(0)|$ (or $\hbar v_\mathrm{F}/k_\mathrm{B}T_\mathrm{c}$), in strong contrast to the so-called Ginzburg–Landau healing length which, as we shall see in Section 5.7, diverges as $(T_\mathrm{c} - T)^{-1/2}$. From now on I shall use the notation ξ_o for the Cooper pair radius, and estimate it as generically $\sim\hbar v_\mathrm{F}/k_\mathrm{B}T_\mathrm{c}$.

It is finally worth noting that the quantity N_p occurring in (5.5.10) has no particular physical significance, at least in the usual BCS limit.[21] To see this, we note that $(N - 2N_\mathrm{p})$ is the total number of BP states, which even in the limit of the normal state is only a fraction $\sim(T/\epsilon_\mathrm{F})$ of the Fermi sea.

5.6 The two-fluid model for superconductors: the Meissner effect

In Section 5.4 I defined a class of what I called "generalized (completely paired) BCS states" by the prescription given in Eqn. (5.4.3). While in the bulk of the last two sections we have concentrated on the simple case in which the "pseudo-molecular" wave function $\phi(\boldsymbol{r}_1, \boldsymbol{r}_2, \sigma_1, \sigma_2 : t)$ has center of mass at rest, this is not the most general case, and it is now necessary to consider the case in which $\phi(\boldsymbol{r}_1\boldsymbol{r}_2\sigma_1\sigma_2 : t)$, and hence the associated "pair wave function" $F(\boldsymbol{r}_1\boldsymbol{r}_2 : \sigma\sigma' : t)$ (the eigenfunction of the two-particle density matrix associated with the macroscopic eigenvalue, see Chapter 2,

[21] It may play a role in the "BCS-BEC crossover" problem discussed in Chapter 8.

Section 2.4) depends on the center-of-mass variable $\boldsymbol{R} \equiv \frac{1}{2}(\boldsymbol{r}_1 + \boldsymbol{r}_2)$ as well as on the relative coordinate $\boldsymbol{\rho} \equiv \boldsymbol{r_1} - \boldsymbol{r_2}$. In this situation, as already noted in Chapter 2 (see Eqn. 2.4.19) we can define an "order parameter" $\Psi(\boldsymbol{R}, t)$ by

$$\Psi(\boldsymbol{R} : t) \equiv F(\boldsymbol{R}, \boldsymbol{\rho}, \sigma\sigma')_{\substack{\sigma=-\sigma'=+1 \\ \rho=0}} \equiv |\Psi(\boldsymbol{R}t)| \, e^{i\phi(\boldsymbol{R}t)} \tag{5.6.1}$$

and from it a *superfluid velocity* $\boldsymbol{v}_\mathrm{s}(\boldsymbol{r}t)$ by (Eqn. (2.4.21))

$$\boldsymbol{v}_\mathrm{s}(\boldsymbol{R}t) \equiv \frac{\hbar}{2m} \boldsymbol{\nabla}_R \phi(\boldsymbol{R}t) \tag{5.6.2}$$

where m is the electron mass. This is precisely analogous to the definition (2.2.7) used for the case of liquid ^{4}He and all the considerations given in Chapter 3, Section 3 concerning the topological stability of $\boldsymbol{v}_\mathrm{s}$ should apply equally to the superconducting case. Thus, we would expect to be able to develop a "two-fluid" description of superconductivity analogous to that given in that section for ^{4}He.

However, there are a number of significant differences between the two systems. The most obvious is that in contrast to the electrically neutral atoms of liquid ^{4}He, the electrons in a superconductor are charged and thus interact by the long-range Coulomb and Ampère forces[22]. The effects of the former are much the same in the superconducting as in the normal phase, and are not particularly interesting, but the Ampère interaction has spectacular consequences such as the Meissner effect, see below. A second difference is that unlike bulk ^{4}He, which is an approximately translation-invariant system, the electrons in a metal, whether normal or superconducting, experience momentum-changing collisions with impurities and with lattice phonons, and thus under (stable or metastable) equilibrium conditions the value of the normal velocity $\boldsymbol{v}_n$ is always zero; as a result, there is no analog in a superconductor of the convective thermal flow (infinite thermal conductivity) observed in ^{4}He. Finally, in contrast to the point-like atoms of ^{4}He, the objects in a superconductor which undergo (pseudo-)BEC, namely the Cooper pairs, have a substantial extension in space, (roughly the pair radius $\xi_0 \sim \hbar v_\mathrm{F}/k_\mathrm{B}T_\mathrm{c}$), and this feature, while not altering the consequences of the two-fluid model qualitatively, may have an important effect on the allowed electromagnetic behavior; see below.

In the following, I discuss explicitly the behavior of a superconductor in a static applied magnetic field. Let's start with the case in which the characteristic range of spatial variation of any electromagnetic vector potential $\boldsymbol{A}(\boldsymbol{r})$ is very large compared to the pair radius ξ_0; crudely speaking this corresponds to what is usually called extreme type-II behavior[23] (cf. below). For orientation, consider first a ring (torus) of radius R and thickness d, such that $d \ll R$ and $\xi_0 \ll d \ll \lambda$ (Fig. 5.2). We can define the (time-independent) order parameter $\Psi(\boldsymbol{R})$ by (5.6.1), and in the absence of any vector potential[24] further define the superfluid velocity $\boldsymbol{v}_\mathrm{s}$ by (5.6.2). Then, in exact analogy

[22]In principle there exists a gravitational analog of the Ampère force, namely the Lense-Thirring interaction (see, e.g. Weinberg 1972, Chapter 9 Section 7), but it is so weak that its effects are completely negligible under terrestrial conditions.

[23]It occurs for any superconductor sufficiently close to T_c, see Section 5.7.

[24]The neglect of any vector potential induced by the current itself is self-consistent in the limit $d \ll \lambda$, cf. below.

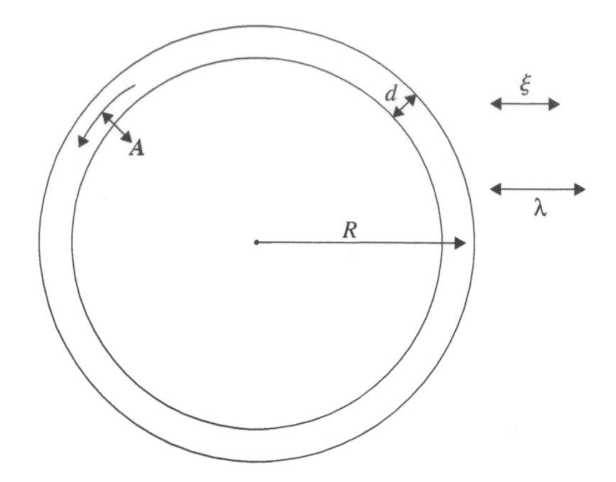

Fig. 5.2 A thin superconducting ring ($\xi \ll d \ll \lambda$) in an applied magnetic vector potential.

to what we did in Chapter 3, Section 3.3, we have the Onsager–Feynman quantization rule[25]

$$\oint \boldsymbol{v}_s(\boldsymbol{r}) \cdot d\boldsymbol{l} = \frac{nh}{2m} \tag{5.6.3}$$

where n is an integer (including zero). Let us now suppose that in (stable or metastable) equilibrium the mass current associated with a finite value of $\boldsymbol{v}_s(\boldsymbol{r})$ is given by

$$\boldsymbol{j}(\boldsymbol{r}) = \rho_s \boldsymbol{v}_s(\boldsymbol{r}) \tag{5.6.4}$$

where ρ_s is a temperature-dependent function whose form will be investigated below: for the moment we anticipate the result that $\rho_s(T) \to 0$ in the limit $T \to T_c$ and (for the simple translation-invariant model used so far) $\rho_s \to \rho$ (the total electronic mass density) for $T \to 0$. We see that for any given temperature the possible values of the circulating current are quantized, that is they are multiples of the quantity $\rho_s(T)(S/R)(\hbar/2m)$, where $S = \pi d^2$ is the cross-sectional area of the ring.

Next, suppose that an external vector potential $\boldsymbol{A}(\boldsymbol{r}) \equiv A_0\hat{\boldsymbol{\phi}}$ is applied to the ring (where $\hat{\boldsymbol{\phi}}$ denotes a unit vector in the tangential direction, see Fig. 5.2) such a potential would result, e.g. from a constant magnetic induction $B = 2A_0/R$ applied through the ring. Now, we know that for a single particle of mass m described by a Schrödinger wave function $\psi(\boldsymbol{r})$, the application of a magnetic vector potential does not change the expression for the probability density, namely

$$\rho(\boldsymbol{r}) = |\psi(\boldsymbol{r})|^2 \tag{5.6.5}$$

but it does change the expression for the probability current density, namely

$$\boldsymbol{j}(\boldsymbol{r}) = \frac{1}{2m}(\psi^*(\boldsymbol{r})(-i\hbar\boldsymbol{\nabla} - e\boldsymbol{A}(r))\psi(\boldsymbol{r}) + \text{c.c.}) \tag{5.6.6}$$

[25]For the rest of this section and the next I use the notation $\boldsymbol{r}$ for the *center-of-mass* coordinate of the Cooper pairs, in distinction to the last two sections, where $\boldsymbol{r}$ denoted the relative coordinate (which plays no role in this section and the next).

Consequently, if we write $\psi(\boldsymbol{r}) \equiv |\psi(\boldsymbol{r})| \exp i\phi(r)$ and define a velocity $\boldsymbol{v}(\boldsymbol{r}) \equiv \boldsymbol{j}(\boldsymbol{r})/\rho(\boldsymbol{r})$, then $\boldsymbol{v}(r)$ contains not just the term $(\hbar/m)\boldsymbol{\nabla}\phi$ but an additional term $eA(\boldsymbol{r})/m$. It is therefore plausible that the correct modification of the definition (5.6.2) of the superfluid velocity in the presence of a vector potential should be similar, with however the proviso that since the "wave function" in question is that of a Cooper pair, not only should m be replaced by $2m$ (as in (5.6.2)) but also e by $2e$. Thus, the generalized definition of the superfluid velocity for a superconductor is

$$\boldsymbol{v}_s(\boldsymbol{r}) = \frac{\hbar}{2m}\left(\boldsymbol{\nabla}\phi(\boldsymbol{r}) - \frac{2e\boldsymbol{A}(\boldsymbol{r})}{\hbar}\right) \tag{5.6.7}$$

The phase ϕ must still be single-valued modulo 2π, and as a result the Onsager–Feynman quantization condition is modified and becomes

$$\oint \boldsymbol{v}_s \cdot d\boldsymbol{l} = \frac{\hbar}{2m}\left(n - \frac{\Phi}{\Phi_0}\right) \tag{5.6.8}$$

where $\Phi \equiv \oint \boldsymbol{A} \cdot d\boldsymbol{l}$ is the flux threading the ring and $\Phi_0 \equiv h/2e$ is the (superconducting) "flux quantum." If we assume that the relation (5.6.4) is unaffected by the presence of the vector potential, then the result (5.6.8) immediately implies that *in the presence of an applied magnetic vector potential the circulating current in the ring cannot be zero.*[26] In fact, if we set $\Phi \ll \Phi_0$ and assume that the stable equilibrium state corresponds to $n = 0$ (i.e. $\phi =$ const), then the local electrical current (the mass current (5.6.4) multiplied by $(2e)/(2m)$) is given by the expression

$$\boldsymbol{j}(\boldsymbol{r}) = -\frac{\rho_s(T)}{m^2} e^2 \boldsymbol{A}(\boldsymbol{r}) \tag{5.6.9}$$

Equation (5.6.9) (or its generalization, see below) expresses the fundamental electromagnetic property of superconductors, namely a diamagnetic response to a (weak) applied vector potential;[27] this property is at the root of more directly observable properties such as the Meissner effect. It was in fact originally written down, on the basis of phenomenological arguments, by F. London in 1938, long before the advent of the BCS theory.

We next consider a "thick" ring (Fig. 5.3), such that $d \gg \lambda$ (we will continue to assume that $\lambda \gg \xi_0$ and that $d \ll R$). Now we must take account of the fact that the circulating electrical current will itself produce, according to Ampère's law, a vector potential which must be added to the externally applied one. Let's assume, first, that the (total) vector potential is everywhere sufficiently weak that the flux threading any relevant circuit is very much less than the flux quantum Φ_0; then it is overwhelmingly plausible that the stable equilibrium state corresponds to the choice $n = 0$ in Eqn. (5.6.8), so that Eqn. (5.6.9) applies. To take into account the self-generated vector potential, we write Ampére's equation $\boldsymbol{\nabla} \times \boldsymbol{H}(r) = \boldsymbol{j}(\boldsymbol{r})$, which

[26] Except of course for the special values such that $\Phi = l\Phi_0$, l integral.

[27] It should be noted that this response occurs even when the vector potential is applied under "Aharanov–Bohm" conditions, i.e. so that the magnetic *field* on the ring is zero (as can be ensured, e.g. by generating the flux by two concentric solenoids.)

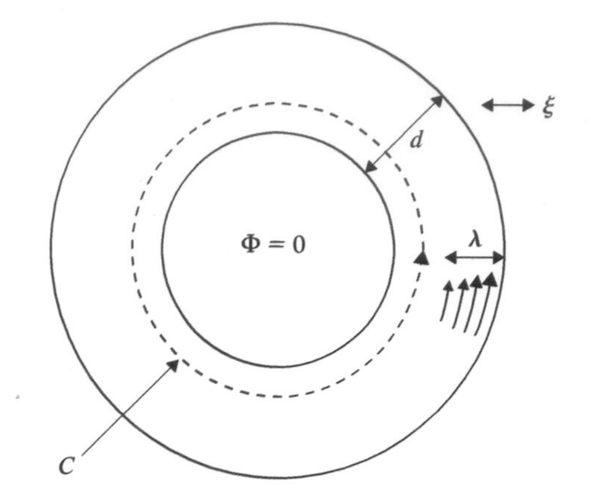

Fig. 5.3 A thick superconducting ring ($d \gg \lambda$) in a weak applied magnetic flux. Both the current and the vector potential A must vanish everywhere on the contour C.

in terms of the (total) vector potential $\boldsymbol{A}(\boldsymbol{r})$ is

$$\boldsymbol{\nabla} \times (\boldsymbol{\nabla} \times \boldsymbol{A}(\boldsymbol{r})) = \mu_0 \boldsymbol{j}(\boldsymbol{r}) \tag{5.6.10}$$

If we choose a gauge such that $\boldsymbol{\nabla} \cdot \boldsymbol{A} = 0$, this becomes

$$\nabla^2 \boldsymbol{A}(\boldsymbol{r}) = -\mu_0 \boldsymbol{j}(\boldsymbol{r}) \tag{5.6.11}$$

and combining this with (5.6.9) yields

$$\nabla^2 \boldsymbol{A}(\boldsymbol{r}) = \lambda_{\mathrm{L}}^{-2}(T) \boldsymbol{A}(r) \tag{5.6.12}$$

where $\lambda_{\mathrm{L}}(T)$ is the *London penetration depth*, defined by

$$\lambda_{\mathrm{L}}(T) \equiv \left(\frac{m^2}{\mu_0 \rho_{\mathrm{s}}(T) e^2} \right)^{1/2} \tag{5.6.13}$$

Anticipating the result that $\rho_{\mathrm{s}}(T)$ tends to nm for $T \to 0$ (where n is the mean electron density) and defining the plasma frequency ω_{p} by $\omega_{\mathrm{p}} \equiv (ne^2/m\epsilon_0)^{1/2}$, we see that the zero-temperature London penetration depth $\lambda_{\mathrm{L}}(0)$ is given by

$$\lambda_{\mathrm{L}}(0) = c/\omega_{\mathrm{p}} \tag{5.6.14}$$

i.e. it is equal to the "high-frequency skin depth" of the normal metal, that is the length over which high-frequency EM radiation incident on the surface is screened out. For the superconductor, Eqn. (5.6.12) implies a similar effect for a *static* vector potential: from the geometry of the sample, we see that the vector potential falls off as $e^{-z/\lambda_{\mathrm{L}}(T)}$, where z is the distance into the metal from the surface (either inner or outer). In view of (5.6.9), the induced electrical current falls off similarly, and so does the magnetic field (the curl of A); thus, once we are at a distance from either surface large compared to $\lambda_{\mathrm{L}}(T)$, not only the current and magnetic field but the vector potential itself are negligibly small.

It is clear that in the above argument no particular role is played by the annular nature of the geometry; if we imagine filling in the center so as to make a solid

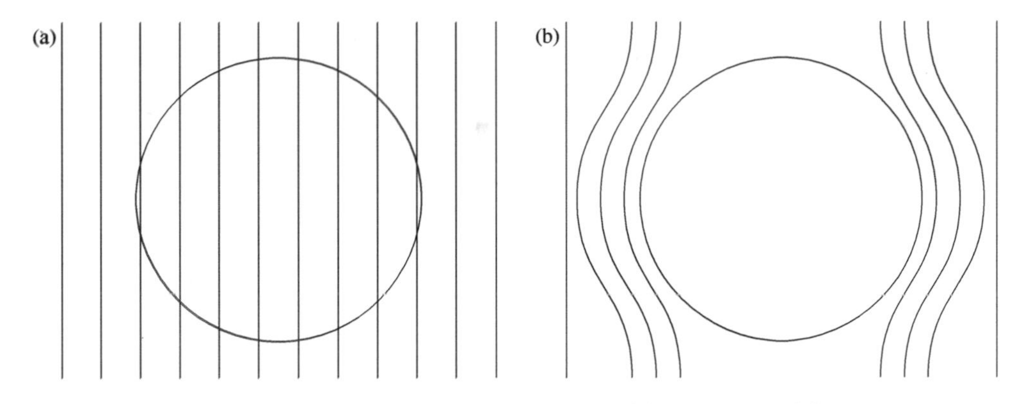

Fig. 5.4 A metal sphere in an applied magnetic field (a) above, and (b) below the superconducting transition.

disk the argument still goes through, with the field, current and vector potential now effectively vanishing in the whole of the interior (i.e. at distances from the surface large compared to the London penetration depth). Moreover, it is clear that a similar argument goes through for any simply connected shape, provided that all the dimensions are large compared to $\lambda_L(T)$. Thus, for any simply connected superconductor,[28] *a sufficiently weak magnetic field is completely expelled*: this is the celebrated Meissner effect. We shall investigate below what "sufficiently weak" means: however, it is clear that a sufficient condition for the above argument to go through is that the flux through any circuit lying entirely within the body is small compared to the flux quantum $\Phi_0 \equiv h/2e$. Because of the need, enforced by Maxwell's equations, to conserve the lines of force, the latter must distort as shown in Fig. 5.4; this costs field energy, so that a superconductor tends to "repel" the source of a magnetic field. For example, a small magnetic pellet, when lowered towards the surface of a superconducting slab, will hover above it (an effect which is nowadays easy to demonstrate using a sample of a slab of YBCO (Chapter 7) doused with liquid nitrogen).

Returning now to our annular ($d \gg \lambda_L$) geometry, let us enquire about the possibility of more general solutions, corresponding to $n \neq 0$ in Eqn. (5.6.8). Since the kinetic energy associated with the supercurrent is proportional to v_s^2, it is plausible that when Φ is close to the value $l\Phi_0$ then the nonzero value $n = l$ will be energetically favored. Let us suppose for simplicity that this is indeed the case throughout the ring, and define the quantity

$$\boldsymbol{A}'(\boldsymbol{r}) \equiv \boldsymbol{A}(\boldsymbol{r}) - \frac{n\Phi_0}{2\pi r}\hat{\boldsymbol{\phi}} \tag{5.6.15}$$

Then Eqn. (5.6.9) is satisfied provided $\boldsymbol{A}(\boldsymbol{r})$ is replaced by $\boldsymbol{A}'(\boldsymbol{r})$; moreover, since $\boldsymbol{\nabla} \times \boldsymbol{A}' \equiv \boldsymbol{\nabla} \times \boldsymbol{A}(\equiv \boldsymbol{B}(\boldsymbol{r}))$, Eqn. (5.6.10) is satisfied with the same replacement. Consequently, for the annular geometry under consideration, all statements made above about $\boldsymbol{A}(r)$, $\boldsymbol{j}(\boldsymbol{r})$, and $\boldsymbol{B}(\boldsymbol{r})$ remain true provided only that $\boldsymbol{A}(r)$ is replaced

[28]We have demonstrated this above only under the assumption that the superconductor is extreme Type-II, but will see below that the result is much more general.

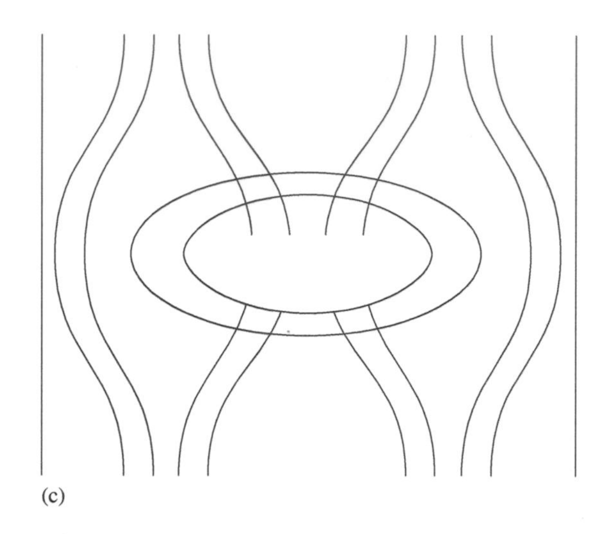

Fig. 5.5 A superconducting ring in an external flux $> \Phi_0$.

by $\boldsymbol{A}'(r)$; in particular, at any point distant by an amount $\gg\lambda_L(T)$ from *either* surface, the current and magnetic field effectively vanish. However, it is now not the actual vector potential $\boldsymbol{A}(\boldsymbol{r})$ but the quantity $\boldsymbol{A}'(\boldsymbol{r})$ which vanishes, and thus the actual flux through the circuit shown in Fig. 5.3 is not zero but takes the value

$$\Phi = n\Phi_0 \tag{5.6.16}$$

Thus, the flux through a multiply connected superconductor is not necessarily zero, but is *quantized* in units of $\Phi_0 \equiv h/2e$; the field lines then distort roughly as shown in Fig. 5.5, and in thermodynamic equilibrium the system will tend to choose the value of n which makes the associated energy least (which, crudely speaking, is the integer closest to Φ_n/Φ_0 where Φ_n is the flux through the ring in the normal state). The observation of this flux quantization effect, with the factor of $2e$ rather than e in the flux quantum Φ_0, is one of the most convincing pieces of evidence for the BCS "pairing" hypothesis.

Let's now ask: Is it possible to extend this argument to the case of a simply connected body as we did in the $n = 0$ case? Prima facie, the answer is no; the easiest way to see this is to note that if the magnetic field is everywhere zero in the bulk of the system, as the extension of the argument would require, then there is no way of getting a nonzero value of n in Eqn. (5.6.16)! An alternative argument, which brings out more clearly the relation to the order parameter, goes as follows: Imagine that Eqn. (5.6.8) is satisfied, with $n \neq 0$, for a "large" circuit which lies entirely within the superconductor. Then, *provided that the order parameter $\Psi(\boldsymbol{r})$ (and thus $\boldsymbol{v}_s(\boldsymbol{r})$) is well-defined everywhere within the superconductor*, it must be satisfied for any circuit, including an arbitrarily small one (radius R, $R \to 0$). But since the magnetic field cannot diverge as R^{-2} (this would cost infinite field energy!) the flux through such a circuit must tend to zero and thus from (5.6.8) $\boldsymbol{v}_s$ must be proportional to r^{-1}, leading to a divergence of the kinetic energy.

Suppose however that it should turn out to be possible to turn one or more small tubes in the body of the superconductor "normal", that is, to ensure that the condensate wave function (order parameter) $\Psi(\boldsymbol{r})$ vanishes within them. Then within those regions $\boldsymbol{v}_s(\boldsymbol{r})$ is not defined, so that from the point of view of the arguments of this section they play exactly the same role as physical holes (compare the discussion in Chapter 3, Section 5). Our arguments concerning the possibility of nontrivial flux quantization ($n \neq 0$ in Eqn. (5.6.8)) then go through for the regions containing these "holes" just as for the physical annulus; the flux through a circuit embracing the normal tube need not be zero but can take the quantized values $n\Phi_0$. Such a configuration is called a *vortex line*, and the normal region is called the core; except close to T_c, the size $\xi(T)$ of the core is of the order of the Cooper pair radius; it is clear that it is very similar to the vortices we already encountered in liquid ^{4}He and the alkali gases. However, note one important difference: While at short distances ($r \ll \lambda_L$, where r is the perpendicular distance from the core) the flow pattern $\boldsymbol{v}_s(\boldsymbol{r})$ ($\propto r^{-1}$) is identical to that of the neutral ^{4}He vortices, at large distances ($r \gg \lambda_L$) it is no longer proportional to r^{-1} but has fallen exponentially to zero, just as in the physical annulus. It turns out that, just as in ^{4}He, vortices with $|n| > 1$ are unstable against dissociation into "singly-quantized" ($|n| = 1$) vortices.

Is it energetically favorable to introduce such vortices? The following argument gives at least a crude order of magnitude of the minimum value of the external magnetic induction B_{ext} for which this is so: The energy per unit length necessary to set up the circulating currents in the vortex is, up to logarithmic factors, of order $\rho_s(\hbar/m)^2$, which by Eqn. (5.6.13) is of order $\mu_0^{-1}\Phi_0^2\lambda_L^{-2}$. On the other hand, the magnetic moment of the vortex per unit length is of order Φ_0, and thus the decrease of the magnetic field energy resulting from its creation is of order $\mu_0^{-1}\Phi_0 B_{\text{ext}}$. Thus there is a minimum value of B_{ext}, usually called the *lower critical field*, above which it is energetically advantageous to create vortex lines; the order of magnitude of B_{c1} is given according to the above argument by

$$B_{c1}(T) \sim \frac{\Phi_0}{\lambda_L^2(T)} \tag{5.6.17}$$

i.e. it is of the order of the field, which when applied to an area of the square of the London penetration depth yields one quantum of flux. (The correct expression for H_{c1} involves other factors, see e.g. de Gennes (1966), Eqn. 3.5.6, but these are generally of order unity). For external fields below B_{c1}, the field is completely expelled as described earlier (the system is said to be in the "Meissner state"). For $B_{\text{ext}} > B_{c1}$, the field begins to penetrate the superconductor by punching vortex lines, and when B_{ext} is substantially greater than B_{c1} the fraction which penetrates is of order one, so that the area per line is of order $\Phi/B_{\text{ext}} \sim (B_{c1}/B_{\text{ext}})\lambda_L^2$. In this regime the screening currents of the different vortices overlap considerably; however, the cores still do not overlap. Eventually, as the field is increased further, the cores begin to overlap, and at this point it becomes energetically advantageous for the whole system to become normal.[29] Thus

[29]Like some other results quoted in this section, this statement, while perhaps plausible, is not completely obvious. A formal proof requires the apparatus of the quantitative Ginzburg–Landau theory to be introduced in the next section.

there is an *upper critical field* B_{c2} given by

$$B_{c2}(T) \sim \frac{\Phi_0}{\xi^2(T)} \tag{5.6.18}$$

beyond which superconductivity is not possible (here $\xi(T)$, as above, is the core radius). The regime between B_{c1} and B_{c2}, where superconductivity exists but full field expulsion does not occur, is called the "mixed state".

The above discussion is intended to emphasize the analogy between the behavior of (some) superconductors in a magnetic field and that of neutral systems such as ^{4}He under rotation. However, while it is a relatively good description, at a qualitative level, of the experimentally observed behavior of many "classic" superconductors which are disordered alloys, and also of most of the "exotic" superconductors to be discussed in Chapter 7, it bears no relation to the magnetic behavior of most of the classic elemental superconductors such as Al, Sn or Pb, which as we shall see is actually considerably simpler than described above.

In fact, it is clear from (5.6.17) and (5.6.18) that the existence of a substantial "mixed" regime requires the condition $\lambda_L(T)/\xi(T) \gg 1$. It actually turns out that not all superconductors satisfy this condition; in fact, in some of the classic elemental superconductors the ratio is $\ll 1$. Following the original work of Ginzburg and Landau, it is conventional to call a superconductor "Type-II" if the ratio $\kappa \equiv \lambda(T)/\xi(T)$ is greater than $1/\sqrt{2}$, and "Type-I" if it is less; as we will see in the next section, although the temperature-dependence of $\lambda(T)$ and $\xi(T)$ is not identical, it is similar enough that most superconductors can be assigned to one type or the other independently of temperature.

A detailed theoretical calculation (based on the Ginzburg–Landau theory to be introduced in the next section, or some generalization of it) yields the result that all Type-II superconductors behave qualitatively as described above, with a finite "mixed-state" regime. A Type-I superconductor behaves quite differently: if we consider the simplest specimen geometry (a long cylinder parallel to the external magnetic field) then such a superconductor shows a complete Meissner effect up to a critical field $B_c(T)$, at which point it simply becomes completely normal and admits the field in its entirety. The value of B_c is simply given by equating the magnetic energy gained by admitting the field, $\frac{1}{2}\mu_0^{-1}B^2$, to the difference in free energy $F_n(T) - F_s(T)$ between the normal and superconducting states. In geometries other than the above we get "demagnetizing" effects, and it is often energetically advantageous in a certain field range for the sample to become normal in some regimes while remaining superconducting in others. This "intermediate" state should not be confused with the "mixed" state of a Type-II superconductor; in distinction to the latter, the normal regions are typically of a size much greater than the microscopic length $\xi(T)$ (or $\lambda_L(T)$).

5.7 The Ginzburg–Landau theory

The Ginzburg–Landau (GL) approach to superconductivity, which was developed some years before the microscopic work of BCS, is a spectacular example of the power of phenomenology even in the absence of a microscopic description. It is particularly useful in discussing the behavior in high magnetic fields, where it allows us to justify and go beyond the "London" phenomenology used in the last section. In this section

I shall first introduce the GL theory in the original phenomenological terms, and then sketch briefly how it can be derived from the microscopic considerations of BCS.

The GL theory is actually a special case of the general Landau–Lifshitz theory of second-order phase transitions, in which one introduces an "order parameter" $\eta(\boldsymbol{r})$ which is zero above the transition temperature T_c but takes a finite value for $T < T_c$, and uses the symmetry of the relevant Hamiltonian to restrict the form of the free energy as a functional of η. In the case of superconductivity, Ginzberg and Landau made the brilliant guess that $\eta(\boldsymbol{r})$ should have the nature of what they called a "macroscopic wave function" and thus denoted $\Psi(\boldsymbol{r})$; following the work of BCS, we now know that $\Psi(\boldsymbol{r})$ is indeed (up to normalization) nothing but the center-of-mass wave function of the Cooper pairs, i.e. the quantity $F(\boldsymbol{r}_1, \boldsymbol{r}_2, \sigma_1, \sigma_2)$ (Eqn. 5.6.1) evaluated, for $\sigma_1 = -\sigma_2$ ($= \uparrow$, say) and relative coordinate $\boldsymbol{\rho} \equiv \boldsymbol{r}_1 - \boldsymbol{r}_2 = 0$, as a function of the center-of-mass coordinate $\boldsymbol{r} \equiv (\boldsymbol{r}_1 + \boldsymbol{r}_2)/2$. It is "macroscopic" in the sense that it is that (unique) eigenfunction of the two-particle density matrix which is associated with a macroscopic eigenvalue. The two properties which will be essential in the ensuing development is that it is a complex scalar and behaves like a true wave function under local gauge transformations.

For a reason which will become clear below I will not immediately assume, as did Ginzburg and Landau, that we are interested in the case that $\Psi(\boldsymbol{r})$ is "small". However, I *will* assume that it is slowly varying over any relevant microscopic length scale. (We will find below that the relevant scale is the Cooper-pair radius ξ_0.) Also, for the moment I will set the electromagnetic vector potential $\boldsymbol{A}(\boldsymbol{r})$ equal to zero. Then it is reasonable to assume that $F\{\Psi(\boldsymbol{r}), T\}$ is the space integral of a free energy density $\mathcal{F}$ which is a function only of $\Psi(\boldsymbol{r})$ and its space derivatives, plus their complex conjugates, and moreover that terms involving more than two powers of the gradient may be neglected. To proceed further we need to invoke various symmetry and analyticity properties, as follows: (A) $\mathcal{F}$ should be invariant under spatial inversion; (B) $\mathcal{F}$ should also be invariant under the "global gauge transformation" $\Psi(\boldsymbol{r}) \to \Psi(\boldsymbol{r}) \exp(i\phi)$; (C) F should be an *analytic* functional of $\Psi(\boldsymbol{r})$ and its complex conjugate $\Psi^*(\boldsymbol{r})$. Assumption (A) follows from the reflection-invariance (in the absence of a vector potential) of the Hamiltonian, while (B) is a natural consequence of the assumption that $\Psi(\boldsymbol{r})$ has the significance of a Schrödinger-like wave function. Assumption (C) is however less obviously justified, and must at the present stage be regarded as an inspired guess. Conditions (A) and (C) together eliminate any item in $\mathcal{F}(\boldsymbol{r})$ linear in the gradient, while (B) and (C) restrict the terms of zeroth order in the gradient to have the form $|\Psi(\boldsymbol{r})|^{2n}$. The terms of second order in the gradient have in general a nontrivial structure which we will examine below; for the moment it is sufficient to notice that the term of lowest (second) order[30] in $\Psi(\boldsymbol{r})$ is constrained by conditions (B) and (C) to be of the form $|\boldsymbol{\nabla}\Psi(\boldsymbol{r})|^2$.

Let us now assume, with Ginzburg and Landau, that when we are close to the second-order transition $\Psi(\boldsymbol{r})$ is "small", and expand, according to assumption (C), F in powers of Ψ. As regards the "bulk" (nongradient) terms, we must obviously keep the term of order $|\Psi|^2$, and also, in order to guarantee stability below T_c, the term of

[30] Any term of the form $\Psi^* \boldsymbol{\nabla}^2 \Psi + \text{c.c.}$ can be converted, by integration by parts, into one in $|\nabla\Psi|^2$, so does not have to be considered separately.

order $|\Psi|^4$. However, in the case of the gradient terms it is sufficient to keep only the term of lowest (second) order in Ψ; any term of the form (e.g.) $|\Psi|^2|(\nabla\Psi)|^2$ or higher will only have the effect of renormalizing the coefficient of $|\nabla\Psi|^2$ by an amount which is itself of order $|\Psi|^2$ and hence negligible for $\Psi \to 0$. Thus, in the absence of a vector potential the form of the free energy functional appropriate to the regime of small Ψ is

$$F = \int \mathcal{F}[\Psi(\boldsymbol{r})]d\boldsymbol{r} \tag{5.7.1}$$

$$\mathcal{F}[\Psi(\boldsymbol{r})] \equiv \mathcal{F}_0(T) + \alpha(T)|\Psi(\boldsymbol{r})|^2 + \tfrac{1}{2}\beta(T)|\Psi(\boldsymbol{r})|^4 + \gamma(T)|\nabla\Psi(\boldsymbol{r})|^2 \tag{5.7.2}$$

In the above expression the normalization of the order parameter $\Psi(\boldsymbol{r})$, and hence the absolute values of the coefficients $\alpha(T)$, $\beta(T)$ and $\gamma(T)$, are arbitrary; however, we see that the ratios $\gamma(T)/\alpha(T)$ and $\alpha^2(T)/\beta(T)$, which as we shall see have a physical meaning, are independent of the normalization.

In order to guarantee that $\Psi(\boldsymbol{r})$ should be zero above the transition temperature T_c and take a uniform non-zero value for $T < T_c$, it is necessary that $\beta(T)$ and $\gamma(T)$ be positive and that $\alpha(T)$ changes sign at T_c (being positive for $T > T_c$). If the temperature-dependence is analytic at T_c, we may take β and γ to have their values at T_c and expand α to lowest order around T_c; thus, we set

$$\beta(T) \cong \beta(T_c) \equiv \beta \tag{5.7.3a}$$

$$\gamma(T) \cong \gamma(T_c) \equiv \gamma \tag{5.7.3b}$$

$$\alpha(T) \cong \alpha_0(T - T_c) \tag{5.7.3c}$$

If now we minimize the free energy $F(T)$ as a functional of $\Psi(\boldsymbol{r})$ with no particular boundary conditions, call the result the "superconducting" free energy $F_s(T)$ and compare it with the "normal-state" free energy $F_n(T)$ (defined to be the value of $F(T)$ obtained for $\Psi = 0$) we find that the free energy difference per unit volume (a measurable quantity) is given for $T < T_c$ by

$$\frac{F_s(T) - F_n(T)}{\Omega} = -\frac{\alpha^2(T)}{2\beta} = -\frac{\alpha_0^2}{2\beta}(T_c - T)^2 \tag{5.7.4}$$

The transition at T_c is therefore second order (as expected!) with a specific heat jump $\alpha_0^2 T_c/\beta$. The (uniform) value of Ψ is zero above T_c, while below T_c it is $(\alpha(T)/\beta)^{1/2}$ $(\propto(T_c - T)^{1/2})$. We may obtain another useful relation by considering the case of uniform superflow, when the amplitude $|\Psi(\boldsymbol{r})|$ of the order parameter is constant at its equilibrium value but its phase $\phi(\boldsymbol{r})$ is spatially varying. The bending (flow) energy per unit volume is then given by

$$E_{\text{bend}} = \gamma|\Psi|^2(\nabla\phi)^2 = \frac{\gamma\alpha(T)}{\beta}(\nabla\phi)^2 \tag{5.7.5}$$

Comparing this expression with the expression (cf. 5.6.4)

$$E_{\text{bend}} = \tfrac{1}{2}\rho_s {\boldsymbol{v}_s}^2 \tag{5.7.6}$$

and using the definition (5.6.2) of $\boldsymbol{v}_s$, we obtain the relation

$$\rho_s(T) = \frac{8m^2}{\hbar^2}\frac{\gamma\alpha(T)}{\beta} \tag{5.7.7}$$

which with (5.7.4) fixes the values of α, β and γ up to the arbitrary scale factor involved in the normalization of Ψ.

It is interesting to ask the question: Over what length do we have to "bend" the phase of the order parameter (say through π) in order that the bending energy should be equal to the superconducting condensation energy $F_s - F_n$? It is clear from (5.7.4) and (5.7.5) that, to within a factor of order unity, the answer is the characteristic length $\xi(T)$ defined by

$$\xi(T) \equiv \left(\frac{\gamma}{\alpha(T)}\right)^{1/2} \sim (T_c - T)^{-1/2} \tag{5.7.8}$$

The length (5.7.8) is usually called the *Ginzburg–Landau correlation (or coherence) length*, though since it plays for the superconducting system a role closely analogous to that played by the quantity (4.3.16) in a dilute Bose-condensed gas (cf. below), it might be more natural to think of it as a "healing length". In any case it should be emphasized that it is *not* in general equal to the "Cooper pair radius" ξ_0, which remains finite as $T \to T_c$, see Section 5.5. (though we shall see below that for T not too close to T_c, $\xi(T)$ and ξ_0 are of the same order of magnitude).

The GL theory really comes into its own when we allow for coupling to an electromagnetic field. Before introducing this subject, however, let us take a moment to enquire whether we should expect an expression of the general type (5.7.2) to remain valid when the order parameter $\Psi(\boldsymbol{r})$, while still slowly varying in space, is not small (as we might expect to be the case for $T/T_c \sim 0.5$, say). From the arguments given above, we see that provided assumptions (A)–(C) continue to hold, the generalization as regards the "bulk" terms is in principle relatively trivial; we simply need to keep terms of all order in $|\Psi|^2$, and we will find that in principle the relevant coefficients can be evaluated from BCS theory. Moreover, assumptions (A) and (C) still exclude a term linear in the gradient of Ψ. As regards the terms of second order in the gradient, the situation becomes more interesting: symmetry now allows the possibility of terms which are not of the simple form $f(|\Psi|^2)|\nabla\Psi|^2$, e.g. at fourth order we can have a term of the form

$$\text{const.}(\Psi^{*2}(\nabla\Psi)^2 + \text{c.c.})$$

As a result, the general form of the gradient terms is

$$E_{\text{bend}} = f_1(|\Psi|^2)(\nabla|\Psi|)^2 + f_2(|\Psi|^2)(|\Psi|^2(\nabla\phi)^2) \tag{5.7.9}$$

where, unlike in the limit of small Ψ (where $f_1 = f_2 = \gamma$) the coefficients f_1 and f_2 are in general not equal. Fortunately, neither this effect nor that of the bulk terms of order $|\Psi|^6$ and higher appears to make much of a qualitative difference,[31] and as a result the simple GL expression (5.7.2), with the coefficient $\gamma(T)$ related to the experimental (or BCS-derived) value of $\rho_s(T)$ is often used for a qualitative analysis throughout the whole range $0 < T < T_c$.

[31] For a complete analysis see Werthamer (1969) or Kosztin et al. (1998).

Let us now turn to the question of the incorporation of a vector potential $\boldsymbol{A}(\boldsymbol{r})$. If the order parameter really does represent something like a Schrödinger wave function, then the kinetic energy (gradient) term should satisfy the standard gauge-invariant prescription $-i\hbar\boldsymbol{\nabla} \to -i\hbar\boldsymbol{\nabla} - e^*\boldsymbol{A}(\boldsymbol{r})$, where e^* is some effective charge (which Ginzburg and Landau tentatively set equal to the electron charge e). Since we now know, following BCS, that $\Psi(\boldsymbol{r})$ is the center-of-mass wave function of a Cooper *pair*, it is clear that the correct value of e^* is actually $2e$ (cf. Eqn. 5.6.7). Thus, making this replacement and adding the standard expression for the electromagnetic field energy $\frac{1}{2}\mu_0^{-1}B^2$ written in terms of $\boldsymbol{A}(\boldsymbol{r})$, we finally obtain the complete Ginzburg–Landau expression for the free energy F as a functional of $\Psi(\boldsymbol{r})$ and $\boldsymbol{A}(\boldsymbol{r})$:

$$F = \int d\boldsymbol{r}\{F_0(T) + \alpha_0(T - T_c)|\Psi(\boldsymbol{r})|^2 + \tfrac{1}{2}\beta|\Psi(\boldsymbol{r})|^4 \\ + \gamma|(\boldsymbol{\nabla} - 2ie\boldsymbol{A}(\boldsymbol{r})/\hbar)\Psi|^2 + \tfrac{1}{2}\mu_0^{-1}(\boldsymbol{\nabla} \times \boldsymbol{A}(\boldsymbol{r}))^2\} \qquad (5.7.10)$$

The expression (5.7.10) is valid for both equilibrium and nonequilibrium configurations of $\Psi(\boldsymbol{r})$, provided that the spatial variation is slow enough.[32] However, the most obvious application is to obtain the equilibrium configuration subject to appropriate boundary conditions. To obtain this we simply set the functional derivatives of F with respect to $\boldsymbol{A}(\boldsymbol{r})$ and $\Psi(\boldsymbol{r})$ equal to zero; this yields the two equations

$$\mu_0^{-1}\boldsymbol{\nabla} \times (\boldsymbol{\nabla} \times \boldsymbol{A}(\boldsymbol{r})) = \left\{-2ie\hbar\gamma\Psi^* \left(\underline{\boldsymbol{\nabla}} - \frac{2ie}{\hbar}\boldsymbol{A}(\boldsymbol{r})\right)\Psi - \text{c.c.}\right\} \qquad (5.7.11)$$

$$\gamma\left(\boldsymbol{\nabla} - \frac{2ie}{\hbar}\boldsymbol{A}(\boldsymbol{r})\right)^2 \Psi(r) + \alpha_0(T - T_c)\Psi(\boldsymbol{r}) + \beta|\Psi(\boldsymbol{r})|^2\Psi(\boldsymbol{r}) = 0 \qquad (5.7.12)$$

Equation (5.7.11) is simply Maxwell's equation (5.6.10), with the electric current $\boldsymbol{j}(\boldsymbol{r})$ identified with the expression on the right-hand side. Equation (5.7.12) is usually known as the Ginzburg–Landau (GL) equation; it is clear that in the limit $\boldsymbol{A} \to 0$ it is identical in structure to the Gross–Pitaevskii equation, (4.3.12), which we met in the context of a Bose-condensed atomic gas (with $-\alpha(T) \to \mu - V_{\text{ext}}(\boldsymbol{r})$). In particular, we see that just as in that case the quantity $\xi(T) = (\gamma/\alpha(T))^{-1/2}$ indeed plays the role of a "healing length" when $\Psi(\boldsymbol{r})$ is perturbed, e.g. by an appropriate boundary condition, from its bulk equilibrium value.

The manifold applications of Eqns. (5.7.11)–(5.7.12), in particular to discuss the structure and multiplicity of vortices, are well covered in many textbooks and will not be discussed here. However, one important point must be mentioned: It is clear on dimensional grounds that the theory described by these equations is characterized at any given temperature T close to T_c, by two important lengths, namely the healing length $\xi(T)$ (Eqn. (5.7.8)) and the London penetration depth, which from Eqns. (5.6.13) and (5.7.7) is given in the present notation by

$$\lambda_{\text{L}}(T) = \left\{\frac{\hbar^2}{8\mu_0 e^2}\frac{\beta}{\gamma\alpha(T)}\right\}^{1/2} \qquad (5.7.13)$$

[32]There is an implicit assumption here that all other degrees of freedom have come into equilibrium, locally, subject to the given value of $\Psi(\boldsymbol{r})$, see below.

The lengths $\xi(T)$ and $\lambda_{\rm L}(T)$ both diverge as $(T_{\rm c}-T)^{-1/2}$ in the limit $T \to T_{\rm c}$, and as a result in this limit the spatial variation of the order parameter is always slow on the scale of the (nondiverging) Cooper pair radius ξ_0, thus automatically guaranteeing the self-consistency of the GL approach in this limit. The ratio of $\lambda_{\rm L}(T)$ to $\xi(T)$, which is temperature-independent in the limit $T \to T_{\rm c}$, is conventionally denoted κ; it is a famous result of GL theory that if $\kappa > 1/\sqrt{2}$ vortices are energetically stable in a certain field regime, i.e. the behavior is Type-II, while if $\kappa < 1/\sqrt{2}$ it is Type-I. If the superconductor is Type-II, then the radius of the vortex cores is indeed, up to a factor of order unity, $\xi(T)$, as stated in the last section.

I will now briefly sketch a possible "derivation" of the GL theory from the microscopic BCS theory. The original derivation is by Gor'kov using the language of Green's functions; it is nicely reformulated in a language closer to that of this book in the book of de Gennes (Chapter 7). The output of the arguments of Gor'kov and de Gennes is the GL *equations* (5.7.11) and (5.7.12); here I prefer to give a more heuristic argument leading to the GL *free energy* (5.7.10). This has advantages when, for example, we need to consider fluctuations around the free energy minimum described by Eqns. (5.7.11)–(5.7.12). I shall proceed as follows: First, I derive the "bulk" terms in the GL free energy (5.7.10) under the assumption that $\Psi(\boldsymbol{r})$ is constant in space, and find explicit expressions for the parameters α and β under these conditions. Next, I show that for $\Psi(\boldsymbol{r})$ of constant amplitude and constant (slow) spatial variation of the phase (with $\boldsymbol{A}=0$) the gradient term is indeed of the form $\gamma|\boldsymbol{\nabla}\Psi|^2$ and find an explicit expression for the coefficient γ. Finally, I invoke the analyticity and gauge-invariance already outlined, plus the assumption that for sufficiently slow spatial variation the "bulk" terms are locally of the form calculated for uniform Ψ. Thus, the argument below should perhaps be regarded not so much as a derivation of GL theory as a demonstration that the latter is *consistent* with the BCS theory as we have developed it[33], provided that the parameters α, β, γ are correctly chosen.

Our initial definition of the order parameter $\Psi(\boldsymbol{r})$ will be that already used in Chapter 2, (Eqn. 2.4.19) and in the last section (Eqn. 5.6.1), namely as the center-of-mass wave function of the Cooper pairs (subsequently, we shall find that a different normalization simplifies the expressions for the coefficients somewhat). That is, we define

$$\Psi(\boldsymbol{r}) \equiv F(\boldsymbol{r}_1\boldsymbol{r}_2,\sigma_1\sigma_2)_{\boldsymbol{r}_1=\boldsymbol{r}_2,\sigma_1=-\sigma_2=\uparrow} \quad \boldsymbol{r} \equiv \tfrac{1}{2}(\boldsymbol{r}_1+\boldsymbol{r}_2) \tag{5.7.14}$$

For the spatially uniform case this reduces to the simple expression

$$\Psi = \sum_{\boldsymbol{k}} F_{\boldsymbol{k}} \tag{5.7.15}$$

where $F_{\boldsymbol{k}}$ is the quantum-mechanical *and thermal* quasi-average of $a_{k\downarrow}a_{-k\uparrow}$ as in Section 5.

We first consider the bulk terms. We make the crucial assumption that while Ψ does not necessarily have its equilibrium value, all the other degrees of freedom have come to equilibrium subject to the given value of Ψ; the resulting "constrained" free

[33]Recall that so far this development has been only for a spatially uniform state (cf. Eqn. (5.4.5).)

energy will be denoted simply $F(\Psi : T)$. We may write, suppressing the T-dependence for clarity,

$$F(\Psi) = F_1(\Psi) + \langle \hat{V} \rangle(\Psi) \tag{5.7.16}$$

where F_1 is a shorthand for the expression $\langle \hat{K} - \mu \hat{N} \rangle - TS$, with $\hat{K}$ the kinetic energy operator. To calculate F we use the BCS ansatz of Section 5.5, and in the interaction term neglect both the Hartree and the Fock terms on the ground that they are not appreciably functions of Ψ. The quantity $\langle \hat{V} \rangle$ is then given entirely by the "pairing" term and has the simple form

$$\langle \hat{V} \rangle(\Psi) = -V_0 |\Psi|^2 \tag{5.7.17}$$

The calculation of the constrained value of F_1 as a function of Ψ is less trivial. Relegating the details to Appendix 5E, I simply quote the result that up to fourth order in Ψ (but for arbitrary T) the expression is

$$F_1(\Psi : T) = F_0(T) + A^{-1}(T)|\Psi|^2 + \tfrac{1}{2}B(T) \cdot (A(T))^{-4}|\Psi|^4 \tag{5.7.18}$$

where the quantities $A(T)$, $B(T)$ are given by

$$A(T) \equiv \frac{1}{2}\frac{dn}{d\epsilon} \ln \left(1.14 \frac{\epsilon_c}{k_B T} \right) \tag{5.7.19a}$$

$$B(T) \equiv \frac{1}{2}\frac{dn}{d\epsilon} \frac{7\,\zeta(3)}{8\pi^2} \left(\frac{1}{k_B T} \right)^2 \tag{5.7.19b}$$

Combining (5.7.17) and (5.7.18), we see that the coefficient of $|\Psi|^2$ in the total constrained free energy $F(\Psi)$ changes sign at the point when $V_0 A(T) = 1$, i.e. at a temperature T_c given (unsurprisingly!) by (5.5.18), i.e.

$$k_B T_c = 1.14 \epsilon_c \exp - \left(\frac{1}{2}\frac{dn}{d\epsilon} V_0 \right)^{-1} \quad (dn/d\epsilon \equiv 2N(0)) \tag{5.7.20}$$

The quantity V_0 now drops out of the problem in favor of T_c. In the limit $T \to T_c$ we can expand the coefficient of $|\Psi|^4$ to zeroth order in $T - T_c$ and the coefficient of $|\Psi|^2$ to first order, thereby obtaining

$$F(\Psi : T) = F_0(T) + \left(\frac{dA^{-1}}{dT} \right)_{T=T_c} (T - T_c)|\Psi|^2 + \frac{1}{2} B(T_c) A^{-4}(T_c)|\Psi|^4 \tag{5.7.21}$$

This is precisely of the form of the bulk terms in the GL expression (5.7.2), with the temperature dependences (5.7.3a–b); however, the explicit expressions for α_0 and β are rather messy (cf. 5.7.18). They are simplified by changing the normalization of $\Psi : \Psi \to \Psi' \equiv V_0 \Psi = A^{-1}(T_c)\Psi$. Since the new order parameter is of the dimensions of energy, i.e. of the "gap" Δ, and reduces to it in the limit of homogeneous equilibrium (cf. Eqn. 5.8.5), I shall use for it the notation $\tilde{\Delta}$, with however the caveat that it is not *in general* equal to the quantity Δ used to parametrize the pair wave function (cf. the remarks made in Appendix 5D in connection with the quantity $\tilde{\Delta}_k$, of which $\tilde{\Delta}$ is a special case). In terms of $\tilde{\Delta}$ the constrained free energy takes the simpler form

$$F(\tilde{\Delta} : T) = F_0(T) + \frac{1}{2}\frac{dn}{d\epsilon} \left\{ - \left(1 - \frac{T}{T_c} \right) |\tilde{\Delta}|^2 + \frac{1}{2} \frac{7\zeta(3)}{8\pi^2} \frac{1}{(k_B T)^2} |\tilde{\Delta}|^4 \right\} \tag{5.7.22}$$

Minimization of the expression (5.7.22) with respect to $\tilde{\Delta}$ leads to the result

$$\tilde{\Delta}(T) = \left(\frac{8\pi^2}{7\zeta(3)}\right)^{1/2} k_{\mathrm{B}}T\left(1-\frac{T}{T_{\mathrm{c}}}\right)^{1/2} \cong 3.06\, k_{\mathrm{B}}T\left(1-\frac{T}{T_{\mathrm{c}}}\right)^{1/2} \tag{5.7.23}$$

which may be verified with some labor to reduce to the $T \to T_{\mathrm{c}}$ limit of the solution of the BCS Eqn. 5.5.20. Thus $\tilde{\Delta}(T)$ indeed reduces in equilibrium to the energy gap $\Delta(T)$.

I now turn to the derivation within BCS theory of an expression for the coefficient $\gamma(T)$, or what is equivalent, for the superfluid density $\rho_{\mathrm{s}}(T)$. The simplest argument is probably one originally introduced by Landau in the context of ^{4}He (cf. Chapter 3, Section 5.6): Imagine that the system is in thermal equilibrium subject to a prescribed uniform value of the superfluid velocity $\boldsymbol{v}_{\mathrm{s}}$. The momentum is then $\rho_{\mathrm{s}}(T)\boldsymbol{v}_{\mathrm{s}}$, and the normal component is at rest in the laboratory frame (due to scattering of the excitations by static impurities, phonons, etc.; as usual we shall take the limits of weak scattering and long times, and thus do not include these scattering mechanisms explicitly). Now imagine we transform to the frame of the moving superfluid. Viewed from this frame of reference, the normal component is moving with velocity $-\boldsymbol{v}_{\mathrm{s}}$ and the total momentum of the system is, by Galilean invariance, $\rho_{\mathrm{s}}(T)\boldsymbol{v}_{\mathrm{s}} - \rho\boldsymbol{v}_{\mathrm{s}} \equiv -\rho_n(T)\boldsymbol{v}_{\mathrm{s}}$, where $\rho_n(T) \equiv \rho - \rho_{\mathrm{s}}(T)$ is the normal density. Now, the correct frame of reference for discussing the excitation spectrum is indeed that in which the superfluid is at rest: in this frame the energies of the GP, BP and EP states of the pair $(\boldsymbol{k}\uparrow, -\boldsymbol{k}\downarrow)$ are given by 0, E_k and $2E_k$ respectively. However, the system is constrained to have a finite total momentum $\boldsymbol{P}$, and thus when determining the relative probabilities of these states we must use the standard Lagrange-multiplier technique and minimize not the free energy F but rather $F - \boldsymbol{v}\cdot\boldsymbol{P}$; the requirement that $\langle\boldsymbol{P}\rangle$ be proportional to $\boldsymbol{v}_{\mathrm{s}}$ then fixes $\boldsymbol{v}$ to be simply $-\boldsymbol{v}_{\mathrm{s}}$. Now the GP and EP states each makes a zero contribution to the total momentum, while the BP states $\boldsymbol{k}\sigma$ and $-\boldsymbol{k}, -\sigma$ contribute $+\hbar\boldsymbol{k}$ and $-\hbar\boldsymbol{k}$ respectively. Consequently the expression for the total momentum (viewed from the frame of the superfluid) is

$$\boldsymbol{P} = \sum_{\boldsymbol{k}} \hbar\boldsymbol{k}(P_k^{\mathrm{BP}\uparrow} - P_k^{\mathrm{BP}\downarrow}) = \sum_k \hbar\boldsymbol{k}\frac{e^{-\beta\hbar\boldsymbol{k}\cdot\boldsymbol{v}_{\mathrm{s}}} - e^{+\beta\hbar\boldsymbol{k}\cdot\boldsymbol{v}_{\mathrm{s}}}}{e^{+\beta E_k} + 2(e^{-\beta\hbar\boldsymbol{k}\cdot\boldsymbol{v}_{\mathrm{s}}} + e^{+\beta\hbar\boldsymbol{k}\cdot\boldsymbol{v}_{\mathrm{s}}}) + e^{-\beta E_k}} \tag{5.7.24}$$

For $v \ll k_{\mathrm{B}}T/\hbar k_{\mathrm{F}}$ this reduces approximately to $(E \equiv (\epsilon^2 + \Delta^2)^{1/2}$: note that for fermions $-\partial n(E)/\partial E = \beta/4\mathrm{sech}^2(\beta E/2)$, so this is consistent with (3.6.8))

$$\begin{aligned}
\boldsymbol{P} &= -\sum_k \hbar\boldsymbol{k}(\hbar\boldsymbol{k}\cdot\boldsymbol{v}_{\mathrm{s}})\frac{\beta}{4}\mathrm{sech}^2\beta E_k/2 \\
&= -\left(\frac{1}{3}\hbar^2 k_{\mathrm{F}}^2 \frac{dn}{d\epsilon}\frac{\beta}{2}\int_0^\infty d\epsilon\ \mathrm{sech}^2\beta E/2\right)\boldsymbol{v}_{\mathrm{s}} \\
&= -\left(\rho\frac{\beta}{2}\int_0^\infty d\epsilon\ \mathrm{sech}^2\beta E/2\right)\boldsymbol{v}_{\mathrm{s}}
\end{aligned} \tag{5.7.25}$$

Thus the normal density $\rho_n(T)$ is given by

$$\frac{\rho_n(T)}{\rho} = Y(T) \tag{5.7.26}$$

where the Yosida function $Y(T)$ is defined by

$$Y(T) \equiv \beta/2 \int_0^\infty \text{sech}^2\beta/2E/2d\epsilon \tag{5.7.27}$$

It is easy to demonstrate that in the limit $T \to T_c$ the superfluid fraction $\rho_s(T)/\rho = 1 - Y(T)$ is given by the expression

$$\frac{\rho_s(T)}{\rho} \cong \frac{7\zeta(3)}{4\pi^2 k_B^2 T_c^2}\Delta^2(T) \tag{5.7.28}$$

On the other hand we have the relation, valid for any normalization of Ψ (now taken to have its equilibrium value)

$$\gamma(T) = \frac{\hbar^2 \rho_s(T)}{8m^2|\Psi(T)|^2} \tag{5.7.29}$$

Then if we choose the normalization so that $\Psi = \tilde{\Delta}$ ($\equiv\Delta$ in homogeneous thermal equilibrium) and use (5.7.26), we finally find that with this normalization

$$\gamma = \frac{n\hbar^2}{4m}\frac{7\zeta(3)}{8\pi^2(k_B T_c)^2} \tag{5.7.30}$$

Note that since with this normalization $\alpha_0 = \frac{1}{2}(dn/d\epsilon)$, the prefactor of $(1-T/T_c)^{-1/2}$ in the expression for the healing length $\xi(T)$ is, up to a factor of order unity

$$\left[\frac{n\hbar^2}{m}\left(\frac{dn}{d\epsilon}\right)^{-1}(k_B T_c)^{-2}\right]^{1/2} \sim \frac{\hbar v_F}{k_B T_c}$$

i.e. precisely of the order of the Cooper pair radius ξ_0.

Once we have the expression for γ, we can use the assumptions of analyticity and locality to generalize Eqn. (5.7.5) to arbitrary slow variations of $\Psi(\boldsymbol{r})$, including variations in the amplitude; and then finally use the gauge-invariance arguments we have already employed several times to justify the general expression (5.7.10) for the GL free energy.

It is however necessary to make one caveat: While we have obtained a general expression, $(1-Y(T))\rho$ (cf. (5.7.24)), for the "superfluid density" $\rho_s(T)$, the latter may not always be a physically meaningful quantity. In fact, in general one would expect the relationship of the (superfluid) current $\boldsymbol{j}(\boldsymbol{r})$ to the superfluid velocity $\boldsymbol{v}_s(\boldsymbol{r}) \equiv (\hbar/2m)\boldsymbol{\nabla}\phi(\boldsymbol{r})$ not to be given by the simple "local" expression (5.6.4), but rather by an expression of the form

$$\boldsymbol{j}(\boldsymbol{r}) = \int K(\boldsymbol{r}-\boldsymbol{r}')\boldsymbol{v}_s(\boldsymbol{r}')d\boldsymbol{r}' \tag{5.7.31}$$

where the kernel K has a spatial range of the order of the largest "microscopic" length in the problem; we might guess (and it turns out to be correct) that this length is of the order of the pair radius ξ_0. Thus, the local ("London") approximation embodied in (5.6.4) is correct only if the scale of variation of $\boldsymbol{v}_s$ is large compared to ξ_0. Since we have seen that in the "GL" limit $T \to T_c$ the two characteristic length scales of the GL theory, $\xi(T)$ and $\lambda_L(T)$, both diverge as $(1-T/T_c)^{-1/2}$ (and in fact $\xi(T)$ is precisely of order $\xi_0(1-T/T_c)^{-1/2}$), it follows that the locality condition is always fulfilled in this limit, irrespective of the value of κ. At lower temperatures, however, $\lambda_{\mathrm{L}}(T)$ is of the order of $\lambda_{\mathrm{L}}(0)$ and $\xi(T)$ is (in so far as it can be defined) of order ξ_0, and the value of $\kappa(T) \equiv \lambda_{\mathrm{L}}(T)/\xi(T)$ is not very different from its $T \to T_{\mathrm{c}}$ limit: thus, for a Type-I superconductor, λ_{L} can be less (even much less) than ξ, and application of the local relation would be inconsistent, since it would predict that the range of variation of $\boldsymbol{v}_{\mathrm{s}}$ is small compared to the pair size. This is the so-called "Pippard" limit, and to discuss it one needs more detailed information on the structure of the kernel $K(\boldsymbol{r}-\boldsymbol{r}')$ than I want to introduce here; I merely quote without proof the result that in the extreme Pippard limit ($\lambda_{\mathrm{L}}(T) \ll \xi(T)$) the actual penetration depth is of order $(\lambda_{\mathrm{L}}^2(T)\xi(T))^{1/3}$ (see, e.g. de Gennes (1966), Chapter 2.)

To conclude this section, let us return to a question which was raised, only to be postponed, in Chapter 2, Section 4, namely why in the Fermi case fragmentation is generally disfavored. I shall actually approach this question indirectly, by showing first that an inhomogeneous coherent state is energetically disadvantageous relative to a uniform one, and then using a result analogous to that of Appendix 2B to draw conclusions about fragmentation.

In the GL regime close to T_{c}, where the formulae derived in this section should apply, the fact that an inhomogeneous value of $\Psi(r)$ is energetically unfavorable compared to a uniform one with the same normalization (same value of N_0) follows immediately from the fact that the bulk terms $\mathcal{F}_{\mathrm{b}}$ in the free energy functional (5.7.1) are concave upwards as a function of $|\Psi(r)|^2 \equiv x$, i.e.

$$\mathcal{F}_{\mathrm{b}}(x+y) \geq \mathcal{F}_{\mathrm{b}}(x) + \mathcal{F}_{\mathrm{b}}(y) \tag{5.7.32}$$

The inequality (5.7.32) actually holds for arbitrary temperature, not just in the GL regime near T_{c}; indeed, were it to fail we should expect the system to undergo phase separation.

Now, while the question of explicit construction of a many-body wave function or density matrix which will give a specified inhomogeneous form of the order parameter $\Psi(\boldsymbol{r})$ is not at all trivial, and will not be discussed here, it is clear that (specializing to the zero-temperature case for simplicity) one possible ansatz to describe pairing simultaneously in states with COM momentum zero and $\boldsymbol{K} \neq 0$ is a generalization of the (coherent) state Eqn. 5.4.6 of the form (omitting normalization etc.)

$$\Psi_{\mathrm{coh}}(\Delta\varphi) = \left(\sum_k \{ c_{\boldsymbol{k}}^{(0)} a_{\boldsymbol{k}}^+ a_{-\boldsymbol{k}}^+ + e^{i\Delta\varphi} c_{\boldsymbol{k}}^{(\boldsymbol{k})} a_{\boldsymbol{k}}^+ a_{-\boldsymbol{k}+\boldsymbol{K}}^+ \} \right)^{N/2} |\mathrm{vac}\rangle \tag{5.7.33}$$

Without detailed calculation, it is clear that such a state posesses simultanously nonzero values of the anomalous averages $\langle a_{\boldsymbol{k}}^+ a_{-\boldsymbol{k}}^+ \rangle$ and $\langle a_{\boldsymbol{k}}^+ a_{-\boldsymbol{k}+\boldsymbol{K}}^+ \rangle$ thus a spatially inhomogeneous order parameter $\Psi(r)$, and hence according to the argument of the last paragraph will be energetically disfavored relative to a state with the same value of N_0 but only one type of pairing (COM momentum either zero or $\boldsymbol{K}$).

However, by an argument parallel to that of Appendix B, the "fragmented" (Fock) state

$$\Psi_{\mathrm{Fock}} = \left(\sum_{\boldsymbol{k}} c_{\boldsymbol{k}}^{(0)} a_{\boldsymbol{k}}^+ a_{-\boldsymbol{k}}^+ \right)^M \left(\sum_{\boldsymbol{k}} c_{\boldsymbol{k}}^{(0)} a_{\boldsymbol{k}}^+ a_{-\boldsymbol{k}+\boldsymbol{K}}^+ \right)^{N/2-M} |\mathrm{vac}\rangle \tag{5.7.34}$$

can be written as a superposition of the coherent states with all possible values of $\Delta\varphi$. Consequently, provided that we can neglect terms in the energy nondiagonal in $\Delta\varphi$ (generally true except under e.g. the conditions specified in Chapter 2, Section 6 point (2)),we can immediately argue, similarly to the considerations given in Section 2.3 in connection with Eqn. (2.3.16), that the Fock (fragmented) state must be energetically disfavored (or least not favored) relative to the "best" coherent state, and thus by the argument of the last paragraph also disfavored with respect to either of its "components" (the states with pairs formed only with COM momentum zero or $\boldsymbol{K}$ respectively).

It should be emphasized that the above arguments refer only to the **bulk** terms in the free energy. The gradient terms, which represent the COM kinetic energy of the condensate, must be considered separately, but it is straightforward to do this for any given situation of interest; it is clear that they can never favor either of the states (5.7.33) or (5.7.34) over both its components simultaneously. Consequently our general conclusion is not affected.

5.8 Generalizations of BCS: the "non-pair-breaking" case

The original BCS theory of superconductivity is based on the very simplest possible model: a Sommerfeld model of free electrons, augmented by an interaction V_0 which is constant within a "shell" of width $2\epsilon_c$ around the Fermi energy and zero elsewhere (see Eqn. 5.2.20). Given the fact that even in the normal state the Sommerfeld model has to be augmented by the modifications introduced by Bloch and by Landau and Silin, and moreover that the effective interaction (5.2.17) (which as noted in Section 5.2 is itself not exact) is nothing like a single constant, it is at first sight a major surprise that the quantitative predictions of the BCS theory work so well (for example, for most of the "classic" superconductors the ration $\Delta(0)/k_B T_c$ is within 5% of the BCS prediction of 1.76). Even more surprising is the fact that many of the predictions continue to work even when the superconductor in question is not even a pure metal but a disordered alloy. In this section I shall sketch, mostly at a qualitative level, the reasons for the remarkable success of the original simple BCS model. The salient fact is that none of the complications to be introduced in this section interfere with the basic idea of pairing electrons in time-reversed states near the Fermi energy; for this reason they are often described as "non-pair-breaking" effects. "Pair-breaking" effects will be discussed in Section 5.9.

Let's first discuss the complications which arise from using the full Bardeen-Pines interaction (5.2.17) rather than the simplified BCS form (5.2.20). We continue to assume that the structure of the superconducting state is represented at $T = 0$ by the many-body wave function (5.4.3), with the function ϕ satisfying (5.4.5), and at $T \neq 0$ by the appropriate generalization (cf. Eqn. 5.5.10). Then all the algebra of Sections 5.4 and 5.5 goes through, and we obtain the gap equation (5.5.16)

$$\Delta_{\boldsymbol{k}} = -\sum_{\boldsymbol{k}'} V_{\boldsymbol{k}\boldsymbol{k}'} \frac{\Delta_{k'}}{2E_{k'}} \tanh \frac{\beta E_{k'}}{2} \tag{5.8.1}$$

with the pairing interaction $V_{kk'}$ given by (5.2.17) with $\boldsymbol{q} = \boldsymbol{k} - \boldsymbol{k}'$, that is

$$V_{kk'} \equiv \frac{\kappa_0}{1 + |\boldsymbol{k} - \boldsymbol{k}'|^2/q_{\mathrm{TF}}^2} \left\{ 1 + \frac{\omega_{\mathrm{ph}}^2(\boldsymbol{k} - \boldsymbol{k}')}{\omega^2 - \omega_{\mathrm{ph}}^2(\boldsymbol{k} - \boldsymbol{k}')} \right\} \tag{5.8.2}$$

where ω should now be interpreted as $(\epsilon_k - \epsilon_{k'})/\hbar$.

It is clear that we can parametrize the wave vector $\boldsymbol{k}$ by its direction $\hat{\boldsymbol{k}}$ and the energy ϵ_k and thus write

$$\begin{aligned} \epsilon_k &\equiv \frac{k^2 - k_{\mathrm{F}}^2}{2m} \\ \Delta_k &\equiv \Delta(\hat{\boldsymbol{k}}, \epsilon_k) \end{aligned} \tag{5.8.3}$$

Let us neglect for the moment the fact that the phonon energy ω_{ph} in general depends on the direction as well as the magnitude of the vector $\boldsymbol{k} - \boldsymbol{k}'$ (we will return to this briefly below in the context of the Bloch theory); then the expression (5.8.2) is evidently invariant under simultaneous rotation of $\boldsymbol{k}$ and $\boldsymbol{k}'$. The gap equation (5.8.1) then certainly possesses one or more "*s*-wave" (isotropic) solutions for which $\Delta(\hat{\boldsymbol{k}}, \epsilon_k)$ is independent of the direction $\hat{\boldsymbol{k}}$. It may or may not possess other ("anisotropic") solutions in which Δ is a (nontrivial) function of $\hat{\boldsymbol{k}}$; however, given experimentally realistic values of the ratio $q_{\mathrm{TF}}^2/k_{\mathrm{F}}^2$ and forms of $\omega_{\mathrm{ph}}(\boldsymbol{k})$ it is usually believed that these will be higher in free energy and thus unstable with respect to the (best) s-wave solution. (As we will see, the situation in this respect is different in ^{3}He and some of the "exotic" superconductors, in particular the cuprates.) Thus from now on we set $\Delta(\hat{\boldsymbol{k}}, \epsilon_k) \equiv \Delta(\epsilon_k) \equiv \Delta(\epsilon)$; the gap equation (5.8.1) then reduces to

$$\Delta(\epsilon) = -\int V(\epsilon, \epsilon') N(\epsilon') \frac{\Delta(\epsilon')}{2E(\epsilon')} \tanh \frac{\beta E(\epsilon')}{2} \, d\epsilon' \tag{5.8.4}$$

where $E(\epsilon) \equiv (\epsilon^2 + |\Delta(\epsilon)|^2)^{1/2}$ and $V(\epsilon, \epsilon')$ is the average of $V_{kk'}$ over the shells $\epsilon = \epsilon_k$, $\epsilon' = \epsilon_{k'}$. Evidently the dependence of $V(\epsilon, \epsilon')$ on its arguments is still quite complicated; however, the salient point is that the minimum range over which the variation is substantial is the Debye energy $\epsilon_{\mathrm{D}} \equiv \hbar\omega_{\mathrm{D}}$ (we must keep in mind here that ϵ_{D} is typically $\sim$2 orders of magnitude smaller than both ϵ_{F} and $\hbar^2 k_{\mathrm{FT}}^2/2m$).

We now carry out a renormalization procedure reminiscent of that routinely employed in atomic scattering calculations to replace the true interatomic potential V by a "*t*-matrix", see e.g. Pethick and Smith (2002), Chapter 5.2. Namely, we choose a cutoff energy ϵ_{c} much less than ϵ_{D}, solve for $\Delta(\epsilon)$ in the range $\epsilon > \epsilon_{\mathrm{c}}$ and plug the result back into the equation for $\Delta(\epsilon)$ in the "low-energy" range $\epsilon < \epsilon_{\mathrm{c}}$. Details of the procedure are given in Appendix 5F and I just quote here the result: the "low-energy" $(\epsilon < \epsilon_{\mathrm{c}})$ $\Delta(\epsilon)$ satisfies the equation

$$\Delta(\epsilon) = -N(0)\, t(\beta, \epsilon_{\mathrm{c}} : \{\Delta(\epsilon)\}) \int_{-\epsilon_{\mathrm{c}}}^{\epsilon_{\mathrm{c}}} \frac{\Delta(\epsilon')}{2E(\epsilon')} \tanh \frac{\beta E(\epsilon')}{2} \, d\epsilon' \tag{5.8.5}$$

In general the "*t*-matrix" $t(\beta, \epsilon_{\mathrm{c}} : \{\Delta(\epsilon)\})$ is a complicated functional of $\Delta(\epsilon)$, and so we have got no further. However, let us suppose that the actual solution of the gap equation corresponds to an (order-of-magnitude) value of Δ in the low-energy region which is much less than ϵ_c. Then, as is shown in Appendix 5F, the quantity t is independent of $\Delta(\epsilon)$; moreover, since we are not interested in values of T much larger than T_c (and will see that T_c is of order Δ) t is also independent of β. Thus the low-energy gap reduces to

$$\Delta(\epsilon) = -N(0)\, t(\epsilon_c) \int_{-\epsilon_c}^{\epsilon_c} \frac{\Delta(\epsilon')}{2E(\epsilon')} \tanh \frac{\beta E(\epsilon')}{2} d\epsilon' \tag{5.8.6}$$

This is clearly identical to the original BCS gap equation (5.5.16) with $V_{\boldsymbol{kk}'} \equiv -V_0$ provided we replace V_0 by $-t(\epsilon_c)$. Thus the analysis at the end of Section 5.5 goes through unchanged, and we find as there that $\Delta(T)$ is given by the formula

$$\Delta(T) = k_B T_c f\left(\frac{T}{T_c}\right) \tag{5.8.7}$$

where the critical temperature T_c is now given by the expression

$$k_B T_c = 1.14 \epsilon_c e^{-1/N(0)|t(\epsilon_c)|} \tag{5.8.8}$$

At first sight T_c, a physical quantity, depends on the cutoff ϵ_c, which is entirely arbitrary. However, it follows from the analysis of Appendix 5F that in the limit $k_B T_c \ll \epsilon_c$ the quantity $t(\epsilon_c)$ has exactly the dependence on ϵ_c to cancel this effect and give T_c independent of ϵ_c.

Thus, the conclusion is that provided the ratio Δ/ϵ_D is very small compared to 1 (the so-called "weak-coupling" limit), and provided we express our results as a function of T/T_c (where we may imagine T_c is taken from experiment) then Eqn. (5.5.20) is essentially exact. Moreover, it turns out that most of the experimental properties depend only on Δ (either directly or through the quantities u_k, v_k which can be expressed as functions of Δ), and thus the same conclusion applies to them. Thus, it is natural that most of the "classic" elemental superconductors should behave in almost all respects according to the predictions of the simple BCS model. In the few cases (Hg, Pb, ...) where there are appreciable deviations from those predictions, the ratio Δ/ϵ_D turns out to be anomalously large ($\sim$0.1), so that these are known as "strong-coupling" superconductors. It turns out that in the strong-coupling case it is actually not sufficient to solve the general gap equation (5.8.1) with the full effective interaction (5.8.2); one needs also to take into account effects which were implicitly neglected in the above formulation of the problem, for example that as a result of electron–phonon scattering the electron momentum eigenstates have a finite energy width. All these complications are usually believed to be adequately taken into account by a field-theoretic approach to the problem originally due to Eliashberg (see e.g. Scalapino 1969), which in the last resort relies only on the smallness of the ratio $c_s/v_F (c_s =$ speed of sound, $v_F =$ Fermi velocity). The Eliashberg theory, while giving excellent agreement with experiment for the strong-coupling superconductors (and a fortiori for the weak-coupling ones, for which it reduces to the BCS theory) is not particularly easy to interpret in simple terms, and I shall therefore not attempt to discuss it here.

This is a good point at which to deal with the Fock term (5.4.13) in the expectation value of $\langle V \rangle$, which was discarded in Section 5.4 in an apparently arbitrary way.

The value of this term is in the general case

$$\langle V \rangle_{\text{Fock}} = \frac{1}{2} \sum_{kk'} V_{kk'} \langle n_k \rangle \langle n_{k'} \rangle \tag{5.8.9}$$

which is not necessarily small. However, consider the difference between the value of the expression (5.8.9) in the superconducting and ($T = 0$) normal phases. Using Eqn. (5.4.28) we see that for the contact potential (5.4.23) this difference is zero; more generally, provided the weak-coupling condition $\Delta/\epsilon_{\text{D}} \ll 1$ is satisfied it will be at most of order $(\overline{V}N^2(0)/\epsilon_{\text{D}})\Delta^3 \ln(\epsilon_{\text{D}}/\Delta)$ (where $\overline{V}$ is a typical value of $V_{kk'}$). Since $\overline{V}N(0)$ is never large compared to 1, this means the difference is at most of order $(\Delta/\epsilon_{\text{D}})\ln(\epsilon_{\text{D}}/\Delta)$ relative to the BCS condensation energy (5.4.31), so that is is legitimate to neglect it within the framework of the BCS calculation. For Δ not very small relative to ϵ_{D} the term (5.8.9) can be taken into account by the Eliashberg technique.

I now deal more briefly with the 'Bloch" and "Landau–Silin" complications. With regard to the former, the obvious generalization of the BCS model is simply to pair electrons in Bloch-wave states with opposite quasimomentum $\boldsymbol{k}$ and spin σ. Generally, this prescription is unambiguous if the Fermi surface intersects only one Bloch band; in cases where more than one band does so, the usual assumption is that only "intraband" pairing occurs (i.e. $\langle a^{\dagger}_{k\uparrow,n} a^{\dagger}_{-k\downarrow,n'} \rangle \sim \delta_{nn'}$); "interband" pairing is pair-breaking (see next section) unless the Fermi surfaces happen to be exactly lined up (cf. Chapter 7, Section 7.8). Needless to say, at the Bloch level the properties of translational and rotational invariance which characterize both the electron and the phonon properties in the Sommerfeld model are replaced by invariance under lattice translations and the crystal point group respectively; as a result, for example, the pair wave function $F(\boldsymbol{r}, \boldsymbol{r}')$ is not strictly speaking a function only of $\boldsymbol{r} - \boldsymbol{r}'$ but has oscillations in $\boldsymbol{r}$ and $\boldsymbol{r}'$ separately with the lattice periodicity. Similarly, the possible solutions of the gap equation have to be classified by their transformation properties under the point group of the crystal; all the evidence is consistent with the assumption that in all the classic superconductors the stable solution corresponds to the identity representation,[34] often denoted for short as "s-wave". Although these complications affect the details of the behavior in the superconducting state (for example, one should not be surprised to find that in a crystal lacking cubic symmetry the superfluid density, like the normal-state conductivity, is a second-rank tensor), the overall behavior is little affected, and in particular the BCS gap equation (5.5.16) retains its validity.

The natural generalization of the BCS theory to take account of "Landau–Silin" effects involves pairing not real electrons but Landau quasiparticles. Since a quasiparticle is characterized by the same quasimomentum $\boldsymbol{k}$ and spin σ as the Bloch quasiparticle from which it has adiabatically evolved (see Appendix A) this prescription is quite unambiguous. As we saw in Section 5.1, the main effects of the transition from real particles to quasiparticles is (a) to renormalize the quasiparticle effective mass and (b) to produce various "molecular fields". The effects of (a) are similar to those of the transition from the Sommerfeld to the Bloch model: the only quantity which is

[34]That is, that in which the order parameter is invariant under all symmetry operations of the point group.

(or may be) affected is the transition temperature T_c, and when expressed in terms of T/T_c all quantities are unchanged. As regards the effects of the molecular fields, it has to be borne in mind that these come into play only in the presence of some macroscopic "polarization" (such as a finite value of some spin or momentum); since the formation of Cooper pairs corresponds to no such polarization, it is completely insensitive to molecular-field effects and the gap equation (5.5.16) retains its validity, even in their presence. However, in general quantities such as the spin susceptibility and normal density, which are responses of "polarizations" to applied fields, are sensitive to molecular-field effects, although (in contrast to the case of ^{3}He, see Chapter 6) for a variety of reasons I will not go into such effects are actually quite difficult to detect in the classic superconductors. Finally, as regards the "Silin" effects, that is the effect of the long-range Coulomb field (a special kind of "molecular field"!), the situation is much the same as in the normal state; they can be taken into account by a generalized random-phase approximation and lead, in the case of the Coulomb field, to essentially the same results[35] (screening of a fixed charge, the existence of plasmons, etc.) As we have already seen in Section 5.6, in the case of the long-range Ampère interaction (which has little effect in the normal state) the effects in the superconducting state are much more spectacular.

To summarize, we expect that for a reasonably pure metal, even at the most sophisticated (Landau–Silin) level of description, provided only that the condition $\Delta \ll \epsilon_D$ is satisfied, the dependence of $\Delta(T)$ on T/T_c will be correctly given by the simple BCS model, and the experimental properties will not be qualitatively modified from the predictions of that model.

We now turn to the question of the effects of disorder, or as it is often put, of superconducting alloys. In this section I confine myself to the case of nonmagnetic disorder, such as might be realized for example in a metallic glass; the electrons are imagined to move in a static, spin-independent potential $U(\boldsymbol{r})$ which, rather than possessing the periodic symmetry characteristic of a crystal, is disordered. Complete disorder is most commonly realized in the case of alloys of two or more different elements (a typical example is the metallic glass PdSiCu); the case of a "dirty metal", that is a crystal with a substantial concentration of (nonmagnetic) impurities, may be treated as a special case of an alloy, with an anomalously long mean free path.[36] The form of the interelectron interaction will be specified below.

Consider the eigenstates of the single-electron Hamiltonian

$$\hat{H}_0 \equiv -\frac{\hbar^2}{2m}\nabla^2 + U(\boldsymbol{r})$$

Since $\hat{H}_0$ is spin-independent, the eigenstates can be chosen to be of the form $\phi_n(\boldsymbol{r})\,|\sigma\rangle$, and the energy ϵ_n is then independent of the spin projection σ. Moreover, since $\hat{H}_0$ is also invariant under time reversal, the time-reverse of $\phi_n(\boldsymbol{r})$, namely $\phi_n^*(\boldsymbol{r})$, is also an

[35]This is because of the nontrivial but standard result that κ_0^{-1} (cf. Eqn. 5.2.6) is very little changed by the superconducting transition (the pairing process essentially "rides up" with the Fermi surface).

[36]If the impurities are not too concentrated, the relevant one-electron states $\phi_n(\boldsymbol{r})$ (see below) will be approximately linear combinations of Bloch waves from the lowest band. As the concentration increases, states from higher bands are mixed in, and eventually, in the limit of strong disorder, the Bloch picture fails completely.

eigenstate,[37] with the same energy ϵ_n as $\phi_n(\boldsymbol{r})$. It is this property which is crucial for the results below. Just as for the translation-invariant case, we can evidently define a Fermi energy ϵ_F (though not, of course, a Fermi surface, since the $\phi_n(\boldsymbol{r})$ are nothing like momentum eigenstates), and as in that case we shall measure the energies ϵ_n from ϵ_F. We note that we can also define a "Fermi velocity", at least to an order of magnitude, by $v_F \equiv (2m\epsilon_F)^{1/2}$.

At this point we need to specify the interelectron interaction. In general we would expect it to be a function of $\boldsymbol{r}$, $\boldsymbol{r}'$ and (for the same reasons as in the crystalline case) the energy difference $\epsilon_n - \epsilon_{n'}$ of the initial and final states. However, by arguments similar to those given earlier in this section we may plausibly replace this by the analog of the simple BCS ansatz (5.2.20), namely

$$V(\boldsymbol{r}, \boldsymbol{r}' : \epsilon_n, \epsilon_{n'}) = \begin{cases} -V_0\delta(\boldsymbol{r} - \boldsymbol{r}') & |\epsilon_n|, |\epsilon_{n'} < \epsilon_c \\ 0 & \text{otherwise} \end{cases} \tag{5.8.10}$$

Let us introduce the notation

$$|n\rangle \equiv |\phi_n(\boldsymbol{r}), \uparrow\rangle, \quad |\bar{n}\rangle \equiv |\phi_n^*(\boldsymbol{r}), \downarrow\rangle \tag{5.8.11}$$

so that the single-particle states n and $\bar{n}$ are time-reverses of one another, and moreover correspond to the same energy ϵ_n. Now, confining ourselves for the moment to the limit $T = 0$, we make for the many-body ground state wave function the ansatz (of which (5.4.6) is just a special case)

$$\Psi_N = \mathcal{N}'' \left(\sum_n c_n a_n^\dagger a_{\bar{n}}^\dagger \right)^{N/2} |\text{vac}\rangle \tag{5.8.12}$$

We can now easily generalize the arguments of Section 5.4 and Appendix 5C: in particular, because n and $\bar{n}$ possess the same single-particle energy ϵ_n, the expectation value of the single-particle energy has the form (cf. 5.4.17)

$$\langle H_0 \rangle = \sum_n |\epsilon_n|(1 - \sqrt{1 - 4|F_n|^2})$$

$$F_n \equiv \frac{c_n}{1 + |c_n|^2} \tag{5.8.13}$$

The potential energy needs a little more care. For the same reasons as in the crystalline case, we assume that we may disregard the Hartree and Fock terms. The pairing term is now given by the expression

$$\langle V \rangle = -V_0 \int d\boldsymbol{r} |F(\boldsymbol{r}, \boldsymbol{r})|^2 = -\sum_{nn'} V_{nn'} F_n F_{n'}^* \tag{5.8.14}$$

where we have defined

$$F(\boldsymbol{r}, \boldsymbol{r}') \equiv \sum_n F_n \phi_n(\boldsymbol{r}) \phi_{\bar{n}}(\boldsymbol{r}') = \sum_n F_n \varphi_n(\boldsymbol{r}) \varphi_n^*(\boldsymbol{r}') \left(\equiv \left\langle \psi_\uparrow^\dagger(\boldsymbol{r}) \psi_\downarrow^\dagger(\boldsymbol{r}') \right\rangle \right) \tag{5.8.15}$$

$$V_{nn'} \equiv V_0 \int d\boldsymbol{r} |\phi_n(\boldsymbol{r})|^2 |\phi_{n'}(\boldsymbol{r})|^2 \tag{5.8.16}$$

(the general form of $F(\boldsymbol{r}, \boldsymbol{r}')$ will be discussed below).

[37] Generally, in the presence of spin-independent disorder, the $\phi_n(\boldsymbol{r})$ are real, so that $\phi_n^*(\boldsymbol{r}) \equiv \phi_n(\boldsymbol{r})$. However, with a view to possible generalizations it is convenient to leave the formulation general.

The crucial point, now, is that while the individual $V_{nn'}$ may possibly oscillate wildly, when averaged over a very small ($\ll\Delta$) range of energies ϵ_n, $\epsilon_{n'}$ the mean value of $V_{nn'}$ is very close to V_0. Hence we can replace Eqn. (5.8.14) by the expression

$$\langle V\rangle = -V_0 \sum_{nn'} F_n F^*_{n'} \tag{5.8.17}$$

All subsequent maneuvers now go through (both at $T = 0$ and at finite T) in exact analogy to the calculations of Sections 5.4 and 5.5; in particular, we assume in view of the above considerations on $V_{nn'}$ that the gap Δ_n is a function only of ϵ_n, and we end up with a gap equation which is a simple generalization of (5.8.6), namely

$$\Delta(\epsilon) = -V_0 \int_{-\epsilon_c}^{\epsilon_c} N_0(\epsilon') \frac{\Delta(\epsilon')}{2E(\epsilon')} \tanh \frac{\beta E(\epsilon')}{2} d\epsilon' \tag{5.8.18}$$

where the single-particle density of states (per spin) $N_0(\epsilon)$ is defined by

$$N_0(\epsilon) \equiv \sum_n \delta(\epsilon - \epsilon_n) \tag{5.8.19}$$

Now, except in very pathological cases, the density of states $N_0(\epsilon)$, at least when averaged over very small ($\ll\Delta$) energies, is approximately independent of ϵ, so we may set it equal to the value $N(0)$ at the Fermi energy and take it out of the integrand. The form of the gap equation is thus precisely identical to that realized for a crystalline system, and consequently all the thermodynamics (when expressed in terms of the experimentally measured T_c) is identical; this applies however disordered the alloy, provided only that the approximation of the inter-electron potential by the δ-function form (5.8.10) is justified. In the case of a weakly disordered system we can actually draw a further conclusion: since in this case neither the singe-particle density of states $N(0)$ nor the effective interaction V_0 is much different from its value in the pure crystal, the transition temperature T_c should be little changed by a modest concentration of nonmagnetic impurities. This indeed seems to be consistent with the experimental data.[38]

While the thermodynamics of superconducting alloys is thus closely similar to that of pure metallic superconductors, some other properties, and in particular the behavior of the Cooper pair wave function, are substantially different. The pair wave function is given according to (5.8.15), (5.8.11) and the obvious generalization of (5.4.21) by

$$F(\boldsymbol{r}_1, \boldsymbol{r}_2) = \sum_n F_n \phi_n(\boldsymbol{r}_1)\phi^*_n(\boldsymbol{r}_2) = \Delta \sum_n \frac{\tanh \beta E_n/2}{2E_n} \phi_n(\boldsymbol{r}_1)\phi^*_n(\boldsymbol{r}_2) \tag{5.8.20}$$

(where we assume $\Delta(\epsilon_n) = \text{const.} = \Delta$ in accordance with the above considerations). Let's write $\boldsymbol{R} \equiv \frac{1}{2}(\boldsymbol{r}_1 + \boldsymbol{r}_2)$, $\boldsymbol{r} \equiv \boldsymbol{r}_1 - \boldsymbol{r}_2$ and average over $\boldsymbol{R}$ (the average is represented by a bar):

$$\overline{F(\boldsymbol{r}) = \Delta \sum_n \frac{\tanh \beta E_n/2}{2E_n} \phi_n(0)\phi^*_n(\boldsymbol{r})} \tag{5.8.21}$$

[38]In some cases (e.g. Al) the effect of (substantial) disorder is actually to raise T_c somewhat. This is because, by averaging out the crystalline anisotropy, it allows the gap to be more isotropic, an effect which is usually "advantageous" to Cooper pairing.

We have seen that for $\boldsymbol{r} = 0$ the value of $\overline{F}$ is quite comparable to what it is in the pure metal (namely of order Δ/ϵ_F). How does it fall off with $\boldsymbol{r}$? In the case of a pure metal we saw in Section 5.5 that it falls off exponentially with a range of order $\xi_0 \sim \hbar v_F/k_B T_c$ (or $\hbar v_F/\Delta(0)$). To discuss the behavior in the dirty limit, we use a semiclassical argument: $F(\boldsymbol{r})$ will drop substantially below its $\boldsymbol{r} = 0$ value as soon as the *difference* in the phase difference $(\arg \phi_n(\boldsymbol{r}) - \arg \phi_n(0))$ for different n's which are well represented in the sum becomes $\sim 2\pi$. Now, from a semiclassical point of view a wave packet which starts from $\boldsymbol{r} = 0$ at time zero with energy spread ΔE will be dephased by $\sim 2\pi$ in a time $t \sim \hbar/\Delta E$. In this case $\Delta E \sim \Delta$ (or $\sim k_B T_c$ near T_c), so $\Delta t \sim \hbar/\Delta$. Thus, the order of magnitude of the "size" r_p of the pairs (i.e. the value of r beyond which $F(r)$ falls off substantially) should be the radius which the wave packet has reached in a time $\sim\hbar/\Delta$. In the case of a pure metal, where the propagation is ballistic, r_p is given simply by $v_F \Delta t \sim \hbar v_F/\Delta$, (or $\hbar v_F/k_B T_c \equiv \text{const.}\xi_0$), so we recover our earlier result. However, in a dirty metal with (elastic) mean free path l the propagation for $r \gtrsim l$ is *diffusive*, so in the limit $r \gg l$ we have $r \sim \sqrt{D\Delta t}$ where $D = 1/3 v_F^2 l$ is the diffusion coefficient. Thus the pair radius r_p in a "dirty" metal (defined as one where $l \ll \xi_0$) is given by the estimate

$$r_p^2 \sim \frac{D\hbar}{\Delta} \sim \frac{\hbar v_F}{\Delta} l \tag{5.8.22}$$

i.e. if ξ_0 is the pair radius in a (possibly hypothetical) pure metal with the same v_F and T_c, then

$$r_p \sim (\xi_0 l)^{1/2} \tag{5.8.23}$$

In a disordered alloy, where l is only a few times the interatomic spacing, the reduction can be very substantial.

The above argument is essentially equivalent to the observation that the actual path length which an electron traverses in propagating out from the origin to radius r is (for $r \gg l$) of order $r^2/l \equiv R$. This suggests the following consideration: Suppose that we wish to bend (say) the phase of the $T = 0$ wave function through π over some distance r, which we will assume to be of the order of the pair radius (or at least not much larger). At what (order -of-magnitude) value of r does the energy required for this bending equal the condensation energy? For a pure metal, the results of Section 5.7 imply that the answer is of order of the pair radius ξ_0. Now for a dirty metal the equivalent length available to put the necessary "phase kinks" in the single-particle wave functions is not r but r^2/l; hence the answer should be that value of r for which $r^2/l = \xi_0$, i.e. the actual dirty-limit pair radius (5.8.23). In other words, the $T = 0$ superfluid fraction, which is unity for a pure superconductor in the Sommerfeld model, should be reduced in a dirty metal to a value of order $l/\xi_0 \ll 1$. This argument might also suggest (though it certainly does not prove) that a similar reduction should take place in the GL region near T_c, i.e. that the healing length $\xi(T)$ should be decreased, relative to its pure-limit value, by a factor of order $(l/\xi_0)^{1/2}$ (while $\lambda_L(T)$, which is proportional to $\rho_s^{-1/2}$, should increase by a factor $\sim(\xi_0/l)^{1/2}$). A more microscopic argument (see e.g. deGennes 1964, Chapter 7) shows that this conclusion is in fact correct, so that dirty metals, and in particular superconducting alloys, are much more likely to be Type-II than their crystalline counterparts–a result which is well supported by experiment.

To conclude this discussion of "non-pair-breaking" disorder, I add two notes. The first concerns the case of spin–orbit scattering, which does not conserve spin but is nevertheless invariant under time reversal. In this case, while we cannot choose the degenerate eigenfunctions of the single-particle Hamiltonian to be spin eigenstates (i.e. of the form (5.8.11)), we can still choose them to be pairs of states related by time reversal. Then, provided only that the range of the interelectron potential is small compared to the mean free path l_{SO} against spin–orbit scattering, the mean value of the pairing potential is still comparable to that for the pure case, and thus just as in the case of nonmagnetic disorder the zero-field thermodynamics is identical to that of the pure case. However, as one might guess, for $l_{SO} \lesssim \xi_0$ the spin dependent properties are modified, since the Cooper pairs now have a triplet as well as a singlet component. See, e.g. Ferrell (1959).

The second note concerns the question of what is sometimes called "localized superconductivity". It is well known that for noninteracting electrons moving, in 3D, in a sufficiently strongly disordered potential the single-electron eigenstates can be localized over a range of energy which may include a wide band around the Fermi energy ("Anderson localization"). Now, if one goes carefully through the arguments of this section concerning nonmagnetic disorder, in particular those which rest on averaging over energy regions $\ll\Delta$, one sees that the case of Anderson localization is in no way excluded; Cooper pairs can form, and the thermodynamics is similar to that of a pure metal. What is at first sight much more surprising is that despite the fact that in the normal state all the single-particle states are localized and the system is thus an insulator, in the superconducting state the Cooper pairs can nevertheless carry a current! Thus, in such a system one would expect a transition from insulator to superconductor to occur as the temperature is lowered. Whether such behavior has been seen experimentally is currently controversial (see, e.g. Sadovskii, 1997).

5.9 Pair-breaking effects

If we consider a general many-body wave function in the class (5.4.3) ("generalized completely paired BCS wave function"), then it is always possible to write it in the form

$$\Psi_N = \mathcal{N}'' \left(\sum_n c_n a_n^\dagger a_{\tilde{n}}^\dagger \right)^{N/2} |\text{vac}\rangle \tag{5.9.1}$$

where $\tilde{n}$ is a unique function of n; clearly, Eqns. (5.4.6) and (5.8.11) are special cases of (5.9.1). What distinguishes these special cases is that the pairs n, $\tilde{n}$ can be chosen so that; (a) they are eigenfunctions of the single-particle Hamiltonian with the same eigenvalue $\epsilon_n \equiv \epsilon_{\tilde{n}}$; (b) $\tilde{n}$ is the state obtained from n by the operation of time reversal. Clearly, except perhaps in special cases, the ability to satisfy simultaneously conditions (a) and (b) requires that the single-particle Hamiltonian be invariant under time reversal. Any effects which destroy this invariance are said to be "pair-breaking";[39]

[39] This definition is adequate for the simple case of s-wave (isotropic) pairing believed to be realized in the classic superconductors. In the case of anisotropic pairing the definition has to be generalized, see Chapter 7, Appendix B.

in general such effects lead to the suppression and, if sufficiently strong, to the complete destruction of superconductivity. In this section I examine some simple examples of pair-breaking effects. I will throughout assume that the potential is of the simple contact form (5.8.10).

Before embarking on specific examples, let us briefly examine why the simultaneous satisfaction of conditions (a) and (b) is necessary for "robust" superconductivity. Violation of condition (a) means that the minimum excitation energy of a pair relative to the Fermi energy is not arbitrarily small and thus the Cooper instability does not occur for arbitrarily small attraction. As regards condition (b), we note that quite generally, for any many-body wave function of the form (5.9.1) the expectation value of the "pairing" term in the potential energy is given by an expression of the general form

$$\langle V\rangle_{\text{pair}} = -V_0 \int d\boldsymbol{r}\ |F(\boldsymbol{r},\boldsymbol{r})|^2 \tag{5.9.2}$$

where the quantity $F(\boldsymbol{r},\boldsymbol{r}')$ is given by[40]

$$F(\boldsymbol{r},\boldsymbol{r}) = \sum_n F_n \phi_n(\boldsymbol{r})\phi_{\tilde{n}}(\boldsymbol{r}) \quad (F_n \equiv c_n/(1+|c_n|^2)) \tag{5.9.3}$$

A "large" value of $\langle V\rangle_{\text{pair}}$ clearly requires a "large" value of $F(\boldsymbol{r},\boldsymbol{r})$, which in turn requires that the contributions from different n not undergo destructive interference. A sufficient condition for this to be so is that $\phi_{\tilde{n}}(\boldsymbol{r}) = \phi_n^*(\boldsymbol{r})$, i.e. that condition (b) is satisfied. To be sure, it is not absolutely guaranteed that condition (b) is also necessary, but in practice it seems very difficult to find counterexamples (one of the few known ones is the LOFF state, see below).

The simplest example of a pair-breaking effect is that of a fictitious[41] static uniform magnetic field B which couples only to the electron spins, via the Zeeman energy $-\mu_B\sigma B$. This is a rather special case, in that the eigenstates of the single-particle Hamiltonian, including the Zeeman term, can still be sorted into pairs n and $\tilde{n}$ related by time reversal, although now $\epsilon_n \neq \epsilon_{\tilde{n}}$. At zero temperature we have two obvious options, namely: (A) pair states of opposite spin and equal single-particle energy, thereby satisfying condition (a). Since in a finite field a free-electron system is polarized so that the Fermi surfaces (the loci $\epsilon_n = \epsilon_{\tilde{n}} = \epsilon_F$) do not coincide (see Fig. 5.6), this is roughly equivalent to starting from the polarized normal ground state and forming one's pairs by taking electrons close to the relevant Fermi surface. From the geometry of Fig. 5.6 it is clear that in general it is not possible to choose the center-of-mass momentum of the pair to be zero, or indeed to be the same for all pairs (though see below), so that this prescription not only violates condition (b), but involves destructive interference between the different n in Eqn. (5.9.3). Option (B) is to pair time-reversed states ($n \equiv (\boldsymbol{k},\uparrow), \tilde{n} \equiv (-\boldsymbol{k},\downarrow)$) just as in zero field; in effect, this is equivalent to refusing to let the system polarize, i.e. to pretending the magnetic

[40]Since the argument at this point is intended to be schematic rather than rigorous, I simplify by omitting in (5.9.3) the spin degrees of freedom. The latter are also not written out explicitly in (5.9.13) below, because the time-reversed pairs introduced by the transformation (5.9.16) automatically have opposite spin at the same point $\boldsymbol{r}$.

[41]This situation may be approximately realized in very thin slabs with the magnetic field in the plane of the slab, since then the effects of the coupling to orbital motion are negligible.

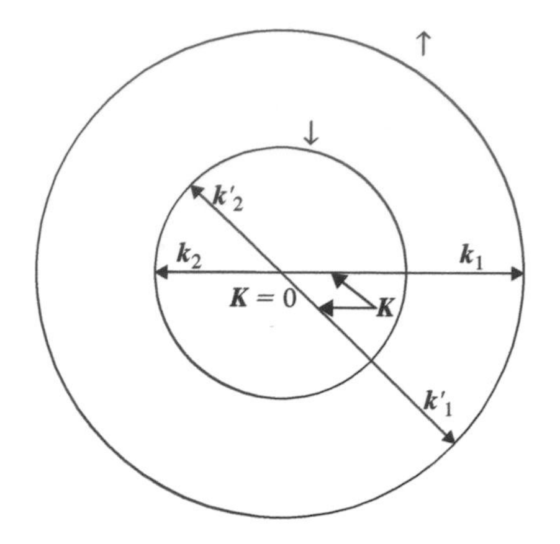

Fig. 5.6 Attempting to pair states $\boldsymbol{k}_{1\uparrow}$ and $\boldsymbol{k}_{2\downarrow}$ close to the Fermi surface in a Zeeman field yields a nonzero centre-of-mass momentum $\boldsymbol{K}$ which is different for different choices of $\boldsymbol{k}_{1\uparrow}$, $\boldsymbol{k}_{2\downarrow}$.

field does not exist and sacrificing the relevant polarization energy. On the other hand, condition (b) is satisfied and thus the potential energy can be as large as in the simple BCS case; indeed, apart from the polarization energy the thermodynamics is identical to that in zero field. A variant (B′) of option B is to allow complete polarization but nevertheless pair time-reversed (and hence unequal-energy) states.

If we assume that for option (A) the destructive interference is so effective that $\langle V\rangle_{\text{pair}}$ is negligibly small, then in this option there is no energy incentive to form pairs, i.e. it corresponds to the normal ground state. For the moment we ignore option (B'). Relative to option (A), option (B) has gained the full BCS superconducting condensation energy $\frac{1}{2}N(0)\Delta^2$, but has sacrificed the polarization energy $N(0)\mu_{\text{B}}^2B^2$. Thus,

$$E_{\text{B}} - E_{\text{A}} = -\frac{1}{2}N(0)\Delta^2 + N(0)\mu_{\text{B}}^2B^2 \tag{5.9.4}$$

Thus, we expect a transition, at $T = 0$, between the normal (N) and superconducting (S) phases at a field (the "thermodynamic critical field") $B_{\text{c}} = \Delta/\sqrt{2}\mu_{\text{B}}$. This actually turns out to be a first-order transition; the second-order transitions corresponding to the "supercooling" (N → S) and "superheating" (S → N) critical fields occur at $B_{\text{SC}} = \Delta/2\mu_{\text{B}}$ and $B_{\text{SH}} = \Delta/\mu_{\text{B}}$ respectively. At finite temperatures the picture is somewhat modified, and above a temperature $T = 0.56T_{\text{c}}$ there is only a single second-order transition (see Maki and Tsuneto 1964).

One might ask: Can one exclude the possibility of a third type of solution, in some sense intermediate between (A) and (B), which while adequately maintaining condition (a) allows a nonzero value of $\langle V\rangle_{\text{pair}}$? The answer is no: as shown independently by Larkin and Ovchinnikov and by Fulde and Ferrell, one can start from the polarized ground state and form pairs over only a part of the Fermi surface and with a common nonzero center-of-mass momentum. This prescription adequately maintains condition (a), and while violating condition (b) nevertheless achieves a nonzero value of $\langle V\rangle_{\text{pair}}$.

The resulting "LOFF" state[42] turns out to have, over a small range of B around B_c, an energy which is lower than that of either the N or S states discussed above. However, unlike these states it is very fragile in the presence of any kind of disorder (even of the nonmagnetic variety), and it is a matter of controversy whether there is at present any direct experimental evidence for its existence (see, e.g. Steglich et al. 1996).

The best-known kind of pair-breaking effect (for a simple s-wave-paired superconductor) is that of random magnetic impurities. Such impurities may be regarded as coupling to the conduction-electron system through an interaction of the form .

$$\Delta\hat{H}_0 \equiv \sum_m J_m \boldsymbol{S}_m \cdot \boldsymbol{\sigma}(\boldsymbol{r}_m) \tag{5.9.5}$$

where $\boldsymbol{\sigma}(\boldsymbol{r}_m)$ is the conduction-electron spin density at the site $\boldsymbol{r}_m$ of the m-th impurity, J_m is the associated coupling constant and $\boldsymbol{S}_m$ is the spin of the impurity, which is regarded as a fixed c-number; the quantities J_m, $\boldsymbol{r}_m$ and the direction of $\boldsymbol{S}_m$ are regarded as random variables. It is clear that in the presence of the term (5.9.5) the single-particle Hamiltonian is no longer invariant under time reversal (indeed, unlike the case of uniform "Zeeman" field, the time reverse of an energy eigenstate is in general not itself an eigenstate); thus magnetic impurities are indeed "pair-breaking". It is convenient in the following not to confine ourselves to the specific Hamiltonian (5.9.5) but to consider a more general class of situations with the above properties, namely that the time-reversal operator $\hat{K}$ does not commute with the single-particle Hamiltonian and moreover the time-reverse of an energy eigenstate is in general not itself an energy eigenstate.

A very beautiful treatment of the problem of the formation of Cooper pairs in the presence of pair-breaking effects, adapted from the original Green's function treatment of Abrikosov and Gor'kov, is given in Chapter 8 of the book of De Gennes (1966), and what follows is essentially a "poor man's version" of this.[43] We start with a qualitative consideration: Analogously to the Zeeman problem discussed above, at $T = 0$ we have two "extreme" choices. (A) we can try to form pairs from energy eigenstates (B); we can readjust the single-particle distribution so that whenever a particular single-particle state is occupied, so is its time reverse, and then form Cooper pairs out of pairs of time-reversed states. Assuming for the moment that choice (A) makes $F(\boldsymbol{r}, \boldsymbol{r})$ zero, this choice gives the normal ground state. Choice (B) saves a condensation energy approximately equal to that of the pure BCS state, $\frac{1}{2}N(0)\Delta_0^2$ where Δ_0 is the BCS gap for the corresponding time-reversal-invariant problem; however, to perform the appropriate rearrangement of the single-particle distribution costs an energy of order $\frac{1}{2}N(0)\Gamma_K^2$, where Γ_K is the characteristic energy width (inverse lifetime) associated with the time reversal operation (see below). Thus, we should expect the superconducting state (B) to be energetically favored at zero temperature only so long as $\Gamma_K \lesssim \Delta_0$. As we shall see, this conclusion is qualitatively correct, although the argument as it stands is too naive.

[42]Or rather class of states: the specific solutions found by LO are not identical to those of FF.

[43]The derivation given by De Gennes is based on the Bogoliubov-De Gennes equations, which I do not introduce in this book.

Now for a more quantitative treatment. For pedagogical convenience I confine myself for the moment to the case $T = 0$. Let's work in the energy basis $|m\rangle$ and define the matrix elements K_{mn} of the time-reversal operator $\hat{K}$ in this basis; in view of the antiunitary character of $\hat{K}$ we must obviously have

$$\sum_n |K_{mn}|^2 = 1 \ \forall \ m. \tag{5.9.6}$$

Without yet specifying a detailed form for the K_{mn}, we make the following two assumptions:

1. The range of $|\epsilon_m - \epsilon_n|$ for which K_{mn} is appreciable is of the order of some quantity Γ_K.
2. K_{mn} is "random", in the sense that we may set

$$\overline{\sum_{nn'} K^*_{mn} K_{mn'}} \cong \overline{\sum_n |K_{mn}|^2} \tag{5.9.7}$$

provided the average over ϵ_n indicated by the bar extends over some minimum scale δ which we assume to be very much less than Γ_K.

It is convenient to seek the many-body groundstate Ψ_N in the form of a generalization not of Eqn. (5.4.6) but of the "alternative" form (5.4.37). Thus we define the normal Fermi sea $|\text{FS}\rangle$ by

$$|\text{FS}\rangle \equiv \prod_{\epsilon_n < 0} a_n^+ |\text{vac}\rangle \tag{5.9.8}$$

(where ϵ_n denotes the eigenvalues of the single-particle Hamiltonian relative to the Fermi energy) and postulate that apart from normalization and the operation analogous to Eqn. (5.4.41) we have

$$\Psi_N = \left(\sum_{\substack{mn \\ (\epsilon_m, \epsilon_n > 0)}} c_{mn} a_m^+ a_n^+ \right)^{N_+} \left(\sum_{\substack{mn \\ (\epsilon_m, \epsilon_n < 0)}} d_{mn} a_m a_n \right)^{N_-} |\text{FS}\rangle \tag{5.9.9}$$

where the normalization is such that

$$\sum_{mn} |c_{mn}|^2 \equiv N_+ = \sum_{mn} |d_{mn}|^2 \equiv N_- \tag{5.9.10}$$

In general, the analysis of a wave function of the type (5.9.9) is cumbrous and messy. However, we can simplify it enormously if we assume that the actual "gap" $\Delta \equiv -V_0 \Psi$ which will emerge from the calculation is small compared to Γ_K (and hence, in general, small compared to Δ_0, the value of Δ for the system in the absence of pair-breaking); this limit will certainly be adequate for the calculation of the critical value of Γ_K (provided that there is a single second-order transition, cf. below). Under this condition it will turn out, as we shall verify, that the probability of excitation of any particular particle state above the Fermi energy, or of any particular hole state below, is very small compared to unity; as a result, the effects of the Pauli principle are negligible, and the problem reduces in effect to a generalization of the Cooper problem, the only difference being that we now explicitly consider the excitation of pairs of holes (described by the coefficients d_{mn}) as well as the pairs of particles described by the c_{mn}.

To be specific, we note that the quantity

$$F_{mn} \equiv \langle N-2|a_n a_m|N\rangle \tag{5.9.11}$$

is under the condition of small excitation of any one m, n simply equal to c_{mn} for $\epsilon_m, \epsilon_n > 0$ and to d^*_{mn} for $\epsilon_m, \epsilon_n < 0$; for different signs of ϵ_m, ϵ_n F_{mn} is identically zero, cf. the form of Eqn. (5.9.9). Thus the kinetic energy can be written in the form

$$KE = \sum_{mn}(|\epsilon_m| + |\epsilon_n|)|F_{mn}|^2 \tag{5.9.12}$$

Consider now the potential energy, neglecting as always the Hartree and Fock terms. According to (5.9.2), this is given by[40]

$$\langle V\rangle = -V_0 \int |F(\boldsymbol{r},\boldsymbol{r})|^2 d\boldsymbol{r} \tag{5.9.13}$$

where

$$F(\boldsymbol{r},\boldsymbol{r}) \equiv \sum_{mn} F_{mn}\varphi_m(r)\varphi_n(r) \tag{5.9.14}$$

Now we make the crucial assumption that only states which are time reverses of one another contribute (on average) to the RHS of Eqn. (5.9.14). Then we can rewrite (5.9.13) as

$$\langle V\rangle = -V_0|\Psi|^2 \tag{5.9.15}$$

where

$$\Psi \equiv \sum_{mn} K_{mn}F_{mn} \tag{5.9.16}$$

where K_{mn} is the matrix element of the time-reversal operator $\hat{K}$ introduced above. Combining (5.9.15) with (5.9.12), we find an expression for the expectation value of the Hamiltonian which is quadratic in the F_{mn}:

$$\langle H\rangle = \sum_{mn}(|\epsilon_m| + |\epsilon_n|)|F_{mn}|^2 - V_0\left|\left(\sum_{mn} K_{mn}F_{mn}\right)\right|^2 \tag{5.9.17}$$

It is now straightforward to obtain the optimum values of the quantity F_{mn} (i.e. of the coefficients c_{mn}, d_{mn} in (5.9.9)) by minimizing the expression (5.9.17) with respect to F^*_{mn}; the only point to remember is that by construction F_{mn} is zero whenever the signs of ϵ_m and ϵ_n are different, and we take this into account in what follows by defining a symbol

$$\Theta_{mn} \equiv \tfrac{1}{4}(sgn\epsilon_m + sgn\epsilon_n)^2 \tag{5.9.18}$$

Then, defining

$$\Delta \equiv -V_0\Psi \equiv -V_0\sum_{mn} K_{mn}F_{mn} \tag{5.9.19}$$

we find

$$F_{mn} = \frac{\Delta K_{mn}}{|\epsilon_m| + |\epsilon_n|}\Theta_{mn} \tag{5.9.20}$$

Now the probability of excitation p_m of the particular state m is given by $\sum_n |F_{mn}|^2$; given that K_{mn} satisfies (5.9.6) and (5.9.7) and that the range of $|\epsilon_m - \epsilon_n|$ for which it is appreciable is of order Γ_K, Eqn. (5.9.20) implies that p_m is at most of order $(\Delta/\Gamma_K)^2$, which by hypothesis is very small compared to unity. Hence our procedure of neglecting the Pauli principle and thus equating F_{mn} to c_{mn} (or d^*_{mn}) is self-consistent.

Substituting the solution (5.9.20) in (5.9.12) and using (5.9.16) and (5.9.19), we find that the expectation value of $\langle H\rangle$ as a function of Δ is

$$\langle H\rangle_\Delta = \Delta^2 \left(\sum_{mn} \frac{|K_{mn}|^2}{|\epsilon_n| + |\epsilon_n|} \Theta_{mn} - V_0^{-1} \right) \tag{5.9.21}$$

so that the condition for Δ to be nonzero[44] in equilibrium (i.e. for Cooper pairing to occur) is

$$\sum_{mn} \frac{|K_{mn}|^2 \Theta_{mn}}{|\epsilon_m| + |\epsilon_n|} < V_0^{-1} \tag{5.9.22}$$

We can rewrite this by using the zero-temperature gap equation for the pure system (with energy gap Δ_0)

$$Q \equiv \sum_{mn} \frac{|K_{mn}|^2 \Theta_{mn}}{|\epsilon_m| + |\epsilon_n|} < N(0) \ln \frac{2\epsilon_c}{\Delta_0} \tag{5.9.23}$$

To proceed further, we need to make some definite assumptions about the form of the matrix elements K_{mn}. The simplest assumption is that $\hat{K}$ relaxes exponentially with a time constant $\tau_K \equiv \Gamma_K^{-1}$ (which for the simple model considered (only) is equal to the (normal-state) conduction-electron spin relaxation time, see de Gennes (1966), Section 8.2). If this is correct, then it is easy to show that "on average" (over ϵ_n, cf. 5.9.7) the behavior of the quantity $K(\omega) \equiv \sum_n |K_{mn}|^2 \delta(\omega - (\epsilon_n - \epsilon_m))$ is

$$K(\omega) \equiv \frac{1}{\pi} \frac{\Gamma_K}{\omega^2 + \Gamma_K^2} \tag{5.9.24}$$

With this ansatz, and the usual assumption of particle–hole symmetry, we find that the quantity Q of (5.9.23) is given by

$$Q = \frac{2N(0)}{\pi} \int_0^{\epsilon_c} d\epsilon \int_0^{\epsilon_c} d\epsilon' \frac{\Gamma_\kappa}{(\epsilon - \epsilon')^2 + \Gamma_\kappa^2} \frac{1}{\epsilon + \epsilon'} \tag{5.9.25}$$

In the limit $\Gamma_K \ll \epsilon_c$ this expression tends to $N(0) \ln 2\epsilon_c/\Gamma_K$. Inserting this expression into (5.9.23), we see that the criterion for the onset of superconductivity is simply $\Delta_0 > \Gamma_K$, in agreement with the conclusion of our earlier qualitative argument.

The generalization of the above zero-temperature argument to finite T is immediate: since the quantity $\langle a_m a_n\rangle$ is given by F_{mn} for the GP state but by $-F_{mn}$ for the EP one (and is of course zero for the BP states), we should multiply the summand in expression (5.9.22) by $P_{\mathrm{GP}} - P_{\mathrm{EP}}$. But since the "perturbation" of the normal-state distribution by the pairing is small, P_{GP} is just given by $\langle n_m\rangle_T \cdot \langle n_n\rangle_T$ where $\langle\ \rangle_T$ is the normal-state expectation value at temperature T, and P_{GP} similarly by $(1-\langle n_m\rangle)\cdot(1-\langle n_n\rangle)$. Thus the factor $P_{\mathrm{GP}} - P_{\mathrm{EP}}$ is $\frac{1}{2}(\tanh\beta|\epsilon_m|/2 + \tanh\beta|\epsilon_n|/2)$, and

[44] Clearly, to determine the actual value of Δ we would need to keep the neglected terms of higher order in Δ/Γ_K.

correspondingly the integral of (5.9.25) is multiplied by $\frac{1}{2}(\tanh\beta|\epsilon|/2 + \tanh\beta|\epsilon'|/2)$. Expressing V_0 now in terms of $T_{co} \equiv 1/k_B\beta_{co}$, the transition temperature for the pure system, rather than Δ_0, we find for the dependence of the critical temperature $T_c \equiv 1/k_B\beta_c$ of the "dirty" system on Γ_K the implicit relation

$$\frac{1}{\pi}\int_0^{\epsilon_c}\int_0^{\epsilon_c} d\epsilon\, d\epsilon' \frac{(\tanh\beta_c|\epsilon|/2 + \tanh\beta_c|\epsilon'|/2)\Gamma_K}{[(\epsilon-\epsilon')^2 + \Gamma_K^2](\epsilon+\epsilon')} = \int_0^{\epsilon_c} \frac{\tanh(\beta_{co}|\epsilon|/2)d\epsilon}{|\epsilon|} \tag{5.9.26}$$

which yields $\beta_c(\Gamma_K \to 0) = \beta_{co}$, as of course it must. Inspection of Eqn. (5.9.25) reveals that for $T_{co} \ll \epsilon_c$ the cutoff ϵ_c disappears from the solution; in the literature it has become conventional to transform the integrals by standard mathematical operations so as to express (5.9.26) in terms of the digamma function $\psi(z) \equiv \Gamma'(z)/\Gamma(z)$ (where $\Gamma(z)$, no relation to Γ_K, is the standard Euler Γ-function):

$$\ln(\mathrm{T_{co}}/\mathrm{T_c}(\Gamma_\kappa)) = \psi\left(\frac{1}{2} + \frac{\Gamma_\kappa}{4\pi \mathrm{k_B T_c}(\Gamma_\kappa)}\right) - \psi\left(\frac{1}{2}\right) \tag{5.9.27}$$

However, this form (which is known as the Abrikosov-Gor'kov formula) obscures the simple physics of pairing of time-reversed states which is manifest in (5.9.26).

While the above discussion is consistent with the standard theory of superconductivity in the presence of magnetic impurities (see De Gennes 1966, Chapter 8), one caveat is in order: Eqn. (5.9.26) is in effect an equation for the instability of the normal groundstate against creation of a small value of Δ, i.e. it defines a "supercooling" critical value of the ratio Γ_K/Δ_o. Now, in the case of simple Zeeman depairing discussed earlier in this section the analogous calculation would be that denoted there as "option (B′)", and it turns out to give the supercooling field $\Delta/2\mu_B$. However, we saw that the **thermodynamic** critical field was larger than this, and corresponded to a first-order rather than a second-order transition. Is it conceivable that a similar situation occurs in the magnetic-impurity problem, i.e. that it might be energetically advantageous to "rearrange" the Fermi sea partially before turning on the pairing interaction (this is in fact "option (B)" of the qualitative argument leading to $\Gamma_K^{(\mathrm{int})} \sim \Delta_o$)? To the best of my knowledge there has been no detailed investigation of this question in the literature.

Finally, I want to emphasize that irrespective of the caveat just raised, the discussion given above merely scratches the surface of a rich and complex subject; for example, I have not even mentioned the "gapless superconductivity" which can occur in a part of the parameter space (on this, see De Gennes 1966, Chapter 8, or Maki 1969).

5.10 The Josephson effect

The term "Josephson effect" is nowadays used in a loose sense, to describe phenomena which occur when a pair of superconducting (or superfluid) systems are connected together by some kind of "weak link", that is, a region where superconductivity (or superfluidity) can exist but is in one way or other suppressed. In the classical superconductors such a region can be a standard tunnel oxide barrier (see below), a point contact, a thin superconducting wire ("microbridge") or some other kind of link; in liquid ^{4}He it is most commonly one or more capillaries connecting two vessels, and in the alkali gases (in so far as it has been realized) a region of high potential energy as provided, e.g. by a blue-detuned laser. While all of these examples involve two spatially separated "bulk" regions and some kind of physical barrier or constriction between them, it is also possible to think of certain phenomena in which particles are transferred between different internal states as a sort of "internal Josephson effect", with the role of the barrier or constriction being replaced by the weakness of the transferring mechanism; examples include Raman-laser-assisted transitions between different hyperfine states in the alkali gases, and NMR phenomena in superfluid ^{3}He (see Chapter 6, Section 5.4). In this section I shall discuss the Josephson effect as it is realized in classic superconductors, and will concentrate mainly on the case considered by Josephson in his original work, namely that of a simple tunnel oxide barrier.

Let's start by considering a general situation in which we have two different metallic regions which would each if isolated be superconducting, connected by some kind of weak link (for the moment of unspecified nature). Then the essential point is that it is necessary, at $T = 0$, to describe the whole combined system by a wave function of the generalized BCS type (5.4.3), where however the arguments $\boldsymbol{r}_1$ and $\boldsymbol{r}_2$ of the "pseudo-molecular" wave function range throughout both bulk systems and also through the weak link (though see below). As in the case of a single bulk system, it is actually more convenient to frame our discussion in terms of the "pair wave function" $F(\boldsymbol{r}\sigma, \boldsymbol{r}'\sigma')$; since we are dealing in this section with the classic superconductors, we shall make the standard assumption that the internal pair structure is s-wave spin singlet, i.e. that F has, at least approximately, the structure

$$F(\boldsymbol{r}\sigma, \boldsymbol{r}'\sigma') = 2^{-1/2}\delta_{\sigma,-\sigma'}f(|\boldsymbol{r}-\boldsymbol{r}'|)\Psi(\boldsymbol{R}) \tag{5.10.1}$$

where $\boldsymbol{R} \equiv \frac{1}{2}(\boldsymbol{r}+\boldsymbol{r}')$ is the position of the pair CoM and the relative wave function f is fixed by the energetics (see Section 5.4). Thus the complex scalar quantity $\Psi(\boldsymbol{R})$ is, up to possible normalization, the CoM wave function of the pairs (GL order parameter).

If the link between the two bulk regions 1, 2 is indeed "weak" as we are supposing, then we should expect that, to a good approximation in each bulk region separately, the magnitude of the order parameter Ψ should be fixed at the equilibrium value imposed by the energetics, and its phase should be constant. Thus

$$\Psi(\boldsymbol{R}) = \Psi_1, \boldsymbol{R} \in 1: \quad \Psi(\boldsymbol{R}) = \Psi_2, \boldsymbol{R} \in 2 \tag{5.10.2}$$

(If the two bulk metals are of identical composition, then evidently $|\Psi_1| = |\Psi_2|$, but we do not need this property for the subsequent discussion: all that matters is that $|\Psi_1|$ and $|\Psi_2|$ are uniquely fixed by the energetics.) The crucial point is that up to this point the *relative phase* $\Delta\phi$ of Ψ_1 and Ψ_2 has not been fixed by the energetics; it is principally the dynamics of this phase which give rise to the complex of phenomena known as the Josephson effect.

At this point it is necessary to make an important distinction. We will see below that the absolute ground state of the combined system normally corresponds to $\Delta\phi = 0$; under that condition the natural assumption is that the phase $\phi(\boldsymbol{R})$ of $\Psi(\boldsymbol{R})$ is constant throughout the whole system, including the weak link. Now imagine that we start "cranking up" $\Delta\phi$, e.g. by applying a voltage across the weak link (see below). As $\Delta\phi$ increases and approaches the value π, there are two possible scenarios:

(a) The magnitude of $\Psi(\boldsymbol{R})$ remains nonzero throughout the weak link, so that $\phi(\boldsymbol{R})$ (and its gradient) is everywhere defined. In this case we can define $\Delta\phi$ by the unique prescription

$$\Delta\phi = \int \boldsymbol{\nabla}\phi(\boldsymbol{R}) \cdot d\boldsymbol{l} \tag{5.10.3}$$

and there is nothing to prevent $\Delta\phi$ from taking a value larger than π (indeed, under appropriate conditions, $\Delta\phi$ may take values $\gg 2\pi$).[45] This is the case of a "hysteretic" weak link: the values of the arguments ϕ_1 and ϕ_2 of Ψ_1 and Ψ_2 (which are defined only modulo 2π) do not uniquely characterize the state of the combined system and thus its energy (the difference between this energy for (e.g.) $\Delta\phi = 2\pi$ and $\Delta\phi = 0$ comes of course from the bending term within the weak link).

(b) On the other hand, the energetics may be such that when $\Delta\phi$ equals π, the magnitude of $\Psi(R)$ vanishes across some surface spanning the weak link, so that $\Delta\phi$ can no longer be defined by (5.10.3). When the quantity $\phi_1 - \phi_2$ is further increased, say to $\pi + \epsilon$, the state of the system is identical to that for $\phi_1 - \phi_2 = -\pi + \epsilon$. The system is said to have undergone a "phase slip". Thus in this ("nonhysteretic") case the quantity $\Delta\phi$ can only be defined modulo 2π ($\Delta\phi \equiv \phi_1 - \phi_2$), and the energy is a unique function of the bulk phases.

A rough-and-ready criterion for whether a particular type of weak link is likely to be hysteretic is whether the condensation energy density *of the weak link* is larger or smaller than the "bending" energy density arising from the variation of the phase $\phi(\boldsymbol{R})$ by π across the weak link. If the weak link is of the same material as the bulk metals (which is usually the case for microbridges) and we are in the GL regime, we can use the result obtained in Section 5.7, namely that the above two energies are equal in order of magnitude when the bending through 2π is over the GL healing length $\xi(T)$. Thus, in such a case we would tend to conclude that a weak link is likely to be hysteretic (nonhysteretic) if its length L is $\gg \xi(T)$ ($\ll \xi(T)$); in particular, any such link should be nonhysteretic in the limit $T \to T_c$ (since as shown in Section 5.7, $\xi(T)$ diverges in that limit). On the other hand, in some kinds of weak link (notably in a standard tunnel oxide barrier) the density of electrons, and hence (in so far as it can be defined) the condensation energy, is very much smaller than in bulk, and consequently (since the "length" (thickness) of such junctions is never more than a few hundred Å) the bending energy always overwhelms the condensation energy, and such junctions are always nonhysteretic. In the literature the term "Josephson junction" is sometimes reserved for the nonhysteretic class of weak links,[46] and I shall confine myself to this subclass for the rest of this section.

Consider then two bulk superconductors characterized by order parameters Ψ_1 and Ψ_2 with the equilibrium magnitudes $|\Psi_1|$ and $|\Psi_2|$, and a phase difference $\Delta\phi \equiv \arg(\Psi_2/\Psi_1)$, $-\pi < \Delta\phi < \pi$. In general, at $T = 0$ under these conditions the energy will consist of a "bulk" term, which is independent of $\Delta\phi$ and of no interest to us in the present context, and a term associated with the junction, which we should expect

[45] Eventually, for large enough $\Delta\phi$, Ψ will go to zero in the middle of the link as in case (b) and we will see a phase slip.

[46] In the case of the "internal" Josephson effect the whole concept of a spatial variation of $\phi(\boldsymbol{R})$ is inapplicable, and the behavior is always nonhysteretic.

to depend on $\Delta\phi$. Without some information about the detailed nature of the junction it is difficult to say much rigorously about the form of the dependence on $\Delta\phi$, but we shall see below that in a large class of junctions, including the original tunnel oxide barriers, it is given, apart from a constant, by an expression of the form

$$E_J(\Delta\phi) = -J\cos\Delta\phi \tag{5.10.4}$$

where J is a positive constant. Below I will use this form for definiteness: it is easy to see how the ensuing discussion may be generalized to forms of $E_J(\Delta\phi)$ different from (5.10.4).

We now invoke a crucial and generic result, which is found in Appendix 2B, namely that the variable which is canonically conjugate to the relative phase $\Delta\phi$ is the *number imbalance*[47] ΔN between the two bulk superconductors, or more accurately $\Delta N/2$: when ΔN and $\Delta\phi$ are regarded as operators (see Appendix 2B)

$$[\hat{\Delta N}, \hat{\Delta\phi}] = -2i \tag{5.10.5}$$

Suppose that the Hamiltonian $\hat{H}$ is given in terms of the two canonically conjugate operators $\hat{\Delta N}$ and $\hat{\Delta\phi}$; then by commuting these operators with $\hat{H}$, using (5.10.5), we obtain their equations of motion. We will verify a posteriori that in a typical situation involving Josephson junctions the conditions for a semiclassical approximation to be valid are well satisfied, so we can make this approximation and obtain the semiclassical equations of motion

$$\frac{\partial}{\partial t}(\Delta N) = -\frac{2}{\hbar}\frac{\partial E}{\partial(\Delta\phi)} \tag{5.10.6a}$$

$$\frac{\partial}{\partial t}(\Delta\phi) = \frac{2}{\hbar}\frac{\partial E}{\partial(\Delta N)} \tag{5.10.6b}$$

What is the dependence of E on ΔN? Rather than considering the (physically realistic) case of a junction biased by a constant externally supplied current, it is convenient for the moment to consider the whole system $(1 + 2 +$ weak link) as isolated. Then, if for the moment we neglect capacitance effects, the only obvious source of dependence of E on ΔN is an external voltage V applied across the junction, giving a term

$$E_V(\Delta N) = -\Delta N \cdot e \cdot V(t) \tag{5.10.7}$$

Thus, substituting (5.10.4) and (5.10.7) in (5.10.6), we obtain the two celebrated *Josephson equations*

$$\frac{d}{dt}\Delta N(t) = \frac{2J}{\hbar}\sin\Delta\phi(t) \tag{5.10.8a}$$

$$\frac{d}{dt}\Delta\phi(t) = -\frac{2eV(t)}{\hbar} \tag{5.10.8b}$$

Since the current $I(t)$ flowing between the bulk superconductors 1 and 2 is just $ed/dt\ (\Delta N)$, the first Josephson equation (5.10.8a) is more commonly written in

[47]While the validity of (5.10.5) is clearly independent of the choice of zero for ΔN, it is convenient to take the latter to be such that $\langle\Delta N\rangle$ is zero in thermal equilibrium with no applied voltage.

the form

$$I(t) = I_c \sin \Delta\phi(t) \tag{5.10.9}$$

where the critical current I_c (evidently the maximum possible value of I obtainable from the Josephson mechanism) is given by

$$I_c = \frac{2eJ}{\hbar} \tag{5.10.10}$$

(or equivalently $J = I_c/2\pi\Phi_0$, where $\Phi_0 \equiv h/2e$ is the superconducting flux quantum as in Section 5.5). Equations. (5.10.8a–b) have many implications, of which one of the most fascinating is that provided a sink is provided for the current flowing across the junction (e.g. by external current leads) a finite current[48] $I = I_c \sin \Delta\phi$ can flow even in the absence of a voltage applied across the junction (the "d.c. Josephson effect"). In addition, by applying a dc voltage V and Eqn. (5.10.8b), we can again produce a dc current ('Shapiro steps"). For these and many other applications see Tinkham (1996), Chapter 6.

I now return briefly to the question of the justification of the form (5.10.4) for the Josephson energy, and of the magnitude of the constant J (or equivalently of the critical current I_c, see (5.10.10). In the limit of small $|\Psi|$ (e.g. the GL limit) one may argue from analyticity and gauge invariance considerations that the lowest-order term must be of the form (apart from possible terms in $|\Psi_1|^2$, $|\Psi_2|^2$ which are of no interest to us)

$$E_J = \text{const.}(\Psi_1^* \Psi_2 + \text{c.c.}) \tag{5.10.11}$$

which then yields the form (5.10.4), but with a constant J which is not obviously positive. To justify a positive value of J, one might use the qualitative argument that the ground state should have $\Delta\phi = 0$ so as to avoid "bending" of the order parameter in the junction; however, this is not rigorous, and indeed it is believed that in certain special cases (typically involving magnetic impurities in the junction) J may be negative, giving equilibrium at $\Delta\phi = \pi$ ("π-junction").[49]

In the important case of a tunnel oxide barrier, the classic calculation of the Josephson coupling energy was given by Ambegaokar and Baratoff (AB) shortly after Josephson's original work; I will just sketch the principle of this calculation without going into the details. AB modeled the tunnel oxide barrier by the Bardeen–Josephson Hamiltonian, in which it is conceived as a region of high potential in which the single-electron wave functions are much attenuated, and parametrized by the single-electron transmission amplitude $t_{\boldsymbol{k}\boldsymbol{k}'}$ from a plane wave state $\boldsymbol{k}$ in 1 to the plane-wave state $\boldsymbol{k}'$ in 2. In the normal state this model when treated to second order in $t_{\boldsymbol{k}\boldsymbol{k}'}$, leads, inter alia, to a junction conductance R_n^{-1} which is proportional to the average of $|t_{\boldsymbol{k}\boldsymbol{k}'}|^2$; it also leads to a reduction in energy of the total system by a similar amount. AB now use Eqns. (5.5.8a–b) to express the single-electron operators occurring in the Bardeen–Josephson Hamiltonian in terms of the Bogoliubov quasiparticle operators, and evaluate the ground state energy (or more generally the free energy) to second order in $t_{\boldsymbol{k}\boldsymbol{k}'}$. The squared matrix elements which occur in the resulting expression

[48]In such a (realistic) situation the phase $\Delta\phi$ is fixed by the condition that I is equal to the current flowing through the external leads, something that is typically controlled by the experimenter.

[49]Cf. Chapter 7, Section 7.10 for comment on the use of this term in the literature.

involve, inter alia, factors such as $v_{\boldsymbol{k}1}v^*_{\boldsymbol{k}'2}$; because the phases of the v's are just these of the corresponding Ψ's, these factors are sensitive to $\Delta\phi$. As a result, the final expression for the free energy contains a term which is exactly of the form (5.10.4), with a value of J which can be evaluated explicitly (cf. below). However, it is interesting to note that the calculation also produces a $\Delta\phi$-independent term which is different from that in the normal state by precisely J; thus, the absolute ground state energy (or free energy) of the system has a contribution from tunneling which is neither increased nor decreased by the formation of Cooper pairs in the bulk metals.[50]

In the limit of identical bulk metals and $T = 0$ the AB calculation produces a celebrated relation between the critical current I_c and the resistance R_n of the junction in the normal state, namely

$$I_c = \frac{\pi\Delta}{2eR_n} \tag{5.10.12}$$

i.e. the critical current is, up to a numerical factor, the current which would be generated in the normal state by a potential bias eV equal to the bulk energy gap Δ. It is perhaps at first sight surprising that I_c (hence J) is *linear* in Δ and hence approximately in $|\Psi|$; however, the generalization of the calculation to finite T shows that in the limit $T \to T_c$, I_c is proportional to $|\Psi|^2$ in accordance with (5.10.11).

Appendices

5A Landau Fermi-liquid theory

In this appendix I shall sketch the barest bones of the theory of Fermi liquids, which was originally conceived by Landau for (normal) liquid ^{3}He and subsequently applied by Silin and others to normal metals; for a much more detailed treatment, see Baym and Pethick (1991). I shall start by introducing the theory in the original form, applied to a neutral translation-invariant system such as liquid ^{3}He, and then indicate the modifications necessary to apply it to a metal. In the latter context, one could say that the main function of the theory is to explain why the modifications it introduces are relatively unimportant, so that for most purposes the electrons, which are in reality strongly interacting, can nevertheless for many purposes be treated as a noninteracting degenerate Fermi gas.

Consider then a set of N neutral Fermi particles of mass m and spin 1/2, moving freely in a volume Ω at zero temperature. For the moment let us forget about their interactions. Then the single particle states are labeled by momentum $\mathbf{p}$ (or wavevector $\boldsymbol{k} \equiv \boldsymbol{p}/\hbar$) and "spin" $\sigma = \pm 1$; the orbital wave-functions are plane-wave states, $\psi_{\boldsymbol{p}}(\boldsymbol{r}) = \Omega^{-1/2} \exp i\boldsymbol{p}\cdot\boldsymbol{r}/\hbar$, and the associated energy is given by

$$\epsilon(\boldsymbol{p},\sigma) = \epsilon(|\boldsymbol{p}|) = p^2/2m.$$

In the groundstate the single-particle states are occupied up to the Fermi momentum $p_F \equiv \hbar k_F, k_F = (3\pi^2 n)^{1/3}$; thus the occupation number is

$$n_{\boldsymbol{p},\sigma} = 1, \quad |\boldsymbol{p}| \le p_F$$
$$n_{\boldsymbol{p},\sigma} = 0, \quad |\boldsymbol{p}| > p_F$$

[50]In the limit of "particle-hole" symmetry" (single-particle densities of states and tunneling matrix elements constant around the Fermi surface).

The simplest excited states of the N-particle system are obtained by taking a particle out of an occupied state $(\boldsymbol{p}, \sigma)$, below the Fermi surface ($|\boldsymbol{p}| < p_{\mathrm{F}}$) and putting it into an unoccupied state ($\boldsymbol{p}'\sigma'$ with ($|\boldsymbol{p}'| > p_{\mathrm{F}}$). The resulting *change* in the energy of the system i.e. "energy of excitation" is $\epsilon(\boldsymbol{p}) - \epsilon(\boldsymbol{p}')$. If we move more than one particle in this way the excitation energy is just the sum of that due to each process separately. Hence we label the states of the noninteracting gas by the set of quantities $\{\delta n_{\boldsymbol{p}\sigma}\}$ and write

$$E - E_0 = \sum_{\boldsymbol{p},\sigma} \epsilon(p)\delta n_{\boldsymbol{p}\sigma} \tag{5.A.1}$$

with the conditions

$$\delta n_{\boldsymbol{p}\sigma} \equiv \begin{cases} 0 \text{ or } 1, & |\boldsymbol{p}| > p_{\mathrm{F}} \\ 0 \text{ or } -1, & |\boldsymbol{p}| < p_{\mathrm{F}} \end{cases}$$

It is also clear that since the momentum and spin are zero in the groundstate, we have $\boldsymbol{P} = \sum_{\boldsymbol{p}\sigma} \boldsymbol{p}\delta n_{\boldsymbol{p}\sigma}$, $S = \sum_{\boldsymbol{p}\sigma} \sigma n_{\boldsymbol{p}\sigma}$.

It is convenient from now on to measure $\epsilon(p)$ from the chemical potential μ, which we recall is at $T = 0$ equal to the Fermi energy $\epsilon_{\mathrm{F}} \equiv p_{\mathrm{F}}^2/2m$. Then Eqn. (5.A.1) remains true for the N-particle states, since for these the term $\sum_{\boldsymbol{p}\sigma} \mu\delta n_{\boldsymbol{p}\sigma}$ is zero identically; we can also use it for states when the total particle number changes, provided that we then interpret the left-hand side as the change not in E but in $E - \mu N$. (This may seem a rather artificial manoeuvre, but it turns out to be quite useful).

Let us now consider what happens when we gradually turn on the interaction between particles. For all reasonable cases the interaction will conserve the particle number; it *may* also conserve spin, momentum and orbital angular momentum (i.e. be of the form $\frac{1}{2}\sum_{ij} V(|\boldsymbol{r}_i - \boldsymbol{r}_j|)$) (this is true to a good approximation in ^{3}He) and we will assume for the moment that it does so. Irrespective of this, the basic hypothesis of Fermi-liquid theory is that *the ground state of the noninteracting system evolves adiabatically into the ground state of the interacting system, and similarly each excited state of the noninteracting one into a corresponding excited state of the interacting one.* It cannot be over-emphasized that the Landau theory does not *justify* this hypothesis; it simply assumes it and explores the consequences. If it is true, then we can label each state of the *interacting* system by the set of quantities $\{\delta n_{\boldsymbol{p}\sigma}\}$ which labeled the "free" states from which it evolved. We say that in the interacting system we have $\delta n_{\boldsymbol{p}\sigma}$ "quasiparticles" (if $|\boldsymbol{p}| > p_{\mathrm{F}}$) or $-\delta n_{\boldsymbol{p}\sigma}$ "quasiholes" (if $|\boldsymbol{p}| < p_{\mathrm{F}}$) in state $\boldsymbol{p}, \sigma$. Note that in this process the density of states in momentum (not energy!) space of the quasiparticle states is exactly that of the original particles and the Fermi momentum p_{F} is *by construction* unchanged. We may if we wish regard a quasiparticle as physically composed of the original "bare" particle plus a "screening cloud" of other particles and holes; however this picture is not essential to the formal development of the theory. If (and *only* if) the total spin (momentum) is conserved by the (full interacting) Hamiltonian, then a quasiparticle with label $(\boldsymbol{p}, \sigma)$ indeed carries spin σ (momentum $\boldsymbol{p}$); however, the ascription of the label is independent of this.

Having settled the labeling of the states of the interacting system, the next step is to write down the *energy* of the states labeled by a particular set of $\{\delta n_{\boldsymbol{p}\sigma}\}$. Formally, we can make a Taylor expansion of the form (ignoring some complications associated with spin, cf. below)

$$E - E_0 = \sum_{\boldsymbol{p}\sigma} \frac{\delta E}{\delta n_{\boldsymbol{p}\sigma}} \delta n_{\boldsymbol{p}\sigma} + \frac{1}{2} \sum_{\boldsymbol{pp}',\sigma\sigma'} \frac{\delta^2 E}{\delta n_{\boldsymbol{p}\sigma} \delta n_{\boldsymbol{p}'\sigma'}} \delta n_{\boldsymbol{p}\sigma} \delta n_{\boldsymbol{p}'\sigma'} + \frac{1}{6} \sum \cdots \tag{5.A.2}$$

Let's define

$$\delta E / \delta n_{\boldsymbol{p}\sigma} \equiv \epsilon(\boldsymbol{p}\sigma) \quad (\epsilon(\boldsymbol{p}\sigma) \text{ measured from } \mu)$$
$$\delta^2 E / \delta n_{\boldsymbol{p}\sigma} \delta n_{\boldsymbol{p}'\sigma} \equiv f(\boldsymbol{pp}', \sigma\sigma') \quad \text{etc.}$$

Evidently, for an extended system, if the energy is to be extensive we must have

$$\epsilon(\boldsymbol{p}\sigma) \sim 1, \qquad f(\boldsymbol{pp}', \sigma\sigma') \sim \Omega^{-1}, \text{etc.} (\Omega \equiv \text{volume of the system})$$

Suppose the total number of excited quasiparticles and quasiholes is some number N_{exc} which is of order Ω, hence proportional to N, but still $\ll N, N_{\text{exc}}/N \ll 1$. In general, then the *second* term in the expansion will be of order $N_{\text{exc}}^2 f \sim N_{\text{exc}}^2/\Omega \sim N_{\text{exc}}(N_{\text{exc}}/N)$ while the third term will be $\sim N_{\text{exc}}^3 \Omega^{-1} \sim N_{\text{exc}}(N_{\text{exc}}/N)^2$. Thus the third term is always negligible compared to the second and can be dropped (as, obviously, can the higher terms in Eqn. (5.A.2)). One might at first sight think that the second term would be negligible relative to the first, but this is not necessarily so; as we shall see below, there are cases when the contribution of order N_{exc} from the first term vanishes from symmetry, and the remaining contribution is proportional to N_{exc}^2/N. Hence both the first and second terms in the right hand side of Eqn. (5.A.2) must be kept, and we get

$$E - E_0(-\mu(N - N_0)) = \sum_{\boldsymbol{p}\sigma} \epsilon(\boldsymbol{p}\sigma) \delta n_{\boldsymbol{p}\sigma} + \frac{1}{2} \sum_{\boldsymbol{pp},\sigma\sigma'} f(\boldsymbol{pp}, \sigma\sigma') \delta n_{\boldsymbol{p}\sigma} \delta n_{\boldsymbol{p}'\sigma'} \tag{5.A.3}$$

Thus, the energy of the system is entirely parametrized by the quantity $\epsilon(\boldsymbol{p}\sigma)$ (the "single quasiparticle energy") and the quantity $f(\boldsymbol{pp}', \sigma\sigma')$ ("Landau interaction function ").

Symmetry enables us to simplify the form of these quantities considerably:

(a) In the absence of a magnetic field, since the interaction is spin-independent no direction in spin space can be picked out and $\epsilon(\boldsymbol{p}\sigma)$ must therefore be spin-independent. Also, from the spatial isotropy, $\epsilon(\boldsymbol{p}\sigma)$ can be a function only of the *magnitude* of $\boldsymbol{p} : \epsilon(\boldsymbol{p}) \equiv \epsilon(p)$. Let us now expand around the Fermi surface $(p = p_{\text{F}})$; since by our convention for ϵ the minimum value $|\epsilon(p)|$ is zero, (i.e. $\epsilon_{\text{true}}(p_{\text{F}})$ is just the chemical potential μ) we can write

$$\epsilon(p) = \left(\frac{d\epsilon}{dp} \right)_{p_{\text{F}}} (p - p_{\text{F}}) + \mathcal{O}(p - p_{\text{F}})^2$$

The second term is typically of order $\epsilon/\epsilon_{\text{F}}$ relative to the first. If, therefore, we confine ourselves to "low-energy" phenomena (typical energies $\ll\epsilon_{\text{F}}$), as we shall, this term can be neglected. We define

$$(d\epsilon/dp)_{p_{\text{F}}} \equiv v_{\text{F}}, \quad p_{\text{F}}/v_{\text{F}} \equiv m^* \tag{5.A.4}$$

where m^* is usually called the "effective mass"[51] of a quasiparticle. Then the relevant part of the single-quasiparticle energy spectrum is completely characterized by the single number m^*. Note that even for a translation invariant system m^*, while having the dimensions of mass, is *not* in general equal to the mass of the bare particle. With this definition, we easily check that the density of single-particle states per unit volume near the Fermi surface is given by

$$\frac{dn}{d\epsilon} = \frac{p_{\mathrm{F}}^2}{\pi^2 \hbar^2 v_{\mathrm{F}}} = \frac{m^* p_{\mathrm{F}}}{\pi^2 \hbar^2} = \left(\frac{m^*}{m}\right)\left(\frac{dn}{d\epsilon}\right)_{\text{free gas}} \tag{5.A.5}$$

where we used the fact that p_{F} is, by construction, the same for the interacting as for the free system.

(b) Consider next the Landau interaction function $f(\boldsymbol{p}\boldsymbol{p}', \sigma\sigma')$, and suppose that we are interested only in quasiparticles having a definite spin projection σ on the z axis. (For simplicity we will ignore a technical complication associated with the vector nature of the spin.) Then, from symmetry, f can be a function only of $|\boldsymbol{p}|, |\boldsymbol{p}'|\boldsymbol{p} \cdot \boldsymbol{p}'$ and $\sigma\sigma'$ (note by our convention $\sigma, \sigma' = \pm 1$). Since f (unlike ϵ) is finite for $p = p_{\mathrm{F}}$, we need keep only the zeroth term in the Taylor expansion in $(p - p_{\mathrm{F}})$, i.e. evaluate f for $p = p' = p_{\mathrm{F}}$. It can then depend only on the **angle** θ between $\boldsymbol{p}$ and $\boldsymbol{p}'$, i.e. on $\hat{\boldsymbol{p}} \cdot \hat{\boldsymbol{p}}' = \cos\theta$. Since for spin 1/2 any function of the product $\sigma\sigma'$ can be written $A + B\sigma\sigma'$, we can then write

$$f(\boldsymbol{p}\boldsymbol{p}', \sigma\sigma') = f_{\mathrm{s}}(\cos\theta) + f_{\mathrm{a}}(\cos\theta)\sigma\sigma' \tag{5.A.6}$$

The Landau function has the dimensions of energy and is proportional to Ω^{-1} (inverse volume). It is convenient to multiply it by the one-quasiparticle density of states (for the total volume, not per unit volume) which is proportional to Ω and has dimensions of inverse energy, so as to produce a dimensionless quantity. That is, we define

$$F_{\mathrm{s}}(\cos\theta) \equiv \Omega\frac{dn}{d\epsilon} f_{\mathrm{s}}(\cos\theta), \quad F_{\mathrm{a}}(\cos\theta) \equiv \Omega\frac{dn}{d\epsilon} \equiv \Omega\frac{dn}{d\epsilon} f_{\mathrm{a}}(\cos\theta). \tag{5.A.7}$$

Finally we expand F_{s} and F_{a} in Legendre polynomials

$$F_{\mathrm{s}}(\cos\theta) \equiv \sum_{\ell} F_{\ell}^{\mathrm{s}} P_{\ell}(\cos\theta), \quad F_{\mathrm{a}}(\cos\theta) \equiv \sum_{\ell} F_{\ell}^{\mathrm{a}} p_{\ell}(\cos\theta). \tag{5.A.8}$$

The infinite set of dimensionless numbers F_{ℓ}^{s}, and F_{ℓ}^{a} (the so-called "Landau parameters") therefore completely characterizes (the relevant part of) the interquasiparticle interaction; fortunately for many purposes we need only the first few.

Molecular fields

It is often helpful to interpret the Landau parameters F_{ℓ}^{s}, F_{ℓ}^{a} in terms of generalized *molecular fields*. For clarity, consider first the artificial case in which only F_0^{a}

[51]In the original Landau theory (of ^{3}He) this is the only "effective mass". In the theory of metals, where other "masses" (cyclotron infrared etc.) are introduced the one defined by Eqn. (5.A.4) is sometimes called the "thermal effective mass" to distinguish it.

is non-zero, i.e. (considering for notational simplicity the case in which all $\delta n_{\boldsymbol{p}\sigma}$ are diagonal in the given spin axes)

$$f(\boldsymbol{pp}', \sigma\sigma') = \Omega^{-1}\left(\frac{dn}{d\epsilon}\right)^{-1} F_0^{\mathrm{a}}\sigma\sigma'$$

The term in the total energy which involves the Landau interaction function f can then be written

$$\delta E^{(2)} = \Omega^{-1}\left(\frac{dn}{d\epsilon}\right)^{-1} F_0^{\mathrm{a}} \sum_{\boldsymbol{pp}',\sigma\sigma'} \sigma\sigma'\delta n_{\boldsymbol{p}\sigma}\delta n_{\boldsymbol{p}'\sigma'}$$

But we saw that for a spin-conserving interaction the quantity $\sum_\sigma \sigma\delta n_{\boldsymbol{p}\sigma}$ is nothing but the (z-component of) total spin, and hence $\delta E^{(2)}$ has the simple form

$$\delta E^{(2)} = \frac{1}{2}\Omega^{-1}\left(\frac{dn}{d\epsilon}\right)^{-1} F_0^{\mathrm{a}}\boldsymbol{S}^2 \tag{5.A.9}$$

Now, the expression Eqn. (5.A.9) is nothing but the energy of polarization of the total spin $\boldsymbol{S}$ of a free Fermi gas in a "field" given by

$$\boldsymbol{\mathcal{H}}_{\mathrm{mol}} \equiv -\left(\frac{dn}{d\epsilon}\right)^{-1} F_0^{\mathrm{a}}\boldsymbol{S} = \text{“molecular field”}$$

(note the factor of $\frac{1}{2}$ cancels because $E_{\mathrm{pol}} = -\int \boldsymbol{\mathcal{H}}_{\mathrm{mol}}\cdot d\boldsymbol{S}$, not $-\boldsymbol{\mathcal{H}}_{\mathrm{mol}}\cdot\boldsymbol{S}$). This suggests the following simple way of calculating the effects of the Landau interaction term. We calculate the response (possibly space and time dependent) of the spin of the system to an external field *without* the Landau term, thereby defining a "bare" (spin) response function $\chi_0^{\mathrm{sp}}(\boldsymbol{k},\omega)$. Then the effect of the Landau (F_0^{a}) term is explicitly taken into account by saying that the system responds with the *bare* response function to the *total* effective field, that is the sum of the true (external) field and the molecular field. Explicitly, for an external field of wave vector $\boldsymbol{k}$ and frequency ω, we have the set of equations

$$\begin{aligned}
\boldsymbol{S}(\boldsymbol{k},\omega) &= \chi_0^{\mathrm{sp}}(\boldsymbol{k},\omega)\boldsymbol{\mathcal{H}}_{\mathrm{tot}}(\boldsymbol{k},\omega)\\
\boldsymbol{\mathcal{H}}_{\mathrm{tot}}(\boldsymbol{k},\omega) &= \boldsymbol{\mathcal{H}}_{\mathrm{ext}}(\boldsymbol{k},\omega) + \boldsymbol{\mathcal{H}}_{\mathrm{mol}}(\boldsymbol{k},\omega)\\
\boldsymbol{\mathcal{H}}_{\mathrm{mol}}(\boldsymbol{k},\omega) &= -(dn/d\epsilon)^{-1}F_0^{\mathrm{a}}\boldsymbol{S}(\boldsymbol{k},\omega)
\end{aligned}\tag{5.A.10}$$

It is obvious that the structure of these equations is exactly identical to those used in the calculation of screening (see Section 5.2), and is an obvious generalization of the familiar (Weiss) "mean field theory" of ferromagnetism. Correspondingly, the solution for the *true* response $\chi_{\mathrm{true}}^{\mathrm{sp}}(\boldsymbol{k},\omega)$ of the spin to an external field, $\chi_{\mathrm{true}}^{\mathrm{sp}}(\boldsymbol{k},\omega) \equiv \boldsymbol{S}(\boldsymbol{k},\omega)/\boldsymbol{\mathcal{H}}_{\mathrm{ext.}}(\boldsymbol{k},\omega)$ (or $\delta\boldsymbol{S}(\boldsymbol{k},\omega)/\delta\boldsymbol{\mathcal{H}}_{\mathrm{ext}}(\boldsymbol{k},\omega)$) is given in terms of χ^0 by the formula

$$\chi_{\mathrm{true}}^{\mathrm{sp}}(\boldsymbol{k},\omega) = \frac{\chi_0^{\mathrm{sp}}(\boldsymbol{k},\omega)}{1 + (dn/d\epsilon)^{-1}F_0^{\mathrm{a}}\,\chi_0^{\mathrm{sp}}(\boldsymbol{k},\omega)} \tag{5.A.11}$$

(compare the results for the "screened" charge density response, (Section 5.2)).

A few notes on the above argument:

1. Just as in the Weiss theory of ferromagnetism, the "molecular field" $\boldsymbol{\mathcal{H}}_{\mathrm{mol}}$ is a *fictitious* field which is unrelated to real magnetism; for example, a muon introduced into liquid ^{3}He will not feel it at all.

2. The above results are *exact* in the approximation of keeping only F_0^{a} and in the limit $\boldsymbol{k} = \omega = 0$ (provided that the quantity $\chi_0^{\mathrm{sp}}(\boldsymbol{k}, \omega)$ is correctly calculated). For nonzero $\boldsymbol{k}$ and ω an implicit approximation is involved, namely that it is valid to use the Landau form of the energy functional even when the $\delta n_{\boldsymbol{k}\omega}$ are varying in space and time. The approximation should be good, crudely speaking, provided $k \ll k_{\mathrm{F}}$ *and* $\omega \ll \epsilon_{\mathrm{F}}/\hbar$ (see below). It should be emphasized that the Landau theory does not claim to describe excitations with wave vector (frequency) $\gtrsim k_{\mathrm{F}}(\epsilon_{\mathrm{F}}/\hbar)$.
3. We can make an immediate application to the static spin susceptibility $\chi_{\mathrm{true}}(0,0)$ of a Landau Fermi liquid. The "bare" static susceptibility, namely that calculated from the first term in the Landau Hamiltonian, $\sum_{\boldsymbol{p},\sigma} \epsilon(p)\delta n_{\boldsymbol{p},\omega}$, is clearly given, just as in the elementary theory of metals, by the density of states at the Fermi energy, $dn/d\epsilon$; in fact, recalling that $\boldsymbol{S}$ is measured in units of $\hbar/2$, we have simply

$$\chi_0 = dn/d\epsilon$$

Inserting this into Eqn. (5.A.11) we get for the "true" susceptibility χ the simple expression[52]

$$\chi = \frac{dn/d\epsilon}{1 + F_0^{\mathrm{a}}} \tag{5.A.12}$$

This result is exact within the Landau theory (it does not depend on keeping only the F_0^{a} term). Since, as we will see, the specific heat of the Fermi liquid measures $(dn/d\epsilon)$ directly, a measurement of the spin susceptibility determines the Landau parameter F_0^{a}. (For liquid ^{3}He, F_0^{a} is approximately -0.7 for most of the pressure interval in which the liquid exists.)

The molecular fields associated with the other F_l's can be handled in a similar way; see, e.g. Leggett (1975), Section II. We note in particular for future reference the expression, analogous to (5.A.12), for the compressibility (the inverse of the bulk modulus κ):

$$\chi_0 \equiv \kappa^{-1} = \frac{dn/d\epsilon}{1 + F_0^{\mathrm{s}}} \tag{5.A.13}$$

A very important general principle is the following: For any kind of disturbance which does not involve a *net* polarization of Fermi surface at any point (i.e. s.t. $\int(\delta n(\hat{\boldsymbol{n}}, \epsilon, \sigma)d\epsilon = 0$ for all $\hat{\boldsymbol{n}} \equiv \boldsymbol{p}/|\boldsymbol{p}|$) *the effect of the molecular fields* (Landau interaction) *vanishes completely*, and the behavior of the system is *exactly* that[53] of a non-interacting Fermi gas of particles with an effective mass m^*.

An immediate application of this principle is to the specific heat of a Fermi liquid. It is clear that if we start from the ground state, the effect of finite temperature in a noninteracting Fermi gas is to round off the Fermi surface but not to shift its average

[52]For notational simplicity we implicitly included, in the above argument, a factor of the gyromagnetic ration γ in the definition of the magnetic field. When this and the factors of $\hbar/2$ are sorted out the net effect is to multiply Eqn. (5.A.12) by a factor μ_n^2 (μ_n being the magnetic moment).

[53]We implicitly assume as always that $k \ll k_{\mathrm{F}}, \omega \ll \varepsilon_{\mathrm{F}}/\hbar, T \ll T_{\mathrm{F}}$.

value (to order T/T_{F}). Thus, the condition is justified and there is nothing for the molecular fields to "take hold of." Consequently, in the low-temperature limit the thermal energy, and hence specific heat, of a Landau Fermi liquid is exactly that of the corresponding free gas of effective mass m^*.

A second and less well-known application of this principle is to orbital magnetic effects in metals such as the de Haas–van Alphen effect. If we neglect for the moment the effect of the magnetic field on the spins, then although the shape of the electron orbits is changed by the field no *net* polarization[54] of the Fermi surface is involved (at least to the extent that $\mu H \ll \varepsilon_{\mathrm{F}}$: of course, for ultrahigh magnetic fields such that $\mu H \gtrsim$ the whole shape of the Fermi surface is radically changed, but then the conditions for FL theory to remain valid are no longer met). Hence, there can be no "molecular-field" effects on such phenomena.

Quasiparticle collisions (Pines and Noziéres, 1966, Section 1.8)

All the above analysis implicitly assumed that the states labeled by the quasiparticle occupation numbers $\delta n_{p\sigma}$ were in fact exact energy eigenstates of the system. However, this cannot in fact be true: if the quasiparticles have any interaction at all, it must be possible for this to lead to scattering and possibly to "decay" of a quasiparticle Fortunately, it turns out that this effect is rather small at temperatures $T \ll T_{\mathrm{F}}$ and thus does not spoil the basics of the Landau Fermi-liquid picture; however, it is essential to take into account when one does transport calculations, etc.

Let us estimate the "lifetime" τ of a quasiparticle due to collisions with other quasiparticles, for quasiparticle energy $\varepsilon \ll \varepsilon_{\mathrm{F}}$. At $T = 0$ no other quasiparticles are excited and the only possible collision processes are those in which the original quasiparticle knocks one out of the Fermi sea, i.e. creates a quasihole, leaving a state in which there are two excited quasiparticles:

$$qp \rightarrow qp + qp + qh$$

Such collisions must satisfy conservation of momentum (and spin); this means that once two of the final-state momenta are specified the third if fixed. Also, energy must be conserved. For finite T collisions with already excited quasiparticles are also possible. Omitting the rather tedious details of the calculation, I simply quote the result that for $\varepsilon, kT \ll \varepsilon_{\mathrm{F}}$,

$$\tau^{-1} \sim \frac{\varepsilon_{\mathrm{F}}}{\hbar}\left(\frac{\varepsilon^2 + \pi^2 k_B^2 T^2}{\varepsilon_{\mathrm{F}}^2}\right) \tag{5.A.14}$$

It is often stated that a necessary and sufficient condition (as regards temperature, etc.) for Landau Fermi liquid theory to be valid for a normal degenerate Fermi system is that the energy "width" of a typical quasiparticle, $\hbar/\tau$, should be small compared to its excitation energy ε relative to the Fermi surface. Since in thermal equilibrium a "typical" quasiparticle has energy $\sim k_B T$, this leads by Eqn. (5.4.14) to the criterion $T \ll T_{\mathrm{F}}$. Actually, the above criterion may be a bit pessimistic: for liquid ^{3}He Fermi

[54]Although the original spherical surface is replaced by a set of "tubes" (cf. e.g. Harrison (1970), p. 125) they are so closely spaced for $\mu H \ll \varepsilon_{\mathrm{F}}$ that an average over any substantial region of $\vec{n}$ gives zero.

liquid theory appears to give a fairly good account of most properties all the way up to ~100 mK, even though at that point $\hbar/\tau$ is quite comparable to $k_B T$. In metals, as we shall see, the situation is complicated by the need to distinguish between elastic and inelastic scattering.

Finally, we should note that a special case of "collisions" is one in which particles in states $(\boldsymbol{p}, -\boldsymbol{p})$ are scattered into states $(\boldsymbol{p}', -\boldsymbol{p}')$. It is just such "collisions" which are crucial for the theory of superfluidity in ^{3}He and superconductivity in metals: see Section 5.4.

Application of Fermi-liquid theory to metals

The formulation of Fermi-liquid theory we have discussed so far is appropriate to a neutral, pure, translation-invariant system such as liquid ^{3}He. In attempting to generalize it so as to apply it to the electrons in metals, we need to take account of a number of ways in which the latter differ from ^{3}He:

1. Whereas the ^{3}He atoms as a whole move freely in space, the electrons in metals move in the field of a crystalline lattice. The effect is to destroy both (a) the translation invariance (or rather to replace it with the translational symmetry of the lattice) and (b) the rotational invariance of the free system.
2. Whereas ^{3}He atoms are electrically neutral and their interactions therefore short-ranged, the electrons in metals interact by the long-range Coulomb interaction.
3. In general, apart from electron–electron and electron–phonon collisions (see below) the electrons can be scattered by static impurities (chemical impurities, defects, dislocations, etc.); for most metals this is in fact the dominant scattering process at low temperatures.
4. As well as moving in the field of the *static* crystalline lattice, the electrons interact with the vibrations of the ions (phonons). A special feature of this interaction is that the typical energy scale if the ions, $\hbar\omega_{\mathrm{D}}$, is usually very small compared to the Fermi energy. Thus one has a "small energy scale" in the problem which is completely absent in the ^{3}He case.

Let us deal with these points in turn:

1. (a) The absence of translation-invariance means, of course, that momentum is no longer conserved. However, just as in standard one-electron theory, we can introduce a *quasimomentum* which labels the irreducible representations of the translation group. In fact, it may be helpful to think of going from the completely free electron gas to the real system in two stages: we first turn on the lattice potential, and then later the interelectron interactions. It is amusing that we can in fact describe not only the second stage but also the first in terms of Fermi liquid concepts: in fact, in the extended-zone representation, the Bloch state labeled by "quasimomentum" $\boldsymbol{k}$ is nothing but the state which evolves adiabatically as the lattice potential is turned on, from the original free-electron state $\boldsymbol{k}$! The only important difference is that because of the periodicity of the crystalline potential, the spectrum $\varepsilon(\boldsymbol{k})$ is not continuous as in the ^{3}He case but has gaps at

the Brillouin zone boundaries. Because the lattice potential does not conserve momentum, it is not surprising that the state labeled by $\boldsymbol{k}$ does not have true momentum $\hbar\boldsymbol{k}$. When at stage 2 we switch on the interparticle interactions, real momentum is not conserved (although the interelectron interaction conserves it, the "noninteracting" (one-electron-in-lattice) Hamiltonian $\hat{H}_0$ does not). However, *quasimomentum* is conserved (mod $\boldsymbol{K}$) both by $\vec{H}_0$ and $\hat{V}$, and hence we can say that the label $\boldsymbol{k}$ of the Landau quasiparticle produced at stage 2 is indeed the "quasiparticle quasimomentum". However, in considering electron–electron collisions, etc., it should be remembered that the quasimomentum is only defined (and only conserved, in view of Umklapp processes) modulo $\boldsymbol{K}$, where $\boldsymbol{K}$ is a reciprocal lattice vector.

(b) The absence of rotational invariance leads mainly to complications of notation rather than of substance. It is clear that we can, for example, no longer assume that the energy spectrum $\varepsilon(\boldsymbol{k})$ is a function only of $|\boldsymbol{k}|$, so just as in the one-electron theory we must allow the Fermi velocity $v_{\mathrm{F}}(\hat{\boldsymbol{n}})$ to be a function of position $\hat{\boldsymbol{n}}$ on the Fermi surface, and generalize the formulae for the one-qp density of states, etc., accordingly. Also, just as in the one-electron theory, one would no longer expect the various effective masses (thermal, cyclotron, etc.) to be identical, even when they involve no explicit "Fermi-liquid" effects. Similarly, in trying to simplify the Landau interaction term one can no longer assume that the function $f(\boldsymbol{p}\boldsymbol{p}'\sigma\sigma')$ depends (for $|\boldsymbol{p}| = |\boldsymbol{p}'| = p_{\mathrm{F}}$) only on the relative angle $\boldsymbol{p}\cdot\boldsymbol{p}'$, and hence can be expressed as $f(\cos\theta)$; rather, we must make an expansion in the irreducible representations of the point group of the crystal.

(c) Finally, it should be mentioned that in crystals containing heavy elements there is another complication, namely the interaction of the spin of the moving electron with the electric field of the nucleus (spin–orbit interaction). To handle this properly one really needs a relativistic generalization of Fermi liquid theory; however, one can handle the lowest-order effects in v_{F} by an appropriate generalization of the forms of $\varepsilon(\boldsymbol{p}\sigma)$ and $f_{\boldsymbol{p}\boldsymbol{p}'\sigma\sigma'}$ (cf. e.g. Fujita and Quader 1987).

2. The Coulomb interaction: There is a standard method of handling this in Fermi liquid theory, namely, the simple "screening" idea which was already introduced in Section 5.2. Just as there, one simply separates out the long-range part of the Coulomb interaction and treats it as an "induced" potential which is found from Poisson's equation in terms of the charge density. The behavior of the latter is determined by an "unscreened" density response function $\chi_0(\boldsymbol{q},\omega)$ which contains, inter alia, all the *short-range* effects of the Coulomb field. That is (cf. the Fourier-transformed forms of Eqns. (5.2.2–4))

$$\delta\varphi(q\omega) = \delta\varphi_{\mathrm{ext}}(q\omega) + \delta\varphi_{\mathrm{ind}}(q\omega)$$

$$q^2\varphi_{\mathrm{ind}}(q\omega) = \delta\rho(q\omega)/\varepsilon_0$$

$$\delta\rho(q\omega) = -\chi_0(q\omega)\delta\varphi(q\omega)$$

which gives for the true density response $\chi(q\omega) \equiv \delta\rho(q\omega)/\delta\varphi_{\text{ext}}(q\omega)$ the by now familiar result

$$\chi(q,\omega) = \frac{\chi_0(q,\omega)}{1 + (e^2/\varepsilon_0 q^2)\chi_0(q,\omega)} \tag{5.A.15}$$

The only way in which this result differs from that of Section 5.2 is that χ_0 is now not to be calculated by any approximation (Lindhard, Hartree-Fock, etc.) but is the *exact* response of the "neutral" Fermi liquid, i.e. the system described by a Hamiltonian with only the term $e^2/\epsilon_0 q^2$ left out: Thus, χ_0 is *not* the response of the free quasiparticle gas (which for notational consistency we should probably now denote χ_{00}!) but contains the effects of the Landau parameters F_l^{s} through the molecular-field effects, cf. Eqn. (5.A.11). Thus, we must do the calculation in three stages:

(i) Calculate the density response $\chi_{00}(q,\omega)$ of the free quasiparticle gas.
(ii) Put in the (short range) "molecular field" corrections to get a "bare" density response function $\chi_0(q,\omega)$ for the Fermi liquid (e.g. if only F_0^{s} is taken into account, this has a structure similar to (5.A.11) with $F_0^{\text{a}} \to F_0^{\text{s}}$ (cf. Eqn. 5.A.13).
(iii) Put in the long-range ("screening") interactions via Eqn. (5.A.15) above. It is clear that stage (ii) will not usually introduce qualitative corrections, though it may change the numbers somewhat.

3. Impurity scattering: An important feature of this is that, generally speaking, the energy "width" of a state *of definite quasimomentum* $\boldsymbol{k}$ will not tend to zero as $\varepsilon \to 0$. This is because the scattering is elastic, and therefore, if ε is initially above the Fermi surface, the energy of the final state into which the qp is scattered is automatically also >0, and hence the Pauli principle puts no restrictions on the scattering. Thus, generally speaking, τ^{-1} is finite as $T \to 0, \varepsilon \to 0$ and can be expressed in terms of a "mean free path" $l \equiv v_{\text{F}}\tau$, which can often be estimated from naive kinetic-theory considerations: typically, $l \sim 1/n_{\text{imp}}\sigma$ where n_{imp} is the number of impurities per unit volume and σ is a cross-section which for chemical impurities is typically of the order of a_0^2 (or say 1Å^2).
   If the criterion for the validity of Landau theory were simply that the energy width $(\hbar/\tau)$ of a "typical" quasiparticle with definite momentum $\boldsymbol{k}$ is small compared to its energy ($\sim k_B T$), then clearly Landau theory would fail for metals below some finite temperature ($\sim \hbar/k_B\tau$). Experimentally this is clearly not the case: the theory actually seems to work *better* in the limit $T \to 0$. An intuitive way of seeing the reason for this is that, as long as the scattering is elastic and involves only an external *c*-number potential, it is always possible to reformulate the FL theory in terms of the exact one-electron eigenstates (which in general are nothing like Bloch waves). Many of the results of the theory (such as the insensitivity of the specific heat to molecular-field corrections, etc.) are quite insensitive to this reformulation. However, for many purposes it is simpler just to work in terms of the Bloch-wave quasiparticle states and handle the scattering, both impurity and other, in a phenomenological way.
4. Finally we comment briefly on the effects of the electron–phonon interaction. We will discuss some consequences of this in Appendix 5B, but now just summarize the implications of the FL description: It turns out that there are *two* regions of energy

and/or temperature where the FL theory is valid, but with different values for the parameters (m^*, F_l^s, etc.): a "high-energy" region ($\epsilon \gg \hbar\omega_D$, or $T \gg T_D$ (but still $\varepsilon \ll \varepsilon_F, \ll T_F$)) where the phonons have essentially no effect, and a "low energy" region ($\epsilon \ll \hbar\omega_D$ *and* $T \ll T_D$) where everything is strongly renormalized by the electron–phonon interaction. (In the intermediate regime ($\epsilon \sim \hbar\omega_D$ or $T \sim T_D$) it is believed that no simple Landau-type theory works.) In the context of superconductivity we are usually interested in the low-energy regime.

We can now begin to see why the "naive" one-electron picture of a normal metal, corrected for screening, gives such impressively good results. In reality, the quasiparticles of the Landau theory are not very much like free electrons, but for most experimentally measurable quantities the difference is absorbed in the renormalization of the effective mass (which is not easy to calculate in band theory anyway!) and, sometimes, in the effect of the molecular fields, which for most experimentally interesting phenomena in the normal phase simply changes the result by a numerical factor. In fact, one has to work quite hard to bring out spectacular effects of specifically "Fermi-liquid" features for a normal metal!

Note added: for a correct treatment of the spin degree of freedom, see e.g. Leggett 1975, section II.A.

5B The effective electron–electron interaction in the presence of phonons

In metals the ions carry a charge density when averaged over the unit cell. If, therefore, we kept the conduction electrons frozen and allowed the ionic system to vibrate, then by a standard argument they would sustain a plasma oscillation at the "ionic plasma frequency" Ω_p given by

$$\Omega_p^2 = \frac{nZ^2e^2}{M\epsilon_0} = Z^2\left(\frac{m}{M}\right)\omega_p^2 \sim (10^{13}\text{sec}^{-1})^2 \tag{5.B.1}$$

where N is the ionic number density, Z the ionic charge and M the (average) mass. In reality, the electrons are not frozen and can move to screen the ionic charge at frequencies of order of Ω_p. Thus it is necessary to carry out a calculation in which *all* long-range interactions (ion–ion, electron–ion, electron–electron) are consistently taken into account. When this is done, it is found that the resultant picture is one in which a first approximation we have screened electrons (as in the simple "screening" picture of Section 5.2) and sound-wave-like oscillations (phonons) of *neutral* atoms (i.e. completely screened ions) just as in an insulator; the lack of complete charge compensation appears as a residual electron-phonon interaction, which in turn generates the effective electron–electron interaction which is believed to be responsible for the formation of Cooper pairs and hence for superconductivity.

We proceed by an obvious generalization of the technique of Section 5.2. Suppose an external electrostatic potential $\varphi_{\text{ext}}(\boldsymbol{r}t)$ is applied to the system. Denoting the electronic and ionic *charge* densities by $\rho_{\text{el}}(\boldsymbol{r}t)$ and $\rho_{\text{ion}}(\boldsymbol{r}t)$ respectively, we write schematically (the χ_0's on number density responses)

$$\begin{aligned}\delta\rho_{\text{el}} &= -e\chi_{\text{el}}^0\varphi_{\text{tot}} \\ \delta\rho_{\text{ion}} &= -Ze\chi_{\text{ion}}^0\varphi_{\text{tot}}\end{aligned} \tag{5.B.2}$$

where the total electrostatic potential $\varphi_{\text{tot}}(rt)$ is the external potential $\varphi_{\text{ext}}(\boldsymbol{r})$ plus an *induced potential* $\varphi_{\text{ind}}(rt)$ which satisfies Poisson's equation:

$$\nabla^2 \varphi_{ind}(r) = -\delta\rho/\epsilon_0, \quad \delta\rho \equiv \delta\rho_{\text{el}} + \delta\rho_{\text{ion}} \tag{5.B.3}$$

Taking Fourier transforms and solving Eqns. (5.B.2) and (5.B.3) self-consistently, we obtain for the true electronic and ionic number density responses $\chi_{\text{el}} \equiv -(\delta\rho_{\text{el}}/\delta\varphi_{\text{ext}})/e^2$ (etc.) the expressions

$$\chi_{\text{el}}(\boldsymbol{q},\omega) = \frac{\chi^0_{\text{el}}(q\omega)}{1 + e^2/\epsilon_0 q^2 \{\chi^0_{\text{el}}(q\omega) + Z^2\chi^0_{\text{ion}}(q\omega)\}} \tag{5.B.4a}$$

$$\chi_{\text{ion}}(\boldsymbol{q},\omega) = \frac{\chi^0_{\text{ion}}(q\omega)}{1 + \frac{e^2}{\epsilon_0 q^2} \{\chi^0_{\text{el}}(q\omega) + Z^2\chi^0_{\text{ion}}(q\omega)\}} \tag{5.B.4b}$$

Equation (5.B.4a) is equivalent to our previous result (5.2.5) in the case that the ionic response can be ignored; as we will see below, this is generally true for $\omega \gg \Omega_{\text{p}}$. From Eqns. (5.B.4) we see that the response of both electrons and ions is screened by the same factor, as we should expect; this factor is often called the dielectric constant $\epsilon(q\omega)(\equiv\varphi_{\text{ext}}(q\omega)/\varphi_{\text{tot}}(q\omega))$. Equations. (5.B.4) are quite general. Let us now specialize to the case in which the *ions*, at least, are described by the jellium model (i.e. regarded as a structureless continuum). In this case we can calculate $\chi_{\text{ion}}(\boldsymbol{q}\omega)$ very simply: the continuity equation is

$$\frac{\partial\rho_{\text{ion}}}{\partial t} = -\nabla\cdot\boldsymbol{J}_{\text{ion}}$$

and in this model there are no inter-ion forces other than the long-range Coulomb ones, so for *arbitrary* $\boldsymbol{J}_{\text{ion}}(rt)$ the equation of motion is

$$\frac{\partial \boldsymbol{J}_{\text{ion}}}{\partial t}(rt) = \frac{n}{M}\boldsymbol{F} \qquad \boldsymbol{F} = \text{force per ion} \equiv -\nabla V_{\text{ext}}.$$

Thus the quantity $\chi^{(0)}_{\text{ion}}$ (the "bare" ionic number density response) is given by the simple formula

$$\chi^{(0)}_{\text{ion}}(\boldsymbol{q}\omega) = -\frac{nq^2}{M\omega^2} \tag{5.B.5}$$

The result (5.B.5) actually holds beyond the jellium model, provided that the frequency ω is large compared to any "fictitous" frequencies ω_q associated with the short-range ionic forces (see below). Inserting it into Eqn. (5.B.4), we obtain (using the definition (1) of Ω_{p})

$$\chi_{\text{el}}(q\omega) = \frac{\chi^0_{\text{el}}(q\omega)}{(1 - \Omega^2_{\text{p}}/\omega^2) + (e^2/\epsilon_0 q^2)\chi^0_{\text{el}}(q\omega)} \tag{5.B.6a}$$

$$\chi_{\text{ion}}(q\omega) = \frac{-nq^2/M\omega^2}{(1 - \Omega^2_{\text{p}}/\omega^2) + \frac{e^2}{\epsilon_0 q^2}\chi^0_{\text{el}}(q\omega)} \tag{5.B.6b}$$

Thus for $\omega \gg \Omega_{\text{p}}$ the electronic response (5.B.6a) is just that which was already calculated in Section 5.2 ignoring the ions. This result is independent of the jellium approximation, since certainly the fictitious frequencies ω_q are small compared to Ω_{p}

and hence the condition $\omega \gg \Omega_\mathrm{p}$ justifies Eqn. (5.B.5) independently of the model. We will see below that the quantity Ω_p (which has no direct physical significance for the real system) is of order of the "Debye frequency" ω_D which emerges from the screening calculation and *is* physically significant (at least in order of magnitude). Thus we can conclude that quite independently of the microscopic details of the model (and irrespective of the value of the wave vector $\boldsymbol{q}$) *the ions are irrelevant* to the electron behavior for frequencies $\omega \gg \omega_\mathrm{D}$.

In principle, we could proceed, at least in the long-wave length, low-frequency regime ($q \ll k_\mathrm{F}, \omega \ll \epsilon_\mathrm{F}/\hbar$) by substituting for $\chi^0_\mathrm{el}(q\omega)$ the full form calculated from Fermi-liquid theory. However, the resultant expression is messy and not very illuminating. Let us therefore for the moment make a further approximation, i.e. regard the electrons as a free Fermi gas. In that case, if we ignore collisions, the quantity $\chi^0_\mathrm{el}(q\omega)$ is just the Lindhard function (for all values of q and ω), see, e.g. Ziman (1964), Section 5.1. Let us specialize to the long-wavelength, low-frequency limit ($q \ll k_\mathrm{F}, \omega \ll \epsilon_f/\hbar$ but not necessarily $\omega \ll \Omega_\mathrm{p}$). We can simplify the formula (5.B.6a) considerably if we restrict ourselves to the regime $\omega \ll v_\mathrm{F} q$. Actually, for most values of q, this is the only region where we expect the phonons to have an important effect: as we saw, phonon effects are negligible for $\omega \gg \Omega_\mathrm{p}$, and since $q_\mathrm{F} v_\mathrm{F} \sim \epsilon_\mathrm{F}$, which is comparable to the *electron* plasma frequency ω_p for most metals, the condition $q v_\mathrm{F} \lesssim \omega \lesssim \Omega_\mathrm{p}$ can be met only for $q/q_\mathrm{F} \lesssim \Omega_\mathrm{p}/\omega_\mathrm{p} \sim (m/M)^{1/2} \lesssim 10^{-2}$. Thus there is a large region ($10^{-2} \ll q/q_\mathrm{F} \ll 1$) over which we expect the general framework of Fermi-liquid theory to be valid for the electrons and at the same time can take $s \equiv \omega/q v_\mathrm{F} \ll 1$.

We work to zeroth order in s. Then the quantity $\chi^0_\mathrm{el}(q\omega)$ is simply the neutral compressibility χ_0, which as we saw (Appendix 5A) was given by[55]

$$\chi_0 = \left(\frac{dn}{d\varepsilon}\right)\frac{1}{1+F^\mathrm{s}_0}$$

Therefore, if we introduce the Thomas–Fermi wave vector k_FT by

$$k^2_\mathrm{FT} \equiv \frac{e^2}{\varepsilon_0}\frac{1}{1+F^\mathrm{s}_0}\left(\frac{dn}{d\varepsilon}\right) \tag{5.B.7}$$

we can write the "dressed" electronic and ionic responses in the form

$$\chi_\mathrm{el}(q,\omega) = \frac{\chi_0}{1 + k^2_\mathrm{FT}/q^2 - \Omega^2_\mathrm{p}/\omega^2} \tag{5.B.8a}$$

$$\chi_\mathrm{ion}(\boldsymbol{q},\omega) = \frac{-n_\mathrm{ion} q^2/M\omega^2}{1 + k^2_\mathrm{FT}/q^2 - \Omega^2_\mathrm{p}/\omega^2} \tag{5.B.8b}$$

Since we have set $q \ll k_\mathrm{F}$ in order to get the results (5.B.8), it is (usually) consistent to set also $q \ll k_\mathrm{FT}$ (recall that k_FT is comparable to k_F for most metals). Then we can neglect the 1 in the denominator, and write the ionic response as

$$\chi_\mathrm{ion}(q\omega) = \frac{-nq^4/Mk^2_\mathrm{FT}}{\omega^2 - (\Omega^2_\mathrm{p}/k^2_\mathrm{FT})q^2}$$

[55] We go back at this point to the more general Fermi-liquid picture, since it does not complicate the formulae appreciably.

It is convenient to introduce the velocity

$$c_s \equiv \Omega_p / k_{FT}$$

and write (5.B.8b) in the form

$$\chi_{ion}(q\omega) = -(q^2/k_{FT}^2)\frac{nq^2/M}{\omega^2 - c_s^2 q^2} \tag{5.B.9}$$

Apart from the factor q^2/k_{FT}^2, this is just the density response of a *neutral* system of density N and atomic mass M which can sustain *sound waves* of velocity c_s. The factor of $(q/k_{FT})^2$ simply indicates that because in the limit $q/k_{FT} \to 0$ the electrons completely screen the ions, the renormalized oscillation (phonon) cannot couple to an external electromagnetic field in this limit. (If we imagine a field such as gravity which couples even to electrically neutral excitations, the response of the system to such a field would be given essentially be Eqn. (5.B.9) *without* the factor q^2/k_{FT}^2.) Note that if we keep the 1 in (5.B.8b), the phonon frequency is given by

$$\omega_{ph}(q) = \left[\Omega_p^2/(k_{FT}^2 + q^2)\right]^{1/2} q \tag{5.B.10}$$

We now turn to the question of the electron electron interaction in the presence of coupling to the ionic motion. Just as in Section 5.2, we can obtain this by defining the "effective potential" $V_{eff}(\boldsymbol{r}t)$ acting on an electron due to a second electron as equal to $-e\delta\varphi(\boldsymbol{r}t)$, where $\delta\varphi(\boldsymbol{r}t)$ is the *total* potential induced by the "external" (unscreened Coulomb) potential of the second electron, (which is $e/4\pi\epsilon_0 r \equiv \varphi_{ext}(\boldsymbol{r})$). From our earlier result the general result is

$$V_{eff}^{el-el}(q,\omega) = \frac{e^2}{q^2\epsilon_0\epsilon(q\omega)} = \frac{1}{q^2\varepsilon_0/e^2 + (\chi_{el}^0(q\omega) + Z^2\chi_{ion}(q\omega))} \tag{5.B.11}$$

In the case of most practical interest ($q \ll q_F, \omega \ll \varepsilon_F/\hbar, qv_F \ll \omega$) this simplifies, according to Eqn. (5.B.9) to the expression

$$V_{eff}^{el-el}(q,\omega) = \frac{e^2/\varepsilon q^2}{1 + k_{FT}^2/q^2 - \Omega_p^2/\omega^2} \tag{5.B.12}$$

Note, by the way, that this result, though not in general quantitatively valid for $q \sim q_F$, is valid to the extent that the *static* electronic susceptibility $\chi(q,0)$ is well approximated by the Thomas–Fermi form in this region. Thus there is some sense in keeping the 1, though it should be remembered that for $q \sim q_F$ the form (5.B.11) is only approximate.

It is convenient to rewrite (5.B.12) in a completely equivalent form by introducing the phonon frequency $\omega_{ph}(q)$ given by (5.B.10). This gives

$$V_{eff}^{el-el}(\boldsymbol{q},\omega) = \frac{e^2}{\varepsilon_0}\frac{1}{q^2 + k_{FT}^2}\left(1 + \frac{\omega_{ph}^2(q)}{\omega^2 - \omega_{ph}^2(q)}\right) \tag{5.B.13}$$

This result has a very simple physical interpretation: The first term, which is completely independent of the ionic variables, simply represents the screened Coulomb interaction which we already calculated for the electron gas. The second term represents an *interaction due to the exchange of virtual phonons* (cf. the discussion in

Section 5.2). Because this term contains the frequency ω "exchanged" between the electrons, it is *nonlocal in time*: i.e. an electron at point $\boldsymbol{r}$ at time t will induce a polarization of the ionic lattice which stays around for a time $\sim \omega_{\mathrm{D}}^{-1}$, so that an electron coming by at a *later* time, when the first electron has already gone away, will see this polarization and thereby in effect experience a "retarded" interaction with the first electron. Note that for fixed q the interaction V_{eff} is zero in the static limit, becomes *negative* (attractive) for $\omega < \omega_{\text{ph}}(q)$ and finally turns repulsive for $\omega > \omega_{\text{ph}}(q)$.

5C The BCS problem in the particle-conserving representation

In this appendix it is convenient to change the notation so that N denotes the number of *pairs* (half the number of electrons) and to write $a^{\dagger}_{\boldsymbol{k}\uparrow}a^{\dagger}_{-\boldsymbol{k}\downarrow} \equiv b^{\dagger}_{\boldsymbol{k}}$, $\hat{\Omega} \equiv \sum_{\boldsymbol{k}} c_{\boldsymbol{k}} b^{\dagger}_{\boldsymbol{k}}$. Thus the ansatz (5.4.6) for the many-body ground state is written

$$\Psi_N = \mathcal{N}\left(\sum_{\boldsymbol{k}} c_{\boldsymbol{k}} b^{\dagger}_{\boldsymbol{k}}\right)^N |\text{vac}\rangle \equiv \mathcal{N}\hat{\Omega}^N |\text{vac}\rangle \tag{5.C.1}$$

(where we leave off the double prime on the normalization factor $\mathcal{N}$ for subsequent clarity). It is convenient to choose the normalization of the $c_{\boldsymbol{k}}$ so as to satisfy the condition

$$\sum_{\boldsymbol{k}} \frac{|c_{\boldsymbol{k}}|^2}{1+|c_{\boldsymbol{k}}|^2} = N \tag{5.C.2}$$

Throughout this appendix I shall assume that N is very large, and neglect corrections of relative order $N^{-1/2}$ or smaller.

By far the simplest way of making contact between the ansatz (5.C.1) and the standard BCS formalism is to replace the operator $\hat{\Omega}^N$ by $\exp(\hat{\Omega})$, and I shall explore this approach below. However, the resulting many-body wave function is clearly unphysical, in that it involves a superposition (not a mixture!) of states with different total particle number, so let us first verify that we can get the standard BCS results directly from the particle-conserving ansatz (5.C.1).

Let us select a particular pair state $\boldsymbol{k}$, and write $\hat{\Omega}' \equiv \sum_{\boldsymbol{k}' \neq \boldsymbol{k}} c_{\boldsymbol{k}'} b^{\dagger}_{\boldsymbol{k}'}$; moreover, let the normalized version of the state $\hat{\Omega}'|\text{vac}\rangle$ be denoted Ψ'_N. Then it is immediately obvious[56] that we can write (5.C.1) in the form

$$\Psi_N = \left(1+|\tilde{c}_{\boldsymbol{k}}|^2\right)^{-1/2}\left(\Psi'_N + \tilde{c}_{\boldsymbol{k}} b^{\dagger}_{\boldsymbol{k}} \Psi'_{N-1}\right) \tag{5.C.3}$$

where for the moment $\tilde{c}_{\boldsymbol{k}}$ is an unknown quantity (we will later identify it with $c_{\boldsymbol{k}}$). Further, consider the quantity $\langle n_{\boldsymbol{k}}\rangle \equiv \langle a^{\dagger}_{\boldsymbol{k}\uparrow}a_{\boldsymbol{k}\uparrow}\rangle$ ($\equiv\langle a^{\dagger}_{-\boldsymbol{k}\downarrow}a_{-\boldsymbol{k}\downarrow}\rangle$ for any state of the form (5.C.1)); since it is clearly zero in both Ψ'_N and Ψ'_{N-1}, and is changed by one by the operator $b^{\dagger}_{\boldsymbol{k}}$, we clearly have

$$\langle n_{\boldsymbol{k}}\rangle = \frac{|\tilde{c}_{\boldsymbol{k}}|^2}{1+|\tilde{c}_{\boldsymbol{k}}|^2} \tag{5.C.4}$$

[56] We use the fact that $(b^{\dagger}_{k})^2 \equiv 0$.

Next, consider two pair states $\boldsymbol{k}$, $\boldsymbol{k}'$, and denote by a double prime quantities $\hat{\Omega}$, Ψ in which the sum omits both states. It is then slightly less obvious (but, as we shall verify below, true) that (5.C.1) can be represented in the form

$$\Psi_N = \left(1 + |\tilde{c}_{\boldsymbol{k}}|^2\right)^{-1/2} \left(1 + |\tilde{c}_{\boldsymbol{k}'}|^2\right)^{-1/2} \times \left(\Psi''_N + \left(\tilde{c}_{\boldsymbol{k}} b^\dagger_{\boldsymbol{k}} + \tilde{c}_{\boldsymbol{k}'} b^\dagger_{\boldsymbol{k}'}\right) \Psi''_{N-1} + \tilde{c}_{\boldsymbol{k}} \tilde{c}_{\boldsymbol{k}'} b^\dagger_{\boldsymbol{k}} b^\dagger_{\boldsymbol{k}'} \Psi''_{N-2}\right) \tag{5.C.5}$$

Using the representation (5.C.5), we can calculate the quantity $\langle b_{\boldsymbol{k}} b^\dagger_{\boldsymbol{k}'}\rangle (\equiv \langle \Psi_N | b_{\boldsymbol{k}} b^\dagger_{\boldsymbol{k}'} | \Psi_N\rangle)$ and find

$$\left\langle b_{\boldsymbol{k}} b^\dagger_{\boldsymbol{k}'}\right\rangle = \frac{\tilde{c}_{\boldsymbol{k}} \tilde{c}^*_{\boldsymbol{k}'}}{\left(1 + |\tilde{c}_{\boldsymbol{k}}|^2\right)\left(1 + |\tilde{c}_{\boldsymbol{k}'}|^2\right)} \equiv F_{\boldsymbol{k}} F^*_{\boldsymbol{k}'} \tag{5.C.6}$$

where we have defined

$$F_{\boldsymbol{k}} \equiv \frac{\tilde{c}_{\boldsymbol{k}}}{1 + |\tilde{c}_{\boldsymbol{k}}|^2} \tag{5.C.7}$$

Using (5.C.3) and the fact that $\langle \Psi'_N | b_{\boldsymbol{k}} b^\dagger_{\boldsymbol{k}} | \Psi'_N\rangle \equiv \langle \Psi'_N | \Psi'_N\rangle \equiv 1$, we see that $F_{\boldsymbol{k}}$ can be written in the form (which is sometimes called an "anomalous average" and denoted $\langle b_k\rangle$)

$$F_{\boldsymbol{k}} = \langle \Psi_{N-1} | \, b_{\boldsymbol{k}} \, | \Psi_N\rangle \tag{5.C.8}$$

in agreement (after the identification of $\tilde{c}_{\boldsymbol{k}}$ with $c_{\boldsymbol{k}}$, see below) with Eqn. (5.4.8) of the text (where, recall, N is the number of *electrons*). We will see below that the Fourier transform of $F_{\boldsymbol{k}}$ can be regarded as the "relative wave function of the Cooper pairs."

As a matter of fact, Eqns. (5.C.4) and (5.C.6) are sufficient as they stand to permit the derivation of the standard BCS results given in Section 5.4, since we can eliminate $\tilde{c}_{\boldsymbol{k}}$ from (5.C.4) to get (5.4.9), and (5.C.6) leads directly to (5.4.14); we do not need to know that $\tilde{c}_{\boldsymbol{k}}$ is actually identical to the $c_{\boldsymbol{k}}$ occurring in the ansatz (5.C.1). However, for completeness I shall now demonstrate this (and in the process verify the correctness of Eqn. (5.C.5), or more precisely of (5.C.6) from which we can work back to (5.C.5)).

We simply consider, for purely formal purposes, the (so far unnormalized) wave function introduced by BCS, namely

$$\Psi_{\text{BCS}} \equiv \mathcal{N}_{\text{BCS}} e^{\hat{\Omega}} \, |\text{vac}\rangle \equiv \mathcal{N}_{\text{BCS}} \sum_{n=0}^{\infty} (n!)^{-1} \, \hat{\Omega}^n \, |\text{vac}\rangle \equiv \sum_n a_n \Psi_n \tag{5.C.9}$$

where the Ψ_n are normalized. From an examination of the form of Ψ_{BCS} expressed explicitly in terms of the coefficients c_k (Eqn. (5.C.11) below) it will turn out that, given the choice (5.C.2), the coefficients a_n are strongly peaked around $n \cong N$. Moreover, it is clear that the expectation value of any particle-conserving operator $\hat{K}$, such as $\hat{n}_{\boldsymbol{k}}$ or $b_{\boldsymbol{k}} b^\dagger_{\boldsymbol{k}'}$, on Ψ_{BCS} is simply the average of its expectation value on the Ψ_n, weighted with the corresponding probabilities $|a_n|^2$. Assuming[57] that this expectation

[57]This assumption may be justified from the consideration that a single application of the operator $\hat{\Omega}$ affects the occupation of any given pair state k only to order N^{-1} (it is of course implicit here that $\hat{K}$ is a "few-particle" operator).

value is only weakly sensitive to n, we can thus draw the conclusion that

$$\langle\Psi_N|\,\hat{K}\,|\Psi_N\rangle = \langle\Psi_{\text{BCS}}|\,\hat{K}\,|\Psi_{\text{BCS}}\rangle \tag{5.C.10}$$

Now we are home, since it is very easy to calculate the RHS of Eqn. (5.C.10). We have explicitly

$$\begin{aligned}\Psi_{\text{BCS}} &= \mathcal{N}_{\text{BCS}} e^{\sum_{\boldsymbol{k}} c_{\boldsymbol{k}} b_{\boldsymbol{k}}^\dagger}\,|\text{vac}\rangle = \mathcal{N}_{\text{BCS}} \prod_{\boldsymbol{k}} e^{c_{\boldsymbol{k}} b_{\boldsymbol{k}}^\dagger}\,|\text{vac}\rangle \\ &= \mathcal{N}_{\text{BCS}} \prod_{\boldsymbol{k}} \left(1 + c_{\boldsymbol{k}} b_{\boldsymbol{k}}^\dagger\right)|\text{vac}\rangle \equiv \mathcal{N}_{\text{BCS}} \prod_{\boldsymbol{k}} \Phi_{\boldsymbol{k}}^{(\text{un})}\end{aligned} \tag{5.C.11}$$

where $\Phi_{\boldsymbol{k}}^{(\text{un})}$ refers only to the "occupation state" of the pair of plane-wave states $(\boldsymbol{k}\uparrow, -\boldsymbol{k}\downarrow)$: in an obvious notation

$$\Phi_{\boldsymbol{k}}^{(\text{un})} \equiv |00\rangle_{\boldsymbol{k}} + c_{\boldsymbol{k}}\,|11\rangle_{\boldsymbol{k}}\,. \tag{5.C.12}$$

It is now obvious that in order for the many-body wave function to be normalized we must have

$$\mathcal{N}_{\text{BCS}} = \prod_{\boldsymbol{k}} \left(1 + |c_{\boldsymbol{k}}|^2\right)^{-1/2} \tag{5.C.13}$$

and so we can rewrite (5.C.10) in the form

$$\Psi_{\text{BCS}} = \prod_{\boldsymbol{k}} \Phi_k, \qquad \Phi_k \equiv \left(1 + |c_{\boldsymbol{k}}|^2\right)^{-1/2} \left(|00\rangle_{\boldsymbol{k}} + c_{\boldsymbol{k}}\,|11\rangle_{\boldsymbol{k}}\right) \equiv u_{\boldsymbol{k}}\,|00\rangle + v_{\boldsymbol{k}}\,|11\rangle \tag{5.C.14}$$

where the last form is the standard notation used in most textbooks. From (5.C.14) it is immediately obvious that the expectation value of $\hat{n}_{\boldsymbol{k}}$ in the BCS state, and thus by the above argument also in the particle-conserving state (5.C.1), is

$$\langle n_{\boldsymbol{k}}\rangle = \frac{|c_{\boldsymbol{k}}|^2}{1 + |c_{\boldsymbol{k}}|^2} \tag{5.C.15}$$

Comparing this result with (5.C.7), we see that $c_{\boldsymbol{k}} = \tilde{c}_{\boldsymbol{k}}$ as claimed.[58] Further, we easily find from (5.C.14) that

$$\left\langle b_k b_{k'}^\dagger \right\rangle = \frac{c_k c_{k'}^*}{\left(1 + |c_k|^2\right)\left(1 + |c_{k'}|^2\right)} \equiv \frac{\boldsymbol{c}_k \boldsymbol{c}_{k'}^*}{\left(1 + |\boldsymbol{c}_k|^2\right)\left(1 + |\boldsymbol{c}_{k'}|^2\right)} \tag{5.C.16}$$

thereby justifying (5.C.6); this concludes the argument for Eqns. (5.4.9) and (5.4.14) of the text. For completeness I note the relation between the parameters c_k and F_k used here and the more common notation (u_k, v_k): taking u_k to be real by convention,

$$c_k = \frac{v_k}{u_k}, \quad F_k = u_k v_k, \quad |u_k|^2 + |v_k|^2 = 1. \tag{5.C.17}$$

Finally, let us verify that the Fourier transform $F(\boldsymbol{r})$ of F_k is indeed the "pair wave function", that is the unique (unnormalized) eigenfunction of the two-particle density

[58] Strictly speaking, the argument as stated allows $c_{\boldsymbol{k}} = c_{\boldsymbol{k}} e^{i\phi_{\boldsymbol{k}}}$ where $\phi_{\boldsymbol{k}}$ is an arbitrary phase. However, provided the (arbitrary) relative phases of the Ψ_N are chosen in the "natural" way, a comparison of (the first equality of) (5.C.16) and (5.C.6) constrains all $\phi_{\boldsymbol{k}}$ to be zero.

matrix associated with an eigenvalue of order N. Consider the general expression (cf. Chapter 2, Eqn. 2.4.2)

$$\rho_2(\boldsymbol{r}_1\sigma_1, \boldsymbol{r}_2\sigma_2 : \boldsymbol{r}_1'\sigma_1', \boldsymbol{r}_2'\sigma_2') \equiv \langle \psi^\dagger_{\sigma_1}(\boldsymbol{r}_1)\psi^\dagger_{\sigma_2}(\boldsymbol{r}_2)\psi_{\sigma_2'}(\boldsymbol{r}_2')\psi_{\sigma_1'}(\boldsymbol{r}'_1)\rangle$$
$$\equiv \sum_{\boldsymbol{k}_1\boldsymbol{k}_2\boldsymbol{k}_3\boldsymbol{k}_4} \left\langle a^\dagger_{\boldsymbol{k}_1\sigma_1} a^\dagger_{\boldsymbol{k}_2\sigma_2} a_{\boldsymbol{k}_3\sigma_2'} a_{\boldsymbol{k}_4\sigma_1'} \right\rangle$$
$$\times\, e^{-i\left(\boldsymbol{k}_1\cdot\boldsymbol{r}_1 + \boldsymbol{k}_2\cdot\boldsymbol{r}_2 - \boldsymbol{k}_3\cdot\boldsymbol{r}'_2 - \boldsymbol{k}_4\cdot\boldsymbol{r}_1'\right)} \tag{5.C.18}$$

The terms corresponding to nonzero center-of-mass momentum ($\boldsymbol{k}_1 \neq -\boldsymbol{k}_2$, etc.) are of no particular interest to us, so we set $\boldsymbol{k}_2 = -\boldsymbol{k}_1$, $\boldsymbol{k}_4 = -\boldsymbol{k}_3$, $\sigma_2 = -\sigma_1$. Also for definiteness[59] let us set $\sigma_1 = \uparrow$. The resulting term, which we call ρ_2(pair) is

$$\sum_{\boldsymbol{k}\boldsymbol{k}'} \langle a^\dagger_{\boldsymbol{k}\uparrow} a^\dagger_{-\boldsymbol{k}\downarrow} a_{-\boldsymbol{k}'\downarrow} a_{\boldsymbol{k}'\uparrow}\rangle e^{-(i\boldsymbol{k}\cdot(\boldsymbol{r}_1-\boldsymbol{r}_2) - i\boldsymbol{k}'\cdot(\boldsymbol{r}_2'-\boldsymbol{r}_1'))} \equiv \sum_{\boldsymbol{k}\boldsymbol{k}'} \langle b^\dagger_{\boldsymbol{k}} b_{\boldsymbol{k}'}\rangle e^{-i\boldsymbol{k}\cdot\boldsymbol{r}} e^{i\boldsymbol{k}'\cdot\boldsymbol{r}'} \tag{5.C.19}$$

which by (5.C.6) is

$$\rho_2(\text{pair}) = \sum_{\boldsymbol{k}\boldsymbol{k}'} F^*_{\boldsymbol{k}} F_{\boldsymbol{k}'} e^{-i\boldsymbol{k}\cdot\boldsymbol{r}} e^{i\boldsymbol{k}'\cdot\boldsymbol{r}'} \equiv F^*(\boldsymbol{r})F(\boldsymbol{r}') \tag{5.C.20}$$

The associated eigenvalue N_0 is given by Eqn. (5.4.32), and is a fraction of order $\Delta/\epsilon_{\mathrm{F}}$ of N. Since ρ_2(pair) does not depend at all on the center-of-mass variables of $\boldsymbol{r}_1$ and $\boldsymbol{r}_2$ ($\boldsymbol{r}'_1$ and $\boldsymbol{r}'_2$), we see that the expression (5.C.20) corresponds to a very simple form of off-diagonal long-range order. Of course, it has this simple form only because of our original assumption that the fraction $\phi(\boldsymbol{r}, \boldsymbol{r}' : \alpha\beta)$ had the form (5.4.5), i.e. is independent of the COM variable.

5D Derivation of the finite-temperature BCS gap equation

According to the considerations of Section 5.5, the total free energy of a system described by a density matrix of the BCS form (cf. Eqn. 5.5.10) consists of the kinetic energy (with the usual subtraction of μN), the entropy term $-TS$, and the pairing term.[60] The first two terms can be written as a sum of contributions $Q_{\boldsymbol{k}}$ from the individual pairs of states ($\boldsymbol{k}\uparrow, -\boldsymbol{k}\downarrow$): explicitly, we have from (5.5.13) and (5.5.15), omitting terms independent of the parameter Δ_k,

$$Q_{\boldsymbol{k}} \equiv \langle K - \mu N\rangle_{\boldsymbol{k}} - TS_{\boldsymbol{k}}$$
$$= -\frac{\epsilon^2_{\boldsymbol{k}}}{E_{\boldsymbol{k}}} \tanh\frac{\beta E_{\boldsymbol{k}}}{2} + \left(E_{\boldsymbol{k}}\tanh\frac{\beta E_{\boldsymbol{k}}}{2} - 2\beta^{-1}\ln\cosh\frac{\beta E_{\boldsymbol{k}}}{2}\right) \tag{5.D.1}$$

The pairing term in the potential is

$$\langle V\rangle_{\text{pair}} = \sum_{\boldsymbol{k}\boldsymbol{k}'} V_{\boldsymbol{k}\boldsymbol{k}'} F^*_{\boldsymbol{k}} F_{\boldsymbol{k}'} \equiv \sum_{\boldsymbol{k}} F^*_{\boldsymbol{k}} \left(\sum_{\boldsymbol{k}'} V_{\boldsymbol{k}\boldsymbol{k}'} F_{\boldsymbol{k}'}\right) \tag{5.D.2}$$

(where $F_{\boldsymbol{k}}$ is given by the expression (5.5.14)), and thus (since we regard $\Delta_{\boldsymbol{k}}$ and $\Delta^*_{\boldsymbol{k}}$ (or $F_{\boldsymbol{k}}$ and $F^*_{\boldsymbol{k}}$) as independent variables) can be regarded, to within terms of relative

[59] Strictly speaking the pair wave function is $F(r)$ times the spin singlet state $2^{-1/2}(\uparrow_1\downarrow_2 - \downarrow_1\uparrow_2)$.

[60] As usual we implicitly neglect the Hartree and Fock terms.

order N^{-1}, as also a sum of contributions from the individual pairs $\boldsymbol{k}$. Thus, to obtain the equilibrium values of the parameters $\Delta_{\boldsymbol{k}}$, it is sufficient to solve the equation

$$\frac{\partial Q_{\boldsymbol{k}}}{\partial \Delta_{\boldsymbol{k}}^*} + \frac{\partial F_{\boldsymbol{k}}^*}{\partial \Delta_{\boldsymbol{k}}^*}\left(\sum_{\boldsymbol{k}'} V_{\boldsymbol{k}\boldsymbol{k}'} F_{\boldsymbol{k}'}\right) = 0 \tag{5.D.3}$$

A straightforward calculation shows that

$$\frac{\partial Q_{\boldsymbol{k}}}{\partial \Delta_{\boldsymbol{k}}^*} = \Delta_{\boldsymbol{k}} \left\{ \frac{\epsilon_{\boldsymbol{k}}^2}{E_{\boldsymbol{k}}^3} \tanh \frac{\beta E_{\boldsymbol{k}}}{2} + \frac{\beta \left|\Delta_{\boldsymbol{k}}\right|^2}{2E_{\boldsymbol{k}}^2} \operatorname{sech}^2 \frac{\beta E_{\boldsymbol{k}}}{2} \right\} \tag{5.D.4}$$

On the other hand, using the definition (5.5.14) of $F_{\boldsymbol{k}}$, we find that $\partial F_{\boldsymbol{k}}^*/\partial \Delta_{\boldsymbol{k}}^*$ is simply the expression in curly brackets in (5.D.4). Thus we obtain from (5.D.3)

$$\Delta_{\boldsymbol{k}} = -\sum_{\boldsymbol{k}'} V_{\boldsymbol{k}\boldsymbol{k}'} F_{\boldsymbol{k}'} \equiv -\sum_{\boldsymbol{k}'} V_{\boldsymbol{k}\boldsymbol{k}'} \frac{\Delta_{\boldsymbol{k}'}}{2E_{\boldsymbol{k}'}} \tanh \frac{\beta E_{\boldsymbol{k}'}}{2} \tag{5.D.5}$$

which is the standard form (5.5.16) of the finite-temperature BCS gap equation.

It is worthwhile to sketch an alternative, more intuitive derivation of Eqn. (5.D.5), which is based on the idea, due to P. W. Anderson, of representing the states $|00\rangle_{\boldsymbol{k}}$ and $|11\rangle_{\boldsymbol{k}}$ as respectively the states $\sigma_{\boldsymbol{k}z} = +1 \cdots -1$ of a fictitious spin-$\frac{1}{2}$ system. In this representation the kinetic-energy term in the Hamiltonian becomes, apart from a constant,

$$\hat{K} - \mu\hat{N} = -\sum_{\boldsymbol{k}} \epsilon_{\boldsymbol{k}} \hat{\sigma}_{\boldsymbol{k}z} \tag{5.D.6}$$

and the operator $b_{\boldsymbol{k}}^\dagger b_{\boldsymbol{k}'}$ becomes $\hat{\boldsymbol{\sigma}}_{\boldsymbol{k}\perp} \cdot \hat{\boldsymbol{\sigma}}_{\boldsymbol{k}'\perp}$ where "$\perp$" denotes the xy-plane component; thus the total Hamiltonian is written

$$\hat{H} = -\sum_{\boldsymbol{k}} \epsilon_{\boldsymbol{k}} \hat{\sigma}_{\boldsymbol{k}z} + \sum_{\boldsymbol{k}\boldsymbol{k}'} V_{\boldsymbol{k}\boldsymbol{k}'} \hat{\boldsymbol{\sigma}}_{\boldsymbol{k}\perp} \cdot \hat{\boldsymbol{\sigma}}_{\boldsymbol{k}'\perp} \tag{5.D.7}$$

In the mean-field approximation which corresponds to the BCS ansatz for the density matrix, the Hamiltonian corresponds to the problem of a set of spins $\boldsymbol{\sigma}_{\boldsymbol{k}}$ in a magnetic field $\boldsymbol{\mathcal{H}}_{\boldsymbol{k}}$ given by

$$\boldsymbol{\mathcal{H}}_{\boldsymbol{k}} = \epsilon_{\boldsymbol{k}} \hat{\boldsymbol{z}} - \sum_{\boldsymbol{k}'} V_{\boldsymbol{k}\boldsymbol{k}'} \langle \boldsymbol{\sigma}_{\boldsymbol{k}'\boldsymbol{k}} \rangle \tag{5.D.8}$$

It is intuitively clear that in equilibrium (for not too "pathological" $V_{\boldsymbol{k}\boldsymbol{k}'}$) the transverse components of all the spins will lie parallel, and since rotation of all of them simultaneously turns out to correspond to a simultaneous change of phase of all the $F_{\boldsymbol{k}}$, which as pointed out in Section 5.4 has no physical significance, we may as well take them to lie along the x-axis. Then we can rewrite (5.D.8) in the form

$$\mathcal{H}_{\boldsymbol{k}} = \epsilon_{\boldsymbol{k}} \hat{\boldsymbol{z}} + \tilde{\Delta}_{\boldsymbol{k}} \hat{\boldsymbol{x}} \tag{5.D.9}$$

where

$$\tilde{\Delta}_{\boldsymbol{k}} \equiv -\sum_{\boldsymbol{k}'} V_{\boldsymbol{k}\boldsymbol{k}'} \langle \sigma_{x\boldsymbol{k}'} \rangle \tag{5.D.10}$$

We will see that the quantity $\tilde{\Delta}_{\boldsymbol{k}}$ is identical to $\Delta_{\boldsymbol{k}}$ in thermal equilibrium (but not more generally[61]).

It is clear that the magnitude of the total (fictitious) "magnetic field" acting on the "spin" $\boldsymbol{k}$ is $(\epsilon_{\boldsymbol{k}}^2 + |\tilde{\Delta}_{\boldsymbol{k}}|^2)^{1/2} \equiv E_{\boldsymbol{k}}$, and that it is oriented along at an angle $\tan^{-1}(\Delta_{\boldsymbol{k}}/\epsilon_{\boldsymbol{k}}) = \sin^{-1}(\Delta_{\boldsymbol{k}}/E_{\boldsymbol{k}}) \equiv \theta_{\boldsymbol{k}}$ to the positive z-axis. In calculating the net polarization of the spin $\boldsymbol{k}$, it is necessary to bear in mind that the spin-1/2 representation describes only two of the pair states associated with the pair $(\boldsymbol{k}\uparrow, -\boldsymbol{k}'\downarrow)$, namely the GP and EP states (which correspond to orientation "along" or "against" the field respectively). The BP states are not incorporated in this description and may formally be taken to correspond to spin zero; however, at finite temperature they have a finite occupation probability, and the absolute probabilities of the GP and EP states are thereby reduced. From formulae (5.5.11)–(5.5.12), we see that the polarization $\langle \boldsymbol{\sigma}_{\parallel} \rangle = P_{\mathrm{GP}} - P_{\mathrm{EP}}$ is given by

$$\langle \boldsymbol{\sigma}_{\parallel \boldsymbol{k}} \rangle = \tanh \frac{\beta E_{\boldsymbol{k}}}{2} \tag{5.D.11}$$

(rather than by the formula $(\tanh \beta E_{\boldsymbol{k}})$ which would apply for a true spin-$\frac{1}{2}$ system). Since $\langle \sigma_{\perp \boldsymbol{k}} \rangle = \frac{1}{2} \sin \theta_{\boldsymbol{k}} \langle \sigma_{\parallel \boldsymbol{k}} \rangle = (\tilde{\Delta}_{\boldsymbol{k}}/2E_{\boldsymbol{k}}) \langle \sigma_{\parallel \boldsymbol{k}} \rangle$, we find from (5.D.10) that in equilibrium

$$\tilde{\Delta}_{\boldsymbol{k}} = -\sum_{\boldsymbol{k}'} V_{\boldsymbol{k}\boldsymbol{k}'} \frac{\tilde{\Delta}_{\boldsymbol{k}'}}{2E_{\boldsymbol{k}'}} \tanh \frac{\beta E_{\boldsymbol{k}'}}{2} \tag{5.D.12}$$

It is easy to check that in equilibrium $\tilde{\Delta}_{\boldsymbol{k}}$ is identical to the quantity $\Delta_{\boldsymbol{k}}$ which was used in Sections 5.4 and 5.5 as a convenient parametrization of the $c_{\boldsymbol{k}}$, i.e. of the many-body states occurring in the density matrix. For this reason much of the literature does not distinguish them (cf. footnote 61).

5E Derivation of bulk terms in the Ginzburg–Landau free energy

If we define as in (5.7.15)

$$\Psi = \sum_{\boldsymbol{k}} F_{\boldsymbol{k}} \tag{5.E.1}$$

then the potential terms (ignoring as usual the Hartree and Fock contributions) are simply $-V_0 \Psi^2$. To get the other terms, namely

$$F_1 \equiv \langle K - \mu N \rangle - TS \tag{5.E.2}$$

we use the technique of Lagrange multipliers and minimize the expression

$$\tilde{F}_1 \equiv F_1 - (\lambda \Psi^* + \lambda^* \Psi) = F_1 - 2\lambda\Psi \tag{5.E.3}$$

where in the second step we have assumed for definiteness that Ψ is chosen real. Now the problem posed by the minimization of (5.E.3) is simply the original BCS problem,

[61] In much of the literature a quantity $\Delta_{\boldsymbol{k}}$ is *defined* by the expression $-\sum_{\boldsymbol{k}'} V_{\boldsymbol{k}\boldsymbol{k}'} F_{\boldsymbol{k}'}$, i.e. as what we have here called $\tilde{\Delta}_{\boldsymbol{k}}$.

with $\Delta \equiv V\Psi_0 \to \lambda$; hence we know right away that the solution is the transcription of that, namely

$$\Psi(\lambda) = \frac{\lambda}{2}\sum_{\boldsymbol{k}} E_{\boldsymbol{k}}^{-1}(\lambda)\tanh\frac{\beta E_{\boldsymbol{k}}(\lambda)}{2} \qquad E_{\boldsymbol{k}}(\lambda) \equiv \left(\epsilon_{\boldsymbol{k}}^2 + \lambda^2\right)^{1/2} \tag{5.E.4}$$

We can expand the RHS of (5.E.4) in λ to give (for arbitrary T)

$$\Psi(\lambda) = A(T)\lambda - B(T)\lambda^3 \implies \lambda \approx A^{-1}\Psi + BA^{-4}\Psi^3 \tag{5.E.5}$$

Then since according to (5.E.3) we have $\partial F_1/\partial\Psi = 2\lambda$ and thus

$$F_1(\lambda) = F_0 + 2\int \lambda \, d\Psi \tag{5.E.6}$$

it follows that up to fourth order in Ψ we have

$$F_1(\Psi) = A^{-1}|\Psi|^2 + \frac{1}{2}BA^{-4}|\Psi|^4 \tag{5.E.7}$$

as stated in the text. The quantities $A(T)$ and $B(T)$ are given explicitly (in the limit $\beta\epsilon_c \gg 1$) by

$$\begin{aligned} A(T) &= \sum_{\boldsymbol{k}}\left(\frac{\tanh\beta\epsilon_{\boldsymbol{k}}/2}{2\epsilon_{\boldsymbol{k}}}\right) = \frac{1}{2}\left(\frac{dn}{d\epsilon}\right)\int_{-\epsilon_c}^{\epsilon_c}\frac{\tanh\beta\epsilon/2}{2\epsilon}\,d\epsilon \\ &= \frac{1}{2}\left(\frac{dn}{d\epsilon}\right)\ln\left(1.14\frac{\epsilon_c}{k_BT}\right) \end{aligned} \tag{5.E.8a}$$

$$\begin{aligned} B(T) &= -\sum_{\boldsymbol{k}}\frac{d}{d\left(\epsilon_{\boldsymbol{k}}^2\right)}\left(\frac{\tanh\beta\epsilon_{\boldsymbol{k}}/2}{2\epsilon_{\boldsymbol{k}}}\right) = -\frac{1}{2}\left(\frac{dn}{d\epsilon}\right)\int_{-\epsilon_c}^{\epsilon_c}\frac{d}{d\left(\epsilon^2\right)}\frac{\tanh\beta\epsilon/2}{2\epsilon}\,d\epsilon \\ &= \frac{1}{2}\left(\frac{dn}{d\epsilon}\right)\frac{1}{(\pi k_BT)^2}\frac{7}{8}\zeta(3) \end{aligned} \tag{5.E.8b}$$

as stated in the text (Eqns. 5.7.19).

5F Renormalization of the BCS gap equation

We write the gap equation (5.8.4) as a matrix equation in energy (ϵ-)space:

$$\Delta = -\hat{V}\hat{K}\Delta \tag{5.F.1}$$

where $\Delta \equiv \Delta(\epsilon)$ is a column vector, $\hat{V}$ is a (nondiagonal) matrix with the obvious elements $V(\epsilon,\epsilon')$ and $\hat{K}$ is a diagonal matrix diag$K(\epsilon)$ with elements $K(\epsilon) = N(\epsilon)(2E^{-1}(\epsilon))\tanh\beta E(\epsilon)/2$, $E(\epsilon) \equiv (\epsilon^2 + |\Delta(\epsilon)|^2)^{1/2}$ (thus, $K(\epsilon)$ is itself a function of $\Delta(\epsilon)$). We select a cutoff energy ϵ_c which is small on the scale of variation of both the density of states $N(\epsilon)$ and the interaction $V(\epsilon,\epsilon')$ (in either of its indices). Next, we introduce diagonal projector matrices $\hat{P}_1$, $\hat{P}_2$ which project on to the states $|\epsilon| < \epsilon_c$ and $|\epsilon| > \epsilon_c$ respectively, and rewrite (5.F.1) in the identically equivalent form

$$\Delta = -\hat{V}\hat{K}\left(\hat{P}_1 + \hat{P}_2\right)\Delta \tag{5.F.2}$$

Transposing the term in $\hat{P}_2$, premultiplying by the operator $(1+\hat{V}\hat{K}\hat{P}_2)^{-1}$ and introducing the operator

$$\hat{t} \equiv \left(1+\hat{V}\hat{K}\hat{P}_2\right)^{-1}\hat{V} \tag{5.F.3}$$

we can rewrite (5.F.2) in the form

$$\Delta = -\hat{t}\hat{K}\hat{P}_1\Delta \tag{5.F.4}$$

Although Eqn. (5.F.4) is valid for any ϵ, we shall be interested in it only for $|\epsilon| < \epsilon_c$. Now, $V(\epsilon,\epsilon')$ is by construction independent of ϵ (ϵ') for $|\epsilon|$ ($|\epsilon'|$) $< \epsilon_c$, so from its definition (5.F.3) $t(\epsilon,\epsilon')$ is independent of ϵ and ϵ' when both $|\epsilon|$ and $|\epsilon'|$ are $< \epsilon_c$ (but is in general a function of β and ϵ_c, and a functional of $\{\Delta(\epsilon)\}$ where $|\epsilon| > \epsilon_c$). Since $N(\epsilon)$ is also by construction equal to $N(0)$ for $|\epsilon| < \epsilon_c$, the renormalized gap Eqn. (5.F.4) reduces to (5.8.5) of the text.

Suppose now that, while satisfying the conditions specified above on ϵ_c, we are able to choose this quantity to be large compared to (a typical value of) the low-energy zero-temperature gap, and hence to T_c which we anticipate will be of the same order. Then for any $T < T_c$ the quantity $K(\epsilon)$ is for $|\epsilon| > \epsilon_c$ of the simple form

$$K(\epsilon) = \frac{N(\epsilon)}{2|\epsilon|} \tag{5.F.5}$$

and is thus independent of both β and $\{\Delta(\epsilon)\}$. Under these conditions Eqn. (5.8.5) of the text reduces to (5.8.6), that is, to the equation (for $|\epsilon| < \epsilon_c$)

$$\Delta(\epsilon) = -N(0)t(\epsilon_c)\int_{-\epsilon_c}^{\epsilon_c} \frac{\Delta(\epsilon')}{2E(\epsilon')}\tanh\frac{\beta E(\epsilon')}{2}d\epsilon' \tag{5.F.6}$$

As noted in the text, this is identical to the gap equation obtained for the simple contact potential (5.2.20), with the replacement $V_0 \to t(\epsilon_c)$, so that all the analysis of Sections 5.4 and 5.5 goes through unchanged.

It remains to verify that the expressions for the critical temperature T_c, etc. are independent of the (arbitrary) choice of a cutoff ϵ_c. To do this it is simplest to iterate Eqn. (5.F.3), i.e. to define a $t(\epsilon_c')$ ($\epsilon_c' < \epsilon_c$) by

$$\hat{t}(\epsilon_c') \equiv \left(1+\hat{t}(\epsilon_c)\hat{K}\hat{P}_{12}\right)^{-1}\hat{t}(\epsilon_c) \tag{5.F.7}$$

where $\hat{P}_{12}$ is a projector onto the interval $\epsilon_c' < |\epsilon| < \epsilon_c$. Since $\hat{t}(\epsilon_c)$ is a constant $t(\epsilon_c)$ as a function of its arguments ϵ, ϵ' for all $|\epsilon|$, $|\epsilon'| < \epsilon_c$, Eqn. (5.F.7) reduces to the simple algebraic relation

$$t(\epsilon_c') = \left(1+N(0)t(\epsilon_c)\ln\frac{\epsilon_c}{\epsilon_c'}\right)^{-1} t(\epsilon_c) \tag{5.F.8}$$

which implies that (note $t < 0$!)

$$[N(0)|t(\epsilon_c)|]^{-1} = \ln\frac{\epsilon_c}{\epsilon_0} \tag{5.F.9}$$

where ϵ_0 is some constant. Substituting (5.F.9) into (5.8.8) of the text, we see that T_c is just $1.14\epsilon_0/k_B$, i.e. independent of ϵ_c. QED.

6

Superfluid ^{3}He

Like its boson cousin ^{4}He, the fermion system ^{3}He is stable in liquid form down to zero temperature at low pressures (below $\sim$34 atm). Below 3 mK the liquid possesses, in addition to the normal phase, three different anomalous phases, of which two,[1] the A and B phases, are shown in Fig. 6.1. With the possible exception of systems showing the fractional quantum Hall affect, these low-temperature phases of liquid ^{3}He are probably the most sophisticated many-body systems of which we can today claim a detailed quantitative understanding. While many of their properties can be understood, at least qualitatively, in terms of a relatively simple generalization of the BCS theory of classic superconductivity, there is one major qualitative novelty: Unlike the Cooper pairs in classic superconductors, which are formed in an s-wave state and thus have a structure which at any given temperature and pressure is fixed, those formed in the anomalous phases of liquid ^{3}He are characterized by a nontrivial internal structure, and in particular by an *orientation*, in orbital and spin space, which can vary in space and time. It is this feature which gives rise to many of the spectacular effects seen in these phases.

The reader will notice that in the above paragraph I have avoided using the word "superfluid." In fact, while this term has been used to identify the anomalous phases of liquid ^{3}He ever since their original discovery, its application has been based more on theoretical presumption than on experimental phenomenology: while the experimental evidence for "superfluidity" in all these anomalous phases is by now convincing, the phenomenon is much less strikingly manifest than in ^{4}He-II. I will discuss this subject briefly in Section 6.5 below; in the meantime, I shall revert to the usual convention in referring to the "superfluid" phases of ^{3}He.

6.1 The normal phase of liquid ^{3}He

Both the electronic structure and the "gross" (van der Waals type) interactions of ^{3}He atoms are identical to those for ^{4}He; the only difference is that the ^{3}He nucleus has spin $\frac{1}{2}$, which means, first, that a system of ^{3}He atoms obeys Fermi rather than Bose statistics, and secondly that, unlike ^{4}He, it can be probed by nuclear magnetic resonance (NMR). Indeed, above 3 mK the behavior of liquid ^{3}He is, at a qualitative level, similar to that of the normal (He-I) phase of its heavier cousin (or of any other

[1] The third (A_1) phase is stable only in nonzero magnetic field and will not be explicitly discussed in this chapter.

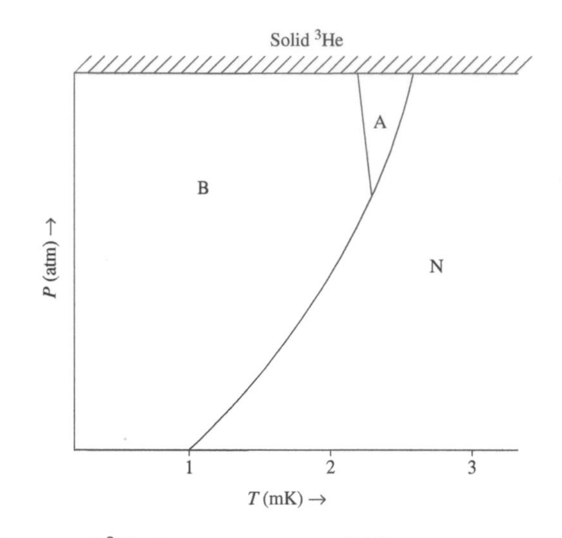

Fig. 6.1 Phase diagram of ^{3}He in zero magnetic field.

ordinary liquid); in particular, it shows no trace of superfluidity, nor of any anomalous NMR behavior.

A noninteracting gas with the density of liquid ^{3}He ($\sim 4 \times 10^{22}$ cm^{-3} at s.v.p.), would have, according to Eqn. (5.1.4), a Fermi wave vector of approximately 1 Å and thus, with the ^{3}He mass, a Fermi temperature of around 7 K; at temperatures well below this we expect, according to the considerations of Chapter 5, Section 5.1, the temperature-dependences of various experimental quantities to have characteristic forms (e.g. specific heat $c_V \sim T$, Pauli spin susceptibility $\chi \sim$ const., etc.). Liquid ^{3}He indeed shows this characteristic "degenerate Fermi gas" behavior, though it is only clearly manifested below $\sim$100 mK; moreover, the quantitative values of c_V, χ, etc. are different from those expected for a degenerate noninteracting gas. This behavior is nicely explained by the "Fermi-liquid" theory of Landau, which was originally developed specifically with respect to liquid ^{3}He and whose application to normal metals we have briefly touched on in Chapter 5, Section 5.1; for details see Appendix 5A. According to this theory, the low-lying states of the many-body system can be characterized by the excitation of fermionic "quasiparticles" (or "quasiholes") which are labeled, like the real atoms, by a momentum $\boldsymbol{p}$ and a spin σ. The net effect of the interatomic interactions is (a) to change the single-particle energy spectrum $\epsilon(\boldsymbol{p}) = p^2/2m$ to a quasiparticle spectrum $\epsilon(\boldsymbol{p}) \cong \epsilon_F + p_F(p - p_F)/m^*$, and (b) to give rise to a set of "molecular fields" which are activated by various kinds of macroscopic polarization of the system. As explained in Appendix 5A, it is conventional to characterize the strengths of these fields by dimensionless quantities F_l^s, F_l^a; the values of these quantities (which may be obtained from experiment) depend on pressure (but not appreciably on temperature, provided $T \ll T_F$) and can be quite large; in particular F_0^s (the molecular-field parameter associated with fluctuations of the total density) is $\sim$100 at pressures near the melting curve. By contrast, the parameter F_0^a (associated with spin polarization) is negative and varies relatively little with pressure (from ~ -0.7 at svp to ~ -0.75 at melting pressure).

As in the case of metals, going over to the Fermi-liquid description, while it effectively takes account of the bulk of the interatomic interactions, does not eliminate the latter entirely; in particular, the residual terms, corresponding to the quasiparticle–quasiparticle scattering processes, are necessary both to give sensible values for transport coefficients such as the thermal conductivity (which in the absence of real scattering processes would be infinite) and to allow the formation of Cooper pairs. We do not know a great deal a priori about these "residual" quasiparticle–quasiparticle interactions; in addition to the effect of the original interatomic van der Waals potential, other effective interactions may be induced by the process of converting a real particle into a Landau quasiparticle. A particularly important class of such terms is thought to be associated with the various molecular fields, and in particular with the one associated with spin polarization. The qualitative argument goes something like this (cf. Leggett 1975, Section IX.A): A single quasiparticle of spin $\boldsymbol{\sigma}$ at point $\boldsymbol{r}$ at time t will produce a molecular field $\boldsymbol{\mathcal{H}}(\boldsymbol{r}t)$ proportioned to $F_0^{\rm a}\boldsymbol{\sigma}$, which will then induce a collective spin polarization $\boldsymbol{S}(\boldsymbol{r}'t')$ at neighboring spacetime points $\boldsymbol{r}'t'$ given by

$$\boldsymbol{S}(r't') \sim \chi(\boldsymbol{r}-\boldsymbol{r}', t-t')\ \boldsymbol{\mathcal{H}}(rt) \sim \chi(\boldsymbol{r}-\boldsymbol{r}', t-t')\ F_0^{\rm a}\boldsymbol{\sigma} \tag{6.1.1}$$

This collective polarization will then produce its own molecular field $\boldsymbol{\mathcal{H}}(r't')$ proportional to $F_{\rm a}^0\boldsymbol{S}(r',t')$, and this will be felt by a second quasiparticle of spin $\boldsymbol{\sigma}'$ at $(\boldsymbol{r}',t')$ as a Zeeman energy $-\boldsymbol{\sigma}'\cdot\boldsymbol{\mathcal{H}}(r't')$. In this way, we generate, between two quasiparticles of spin $\boldsymbol{\sigma},\boldsymbol{\sigma}'$ at spacetime points $\boldsymbol{r},t$ and $\boldsymbol{r}',t'$ respectively a spin-dependent effective interaction of the form

$$V_{\rm eff}(\boldsymbol{r}-\boldsymbol{r}', t-t' : \boldsymbol{\sigma},\boldsymbol{\sigma}') \sim -(F_0^{\rm a})^2\chi(\boldsymbol{r}-\boldsymbol{r}', t-t')\boldsymbol{\sigma}\cdot\boldsymbol{\sigma}' \tag{6.1.2}$$

or in Fourier-transformed form

$$V_{\rm eff}(\boldsymbol{q},\omega : \boldsymbol{\sigma}\boldsymbol{\sigma}') \sim -(F_0^{\rm a})^2\chi(q\omega)\boldsymbol{\sigma}\cdot\boldsymbol{\sigma}' \tag{6.1.3}$$

This induced interaction is evidently somewhat analogous to the effective interaction induced between the electrons in metals by their interaction with the phonon field (in fact, it is often regarded as resulting from "exchange of spin fluctuations"). However, it differs from the latter in two important ways: (a) it is spin-dependent, and in fact in the limit $\omega = 0$, $\boldsymbol{q} \to 0$ it is always[2] attractive for parallel spins and repulsive for antiparallel ones (b) in the electron–phonon case the strength of the induced interaction depends only on properties such as the phonon frequencies which refer to a different system from the electrons themselves (and thus are very little affected by the onset of Cooper pairing in the electron system), in the case of liquid ^{3}He the system which is mediating the induced interaction (the quasiparticles) is the very same system between which that interaction operates. As we shall see in Section 6.3, this consideration provides a very natural explanation, at least at the qualitative level, of the phase diagram of superfluid ^{3}He.

[2]Since in this limit $\chi(\boldsymbol{q},\omega)$ must be positive to guarantee stability.

In the rest of this chapter I shall take the point of view that the low-energy effective Hamiltonian of liquid ^{3}He can be expressed in terms of *quasiparticle* operators $a^\dagger_{p\sigma}$, $a_{p\sigma}$, in the form

$$\hat{H} = \sum_{p\sigma} \epsilon(p) a^\dagger_{p\sigma} a_{p\sigma} + (\text{M.F.}) + \sum_{ijkl} V_{ijkl}\, a^\dagger_i a^\dagger_j a_k a_l \tag{6.1.4}$$

Here (M.F.) stands for the molecular-field terms, which when necessary will be taken into account by the phenomenological technique outlined in Appendix 5A, the quantity $\epsilon(p)$ has the approximate form

$$\epsilon(p) = \epsilon_{\text{F}} + v_{\text{F}}(p - p_{\text{F}}) \equiv \epsilon_{\text{F}} + \frac{p_{\text{F}}}{m^*}(p - p_{\text{F}}) \tag{6.1.5}$$

and in the last term the label i stands for $(\boldsymbol{p}, \sigma)$. The residual interaction V_{ijkl} will be taken to be generic within the constraints imposed by symmetry. Apart from the molecular-field terms, the Hamiltonian is thus in one–one correspondence with that of a weakly interacting Fermi gas, with $m \to m^*$. It is worth emphasizing that this approach (sometimes called the "superfluid Fermi liquid" description) makes sense only because the single-quasiparticle states principally involved in the Cooper-pairing transition have excitation energies $\sim k_{\text{B}} T_{\text{c}} \ll k_{\text{B}} T_{\text{F}}$ and are therefore close enough to the Fermi surface to be adequately described, in the normal phase, by Fermi-liquid type ideas. As we shall see in Chapter 7, this serendipitous state of affairs is not necessarily generic to superfluid Fermi systems; it is our good fortune that it is adequately realized for ^{3}He.

6.2 Anisotropic Cooper pairing

Apart from the direct electromagnetic interaction between the nuclear dipole moments, which even at the distance of closest approach of two atoms is only $\sim 10^{-7}$ K and hence might, at least at first sight, be expected to have negligible effect, the interaction energies between two quasiparticles in liquid ^{3}He arise at a fundamental level from the interplay of Coulomb and kinetic energies, and thus should be invariant under rotations in orbital and spin space separately. Under these circumstances it is natural to conjecture that any Cooper pairs which are formed in the liquid should correspond to a unique behavior under such rotations, i.e. should be characterized by a definite total spin S (=0 or 1) and to a definite value l of their relative angular momentum. As we saw in Chapter 5, for the pairs in classic superconductors the correct assignment is almost certainly $l = S = 0$ (s-wave spin singlet): what is it likely to be for liquid ^{3}He?

For the problem of two identical fermions of spin $\frac{1}{2}$ interacting in free space via a central potential, the energy eigenstates can indeed be classified by their l- and S-values, and the Pauli principle requires that the spin singlet (triplet) states have even (odd) values of l. Moreover, in the absence of explicit spin-dependence of the potential the ground state always corresponds to $l = S = 0$; the easiest way to see this is to note that the effective potential for radial motion always contains, for $l \neq 0$, the extra "centrifugal" term $\hbar^2 l(l+1)/mr^2$, and since there are no a priori restrictions on the relative radial motion the lowest-energy state in this effective potential must always lie higher than that for $l = 0$. For Cooper pairs formed from quasiparticles in a many-body system, the assertion that singlet (triplet) states correspond to even (odd) values of l should prima facie still be valid;[3] however, the conclusion that the ground state must be $l = S = 0$ fails, for two reasons. First, the process of transition from free particles to quasiparticles may introduce an explicit spin-dependence into the effective interquasiparticle interaction (and in fact, as we have seen in Section 6.1, the "spin-fluctuation-exchange-induced" interaction is likely to favor spin triplet states). Secondly, even for spin singlet pairing the above argument may fail, since by their nature the Cooper pairs in a "weak-coupling" system such as ^{3}He are formed from states near the Fermi surface and thus are constrained to have a relative momentum of order $2\hbar k_F$. Indeed, this consideration, coupled to the observation that the hard-core repulsion prevents two ^{3}He atoms from getting closer to one another than a distance $r_0 \sim 2.5$ Å, might suggest intuitively that the relative angular momentum l should be of order $k_F r_0$, i.e. perhaps 2 (this was in fact the majority guess among theorists prior to the experimental discovery of the superfluid phases), but certainly not zero.

On the basis of the above considerations, I will assume in this chapter that a *very schematic* description of the ground state of liquid ^{3}He is given by a generalized "completely paired" BCS wave function of the form (5.4.3), where the function $\phi(\boldsymbol{r}_1, \boldsymbol{r}_2, \sigma_1, \sigma_2)$ corresponds to a definite value of total spin S and of orbital angular momentum l, neither of which is necessarily zero (so that in particular we cannot necessarily make the assumption (5.4.5)). The implicit understanding is that the quantities $\boldsymbol{r}_j$ and $\boldsymbol{\sigma}_j$ refer to quasiparticles rather than real atoms, so that the description (5.4.3) does not really make sense on a spatial scale $\lesssim k_F^{-1}$ (at which the quasiparticles are not well-defined); however, it should be adequate on the scale of the pair radius ξ_0, which in ^{3}He as in classic superconductors turns out to be $\gg k_F^{-1}$.

Although it is almost certainly not explicitly relevant to the case of real superfluid ^{3}He, let us first briefly consider for orientation the possibility that the function $\phi(\boldsymbol{r}\boldsymbol{r}'\alpha\beta)$ of Eqn. (5.4.3) is a spin singlet, i.e. satisfies the condition (5.4.5). Then the whole analysis of Section 5.4 goes through unchanged up to and including Eqn. (5.4.22): we find a pair wave function $F(\boldsymbol{r})$ whose Fourier components F_k are given by

$$F_k = \frac{\Delta_{\boldsymbol{k}}}{2E_{\boldsymbol{k}}}, \quad E_{\boldsymbol{k}} \equiv (\epsilon_{\boldsymbol{k}}^2 + |\Delta_{\boldsymbol{k}}|^2) \tag{6.2.1}$$

where the quantity Δ_k is a solution of the gap equation (5.4.22), namely,

$$\Delta_{\boldsymbol{k}} = -\sum_{\boldsymbol{k}'} V_{\boldsymbol{k}\boldsymbol{k}'} \frac{\Delta_{\boldsymbol{k}'}}{2E_{\boldsymbol{k}'}} \tag{6.2.2}$$

with

$$V_{\boldsymbol{k}\boldsymbol{k}'} \equiv \langle \boldsymbol{k}\uparrow, -\boldsymbol{k}\downarrow |\hat{V}| \boldsymbol{k}'\uparrow, -\boldsymbol{k}'\downarrow\rangle \tag{6.2.3}$$

It is important to bear in mind that in general the quantity $F_{\boldsymbol{k}}$, and thus $\Delta_{\boldsymbol{k}}$, may be (nontrivially) complex, i.e. $\arg(\Delta_{\boldsymbol{k}}/\Delta_{\boldsymbol{k}'})$ is not necessarily zero.

[3]For completeness it should be noted that this conclusion can be avoided if the Cooper pairs are formed in such a way that the quasiparticles are never at the same point at the same time (the so-called "odd-frequency" scenario, see e.g. Abrahams et al. 1995). This hypothesis would require a generalization of the simple ansatz (5.4.6); it seems very unlikely to be relevant to real-life ^{3}He.

We now deviate from the discussion of Section 5.4 by allowing the pair interaction $V_{\boldsymbol{k}\boldsymbol{k}'}$ to have a form more general than the simple BCS form (5.2.20).[4] However, $V_{\boldsymbol{k}\boldsymbol{k}'}$ must still be invariant under simultaneous rotation of the vectors $\boldsymbol{k}$ and $\boldsymbol{k}'$, and thus can always be written as a sum of terms corresponding to given dependence on $\hat{\boldsymbol{k}} \cdot \hat{\boldsymbol{k}}'$:

$$V_{\boldsymbol{k}\boldsymbol{k}'} = \sum_k V_l(k,k')\; P_l(\hat{\boldsymbol{k}} \cdot \hat{\boldsymbol{k}}') \equiv \sum_l V_l(\epsilon_k, \epsilon_{k'}) P_l(\hat{\boldsymbol{k}} \cdot \hat{\boldsymbol{k}}') \tag{6.2.4}$$

where $P_l(\cos\theta)$ $(\cos\theta \equiv \hat{\boldsymbol{k}} \cdot \hat{\boldsymbol{k}}' +)$ are the usual Legendre polynomials. When we substitute (6.2.4) into the gap equation (6.2.2), the same difficulty regarding the high-energy divergence arises as in the s-wave case, but can be solved in a similar way: Provided that the scale of variation of the V_l with ϵ_k is large compared to T_c (a condition almost certainly well-satisfied in real-life ^{3}He) we can introduce a cutoff ϵ_c and replace $V_{kk'}$ by the "generalized BCS" form

$$V_{kk'} = -\sum_l \tilde{V}_l P_l(\cos\theta_{\boldsymbol{k}\boldsymbol{k}'}) \tag{6.2.5}$$

where

$$\tilde{V}_l \equiv \begin{cases} \text{const.} & |\epsilon_k|, |\epsilon_{k'}| \text{ both} < \epsilon_c \\ 0 & \text{otherwise} \end{cases} \tag{6.2.6}$$

When we substitute (6.2.5) in the gap equation (6.2.2) and take into account that a typical value of $|\Delta|$ will turn out to be $\sim k_B T_c$, and thus $\epsilon_c \gg |\Delta|$, we find that the terms corresponding to different l approximately decouple, and that the solution(s) with largest condensation energy correspond to that value of l (call it l_0) for which $\tilde{V}_l$ is largest (most attractive): to a good approximation, in fact, the solution(s) are of the general form[5]

$$\Delta(\boldsymbol{k}) = \Delta(\hat{\boldsymbol{k}}) = \sum_m a_m Y_{l_0 m}(\theta, \phi) \tag{6.2.7}$$

where θ and ϕ are the angles of $\hat{\boldsymbol{k}}$. In general, different choices of the a_m yield different condensation energies E_{cond}, and one of course expects the actual choice made by the system to correspond to the largest possible value of E_{cond}. However, in view of the rotational invariance of the interaction (6.2.5) one knows that if $\Delta(\hat{\boldsymbol{k}})$ is an (optimal) solution then so is $\Delta(R\hat{\boldsymbol{k}})$ where R is any rotation, and the same statement can be made about the Fourier components $F_{\boldsymbol{k}} \equiv \Delta_{\boldsymbol{k}}/2E_{\boldsymbol{k}}$ of the pair wave function $F(\boldsymbol{r})$ and hence about $F(\boldsymbol{r})$ itself. Thus, the "orientation" of the pair wave function is arbitrary: the pairs have a *nontrivial orientational (orbital) degree of freedom.* This is the fundamentally novel feature of the Cooper pairs in superfluid ^{3}He with respect to those in classic superconductors: we will explore it further after we have made the necessary extension to the spin degrees of freedom.

Let's then turn to the possibility of spin triplet pairing, which according to the considerations given above should be associated with an odd-parity (l = odd) orbital

[4]The classic reference is Anderson and Morel (1961).

[5]It is worth noting that for any nonzero value of l_0, any expression of the form (6.2.7) must have (in 3D) at least two nodes on the Fermi surface as a function of $\hat{\boldsymbol{k}}$.

state. The simplest case is that we can choose the spin axes so that the off-diagonal elements (in spin space) of the quantity $\phi(\boldsymbol{r}, \boldsymbol{r}' : \alpha\beta)$ (Eqn. 5.4.3) are zero, so that the $\boldsymbol{k}$-space second-quantized form of the many-body wave function is, analogously to (5.4.6),

$$\Psi_N = \mathcal{N}'' \left(\sum_{\boldsymbol{k}} (c_{\boldsymbol{k}\uparrow} a^\dagger_{\boldsymbol{k}\uparrow} a^\dagger_{-\boldsymbol{k}\uparrow} + c_{\boldsymbol{k}\downarrow} a^\dagger_{\boldsymbol{k}\downarrow} a^\dagger_{-\boldsymbol{k}\downarrow}) \right)^{N/2} |\text{vac}\rangle \tag{6.2.8}$$

(where we must take $c_{\boldsymbol{k}\sigma} \equiv -c_{-\boldsymbol{k}\sigma}$, but where in general for given $\boldsymbol{k}$ there is no particular relation between either the amplitudes or the phases of the complex coefficients $c_{\boldsymbol{k}\uparrow}$, $c_{\boldsymbol{k}\downarrow}$). A state of the form (6.2.8) is known as an "equal-spin-pairing" (ESP) state; we shall see that the usual identification of ^{3}He-A is as a special kind of ESP state. Note carefully that (6.2.8) is not equivalent to the state

$$\Psi^{(F)}_N = \mathcal{N}^{(F)} \left(\sum_{\boldsymbol{k}} c_{\boldsymbol{k}\uparrow} a^\dagger_{\boldsymbol{k}\uparrow} a^\dagger_{-\boldsymbol{k}\downarrow} \right)^{N/4} \left(\sum_{\boldsymbol{k}} c_{\boldsymbol{k}\downarrow} a^\dagger_{\boldsymbol{k}\downarrow} a^\dagger_{\boldsymbol{k}\downarrow} \right)^{N/4} |\text{vac}\rangle \tag{6.2.9}$$

In the language of Chapter 2, Section 2.3, and Appendix 2B, (6.2.8) is a "coherent" state as regards the spin degree of freedom, while (6.2.9) represents a "Fock" (fragmented) state. I return below to the question of why it is thought that the coherent state is a better representation of the physical system ^{3}He-A than the Fock one.

It is clear that in Eqn. (6.2.8) the spin degree of freedom enters on the same footing as the orbital label $\boldsymbol{k}$, and thus the analysis done above for the singlet case is straightforwardly generalized: for example, instead of a single pair amplitude $F_{\boldsymbol{k}}$ we introduce independent amplitudes $F_{\boldsymbol{k}\sigma} \equiv c_{\boldsymbol{k}\sigma}/(1 + |c_{\boldsymbol{k}\sigma}|^2)$ for the "up" ($\sigma = +1$) and "down" ($\sigma = -1$) pairs (and correspondingly "gaps" $\Delta_{\boldsymbol{k}\sigma}$ related to $F_{\boldsymbol{k}\sigma}$ by $\Delta_{\boldsymbol{k}\sigma} \equiv 2E_{\boldsymbol{k}\sigma}F_{\boldsymbol{k}\sigma}$, $E_{\boldsymbol{k}\sigma} \equiv (\epsilon^2_{\boldsymbol{k}} + |\Delta_{\boldsymbol{k}\sigma}|^2)^{1/2}$), and the coordinate-space wave function may be written in the schematic form

$$F(\boldsymbol{r} : \alpha\beta) = F_\uparrow(\boldsymbol{r})|\uparrow\uparrow\rangle + F_\downarrow(\boldsymbol{r})|\downarrow\downarrow\rangle \tag{6.2.10}$$

Moreover, to the extent that the Hamiltonian is invariant under rotation of the spins alone, any matrix elements of the form $\langle \boldsymbol{k}\uparrow, -\boldsymbol{k}\uparrow |V| \boldsymbol{k}'\downarrow, -\boldsymbol{k}'\downarrow\rangle$ must be zero; thus the "up-spin" and "down-spin" gaps $\Delta_{\boldsymbol{k}\uparrow}, \Delta_{\boldsymbol{k}\downarrow}$ are not coupled together, and each separately satisfies an equation of the form (6.2.2) (with $V_{\boldsymbol{k}\boldsymbol{k}'} \equiv \langle \boldsymbol{k}\sigma, -\boldsymbol{k}\sigma|V|\boldsymbol{k}'\sigma, -\boldsymbol{k}'\sigma\rangle$ independent of σ). Note that just as in the spin singlet case, each $\Delta_{\boldsymbol{k}\sigma}$ must have at least two nodes as a function of $\hat{\boldsymbol{k}}$ (since l_0 can now certainly not be zero).

As in the anisotropic singlet case, we can argue that if $\Delta_{\boldsymbol{k}\sigma} \equiv \Delta_\sigma(\boldsymbol{k})$ is an (optimal) solution of the gap equation, then so is $\Delta_\sigma(\hat{R}_\sigma \boldsymbol{k})$, where $\hat{R}_\sigma$ denotes an arbitrary rotation of the orbital coordinates, which may be *different* for the two spin populations. Actually, the degeneracy goes beyond this, since there is nothing in the (rotationally invariant) Hamiltonian to tell us in which direction to choose our spin axes. Thus, in the present simple model (based on a generalized BCS ansatz of the form (5.4.3)), the ground state possesses a very high degree of degeneracy; we shall see below that in more general models some but not all of the degeneracy is removed.

Before going further, let us digress for a moment to discuss the spin susceptibility. Imagine an ESP state with a given (not necessarily unique) pairing axis which we choose as the z-axis. Consider the effect of a weak magnetic field applied along the z-axis. In the normal phase the effect is simply to shift the up-spin Fermi surface relative to the down-spin one, so that $\epsilon(k_{\mathrm{F}}, \sigma) = \epsilon_{\mathrm{F}0} - \mu\sigma H$ where $\epsilon_{\mathrm{F}0}$ is the zero-field Fermi energy. If now we start from this state, we can define $\epsilon_{\boldsymbol{k}\sigma}$ to be the energy of a given single-particle state $(\boldsymbol{k}, \sigma)$ relative to the appropriate Fermi energy $\epsilon(k_{\mathrm{F}}, \sigma)$, and since we pair only parallel spins the argument leading to the gap equations analogous to (6.2.2) goes through word for word, with the field nowhere entering the expressions explicitly. Thus, to the extent that we can neglect the energy-dependence of the single-particle density of states near the Fermi energy[6] (a fairly good approximation for real-life liquid ^{3}He) the superfluid condensation energy is independent of the magnetic field and thus the susceptibility is unchanged from the normal-state value. It is clear that this conclusion holds wherever the field is along an axis with respect to which the pairing is of ESP nature (no antiparallel spins paired); for other axes we should expect the susceptibility to be reduced, similarly to what happens in the spin singlet case.

Let's now consider the most general BCS-type ansatz corresponding to triplet pairing: that is, the many-body ground state is assumed to have the form of Eqn. (5.4.3), which after Fourier transformation reads

$$\Psi_N = \mathcal{N}'' \left(\sum_{\boldsymbol{k}\alpha\beta} c_{\boldsymbol{k}\alpha\beta} a^{\dagger}_{\boldsymbol{k}\alpha} a^{\dagger}_{-\boldsymbol{k}\beta} \right)^{N/2} |\mathrm{vac}\rangle \tag{6.2.11}$$

where the coefficients $c_{\boldsymbol{k}\alpha\beta}$ are odd in $\boldsymbol{k}$ and symmetric under the interchange $\alpha \leftrightarrow \beta$; in general it is not possible (as it was for the special case of an ESP state) to find a single choice of spin axes which makes the off-diagonal elements $c_{\boldsymbol{k}\alpha\beta} \equiv c_{\alpha\beta}(\boldsymbol{k})$ zero for all $\boldsymbol{k}$ simultaneously.

The simplest procedure for dealing with states of the general form (6.2.11) is probably the following: For any given value of $\boldsymbol{k}$, since the matrix $c_{\alpha\beta}(\hat{\boldsymbol{k}})$ is symmetric,[7] it is always possible to find a choice of spin axes (which may not be unique) which will diagonalize it, and proceed to define quantities such as $F_{\boldsymbol{k}\sigma}$ and $\Delta_{\boldsymbol{k}\sigma}$ just as in the ESP case (with $\Delta_{\boldsymbol{k}\sigma} \equiv 2E_{\boldsymbol{k}\sigma}F_{\boldsymbol{k}\sigma}, E_{\boldsymbol{k}\sigma} \equiv (\epsilon_{\boldsymbol{k}}^2 + |\Delta_{\boldsymbol{k}\sigma}|^2)^{1/2}$). We can further write the deviation of the single-particle occupation numbers, $\delta\langle n_{\boldsymbol{k}\sigma}\rangle$, from their normal-state values in this basis in the form (an obvious generalization of (5.4.9))

$$\delta\langle n_{\boldsymbol{k}\sigma}\rangle = \tfrac{1}{2}\left\{1 - \sqrt{1 - 4|F_{\boldsymbol{k}\sigma}|^2}\right\} \mathrm{sgn}(\epsilon_{\boldsymbol{k}}) \tag{6.2.12}$$

so that the total kinetic energy is given by an expression which is the obvious generalization of that in the s-wave case. However, in this basis (with a $\boldsymbol{k}$-dependent choice of spin axes) the form of the potential energy is rather unwieldy. It is therefore convenient to generalize the notation so as to use a ***k*-*independent*** choice of spin axes. When we do so, the quantities $F_{\boldsymbol{k}\sigma}$ and $\Delta_{\boldsymbol{k}\sigma}$ become matrices $F_{\boldsymbol{k}\alpha\beta}$, etc., in spin space; the transformation properties of $F_{\boldsymbol{k}\sigma} \equiv \langle a^{\dagger}_{\boldsymbol{k}\alpha} a^{\dagger}_{-\boldsymbol{k}\beta}\rangle$ are just those of a simple Schrödinger wave function of two spin-$\frac{1}{2}$ particles, but in the general case the transformation properties of $\Delta_{\boldsymbol{k}\alpha\beta}$ are more complicated, and the resulting formalism is rather messy.

[6]And also a possible small effect associated with the handling of the cutoff ϵ_{c}.

[7]That it is not necessarily Hermitian is irrelevant, since $\arg c_{\alpha\beta}(\boldsymbol{k}) \equiv \arg c_{\beta\alpha}(\boldsymbol{k})$.

Fortunately, for many practical applications, in particular for real-life superfluid ^{3}He in zero or low magnetic field, it turns out to be adequate to confine ourselves to a particular subclass of pairing states, the so-called "unitary" states, which are defined[8] by the statement that for any given $\boldsymbol{k}$ the quantity $|F_{\boldsymbol{k}\sigma}|^2$ is independent of σ, so that when regarded as a matrix the excitation energy $E_{\boldsymbol{k}}$ is just proportional to the unit matrix. In that case we can write in a general spin coordinate system

$$F_{\boldsymbol{k}\alpha\beta} \equiv \frac{\Delta_{\boldsymbol{k}\alpha\beta}}{2E_{\boldsymbol{k}}} \tag{6.2.13}$$

$$E_{\boldsymbol{k}} \equiv (\epsilon_{\boldsymbol{k}}^2 + |\Delta_{\boldsymbol{k}}|^2)^{1/2} \tag{6.2.14}$$

where the quantity $|\Delta_{\boldsymbol{k}}|^2$ is defined by

$$|\Delta_{\boldsymbol{k}}|^2 \equiv \sum_{\beta} |\Delta_{\boldsymbol{k},\alpha\beta}|^2 \qquad (= \text{ind. of } \alpha) \tag{6.2.15}$$

With a similar definition of $|F_{\boldsymbol{k}}|^2$, the generalization of the s-wave expression for the kinetic energy $\langle T \rangle$ is simply

$$\langle T \rangle = \sum_{\boldsymbol{k}} |\epsilon_{\boldsymbol{k}}| \left(1 - \sqrt{1 - 4|F_{\boldsymbol{k}}|^2}\right) \tag{6.2.16}$$

while the expression for the pairing terms in the potential energy is (remember, we are assuming invariance under spin rotation!)

$$\langle V \rangle = \sum_{\boldsymbol{k}\boldsymbol{k}'} \sum_{\alpha\beta} V_{\boldsymbol{k}\boldsymbol{k}'} F_{\boldsymbol{k}\alpha\beta} F^*_{\boldsymbol{k}'\beta\alpha} \tag{6.2.17}$$

Differentiation of the sum of $\langle T \rangle$ and $\langle V \rangle$ with respect to $F^*_{\boldsymbol{k}\beta\alpha}$ leads to the (set of) gap equation(s)

$$\Delta_{\boldsymbol{k},\alpha\beta} = -\sum_{\boldsymbol{k}'} V_{\boldsymbol{k}\boldsymbol{k}'} \frac{\Delta_{\boldsymbol{k}',\alpha\beta}}{2E_{\boldsymbol{k}'}} \tag{6.2.18}$$

By an analysis similar to that conducted above for the spin singlet case, the optimal solutions should be of the approximate form

$$\Delta_{\boldsymbol{k},\alpha\beta} = \sum_{m} a_{m,\alpha\beta} Y_{l_0 m}(\hat{\boldsymbol{k}}) \tag{6.2.19}$$

where l_0 is the (odd) l-value corresponding to the most attractive component of the pairing potential $V_{kk'}$. However, there is no general requirement that the $a_{m,\alpha\beta}$ have to be the same for different α, β, and as a result the quantity $|\Delta_{\boldsymbol{k}}|$, which according to (6.2.14) is the minimum value of the excitation energy E_k, need not have any nodes[9] as a function of $\hat{\boldsymbol{k}}$.

To conclude this section, let's introduce a very convenient alternative notation for describing (unitary[10]) spin triplet states. Consider the pair wave function $F(\boldsymbol{r} : \alpha\beta)$, or,

[8]Note that the "ESP" property is neither necessary nor sufficient for unitarity (e.g. the A_1 state is ESP but not unitary, while the BW phase is unitary but not ESP).

[9]For $l = 1$. For $l \geq 3$ there must be nodes (in 3D).

[10]It can also be used to describe nonunitary states, but the vector $\boldsymbol{d}$ is then complex.

better, its Fourier transform $F_{\alpha\beta}(\boldsymbol{k})$. We concentrate on a given value of $\boldsymbol{k}$ and find a set of spin axes in which the matrix $F_{\boldsymbol{k}}$ is diagonal; then, by the definition of a unitary state, $|F_{\boldsymbol{k}}|^2$ is proportional to the unit matrix, so we can write schematically

$$F_{\alpha\beta}(\boldsymbol{k}) = F_\uparrow(\boldsymbol{k})|\uparrow\uparrow\rangle + F_\downarrow(\boldsymbol{k})|\downarrow\downarrow\rangle, \quad |F_\uparrow(\boldsymbol{k})| = |F_\downarrow(\boldsymbol{k})| \equiv |F_{\boldsymbol{k}}| \tag{6.2.20}$$

Now, any spin wave function of two spin-$\frac{1}{2}$ particles of the form (6.2.20) may easily be shown to be equivalent to the state with $S = 1$ and $\boldsymbol{S}\cdot\hat{\boldsymbol{d}} = 0$ where $\hat{\boldsymbol{d}}$ is a real unit vector in the plane perpendicular to the chosen z-axis, and making (with the usual conventions) an angle $\theta = \frac{1}{2}\arg(F_\uparrow/F_\downarrow)$ with the y-axis. Consequently, any state of the form (6.2.20) can be parametrized by a vector $\boldsymbol{d}(\boldsymbol{k})$ which is real up to an overall phase, i.e. $\boldsymbol{d}(\boldsymbol{k}) \times \boldsymbol{d}^*(\boldsymbol{k}) = 0$, with direction along $\hat{\boldsymbol{d}}$ and magnitude equal to $|F_{\boldsymbol{k}}|$.[11] Since for a unitary state the quantity $\Delta_{\boldsymbol{k},\alpha\beta}$ is simply proportional to $F_{\boldsymbol{k},\alpha\beta}$, it can be similarly parametrized if necessary. I just anticipate at this point the result that the usual identification of the A phase of liquid ^{3}He is with the $l = 1$ state known as the ABM state and having a $\hat{\boldsymbol{d}}(\boldsymbol{k})$ given by

$$\hat{\boldsymbol{d}}(\hat{\boldsymbol{k}}) = \hat{\boldsymbol{d}} f(\hat{\boldsymbol{k}}), \quad f(\hat{\boldsymbol{k}}) = \sin\theta e^{i\phi} \tag{6.2.21}$$

(so that it is ESP), while the B phase is identified with the BW state

$$\hat{\boldsymbol{d}}(\hat{\boldsymbol{k}}) = \text{const. } \hat{\boldsymbol{k}} \tag{6.2.22}$$

Further discussion of these states is given below, but we note here that the BW phase has a gap $|\Delta(\hat{\boldsymbol{k}})|$ which is independent of $\hat{\boldsymbol{k}}$, i.e. isotropic, while the ABM-state gap is proportional to $|\sin\theta|$ and thus has two nodes, at $\theta = 0$ and π. Also, the ABM state, being ESP, has a susceptibility equal to the N-state value, on the other hand it is clear that for the BW state it is impossible to find any axis which is perpendicular to $\hat{\boldsymbol{d}}(\hat{\boldsymbol{n}})$ for all $\hat{\boldsymbol{n}}$, so that it is not ESP and must have a reduced[12] susceptibility. The apparent "angular momentum" of the pairs in the ABM state is discussed in Appendix 6.A.

6.3 Generalized Ginzburg–Landau approach: spin fluctuation feedback

The generalization of BCS theory to Cooper pairs formed in a state with l and possibly S nonzero given in the last section, which was essentially the theory of "superfluid ^{3}He" before the experimental discovery of the latter in 1972, ran into a major difficulty as soon as the phase diagram was established: As we shall see in a moment, for the case of $l = 1$ pairing this theory predicts unambiguously that the BW phase is always the most stable solution of the gap equation, independently of pressure or temperature, while the experimental phase diagram, even in zero external field, unambiguously shows two different anomalous phases, the A and B phases.

[11] A formal expression for $\boldsymbol{d}(\boldsymbol{k})$ is given in Eqn. (6.3.4) below.

[12] For any given region of the Fermi surface the state is "antiparallel-spin" as regards the axis parallel to $\boldsymbol{d}(\hat{\boldsymbol{n}})$ and ESP-like for the two axes perpendicular to $\boldsymbol{d}(\hat{\boldsymbol{n}})$. It is therefore plausible to argue (and true, though it of course requires a more rigorous argument to prove it) that the overall susceptibility is isotropic and equal to $\frac{2}{3}$ of the N-state value in the absence of molecular-field effects (which in ^{3}He$_B$ reduce it by a further factor of $\sim$2).

Faced with this difficulty, we realize that just as the phenomenon of BEC, as seen for example in ^{4}He-II, is much more general than its realization in the noninteracting Bose gas, so the phenomenon of Cooper pairing ("pseudo-BEC") may occur under circumstances much more general than can be described by the simple ansatz (5.4.3) for the many-body wave function (even when we interpret the arguments as referring to Landau quasiparticles). In fact, according to the considerations given in Chapter 2, Section 2.4, all that is essential to the Cooper pairing phenomenon is the existence of a single macroscopic ($O(N)$) eigenvalue of the two-particle density matrix. In the following I shall assume that in the superfluid phases of liquid ^{3}He such a unique macroscopic eigenvalue N_0 exists and is associated with an eigenfunction[13] ("order parameter")

$$\langle\psi^\dagger_\alpha(\boldsymbol{r})\psi^\dagger_\beta(\boldsymbol{r}')\rangle \equiv F(\boldsymbol{r},\boldsymbol{r}' : \alpha\beta) \tag{6.3.1}$$

with the "anomalous average" denoted by the pointed brackets having its usual sense (see Appendix 5C).[14] Perhaps surprisingly, it turns out that simply by studying the possible spin- and orbital-space structure of the order parameter (6.3.1) and exploiting the invariance properties of the Hamiltonian, one can obtain some significant generic results, whose validity is independent of that of the ansatz (5.4.3). In the following I shall mostly confine myself for pedagogical simplicity to the case, usually believed to be realized in real-life superfluid ^{3}He, of $l = 1$ pairing, and to the "GL" temperature regime close to T_c where the order parameter is "small," although the results are actually much more general.

We consider a spatially uniform situation, so that F is a function of the relative coordinate $\boldsymbol{r} - \boldsymbol{r}'$, and take the Fourier transform with respect to this variable, which yields the quantity

$$F_{\alpha\beta}(\boldsymbol{k}) \equiv \left\langle a^\dagger_{\boldsymbol{k}\alpha} a^\dagger_{-\boldsymbol{k}\beta} \right\rangle \tag{6.3.2}$$

It is convenient to write $\hat{\boldsymbol{n}} \equiv \boldsymbol{k}/|\boldsymbol{k}|$ and integrate (6.3.2) over the magnitude $|\boldsymbol{k}|$, thus defining[15]

$$F_{\alpha\beta}(\hat{\boldsymbol{n}}) \equiv \int d|\boldsymbol{k}| F_{\alpha\beta}(\boldsymbol{k}) \tag{6.3.3}$$

Finally, we introduce as in the last section a spin-space vector $\boldsymbol{d}(\hat{\boldsymbol{n}})$ by

$$\boldsymbol{d}_i(\hat{\boldsymbol{n}}) \equiv -i \sum_{\alpha\beta} (\sigma_2\sigma_i)_{\beta\alpha} F_{\alpha\beta}(\hat{\boldsymbol{n}}) \tag{6.3.4}$$

The quantity (6.3.4) forms an adequate and convenient description, for our purposes, of the structure of the Cooper pairs. It is a generalization of the complex scalar order parameter Ψ which we used in the case of $l = 0$ pairing (see Chapter 5, Section 5.7).

[13]The definition of F given in (6.3.1) does not, strictly speaking, reduce to that used in the last section under the conditions assumed there, since the operators which occur in (6.3.1) create real ^{3}He atoms rather than Landau quasiparticles; however, from the point of view of the symmetry arguments which are the subject of this section this difference is unimportant.

[14]The normalization is such that $\sum_{\alpha\beta} \iint d\boldsymbol{r} d\boldsymbol{r}' |F|^2 = N_0$.

[15]The normalization is somewhat arbitrary, e.g. we could as well integrate over $d\epsilon_k$.

Now, proceeding as in the $l = 0$ case, we express the free energy F of the system as a functional of $\boldsymbol{d}(\hat{\boldsymbol{n}})$, and assume[16] that this functional is analytic. We also constrain the form of the functional by the consideration that F must be invariant not only, as in the $l = 0$ case, under global gauge transformations, but also under rotations of the spin and orbital coordinate systems separately, i.e. we must have for any given temperature T

$$F\{\boldsymbol{d}(\hat{\boldsymbol{n}}) : T\} = F\{\hat{R}_{\text{spin}}\boldsymbol{d}(\hat{\boldsymbol{n}}) : T\} = F\{\boldsymbol{d}(\hat{R}_{\text{orb}}(\hat{\boldsymbol{n}})) : T\} \tag{6.3.5}$$

(where $\hat{R}_{\text{spin}}$ indicates a rotation of the vector $\boldsymbol{d}$ in spin space, etc.)

The conditions specified in the last paragraph actually constrain the possible forms of $F\{\boldsymbol{d}(\hat{\boldsymbol{n}})\}$ quite severely. Gauge invariance alone forbids terms of odd order, so if we omit the $\boldsymbol{d}$-independent term, which is clearly irrelevant to the present discussion, the lowest-order term is the second-order one; provided $\boldsymbol{d}(\boldsymbol{n})$ corresponds to a unique value of l (cf. below), the above conditions constrain this to have the unique form

$$F_2\{\boldsymbol{d}(\hat{\boldsymbol{n}}) : T\} = \alpha(T) \int |\boldsymbol{d}(\hat{\boldsymbol{n}})|^2 \frac{d\Omega}{4\pi} \tag{6.3.6}$$

As in Chapter 5, Section 5.7, we identify the transition temperature T_c with the temperature at which the coefficient $\alpha(T)$ changes sign (note that this is, trivially, independent of the form of $\boldsymbol{d}(\hat{\boldsymbol{n}})$!) and expand $\alpha(T)$ around T_c:

$$\alpha(T) \cong \alpha_0(T - T_c). \tag{6.3.7}$$

We now consider the fourth-order term. In contrast to the $l = 0$ case of Chapter 5, Section 5.7, where the symmetry allowed only a term of the form $\beta|\Psi|^4$, there are now several allowed terms even in the (simplest) $l = 1$ case. Before discussing the general case, let us return for a moment to the simple BCS-type theory of the last section. The salient point is that in that theory the pairing-potential terms have the simple form

$$\langle V \rangle = - \iint \frac{d\Omega}{4\pi} \frac{d\Omega'}{4\pi} V(\hat{\boldsymbol{n}} \cdot \hat{\boldsymbol{n}}')\boldsymbol{d}(\hat{\boldsymbol{n}}) \cdot \boldsymbol{d}(\hat{\boldsymbol{n}}') \tag{6.3.8}$$

where the quantity $V(\hat{\boldsymbol{n}} \cdot \hat{\boldsymbol{n}}')$ is by hypothesis not itself a functional of $\boldsymbol{d}(\boldsymbol{n})$. Moreover, the kinetic-energy and entropy terms in F are sums of terms arising from the various pairs $(\boldsymbol{k}, -\boldsymbol{k})$ separately, and moreover are functions only of the quantities[17] $\epsilon_{\boldsymbol{k}}$ and $E_{\boldsymbol{k}} \equiv (\epsilon_{\boldsymbol{k}}^2 + |\Delta(\boldsymbol{k})|^2)^{1/2}$; and since (at least to order $(\ln \epsilon_c/\Delta)^{-1}$) the quantity $|\Delta(\boldsymbol{k})|^2$ is uniquely related to $|\boldsymbol{d}(\hat{\boldsymbol{n}})|^2$ where $\boldsymbol{d}(\hat{\boldsymbol{n}})$ is the quantity defined above, we have for those terms

$$\begin{aligned}\langle K \rangle - TS &= \text{const.} + \int \Phi[|\boldsymbol{d}(\hat{\boldsymbol{n}})|^2 : T] \frac{d\Omega}{4\pi} \\ &= \text{const.} + A^{-1}(T) \int |\boldsymbol{d}(\hat{\boldsymbol{n}})|^2 \frac{d\Omega}{4\pi} + \frac{1}{2} B(T) A^{-4}(T) \int |\boldsymbol{d}(\hat{\boldsymbol{n}})|^4 \frac{d\Omega}{4\pi} + O(d^6)\end{aligned} \tag{6.3.9}$$

where $A(T)$ and $B(T)$ are the same quantities as defined for the s-wave case (by Eqns. 5.E.8a and 5.E.8b) respectively.

[16] As in the s-wave case, this assumption is not completely obvious a priori.

[17] To avoid cluttering up the notation I discuss explicitly the unitary case, but it is easily seen that the conclusion does not depend on this.

Equations (6.3.8) and (6.3.9) are general within a BCS-type theory: in deriving them we have not used the assumption that $\boldsymbol{d}(\boldsymbol{n})$ corresponds to a single (odd) value of l. However, it is easily seen that unless two or more harmonics V_l of $V(\hat{\boldsymbol{n}}\cdot\hat{\boldsymbol{n}}')$ are closely degenerate, the optimal value of the second-order terms F_2 in $F \equiv \langle K\rangle - TS + \langle V\rangle$ is obtained by indeed choosing $\boldsymbol{d}(\boldsymbol{n})$ to be composed entirely of harmonics $Y_{l_0 m}$ with l_0 the value of l for which V_l is maximal, and that with that choice the expression for F_2 reduces to (6.3.6), with $\alpha(T) \equiv A^{-1}(T) - V_l \cong \alpha_o(T - T_c), \alpha_o \equiv dA^{-1}/dT)_{T_c}$ as in the s-wave case. Thus, using (6.3.7), we see that the free energy is of the form (apart from a d-independent constant)

$$F\{\boldsymbol{d}(\boldsymbol{n}):T\} = \alpha_0(T-T_c)\int |\boldsymbol{d}(\boldsymbol{n})|^2 \frac{d\Omega}{4\pi} + \frac{1}{2}\beta\int |\boldsymbol{d}(\boldsymbol{n})|^4\frac{d\Omega}{4\pi} + O(d^6)$$

$$\equiv \alpha_0(T-T_c)\langle|\boldsymbol{d}|^2\rangle + \frac{1}{2}\beta\langle|\boldsymbol{d}|^4\rangle + O(d^6) \quad (\beta > 0) \qquad (6.3.10)$$

From Eqn. (6.3.10) it is immediately clear that, given various possible choices of $\boldsymbol{d}(\boldsymbol{n})$ of the form

$$d_i(\hat{\boldsymbol{n}}) = \sum_m a_m^{(i)} Y_{l_0 m} \qquad (6.3.11)$$

the optimum choice is that which minimizes the value of $\langle|\boldsymbol{d}|^4\rangle$ for given $\langle|\boldsymbol{d}|^2\rangle$, i.e. which minimizes the fluctuations of $|\boldsymbol{d}|^2$ over the Fermi surface. For the case $l_0 = 1$ it is clear that the optimal choice is the BW state (6.2.22), since for this choice $|\boldsymbol{d}|^2$ is constant. Thus, we reach the previously announced conclusion that *in the generalized BCS theory of Section 6.2, a state of the BW form (6.2.22) is always the most stable.*[18]

Let us now return to the general case, but specialize to the case of p-wave pairing ($l = 1$). Then since any given component $d_\alpha(\hat{\boldsymbol{n}})$ of the vector $\boldsymbol{d}(\hat{\boldsymbol{n}})$ can be written in the form

$$\sum_{i=1}^{3} d_{\alpha i}\hat{n}_i,$$

the pairing state is completely characterized by the set of nine complex quantities $d_{\alpha i}$, and any fourth-order invariants must be expressible as fourth-order combinations of these, which are contracted with respect to both their spin and their orbital indices. For example,

$$I_1 = \Big(\sum_{\alpha i} |d_{\alpha i}|^2\Big)^2 \qquad (6.3.12)$$

$$I_2 = \sum_{\alpha\beta ij} d^*_{\alpha i} d_{\alpha j} d^*_{\beta i} d_{\beta j} \qquad (6.3.13)$$

[18]We have actually proved this only in the GL regime near T_c, but it is not difficult to extend the proof to general temperatures. On the irrelevance of the Landau molecular fields, see Appendix 5A.

etc. It turns out that there are five such invariant combinations, and so the general form of the GL free energy is written in this notation, apart from a d-independent term

$$F\{d_{\alpha i}, T\} = \alpha_0(T - T_c) \sum_{i\alpha} |d_{i\alpha}|^2 + \frac{1}{2} \sum_{s=1}^{5} \beta_s I_s\{d_{\alpha i}\} \tag{6.3.14}$$

Apart from the constraint imposed by stability (no form of $d_{\alpha i}$ should be able to make the fourth-order term negative) there are no a priori constraints on the coefficients β_s in the general case. Evidently, for any given values of the β_s's the thermodynamically stable form of the $d_{\alpha i}$ (i.e. of $\boldsymbol{d}(\hat{\boldsymbol{n}})$) will be that choice which minimizes the fourth-order term for a given value of the "normalization" $\sum_{i\alpha} |d_{i\alpha}|^2$.

If no constraints at all are imposed on the $d_{\alpha i}$, the problem of finding the choice of them which minimizes the fourth-order term in (6.3.14) is quite messy. However, it simplifies greatly if we impose the constraint of unitarity (which in this notation is $\epsilon_{\alpha\beta\gamma} d^*_{\alpha i} d_{\beta j} = 0$, $\forall i, j, \gamma$); while there is no particular a priori reason to do this, all the evidence is that the states of ^{3}He which are realized in zero magnetic field are in fact unitary, so from a heuristic point of view it is not unreasonable. Within the space of unitary states it is found (see, e.g. Leggett 1975, Section IX B) that there are only four states which can be extrema of the free energy, and one of these (the so-called "axial" or "2D" state) can never be an absolute minimum for any set of values of the β_s's. The remaining three are the BW (or "isotropic") state (6.2.22), the ABM state (6.2.21), and the "polar" state, which is also given by formula (6.2.21) but with $f(\boldsymbol{k}) = \cos\theta$; each of these corresponds to the absolute minimum of the free energy for a region of the space of the β_s's.

Of these three possibly stable phases, only the BW phase has a spin susceptibility which is reduced from the normal-state value, so it seems natural to identify it with the experimentally observed B phase; at this stage the polar and ABM phases (both ESP, thus with unchanged susceptibility from the N phase) are candidates for the experimental A phase (we will see later that the NMR evidence unambiguously favors ABM).

The above considerations, however, are purely formal. What *physical* mechanism could favor an ESP-type state over the BW state (which as we have seen is always stable in the weak-coupling approximation)? The following beautiful idea is due to Anderson and Brinkman: As we saw in Section 6.2, there is a contribution to the effective interaction between Landau quasiparticles from virtual spin polarization of the medium, which is attractive for parallel spins and proportional to the spin susceptibility of the medium, $\chi(\boldsymbol{r} - \boldsymbol{r}', t - t')$. This effective interaction is reminiscent of the electron–electron attraction in classic superconductors which is mediated by charge polarization of the ionic background, with however the crucial difference that in the ^{3}He case the medium which is being polarized is precisely the medium from which the Cooper pairs are themselves being formed. Thus, in general the formation of Cooper pairs modifies the spin susceptibility $\chi(\boldsymbol{r} - \boldsymbol{r}', t - t')$ and hence modifies the very interaction which helped to form them – the "spin fluctuation feedback" mechanism.

Needless to say, to obtain quantitative results a concrete microscopic calculation is needed at this point: see Brinkman et al. (1974). However, it is quite easy to see qualitatively why the spin fluctuation mechanism is likely to favor an ESP-type state, by looking at the behavior of the static uniform spin susceptibility tensor χ_{ij} in the various states. As we have seen, in the BW (isotropic) state χ_{ij} remains isotropic and all its components are reduced from the N-state value, thereby weakening the spin-fluctuation-induced attraction. By contrast, in an ESP state, while the components of χ perpendicular to the pairing axis are (or may be) reduced, that along the pairing axis stays at the N-state value. Since it is just this component which gives rise to the effective spin-fluctuation-induced interaction for the formation of pairs oriented along this axis, the interaction is not reduced from its N-state value. This argument as it stands would say that sufficiently strong spin fluctuation effects would tend to favor *any* ESP state over the BW state; that it is in fact the ABM state which is uniquely favored requires a more detailed argument,[19] as does of course the prediction of the precise region of the P-T plane over which the ABM phase is stable relative to the BW one; see, e.g. Brinkman et al. (1974). However, to summarize, theoretical considerations (as well as the NMR evidence to be discussed in the next section) seem entirely consistent with the by now almost universally accepted identification of the experimentally observed A phase of liquid ^{3}He with the ABM phase and of the B phase with the BW phase.

6.4 Spontaneously broken spin–orbit symmetry and spin dynamics

Perhaps the most spectacular consequence of the onset of Cooper pairing ("pseudo-BEC") in liquid ^{3}He – and certainly the one which was most manifest in the earliest experiments – is the ability to amplify ultraweak effects, in particular those of the nuclear dipole–dipole interaction. As we shall see in this section, this amplification leads to a uniquely rich behavior in nuclear magnetic resonance (NMR) experiments; it may also permit the detection of other kinds of ultraweak effects.

The direct interaction between the nuclear dipole moments of two nearby ^{3}He atoms is of purely electromagnetic origin and has the form

$$\hat{H}_{\mathrm{D}} = \frac{1}{2}\frac{\mu_0\mu_n^2}{4\pi}\sum_{ij} r_{ij}^{-3}\{\boldsymbol{\sigma}_i\cdot\boldsymbol{\sigma}_j - 3\,\boldsymbol{\sigma}_i\cdot\hat{\boldsymbol{r}}_{ij}\boldsymbol{\sigma}_j\cdot\hat{\boldsymbol{r}}_{ij}\} \quad \left(\hat{\boldsymbol{r}}_{ij}\equiv\frac{\boldsymbol{r}_{ij}}{|\boldsymbol{r}_{ij}|}\right) \tag{6.4.1}$$

where μ_n is the nuclear moment ($\sim$$-2{\cdot}1$ nuclear magnetons). Even at the distance of closest approach of two ^{3}He atoms (the hard-core radius, $\sim$2 Å) this interaction is only of order 10^{-7} K, several orders of magnitude smaller than the thermal energy at the temperatures at which the superfluid phases appear ($\sim$$10^{-3}$ K). However, it has one crucial property which distinguishes it from the much stronger forces (van der Waals, hard-core, "exchange") which we have taken into account so far: while it is of course invariant under simultaneous rotation of the spin and orbital coordinate systems, it is not invariant under rotation of either alone. Consequently, if the orbital coordinates are effectively fixed, the dipole interaction can exert a torque on the total spin of the system. However, in normal liquid ^{3}He this torque, while not identically zero, is

[19]Such an argument shows that the "optimal" states from the point of view of the spin-fluctuation mechanism are not merely ESP but have a uniform direction of $\boldsymbol{d}(\hat{\boldsymbol{n}})$; this favors the ABM and polar states over the axial one. The ABM state is already favored over the polar state in the weak-coupling approximation because of its smaller $\langle d^4\rangle/\langle d^2\rangle^2$ ratio.

very tiny indeed, and its only observable effect is to produce a nonzero spin relaxation rate T_1^{-1} which is of the order of inverse hours or even days. It was therefore a major surprise when the earliest NMR experiments on the A phase found that the resonance frequency, while remaining sharp, shifted sharply upwards from its normal-state (Larmor) value $\omega_L \equiv \gamma\mathcal{H}_{ext}$ ($\gamma \equiv \mu_n/\hbar$), apparently indicating the existence in that system of a strong extra torque in addition to that provided by the external magnetic field $\mathcal{H}_{ext}$.

Let's start by examining the problem from a rather general point of view, without necessarily for the moment making the assumption that the superfluid phases of ^{3}He are characterized by the formation of Cooper pairs. A simple sum-rule argument (see Leggett 1972) indicates that if the resonance frequency ω_{res} is unique (as it appears to be, experimentally, for ^{3}He-A) then its value should be given by the formula

$$\omega_{res}^2 = -\chi_0^{-1}\langle[\hat{S}_x, [\hat{S}_x, \hat{H}]]\rangle \equiv \chi_0^{-1}\frac{\partial^2\langle H\rangle}{\partial\theta_x^2} \tag{6.4.2}$$

In formula (6.4.2) $\hat{S}_x$ is the operator of the total x-component of spin of the system, χ_0 is the static spin susceptibility, and the quantity $\partial^2\langle H\rangle/\partial\theta_x^2$ is the second derivative of the energy when the spins (only) are uniformly rotated by an angle θ_x around the x-axis, keeping the orbital coordinates fixed. Now, the bulk of the terms in $\hat{H}$ (van der Waals, hard-core, "exchange", etc.) are, as we have already seen, invariant under such rotation and thus contribute nothing to the RHS of Eqn. (6.4.2); the only terms which are not so invariant are, first, the external-field (Zeeman) term $-\gamma S_z\mathcal{H}_{ext}$, which is easily seen to contribute precisely ω_L^2, and the dipole interaction $\hat{H}_D$. Thus, we obtain the result

$$\omega_{res}^2 - \omega_L^2 = \chi_0^{-1}\frac{\partial^2\langle H_D\rangle}{\partial\theta_x^2} \tag{6.4.3}$$

Barring pathologies, one expects the second derivative to be of order[20] $-\langle H_D\rangle$, so that the deviation of the observed (squared) resonance frequency from the Larmor value is a direct measure (at least up to a factor $\sim$1) of the equilibrium expectation value of the nuclear dipole interaction energy. From the experimental results we then read off that this quantity is unobservably small in the normal phase, but in the A phase acquires a surprisingly large (actually temperature-dependent) value, of order 10^{-3} erg/cm^3.

Now, if we consider two ^{3}He nuclei at their distance of closest approach, then as we have seen the nuclear dipole energy g_D is of order 10^{-7} K; evidently, as with a pair of macroscopic bar magnets, it favors (for parallel spins) the "end-to-end" configuration ($\boldsymbol{r}_{ij} \parallel \boldsymbol{\sigma}_i, \boldsymbol{\sigma}_j$) over the "'side-by-side" one. If we were to assume that the atoms paired off so that each nucleus was paired with a neighbor in the end-to-end configuration, then the dipole energy per unit volume would be of the order of g_D times the atomic density, i.e. $\sim$0.1 erg/cm^3, more than enough to account for the experimental results. However, the problem lies in the thermal disorder: Since the end-to-end configuration is favored over the side-by-side one only by an energy $g_D \lesssim 10^{-7}$ K, and the thermal energy k_BT at the relevant temperatures is $\sim$$10^{-3}$ K, one would expect the former

[20] In fact, it may be shown from the tensorial structure of $\hat{H}_D$ that $\sum_{i=1}^3 \partial^2\langle H_D\rangle/\partial\theta_i^2 = -\langle H_D\rangle$.

configuration to be favored over the latter by *at most*[21] a factor $g_D/k_BT \lesssim 10^{-4}$, giving a shift in the resonance frequency which is unobservably small. This argument is entirely consistent with the data for the normal phase, where no shift is seen.[22]

How can the dipole energy have a value large enough to explain the observed A-phase shift? The key lies in an idea sometimes known as "spontaneously broken spin–orbit symmetry" (SBSOS). Suppose that for some reason the "strong" (van der Waals, etc.) forces in the problem force all the pairs of atoms to behave in *the same* way; in particular, suppose they constrain all pairs to have the same relation of their orbital orientation to their spins. Of course, these "strong" forces cannot themselves effect any preference for the "end-to-end" configuration over the "side-by-side" one. However, all the pairs must now either all *simultaneously* adopt the end-to-end configuration, or all simultaneously adopt the side-by-side one. The difference between these two possibilities is no longer of order g_D, but rather $\sim N_p g_D$ where $N_p \sim 10^{22}$ is the total number of pairs in the system.[23] Since $N_p g_D \gg k_BT$, the preference for the end-to-end configuration is now essentially 100%. Once the system is locked into this equilibrium configuration, it is clear that any deviation of the spins from their initial direction through a small angle θ, such as occurs in the NMR experiments (the orbital configuration remaining fixed) will cost an energy of the order of $N_p g_D \theta^2$; thus we obtain a value of $\partial^2 \langle H_D \rangle / \partial \theta^2$ sufficient to explain, via Eqn. (6.4.3), the experimentally observed resonance shift.

The mechanism by which SBSOS leads to a large value of the dipole energy may be thought of as analogous to the mechanism of ferromagnetism in a system of spins described by the isotropic Heisenberg Hamiltonian, with the van der Waals forces playing the role of the exchange forces in the magnetic problem and the dipole energy that of the Zeeman term due to the external field; for a detailed comparison, see Leggett (2004a). However, an important difference is that while in the ferromagnetic case the ordering produces a macroscopic value of a one-particle quantity (the total spin of the system), which is directly detectable in experiment, in the case of SBSOS the ordering is manifest only in the two-particle correlations and thus shows up only more indirectly, via its effect on the NMR.

The concept of SBSOS is more general than that of (pseudo)-BEC, and can be applied under appropriate conditions even when there is no question of the latter (e.g. in the ordered phases of *solid* ^{3}He, see Osheroff et al. 1980). However, in the context of the superfluid phases of liquid ^{3}He it will be obvious from the discussion of earlier sections that Cooper pairing provides precisely the conditions for SBSOS. In fact, the invariance under spin and orbital rotations separately ($F(\boldsymbol{d}(\hat{\boldsymbol{n}}) = F(\hat{R}_{\text{spin}})\boldsymbol{d}(\hat{\boldsymbol{n}}) = F(\boldsymbol{d}(\hat{R}_{\text{orb}}\hat{n}))$) which we invoked in developing the generalized GL description of Section 6.3 is broken by the dipole energy; a residual invariance, namely that under simultaneous rotation of the spin and orbital coordinate systems ($\hat{R}_{\text{spin}} = \hat{R}_{\text{orb}}$) is left over, but in practice this is almost always broken by further small

[21]The real situation is actually worse than this, since because of the Fermi degeneracy the thermal energy k_BT should by replaced by $k_BT_F \sim 1$ K.

[22]Or rather, was seen in the 1972 experiments. However, see Haard et al. (1994).

[23]Or more precisely the number which are free to readjust, which is a fraction of the order of $T/T_F \sim 10^{-3}$ of the whole (still a very large number!)

perturbations such as the effects of magnetic fields or boundaries. Specifically, the effect of $\hat{H}_D$ on the ABM and BW phases is as follows: In the ABM phase, where we recall (Eqn. 6.2.21) that the order parameter is of the form $\hat{\boldsymbol{d}}f(\hat{\boldsymbol{n}})$ where $f(\hat{\boldsymbol{n}})$ is characterized by an orbital vector $\hat{\boldsymbol{l}}$ (the "intrinsic angular momentum of the pair," cf. Appendix 6A), $\hat{H}_D$ constrains $\hat{\boldsymbol{d}}$ to lie,[24] in equilibrium, along $\hat{\boldsymbol{l}}$; under normal conditions $\hat{\boldsymbol{d}}$ lies perpendicular to the external magnetic field so as to minimize the polarization energy (cf. the discussion in Section 6.2). The effect of $\hat{H}_D$ on the BW phase is more subtle: The "reference" state (6.2.22) originally considered by Vdovin and by Balian and Werthamer is the "$^3P_0$" state, in which the relative orbital angular momentum $\boldsymbol{L}$ and total spin $\boldsymbol{S}$ of any pair satisfies $\boldsymbol{J} \equiv \boldsymbol{L} + \boldsymbol{S} = 0$, and which is, thus by the Wigner–Eckart theorem isotropic in all its properties. The $^3P_0$ state, however, does not correspond to a minimum of the dipole energy; rather, the latter is achieved by starting from this state and performing a *relative* rotation of the spin and orbital coordinate systems through a characteristic angle $\cos^{-1}(-1/4) \cong 104°$ around an arbitrary axis $\hat{\boldsymbol{\omega}}$. The resultant state is isotropic as regards the spin and orbital properties separately, but is anisotropic from the point of view of properties such as NMR which involve *correlations* of the spin and orbital behavior; in particular, while the intrinsic orbital angular momentum $\boldsymbol{L}$ and spin $\boldsymbol{S}$ of the pairs satisfy $\langle \boldsymbol{L} \rangle = \langle \boldsymbol{S} \rangle = 0$, the quantity $\langle \boldsymbol{L} \times \boldsymbol{S} \rangle$ is nonzero (it may be verified to lie along $\hat{\boldsymbol{\omega}}$). For this reason the equilibrium state is often called "pseudo-isotropic." The arbitrariness of the "spin–orbit rotation axis" $\hat{\boldsymbol{\omega}}$ is usually removed up to a sign, in practice, by very small orienting energies associated with the external magnetic field and/or the walls of the system.

Having resolved the issue of how the dipole forces determine the equilibrium configuration of ^{3}He-A and B, let us now turn to their effect on the spin dynamics (NMR), which clearly involves a deviation of the system from equilibrium. For orientation I start with the very simplest case, namely the "longitudinal" resonance of the ABM phase conventionally identified with ^{3}He-A. I take the DC external magnetic field to lie along the z-axis, and for definiteness assume that in equilibrium the $\boldsymbol{d}$-vector lies along the x-axis, so that the initial state, as viewed from the z-axis, is an ESP state. Then the pair wave function is a product of a spin function and a function $F(\hat{\boldsymbol{n}}, r)$ of the orbital coordinates, which I will assume remains effectively fixed throughout the experiment; further, we shall be considering the motion of the total longitudinal component of spin, S_z, so that by symmetry we expect the state of the system to remain unitary and ESP throughout. Thus the spin part of the pair wave function is, throughout the experiment, of the simple form

$$\Psi(t) = (e^{i(\Delta\phi/2)}|\uparrow\uparrow\rangle + e^{-i(\Delta\phi/2)}|\downarrow\downarrow\rangle)^{N/2} \tag{6.4.4}$$

where we took out a possible overall phase factor which has no physical significance. In the $\boldsymbol{d}$-vector notation, this corresponds to $\boldsymbol{d}$ lying always in the xy-plane. Now, according to the considerations of Appendix 2B, the variable canonically conjugate to the relative phase $\hat{\Delta\phi}$ of the pairs is simply the total z-component of spin of

[24]This is intuitively plausible, since the "end-over-end" configuration is formed by $\boldsymbol{l}$ lying perpendicular to the spins involved, which means parallel to $\boldsymbol{d}$.

the system, S_z:

$$[\hat{S}_z, \hat{\Delta}\phi] = 2i \tag{6.4.5}$$

Thus, if we can express the effective Hamiltonian as a function of the pair of conjugate variables S_z and $\Delta\phi$, we will have a complete description of the longitudinal NMR behavior.

A straightforward calculation, which I shall not give here, indicates that the form of the dipole energy is, apart from a constant, simply

$$H_{\mathrm{D}} = -\tfrac{1}{4} g_{\mathrm{D}} \cos \hat{\Delta}\phi \tag{6.4.6}$$

where g_{D} is a constant which in general depends on the way the short-range behavior is handled. From (6.4.6) we see that the equilibrium configuration corresponds to $\Delta\phi = 0$, or in the $\boldsymbol{d}$-vector notation $\boldsymbol{d} \parallel \boldsymbol{l}$, as stated earlier. The S_z-dependent term is a little more tricky; apart from the simple Zeeman term $-S_z\mathcal{H}$, which for a constant external field $\mathcal{H}$ has, as we shall see, only a rather trivial effect, there should also be a term arising (inter alia) from the fact that production of a finite S_z requires us to supply extra kinetic energy to the system. We now invoke the crucial consideration that in real-life ^{3}He-A at temperatures not too small compared to T_c, both the characteristic quasiparticle collision times and the characteristic time, $\sim\hbar/\Delta$, associated with the pairing process, are short compared to the frequency, which we will determine below, associated with the dipole forces. Consequently, it is reasonable to assume that *the system is in equilibrium subject to the given values of the "macroscopic" variables* S_z *and* $\Delta\phi$. But in that case, the intrinsic energy associated with the spin polarization is by definition $S_z^2/2\chi$ where χ is the experimental spin susceptibility (so that minimization of this term plus the Zeeman one gives back the result $S_z = \chi\mathcal{H}$). Thus, the complete effective Hamiltonian for the longitudinal NMR takes the form (where we allow the field $\mathcal{H}(t)$ to be time-dependent)

$$\hat{H}_{\text{eff}} = \frac{\hat{S}_z^2}{2\chi} - \hat{S}_z\mathcal{H}(t) - \frac{1}{4} g_{\mathrm{D}} \cos \hat{\Delta}\phi \tag{6.4.7}$$

with $\hat{S}_z$ and $\hat{\Delta}\phi$ satisfying the canonical commutation relation (6.4.5).

It is clear that the problem described by (6.4.7) and (6.4.5) is nothing but that of a simple quantum pendulum, with the correspondence (up to numerical factors) $\Delta\phi \to \theta$ (angle of pendulum with vertical), $S_z \to L$ (angular momentum), $\chi \to I$ (moment of inertia) and $g_{\mathrm{D}} \to gl$ (gravitational torque). The quantity corresponding to $\mathcal{H}(t)$ is slightly less familiar; it corresponds to a sort of "angular vector potential" related to the external (nongravitational) torque applied to the system in the same way as the familiar electromagnetic vector potential is to the electric field (i.e. the "torque" is $d\mathcal{H}/dt$).

The theory of a fully quantum simple pendulum is, of course, itself not entirely trivial. However, at this point we can exploit the fact (which we already implicitly assumed in writing the many-body wave function in the "coherent" form (6.4.4), cf. below) that for the real-life case of longitudinal NMR in ^{3}He-A we are almost invariably in the extreme semiclassical limit. To see this, we assume (as will turn out to be self-consistent) that to calculate the ground sate we may approximate $\cos \hat{\Delta}\phi$ in (6.4.7) by its $\hat{\Delta}\phi \to 0$ limit, $1 - (\hat{\Delta}\phi)^2/2$. Then the problem reduces to that of a simple harmonic oscillator, and the rms value $(\Delta\phi)_0$ of $\hat{\Delta}\phi$ in the ground state is equal to $(g_D\chi)^{-1/4}$. For a reasonably macroscopic (say $\sim$1 cm^3) sample of liquid ^{3}He this number is tiny ($\sim 10^{-4}$) so that on the scale of 2π the quantity $\Delta\phi$ may be taken, in the ground state and by extension in all the states which are likely to occur in the dynamics, to have a definite value. More precisely, when we calculate the equations of motion of $\hat{S}_z$ and $\hat{\Delta}\phi$ from (6.4.5) and (6.4.7) and take their expectation values, the quantity $\langle \sin \hat{\Delta}\phi(t)\rangle$ may be replaced to a high degree of accuracy by $\sin\langle\hat{\Delta}\phi(t)\rangle$, whereupon writing $\langle\hat{\Delta}\phi(t)\rangle \equiv \Delta\phi(t)$, etc. we recover the equations of motion of a classical pendulum:

$$\frac{dS_z}{dt} = -\frac{1}{2} g_D \sin\Delta\phi \tag{6.4.8a}$$

$$\frac{d}{dt}(\Delta\phi) = 2\left(\frac{S_z}{\chi} - \mathcal{H}(t)\right) \tag{6.4.8b}$$

From (6.4.8) we recover all the results which are standard for a classical pendulum: in particular, for small deviations from equilibrium the motion is harmonic with frequency[25] $\omega_0 \equiv (g_D/\chi)^{1/2}$, while for larger amplitude the frequency decreases and becomes zero at the point where the total energy equals $2g_D$; beyond this point the motion corresponds to a rotation rather than a libration. All these effects have been seen in experiments on the longitudinal NMR of ^{3}He-*A*: see Wheatley 1978. It is amusing that the "nonlinear self-trapping" recently observed (Albiez et al. 2005) in a Bose-condensed alkali gas is exactly the "external" analog, in a single-component system, of the nonlinear ringing discussed by Wheatley.

As already noted, the assumption that $\Delta\phi$ is essentially a classical variable (i.e. that we can ignore its (very small) quantum fluctuations) is already embodied in the coherent-state description (6.4.4). The assumption is justified, as we have seen, by the consideration that the dimensionless parameter $\lambda \equiv (g_D\chi)^{-1/4}$ is very small compared to unity for any reasonably macroscopic sample of liquid ^{3}He. However, both g_D and χ scale with the volume, and it is conceivable that by using (for example) ultrasmall inclusions of liquid ^{3}He trapped in a matrix of solid ^{4}He (Schrenk et al. 1996), we could obtain values of λ comparable to or larger than unity. In such a case one would expect that the correct description of the ground state (in zero field) would not be a coherent state of the form of (6.4.4), but would be more like the "Fock state"

$$\Psi_{\text{Fock}} \sim (\uparrow\uparrow)^{N/4}(\downarrow\downarrow)^{N/4} \tag{6.4.9}$$

The question of the spin dynamics in this limit has been considered in Leggett (2004b).

While the "pendulum" analogy is quite helpful to one's intuition, it is also instructive to view the longitudinal NMR of ^{3}He-*A* as a sort of "internal Josephson effect," with the obvious correspondence of $\Delta\phi$ to the quantity so denoted in Chapter 5, Section 5.10, and S_z the analog of the number difference ΔN. The field $\mathcal{H}(t)$ is the analog of the voltage bias $eV(t)$. The analog, in the original Josephson context, of the polarization term $S_z^2/2\chi$, is the capacitance (charging) energy $(\Delta N)^2e^2/2C$, where C is

[25]The analogous phenomenon realized in superconducting grains (cf. below) is labeled the "Josephson plasma resonance."

the relative capacitance of the two superconductors connected by the junction. While this energy is usually negligibly small in the case of two bulk superconductors originally considered in Josephson's work, it can be appreciable in the case of small metallic grains, and in fact in such systems it is not uncommon to be in the (analog of the) "Fock" regime described by (6.4.9); this requires, crudely speaking, $e^2/C \gg I_c\phi_0$ where I_c is the critical current of the junction and ϕ_0 the flux quantum. See, e.g. Tinkham (1996), Chapter 7, and Chapter 2, section 6, (point 2).

Once we have understood the physics of the longitudinal resonance in ^{3}He-A, it is fairly straightforward to generalize the theory to discuss the spin dynamics in an arbitrary Cooper-paired phase. To make the generalization, it is convenient to introduce for the ABM state (^{3}He-A) instead of the phase difference $\Delta\phi$ between the $\uparrow\uparrow$ and $\downarrow\downarrow$ pairs, the angle θ_z of rotation of the vector $\boldsymbol{d}$ around the z-axis, which is just half of $\Delta\phi$; thus one has the commutation relation

$$[S_z, \theta_z] = i \tag{6.4.10}$$

But there is nothing special about the z-axis from the point of view of the kinematics; thus, if i (=1, 2, 3) denotes an arbitrary Cartesian axis system and if one defines the angle of rotation θ_i of the spin coordinate system (i.e. of the $\boldsymbol{d}(\boldsymbol{n})$ uniformly) around the axis i relative to some initial configuration (e.g. the equilibrium one), and S_i denotes the projection of the total spin $\boldsymbol{S}$ on that axis, one should have

$$[S_i, \theta_j] = i\delta_{ij} \tag{6.4.11}$$

Thus the semiclassical equations of motion take the form[26]

$$\frac{d\boldsymbol{S}}{dt} = \boldsymbol{S} \times \boldsymbol{\mathcal{H}}(t) - \frac{\partial E_{\mathrm{D}}}{\partial \boldsymbol{\theta}} \tag{6.4.12a}$$

$$\frac{d\boldsymbol{\theta}}{dt} = \boldsymbol{\theta} \times \left(\boldsymbol{\mathcal{H}}(t) - \frac{\boldsymbol{S}}{\chi}\right) \tag{6.4.12b}$$

Once the dipole energy $E_{\mathrm{D}}(\boldsymbol{\theta})$ is explicitly given as a function of a general rotation $\boldsymbol{\theta}$, we can solve Eqns. (6.4.12) for a general motion of $\boldsymbol{S}$.

The linear behavior (that is, the small oscillations around the equilibrium configuration) is of special interest, in particular for the two phases (ABM and BW) which are conventionally identified with the A and B phases respectively of superfluid ^{3}He. We already saw that in the ABM phase there is a longitudinal resonance at a frequency $\omega_{\parallel} = \omega_0 \equiv (g_{\mathrm{D}}/\chi)^{1/2}$, and it turns out that it is precisely the same number which determines the shift of the transverse resonance frequency:

$$\omega_{\mathrm{res}}^2 = \omega_{\mathrm{L}}^2 + \omega_0^2 \equiv \omega_{\perp}^2 \tag{6.4.13}$$

The relation $\omega_{\perp}^2 - \omega_L^2 = \omega_{\parallel}^2$ is well verified experimentally in ^{3}He-A.

The case of the BW phase is a little more tricky. It is clear that the dipole energy can depend, in bulk and in zero external field, only on the "spin–orbit rotation angle" θ,

[26] In deriving (6.4.12) we use the fact that while the different components of $\boldsymbol{S}$ fail to commute, the components of the semiclassical variable $\boldsymbol{\theta}$ do so to a high degree of approximation; and we assume that the polarization term has the simple isotropic form $\boldsymbol{S}^2/2\chi$. (The latter is actually not true for the ABM phase, but this does not affect the results: see Leggett 1974, Section 6.)

not on the axis $\boldsymbol{\omega}$ of the rotation. Now, once θ has settled on its equilibrium value $\theta_0 \equiv \cos^{-1}(-\frac{1}{4})$, any further "twisting" *perpendicular* to $\boldsymbol{\omega}$ is, to linear order, simply equivalent to a change in $\boldsymbol{\omega}$ with θ fixed, so cannot give rise to a dipole torque. On the other hand, twisting *parallel* to $\boldsymbol{\omega}$ changes θ from the equilibrium value θ_0 and thus activates a dipole torque. Now, in sufficiently large sample in a weak external DC magnetic field $\mathcal{H}$, there is a very tiny "depairing" effect which, while leaving the original pseudo-isotropic state to all intents and purposes unchanged, tends to orient $\hat{\omega}$ parallel to $\mathcal{H}$. Thus we expect that in a "standard" (transverse) NMR experiment, in which the RF field tends to twist the spin coordinate system around an axis perpendicular to $\mathcal{H}$, the dipole forces would not come into play and we should see the usual resonance at the normal-state Larmor frequency, as indeed we do. On the other hand, in a *longitudinal* NMR experiment, in which the RF field is parallel to $\mathcal{H}(\hat{\omega})$ and thus tends to twist the spin system around $\hat{\omega}$, the dipole forces should come into play and we should get a well-defined resonance just as in the ABM phase. Such a resonance is indeed seen in ^{3}He-B, and in fact its observation was, historically, what finally clinched its by then expected identification with the BW phase.

6.5 Supercurrents, textures and defects

In the last section we saw that even when the order parameter is constant in space, its variation (or more precisely the variation of the spin part of it) in time can give rise to spectacular effects. When we allow the order parameter also to vary in space, the consequences are even more intriguing, and have been the subject of whole books (e.g. Volovik 1992); here I shall have space merely to scratch the surface and whet the reader's appetite for further reading.

In a simple s-wave Fermi superfluid such as the classic superconductors, the internal structure of the Cooper pairs is completely determined, at any given P and T, by the energetics; we may thus regard the order parameter as simply a complex scalar function of the center-of-mass variable $\boldsymbol{R}$. Since the overall magnitude is again fixed by the energetics, the only interesting variable is then the phase $\phi(\boldsymbol{R})$; as we have seen, we can define the superfluid velocity in terms of the gradient of ϕ, $\boldsymbol{v}_s \equiv (\hbar/2m)\boldsymbol{\nabla}\phi$, and then, with the qualifications of Chapter 5, Section 5.7, and apart from the effects associated with the gauge coupling to the electromagnetic field, the system behaves with respect to its flow properties in a way qualitatively similar to ^{4}He-II: see Chapter 5, Section 5.6. Similar behavior would be shown by ^{3}He-B, were it to correspond to the original $^3P_0$ state of Vdovin and Balian and Werthamer. In real life, ^{3}He-B is characterized, in addition to the overall phase of the pair wave function, by a "spin–orbit rotation" whose axis and angle can vary in space; however, this variable is independent of the overall phase, and spatial variation of the latter leads, just as in ^{4}He or classic superconductors, to a simple mass flow, with $\boldsymbol{v}_s \equiv (\hbar/2m)\boldsymbol{\nabla}\phi$, $\boldsymbol{j} = \rho_s(T)\boldsymbol{v}_s$ and the superfluid density $\rho_s(T)$ isotropic.

A state in which, while the overall magnitude of the Cooper-pair order parameter is constant, the orientation varies in space, is called a "texture"; as we shall see, most kinds of texture are characterized by mass or spin currents or both. In the following it is necessary to bear in mind that generally speaking a nontrivial texture will not minimize the local value of the dipole energy; the cost is compensated by the saving in other energies, for example those associated with boundary effects.

At first sight the case ^{3}He-A (the ABM phase) should be rather simple, since in this case the order parameter is a simple product of a spin function, characterized by a real vector $\hat{\boldsymbol{d}}$ (the direction along which the pair spin projection is zero, see Section 6.2), and an orbital function, which at first sight should be characterized by a real vector $\hat{\boldsymbol{l}}$ (the direction of the apparent relative angular momentum of the pairs) and an overall phase ϕ (see however below). Let us first consider a situation where the direction of $\hat{\boldsymbol{l}}$ is constant in space. First, suppose that the spin vector $\hat{\boldsymbol{d}}$ is also constant. Then the only possible spatial variation is that of the overall phase: schematically, taking the $\boldsymbol{d}$-vector to lie along the x-axis for definiteness, we have in the ESP representation,

$$F(\boldsymbol{R}, \sigma\sigma') = e^{i\phi(\boldsymbol{R})}(|\uparrow\uparrow\rangle + |\downarrow\downarrow\rangle)f(\hat{\boldsymbol{n}}) \tag{6.5.1}$$

where $f(\hat{\boldsymbol{n}}) = \sin\theta\, e^{i\chi}$ with θ and χ the azimuthal and polar angles of $\hat{\boldsymbol{n}}$ with respect to the orbital vector $\hat{\boldsymbol{l}}$, and we have suppressed the (fixed) dependence of F on the magnitude $|\boldsymbol{r}|$ of the relative coordinate. For a state of the kind (6.5.1) we can define the superfluid velocity just as for ^{4}He, superconductors or ^{3}He-B, by

$$\boldsymbol{v}_s(\boldsymbol{R}) = \frac{\hbar}{2m}\boldsymbol{\nabla}\phi(\boldsymbol{R}) \tag{6.5.2}$$

However, while a two-fluid model similar to that discussed for ^{4}He in Chapter 3, Section 3.3 is still expected to apply to this kind of situation, the "superfluid density" $\rho_s(T)$ is now in general a *tensor* quantity. To see this, we write the expression (calculated in the absence of molecular-field effects) for the superfluid current $\boldsymbol{J}_s(T)$ (i.e. that induced by a superfluid velocity $\boldsymbol{v}_s$ with normal velocity $\boldsymbol{v}_n = 0$):

$$J_{si}(T) = \rho\, v_{si} - \sum_j \frac{df(E_{\boldsymbol{p}}, T)}{dE_{\boldsymbol{p}}}(p_i p_j)v_{sj} \tag{6.5.3}$$

where f is the Fermi distribution function and $E_{\boldsymbol{p}} = (\epsilon_p^2 + |\Delta(\hat{\boldsymbol{n}})|^2)$ is the excitation energy of a quasiparticle with momentum $\boldsymbol{p}$ (cf. Eqns. 3.6.8 and 5.7.25). Since for the ABM phase $|\Delta(\boldsymbol{n})| \sim \sin\theta$, it is clear that except at $T = 0$ the relation between $\boldsymbol{J}$ and $\boldsymbol{v}_s$ defines a tensor quantity ρ_{sij} by

$$J_i = \rho_{sij}(T)v_{sj} \tag{6.5.4}$$

and from symmetry the axes of the uniaxial tensor ρ_{sij} are defined by $\hat{\boldsymbol{l}}$: formally,

$$\rho_s(\hat{\boldsymbol{v}}_s) = A(T) - B(T)(\hat{\boldsymbol{v}}_s \cdot \hat{\boldsymbol{l}})^2 \tag{6.5.5}$$

where it may be seen that A and B are both positive (at any given nonzero T, there are more quasiparticles in the direction parallel to $\hat{\boldsymbol{l}}$, so the second term in (6.5.3) is larger for $\boldsymbol{v}_s$ in this direction). Fermi-liquid ("molecular-field") effects actually enhance the anisotropy of ρ_s substantially, so that in an intermediate range of temperature ^{3}He-A behaves as "predominately normal" for flow in a direction parallel to $\hat{\boldsymbol{l}}$ but "predominately superfluid" for flow normal to $\hat{\boldsymbol{l}}$: see Fig. 6.2.

Next, let us consider a situation in which both the orbital orientation $\hat{\boldsymbol{l}}$ and the overall phase ϕ are constant in space, but the direction of the spin orientation vector $\hat{\boldsymbol{d}}$ is varying. Quite generally, at any given point in (real) space the direction of $\hat{\boldsymbol{d}}$ and

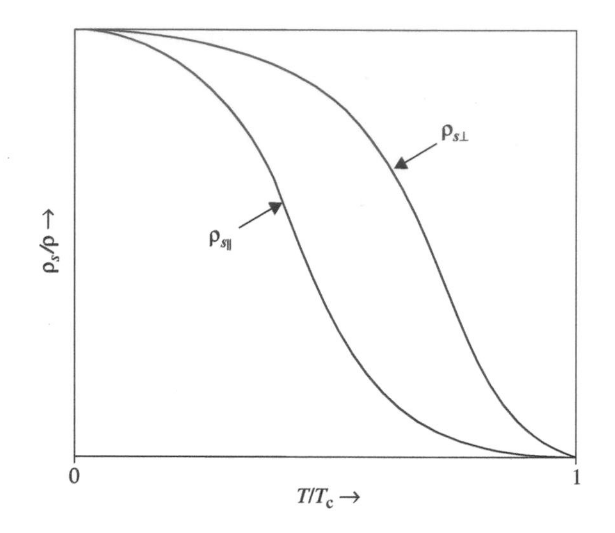

Fig. 6.2 Temperature dependence of the two different eigenvalues of the superfluid density in ^{3}He (qualitative).

its gradient define a plane in spin space, and if we take the normal to this plane as the z-axis for an ESP representation the order parameter has, locally, the form

$$F(\boldsymbol{R} : \sigma_1\sigma_2) = (e^{i((\chi(\boldsymbol{R}))/2)}|\uparrow\uparrow\rangle + e^{-i((\chi(\boldsymbol{R}))/2)}|\downarrow\downarrow\rangle) \times f(\hat{n}) \tag{6.5.6}$$

It is intuitively clear from the representation (6.5.6) that the up-spin Cooper pairs are flowing in the positive direction with "superfluid velocity" $\boldsymbol{v}_{s\uparrow} \equiv (\hbar/2m)\boldsymbol{\nabla}\chi$ and the down-spin pairs are flowing in the opposite direction with $\boldsymbol{v}_{s\downarrow} = -\boldsymbol{v}_{s\uparrow}$; thus, irrespective of the direction of $\boldsymbol{v}_{s\uparrow}$ relative to $\hat{\boldsymbol{l}}$, the configuration (6.5.6) corresponds to a spin supercurrent but no mass current. It is clear that in the situation considered $\hat{\boldsymbol{d}}$ cannot be everywhere parallel to $\hat{\boldsymbol{l}}$, and hence as already remarked the local dipole energy cannot in general be a minimum; it is in fact necessary that this should be so if the texture in question is to be an equilibrium state, since the divergence of the spin current, which is in general nonvanishing, must be compensated by the effect of the dipole torque as given by the second term of Eqn. (6.4.12a). (For details, see Vuorio 1974.)

Before going on to discuss orbital textures in ^{3}He-A, let us go back for a moment to ^{3}He-B and consider a situation where the overall phase is constant but the angle, or axis, or both of the spin–orbit rotation $\hat{R}(\theta, \hat{\omega})$ is spatially varying. It is convenient to regard the orbital coordinates as fixed, so that $\hat{R}$ then rotates the spin vectors $\boldsymbol{d}(\hat{\boldsymbol{n}})$ in the plane perpendicular to $\hat{\omega}$; then, similarly to the analysis done above for the A phase, one expects a counterflow of the spin populations parallel and antiparallel to $\hat{\omega}$, i.e. a supercurrent of the component of spin density parallel to $\hat{\omega}$. If on the other hand the angle θ is constant (e.g. at the value $\cos^{-1}(-\frac{1}{4})$ which minimizes the dipole energy, see Section 6.4) but the axis $\hat{\omega}$ varies, this is equivalent to a local rotation of the spin system around an axis perpendicular to $\hat{\omega}$, and we expect it to generate a supercurrent of the corresponding component of the spin density. In each of these cases, the spin current is "parallel" in spin space to the spin superfluid velocity, but because the spin of the Cooper pairs is correlated to their position on the Fermi surface, the corresponding "'spin superfluid density" is a tensor in orbital space (and different for the two cases). Consequently, in the most general case we can define a "spin superfluid velocity dyad" $v_{i\alpha}^{(\mathrm{s})}$ by

$$v_{i\alpha}^{(\mathrm{s})} \equiv \frac{\hbar}{2m} \partial_i \theta_\alpha \tag{6.5.7}$$

where θ_α denotes a differential rotation of the spin coordinate system around axis α in spin space, and ∂_i is the derivative in orbital space: then the supercurrent of spin density component α in direction i is given by an expression of the form

$$J_{i\alpha} = \sum_{j\beta} \rho_{ij,\alpha\beta}^{(\mathrm{s})} \, v_{j\beta}^{(\mathrm{s})} \tag{6.5.8}$$

where the spin superfluid density $\rho^{(\mathrm{s})}$ is now in general a tensor in both orbital and spin space, whose explicit form depends on the local spin–orbit rotation and hence is in general itself position-dependent: for details see, e.g. Vollhardt and Wölfle 1990, Section 10.1.

We now return to the A phase and turn to the more delicate question of situations where the orbital vector $\hat{\boldsymbol{l}}$, which characterizes the orbital-space orientation of the ABM order parameter, itself varies in real space. In all cases up to now, we have been able to characterize any situation involving mass[27] superflow in terms of a *globally* defined phase $\phi(\boldsymbol{R})$, in terms of which we can define a unique superfluid velocity $\boldsymbol{v}_\mathrm{s}(\boldsymbol{R})$ by Eqn. (6.5.2). The quantity $\boldsymbol{v}_\mathrm{s}(\boldsymbol{R})$ then automatically satisfies (within any simply connected region) the irrotationality condition

$$\boldsymbol{\nabla} \times \boldsymbol{v}_\mathrm{s} = 0 \tag{6.5.9}$$

As a consequence, all the standard results concerning vortices, etc., which we have obtained explicitly for the case of a simple spinless Bose condensate such as ^{4}He, go through (this includes the case of ^{3}He-A for the special case that $\hat{\boldsymbol{l}}$ is constant in space).

However, the orbital part of the order parameter of ^{3}He-A (assumed as always to be the ABM phase) is of a peculiar nature: its dependence on the angle $\hat{\boldsymbol{n}}$ of the relative pair momentum is of the form

$$F(\hat{\boldsymbol{n}} : \boldsymbol{R}) = \hat{\boldsymbol{n}} \cdot \hat{\mathbf{e}}_1(\boldsymbol{R}) + i\hat{\boldsymbol{n}} \cdot \hat{\mathbf{e}}_2(\boldsymbol{R}) \tag{6.5.10}$$

where $\hat{\mathbf{e}}_1(\boldsymbol{R})$ and $\hat{\mathbf{e}}_2(\boldsymbol{R})$ are two mutually perpendicular real unit vectors such that $\hat{\mathbf{e}}_1 \times \hat{\mathbf{e}}_2 \equiv \hat{\boldsymbol{l}}$. From (6.5.10) we see that *a differential change in the overall phase of the pair wave function is equivalent to a differential rotation around* $\hat{\boldsymbol{l}}$. But, if the direction of $\hat{\boldsymbol{l}}$ is varying in space, differential rotations around it are nonholonomic, i.e. we cannot define the "rotation angle" globally. Consequently, while we are still free to define a "superfluid velocity" $\boldsymbol{v}_\mathrm{s}(\boldsymbol{R})$ by Eqn. (6.5.2) (where ϕ is a *differential* phase rotation), it does not in general satisfy the simple irrotationality

[27] The situation for the spin superflow described, e.g. by (6.5.7) is analogous to that discussed below.

condition (6.5.9); rather, it may be shown (see, e.g. Vollhardt and Wölfle (1990), Sections 7.1, 7.3.3) to satisfy the Mermin–Ho identity

$$\nabla \times \boldsymbol{v}_s = \frac{\hbar}{4m} \sum_{ijk} \epsilon_{ijk} \, \hat{l}_i \, \nabla \hat{l}_j \times \nabla \hat{l}_k \tag{6.5.11}$$

As we shall see, this leads to interesting consequences for the structure of possible textures and, in principle, for the decay of superflow.

For the moment we notice a further consequence of a "bending" of $\hat{\boldsymbol{l}}$ in space: Suppose $\hat{\boldsymbol{l}}$ is initially along (say) the z-axis and $\hat{\mathbf{e}}_2$ along the y-axis. Consider now a particular point $\hat{\boldsymbol{n}}$ on the Fermi surface, say for definiteness in the xz-plane. According to Eqn. (6.5.10), the order parameter $F(\boldsymbol{n})$ at this point is purely real. Now suppose $\hat{\boldsymbol{l}}$ tilts slightly into the yz-plane, so that $\boldsymbol{e}_2$ acquires a small z-component. Then according to (6.5.10), $F(\hat{\boldsymbol{n}})$ acquires a small imaginary part: its overall phase[28] has changed! One might thus expect intuitively that if $\hat{\boldsymbol{l}}$ varies in space, so that $F(\hat{\boldsymbol{n}}, \boldsymbol{R})$ acquires a phase gradient, this would lead to a contribution from this part of the Fermi surface to a mass supercurrent; and provided the contributions from different regions of the Fermi surface do not cancel, one would thus get a supercurrent associated with the bending of $\hat{\boldsymbol{l}}$ even if $\boldsymbol{v}_s = 0$. In fact, symmetry under spatial reflection and time reversal allows a current of the general form

$$\boldsymbol{j} = C(T) \nabla \times \boldsymbol{l} - C_0(T) \boldsymbol{l}(\boldsymbol{l} \cdot \nabla \times \boldsymbol{l}) \tag{6.5.12}$$

and a microscopic calculation (which I do not give here) shows that indeed the quantities $C(T)$ and $C_0(T)$ are nonzero (and both positive). It is interesting to note that a term of the form of the first on the RHS of Eqn. (6.5.12) might be expected to occur for a Bose condensate of tightly bound diatomic molecules each possessing intrinsic angular momentum $\boldsymbol{L} = \hbar \hat{\boldsymbol{l}}$. However, it does not appear straightforward to get the correct form of the coefficient $C(T)$ out of this consideration, and in addition the "molecular" system shows no analog of the second term in (6.5.12); cf. Appendix 6A.

A very useful tool for analyzing the possible varieties of texture in ^{3}He-A and B (and in other ordered condensed matter systems) is the branch of mathematics known as homotopy theory. A very beautiful introduction to this subject and its application to various condensed matter systems, including superfluid ^{3}He, is given by Mermin (1979). Before reviewing a few results concerning topological defects (that is, "singular" textures where the order parameter is constrained to vanish, or to change its fundamental structure; in some part of the relevant space), I first mention a very fundamental result concerning the *singularity-free* ^{3}He-A phase: namely, provided the orbital vector $\hat{\boldsymbol{l}}$ is allowed to vary freely as a function of position (or time) it is always possible to change the circulation by multiples of 4π. We already met this theorem in a different context in Chapter 2, Section 2.5 (point 6) and Chapter 4, Section 4.7 (point 2): If we start from a situation, let us say in a toroidal geometry, in which $\hat{\boldsymbol{l}}$ is constant (say $\| \hat{\boldsymbol{z}}$) in space and the condensate phase (which we recall is under this condition well-defined) has a circulation of 4π around the annulus, then by rotating $\boldsymbol{l}$ locally around an axis in the xy-plane which is (e.g.) perpendicular to the radius vector

[28]The relative phase of the two spin components is of course unaffected.

at that point, we achieve a fixed state in which $\boldsymbol{l}$ is everywhere antiparallel to the z-axis and the phase (again well-defined) is constant. Thus we have achieved a reduction in the circulation by 4π, without anywhere leaving the manifold of ABM states. It should be said that while in principle such a mechanism could lead to the decay of supercurrents in ^{3}He-A, in practice there is no evidence for this, probably because in real life the direction of the $\hat{\boldsymbol{l}}$-vector tends to be pinned by dipolar forces and walls.

In the above argument, the direction of $\hat{\boldsymbol{l}}$ at a given point on the annulus was implicitly a function of time. Imagine now that the "annulus" is actually part of a simply connected 3D space, with no "hole" at the origin. By replacing the dependence on time by a corresponding dependence on the radial coordinate, we see that we can achieve a (static) configuration in which $\hat{\boldsymbol{l}}$ is (say) $\|\hat{\boldsymbol{z}}$ on the annular circuit and the phase has a circulation 4π on it, while by the time we get to the origin $\hat{\boldsymbol{l}}$ is $\|-\hat{\boldsymbol{z}}$ and there is no circulation in the phase. Thus, at the origin everything is completely homogeneous and there is no need for a region ("core") in which the order parameter vanishes or substantially changes its structure. In view of the 4π circulation at large r, which in a simple system such as ^{4}He would indicate a (doubly-quantized) line vortex, the texture (see Fig. 6.3) which we have created in this way is therefore sometimes

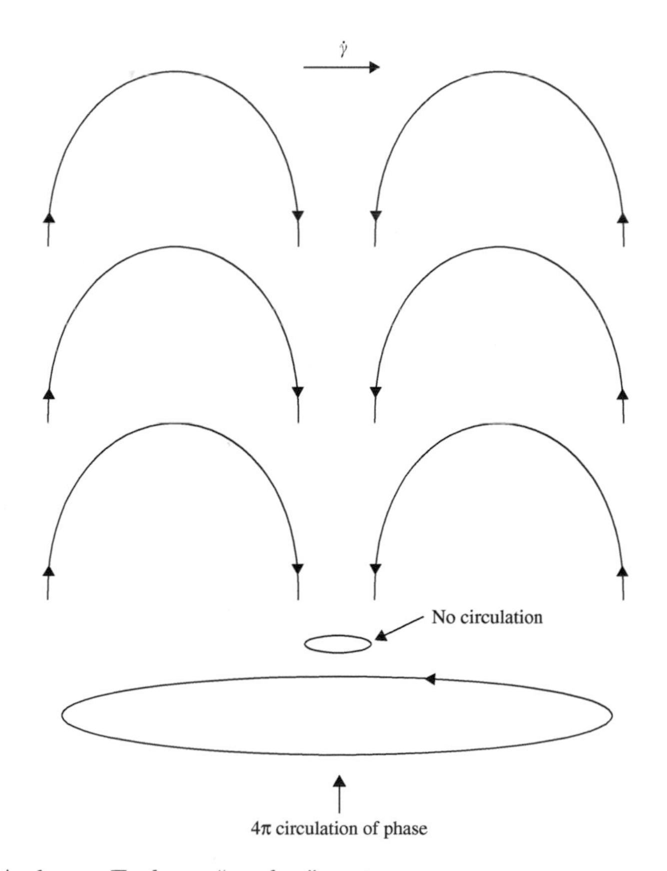

Fig. 6.3 An Anderson-Toulouse "coreless" vortex.

called a "coreless vortex" (or Anderson–Toulouse vortex); it is closely related to the "boojum" texture discussed below.

I now turn to the question of true topological defects in superfluid ^{3}He. We have already met one simple example of a topological defect, namely the (line) vortices which can occur in system such as ^{4}He or classic superconductors where the order parameter is a complex scalar object. For such an object the phase is defined only modulo 2π, and thus as we have seen it is possible to have a circulation of $2\pi n$ around a line region (the vortex "core") on which the order parameter is not defined; such a "vortex line" configuration cannot be deformed into the uniform configuration (phase constant everywhere in space) while staying in the ordered phase everywhere outside the core. In a similar way, in a system such as a Heisenberg ferromagnet where the order parameter is a vector in 3D space, it is possible to contemplate a configuration ("hedgehog") in which, for example, this vector points radially outwards from a particular point in space ($\hat{\boldsymbol{S}}(\boldsymbol{r}) = \hat{\boldsymbol{r}}$ if the point is taken as origin); then it clearly cannot be defined at the origin, so we have a point rather than a line defect. Homotopy theory gives us a systematic way of classifying defects in a system with a given type of order parameter; it has been widely applied to both the A and B phases of superfluid ^{3}He, and a wide variety of possible defects has been predicted in each phase. However, whether a particular type of defect will occur, in equilibrium, in a particular geometry is a matter of energetics; to discuss this we would need to investigate the gradient energy in the ordered state, a rather complicated topic for which there is no space here (see Leggett 1975, Section X.D, or Vollhardt and Wölfle (1990), Section 7.2). In practice, somewhat disappointingly, the only kind of bulk topological defect for which there is substantial experimental evidence in either the A or B phase is a singly-quantized vortex qualitatively similar to the familiar one in ^{4}He; the disappointment is somewhat mitigated by the fact that the cores of these vortices are almost certainly formed not by turning a cylinder of the liquid normal, as in ^{4}He, but by transforming the order parameter to a form which is no longer in the ABM (BW) class, and can thus handle the topological constraints appropriately without anywhere vanishing. This has been a topic of considerable research, both theoretical and (using NMR techniques) experimental; see, e.g. VW, Section 7.8.

By contrast, the experimental evidence for nontrivial *nonsingular* textures in both the A and B phases is substantial. In the equilibrium B phase, there is no reason why the phase of the order parameter should vary in space, and the *angle* of the spin–orbit rotation is usually kept close to its equilibrium value $\cos^{-1}(-\frac{1}{4})$ by the dipole energy. However, the *axis* $\hat{\boldsymbol{\omega}}$ of the rotation is sensitive to small orientational energies connected with the external field, the container walls, etc., and consequently it is not uncommon to find that it varies substantially in space; the effect of this variation shows up spectacularly in the NMR behavior, since by following the argument of Section 6.4 we see that once $\hat{\boldsymbol{\omega}}$ is no longer parallel to the external field the *transverse* resonance frequency can be substantially shifted (an effect which was in fact observed in some of the very earliest NMR experiments on the B phase).

In the A phase textural effects are even more pronounced. In the first place, there is strong theoretical reason to believe that the energetics will always pin the $\hat{\boldsymbol{l}}$-vector normal to the walls of the container (and experiment seems to be consistent with this

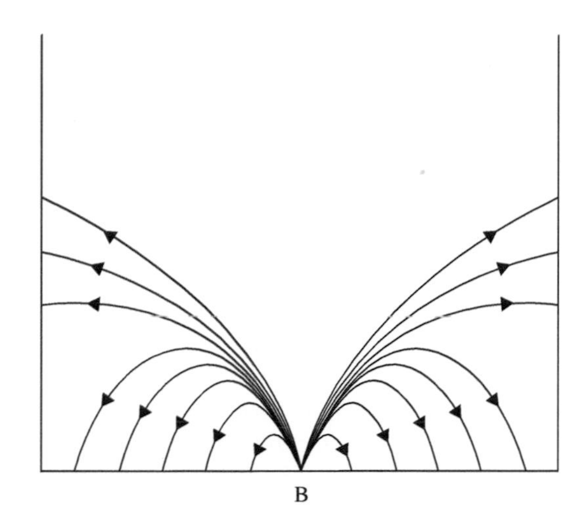

Fig. 6.4 A "boojum" texture.

hypothesis), so that except in the simplest (e.g. flat-slab) geometries this consideration alone enforces a nontrivial texture. However, in addition, it follows from purely topological arguments under this condition that in any simply connected geometry (such as a cylinder or sphere) the ABM-phase order parameter must have at least two topological singularities on the walls of the container. While the analysis of the relative stability of the possible configurations satisfying this constraint is complicated, the general belief is that over at least some part of the phase diagram the most stable would be the texture shown in Fig. 6.4, which can be thought of as half of an Anderson–Toulouse coreless vortex deformed in such a way as to make $\hat{\boldsymbol{l}}$ everywhere normal to the walls; because of the infinite curvature of $\hat{\boldsymbol{l}}$ which would otherwise occur at the point B, a small region near that point must become normal (or at least non-ABM-like), so it is indeed a genuine (surface) topological defect. In the literature this configuration has become known as a "boojum."[29] There is some experimental evidence that in a suitable geometry such a boojum may occur, and even that it may move around the surface and thereby trigger decay of the supercurrent according to the scenario set out above.

A second interesting feature of the A phase is that while in a "simple" geometry such as a flat slab the uniform texture ($\hat{\boldsymbol{l}}$ constant in space) is stable in the absence of a superflow or heat current, this may no longer be true in their presence: in fact, it is theoretically predicted that under certain conditions a helical texture is formed, and the experimental evidence seems consistent with this: see VW, Section 7.10.2.

Finally, one may consider textures in which (for example) the $\hat{\boldsymbol{l}}$-vector is constant in space but the $\boldsymbol{d}$-vector rotates between two 3D regions in which it is respectively parallel and antiparallel to $\hat{\boldsymbol{l}}$ (recall that the dipole energy depends only on $(\hat{\boldsymbol{d}} \cdot \hat{\boldsymbol{l}})^2$).

[29]For an amusing account of the etymology and history of the term "boojum" see Mermin (1990), Chapter 1.

Such a texture is called a $\boldsymbol{d}$-soliton; while it is not the stable equilibrium configuration, it may be created by appropriate NMR techniques, and then detected by its effect on the NMR frequency shifts.

In the last section we looked at (some) consequences of the variation of orientation of the order parameter in time, and in this section at those of its variation in space. It would be logical at this point to consider situations in which the orientation varies in both space and time, i.e. at the question of (low-frequency)[30] collective oscillations. In view of space limitations I just quote the principal results of such a consideration. First, both ^{3}He-B and ^{3}He-A can sustain simple *sound wave* oscillations similar to the phonons of ^{4}He in which the overall phase and the conjugate variable, the total density, oscillate at fixed order-parameter orientation. Since in normal ^{3}He, at T_c and for typical ultrasound frequencies, the density oscillations are not hydrodynamic but of "zero-sound" type (see, e.g. Baym and Pethick 1991, Section 2.3.1) there is a rather complicated crossover behavior as T falls below T_c, but in the limit $T \to 0$ this oscillation is essentially hydrodynamic in both phases, as in ^{4}He. Secondly, *spin waves* exist in both phases, although they are considerably easier to see in the B phase because of the much reduced density of fermion quasiparticles, which act as a damping mechanism. In these excitations the overall phase and the orbital orientation remain fixed, while the pair spin orientation and its conjugate variable, the total spin density, vary in space and time. Since the spin-wave velocity is smaller by more than an order of magnitude than that of sound waves, it is expected that in ^{3}He-B, at temperatures below $\sim$80 μK where the number of excited Fermi quasiparticles is exponentially small, the thermodynamics and transport properties will be dominated by spin waves. Finally, one might ask whether in the A phase[31] one expects to see "orbital waves," in which the orbital orientation (direction of $\hat{\boldsymbol{l}}$) varies in space and time while the spin orientation and, possibly, the overall phase stays constant. The answer turns out in practice to be no,[32] for a number of reasons, of which the most important is that since the energy of a quasiparticle of given wave vector $\boldsymbol{k}$ depends on $|\Delta(\hat{\boldsymbol{k}}|$ and hence on the angle $\hat{\boldsymbol{k}} \cdot \hat{\boldsymbol{l}}$ of $\hat{\boldsymbol{k}}$ relative to $\hat{\boldsymbol{l}}$, then to the extent that the quasiparticle distribution is held fixed a rotation of $\hat{\boldsymbol{l}}$ changes the energy of the quasiparticle system. This energy cost turns out to be enormous at any currently attainable temperature, and in effect locks the direction of $\hat{\boldsymbol{l}}$, even in bulk, over time scales shorter than the quasiparticle relaxation time τ. Over a longer time scale relaxation of the quasiparticle distribution can take place, but its effect is generally to make the motion of $\hat{\boldsymbol{l}}$ strongly overdamped. For further discussion of this topic, which is still somewhat controversial, see VW, Sections 10.3, 11.2.2.

To conclude, I emphasize again that the discussion in this section has barely scratched the surface of a research area which has generated certainly hundreds and probably thousands of papers over the last 30 years.

[30]There is also an interesting class of excitations with ($q \to 0$) frequencies of order Δ: see VW, Section 11.3.

[31]In the B phase there is no distinction between "spin" and "orbital" waves, since they are simply alternative languages to describe a space- and time-varying spin–orbit rotation.

[32]In the sense that there are no observable propagating modes for which $\omega \to 0$ as $q \to 0$. (There does exist a mode – the so-called "normal-flapping" mode – which has (typically) $\omega \sim \Delta$.)

Appendix

6A Note on the "angular momentum" of ^{3}He-A

Let us for the purposes of this appendix proceed in a naive fashion, by neglecting the question of boundary conditions and thus assuming the A phase of superfluid ^{3}He to be described by the spatially uniform many-body wave function (6.2.8), with the $c_{\boldsymbol{k}}$'s given by the form corresponding to the ABM phase order parameter (6.6.21). Then the spin part of the wave function factors out and is irrelevant in the present context, and we can write, apart from normalization

$$\Psi_N = \left(\sum_{\boldsymbol{k}} c_{\boldsymbol{k}} a^\dagger_{\boldsymbol{k}} a^\dagger_{-\boldsymbol{k}}\right)^{N/2} |\text{vac}\rangle \tag{6.A.1}$$

with c_k given by

$$c_{\boldsymbol{k}} = f(|\boldsymbol{k}|)\sin\theta_{\boldsymbol{k}} e^{i\phi_{\boldsymbol{k}}} \tag{6.A.2}$$

Intuitively, Eqn. (6.A.1) with (6.A.2) describes a set of $N/2$ Cooper pairs each of which has angular momentum $\hbar$ around the $z(\hat{\boldsymbol{l}})$-axis, so that the total wave function (6.A.1) is an eigenfunction of total angular momentum $\hat{\boldsymbol{l}}$ with eigenvalue $(N\hbar/2)\hat{\boldsymbol{l}}$. That this conclusion is in fact correct may be verified by writing the operator $\hat{\Omega} \equiv \sum_k c_k a^\dagger_k a^\dagger_{-k}$ in coordinate-space representation, taking its commutator[33] with $\hat{L}_z$, and transforming back into $\boldsymbol{k}$-representation:

$$[\hat{L}_z, \hat{\Omega}] = -i\hbar \sum_{\boldsymbol{k}} \frac{\partial c_{\boldsymbol{k}}}{\partial \phi_{\boldsymbol{k}}} a^\dagger_{\boldsymbol{k}} a^\dagger_{-\boldsymbol{k}} = \hbar\hat{\Omega} \tag{6.A.3}$$

where at the second step we used (6.A.2). Application of (6.A.3) to the wave function (6.A.1) then immediately yields the result stated, namely that the latter is an eigenfunction of $\hat{\boldsymbol{L}}$ with eigenvalue $(N\hbar/2)\hat{\boldsymbol{l}}$. This result is somewhat counterintuitive, in the sense that it implies that the limit of $\langle\hat{\boldsymbol{L}}\rangle$ in the superfluid ground state as the gap tends to zero is macroscopically different from its value (zero) in the normal ground state.

A simple solution to the difficulty is to replace (6.A.1) by the appropriate form of the "alternative" ansatz (5.4.38), namely[34]

$$\Psi_N = (\hat{\Omega}^+)^{N_+}(\hat{\Omega}^-)^{N_-}|\text{FS}\rangle \tag{6.A.5}$$

[33]Since $\boldsymbol{k}$ in (6.4.3) is technically a discrete variable, the derivative with respect to $\phi_{\boldsymbol{k}}$ must of course be treated as the appropriate limit of a difference.

[34]Just as in the s-wave case, we must in fact generalize (6.A.4) to read

$$\Psi_N = \sum_{N_+} c(N_+)(\hat{\Omega}^+)^{N_+}(\hat{\Omega}^-)^{N_+}|\text{FS}\rangle \tag{6.A.4}$$

where the function $c(N_+)$ is slowly varying. (A corresponding generalization is necessary for general N_m, see below.) Just as (6.A.1) is the particle-conserving version of the wave function (5.C.9), so (6.A.9) is the particle-conserving version of the wave function written down by Combescot (1978), who to my knowledge was the first to point out explicitly the possibility exploited here.

where $|\mathrm{FS}\rangle$ denotes the normal ground state and

$$\hat{\Omega}^+ \equiv \sum_{k>k_F} c_{\boldsymbol{k}} a^\dagger_{\boldsymbol{k}} a^\dagger_{-\boldsymbol{k}}, \quad \hat{\Omega}^- \equiv \sum_{k<k_F} d_{\boldsymbol{k}} a_{-\boldsymbol{k}} a_{\boldsymbol{k}} \tag{6.A.6}$$

with

$$d_{\boldsymbol{k}} \equiv c_{\boldsymbol{k}}^{-1} \tag{6.A.7}$$

and where in the limit $\Delta/\epsilon_F \to 0$ (thus $\mu \to \epsilon_F$) we have $N_+ = N_-$. The crucial observation is that since apart from the irrelevant factor of $(1+|c_k|^2)^{-1}$ the operator $\hat{\Omega}^-$ is just the Hermitian conjugate of $\hat{\Omega}^+$, we have $[\hat{L}_z, \hat{\Omega}^-] = -\hbar\hat{\Omega}^-$. Thus, given the condition $N_+ = N_-$, the many-body wave function (6.A.5) is an eigenfunction of $\hat{L}_z$ with eigenvalue exactly zero!

Since we will see in Chapter 8, Section 8.4, that at least in the s-wave case the ansatz (6.A.1) extrapolates smoothly from the BCS to the BEC (molecular) limit, and since for an ABM-type state (6.A.1), with (6.A.2), is indeed very plausibly the correct form in the BEC limit (where we would expect $\langle \boldsymbol{L}\rangle = (N\hbar/2)\hat{\boldsymbol{l}}$), it is natural to ask how the above considerations might be generalized to the case where Δ/ϵ_F is not small (cf. Chapter 8, Section 8.4) and thus the chemical potential μ is not the same as the Fermi energy. A plausible answer (which I shall not attempt to demonstrate) goes as follows: for $\mu > 0$ we define $|\mathrm{FS}(\mu)\rangle$ to be the Fermi sea corresponding to $\hbar^2 k_F^2/2m = \mu$ and define the difference between the total particle number in $|\mathrm{FS}\rangle$ and in $|\mathrm{FS}(\mu)\rangle$ to be N_m ("missing" number). Then we replace (6.A.1) by the ansatz, a generalization of (6.A.5),

$$\Psi_N = (\Omega^+)^{N_+}(\Omega^-)^{N_-}|\mathrm{FS}(\mu)\rangle \tag{6.A.8}$$

where now to recover the correct total number of particles we must set

$$N_+ - N_- = 2N_m \tag{6.A.9}$$

It then follows immediately that (6.A.8) is an eigenfunction of $\hat{L}$ with eigenvalue $N_m\hbar/2$. It is very reasonable to assume that just as in the s-wave case (see Chapter 8, Section 8.4), as we increase the attractive interaction the chemical potential μ will fall from its value ϵ_F in the "extreme BCS limit" and eventually cross zero, so we see that the angular momentum increases gradually from zero to its "BEC limit" value which is attained at $\mu = 0$. For $\mu < 0$ we set $N_- = 0$, so (6.A.9) and (6.A.1) are identical.

The question of what is the true expectation value of the angular momentum of superfluid ^{3}He-A in a given container geometry is one which is more than 30 years old and still has apparently not attained a universally agreed resolution, and I shall not attempt to discuss it here, merely remarking that since the moment of inertia of a macroscopic system is proportional to $N^{5/3}$ rather than N, the energy of states differing in $\langle \boldsymbol{L}\rangle$ by $O(N)$ may itself differ only by $O(N^{1/3})$, so that it is not clear that the problem is even uniquely defined in the sense we are used to in many-body physics. However, it is worth remarking that some theoretical papers (e.g. Volovik and Mineev 1981) have indeed reached, often by quite a circuitous route, the conclusion that in certain geometries $\langle \boldsymbol{L}\rangle = (N_m\hbar/2)\hat{\boldsymbol{l}}$. For a partial list of references on this problem up to 1990, see VW, Section 9.3, and for a more recent discussion Stone and Roy (2004).

7 Cuprate superconductivity

7.1 Introduction

The class of "classic" superconductors, which includes all superconductors discovered prior to 1975 and is generally believed to be well described by the BCS theory outlined in Chapter 5 of this book, is, as already mentioned at the beginning of that chapter, characterized by a number of common properties:

1. With the exception of a few systems discovered since 1975 which have T_c's up to ~40 K, the transition temperature T_c is always below 25 K.
2. The normal state of the metal or alloy in question appears to be well described by Landau Fermi-liquid theory.
3. The superconducting phase does not occur in proximity to other kinds of phase transition.
4. The order parameter is (universally believed to be) of the simple s-wave type (or more precisely corresponds to the unit representation of the symmetry group of the crystal).
5. The behavior seems consistent with the hypothesis that the dominant role in the formation of Cooper pairs is, as postulated in the BCS theory, an attractive interaction resulting from the exchange of virtual phonons.

In addition, two very widely though not universally shared properties of the classic superconductors are that

6. the crystal structure is relatively simple, and in particular not strongly anisotropic, and
7. in the case of alloys, superconductivity is not particularly sensitive to the chemical stoichiometry.

For the purposes of this book I shall define a given superconductor to be "exotic", i.e. not in the class of classic superconductors, if it fails to satisfy at least one of the conditions (1)–(7) and, in addition, there is reason to believe that it may not be well described by the BCS theory of Chapter 5.[1] By this definition, the class of exotic superconductors contains a number of different groups, all discovered since

[1]This definition is somewhat arbitrary; slight modifications of it would permit the inclusion of (e.g.) $BaKBiO_3$ and MgB_2, and perhaps the A-15 compounds, or alternatively the exclusion of the alkali-doped fullerenes. The high-pressure metallic hydrides are usually not regarded as "exotic".

Table 7.1 Differences between the classic superconductors and various classes of "exotic" superconductors.

Property	⟵ Class ⟶					
	Classic	Heavy-fermions	Organics	Ruthenates	Fullerene	Cuprates
$T_c < 25$ K	(✓)	✓	✓	✓	×	×
FL normal state	✓	×	×	×	✓	×
No neighboring phase trans.	✓	×	✓	✓	✓	×
OP s-wave	✓	?	?	×	✓	×
Phonon mechanism	✓	×	?	?	✓	×
Crystal structure simple	✓	✓	×	×	×	×
Stoichiometry-insensitive	✓	✓	✓	✓	×	×

1975: the heavy-fermion superconductors, the organics, the ruthenates, the alkali-doped fullerenes and the cuprate ("high-temperature") superconductors. As will be seen from Table 7.1, all of these fail to satisfy at least one of the criteria (1)–(7), and most violate more than one; most spectacularly, the cuprates violate *all* of the conditions (1)–(7).

At present there does not exist a theory of exotic superconductivity as a whole, or even of any of the individual subgroups, which is comparable to the BCS theory either in the level of quantitative prediction or in the level of acceptance by the community; in particular, the nature and origin of superconductivity in the cuprates is a hotly debated issue. For this reason, my approach in this chapter, which deals with the cuprates, and in Section 8.1 which covers the other "exotics", will be much more descriptive and less quantitative than in Chapter 5.

7.2 The cuprates: composition, structure, and phase diagram

The cuprates are the only class of materials to date in which superconductivity has been (reproducibly) observed to occur above 50 K. They do not occur in nature, and have been synthesized in the laboratory only since the early 1980s. At the time of writing data have been reported on several hundred different cuprate materials, of which a substantial fraction show superconductivity at temperatures of the order of 100 K. However, the value of T_c is usually critically dependent on the chemical stoichiometry, and there exists a nonnegligible subclass of cuprates which apparently cannot be made superconducting under any conditions – a feature which is not often commented on, but which may hold an important clue to the origin of superconductivity in these materials (cf. below).

It is of course possible (and conventional) to specify the composition of the cuprates directly in terms of the chemical composition, e.g. $La_{2-x}Sr_xCuO_4$, $Bi_2Sr_2CaCu_2O_{8+\delta}$

(where x and δ are positive fractional quantities). If one does so, one notices that the compositions at which superconductivity occurs are usually[2] close to an integral value of the stoichiometry (e.g. in the above examples, La_2CuO_4 and $Bi_2Sr_2CaCu_2O_8$) at which the total valence of the formula unit is equal to the number of Cu's, and we shall see below that this is not an accident. In the literature various shorthands are used, usually based on the chemistry of the material in question; for present purposes it is actually more convenient to use a slightly unconventional notation which may better reflect the physically important aspects of the crystal structure. Namely, we write the formula for a given cuprate in the form

$$(CuO_2)_n A_{n-1} X \tag{7.2.1}$$

where n is a positive integer, A is an alkaline earth or rare earth element (or Y) and X is an arbitrary collection of elements, possibly including Cu and/or O, and in general in nonrational stoichiometric proportions. In this notation $La_{2-x}Sr_xCuO_4$ would correspond to formula (7.2.1) with $n = 1$ and $X = La_{2-x}Sr_xO_2$, while Bi_2Sr_2Ca $Cu_2O_{8+\delta}$ would correspond to $n = 2$, A = Ca, $X = Bi_2Sr_2O_{4+\delta}$.

The point of the notation (7.2.1) is that it reflects the characteristic crystal structure of the cuprates. The "naive" unit cell (i.e. the cell corresponding to a formula unit)[3] consists of a group ("multilayer") of n CuO_2 planes (on which more below), spaced by $n-1$ layers of the "spacer" or "intercalant" element A and separated from the next plane or set of planes by the "charge reservoir" unit X. The distance between neighboring CuO_2 planes within a multilayer is approximately 3.1 Å, while the distance between neighboring multilayers (i.e. the c-axis dimension of the "naive" unit cell, cf. Fig. 7.1) depends on X and can be much greater (~15 Å for $Bi_2Sr_2Ca_2Cu_3O_{10}$). This is illustrated in Fig. 7.1 (where we indicate the conventional labeling of the crystal axes) for the case of $Tl_2Ba_2CaCu_2O_8$ ($n = 2$, A = Ca, $X = Tl_2Ba_2O_4$). A series of compounds, such as $Tl_2Ba_2CuO_6$, $Tl_2Ba_2CaCu_2O_8$, $Tl_2Ba_2Ca_2Cu_3O_{10}, \ldots$, for which A and X are the same but n takes increasing values $1, 2, 3, \ldots$ is called a "homologous series". It should be noted that the chemical stoichiometry of the cuprates, at least in the region where superconductivity occurs, always appears to be such as to give rise to close to one hole per CuO_2 unit in the plane (cf. below); if we assume that the valence of the charge reservoir group X is approximately independent of n, this then has the consequence that while for the case where the spacer element A is an alkaline earth (Ca, Sr, Ba,...) homologous series occur in a natural way (since each added A donates the two electrons necessary to leave the corresponding added CuO_2 group with one hole), they cannot occur when the spacer is Y or a rare earth (the valence of the group Y(or RE)CuO_2 is two). Thus, the well-known class of bilayer compounds Y(or RE)$Ba_2Cu_3O_{6+\delta}$ has no one- or three-layer homologues. As an aside, I note that there exists one special system (the so-called "infinite-layer" material $Ca_{1-x}Sr_xCuO_2$) for which the charge-reservoir group X is totally absent, so that the structure is simply repeated CuO_2 planes spaced by Ca or Sr. Some important cuprate superconductors

[2]The case of YBCO is anomalous in this respect, because the number of Cu atoms is different from the number of (CuO_2) plane units, see below.

[3]In some cases the true (crystallographic) unit cell contains two "naive" unit cells because neighboring groups of CuO_2 planes are staggered with respect to one another.

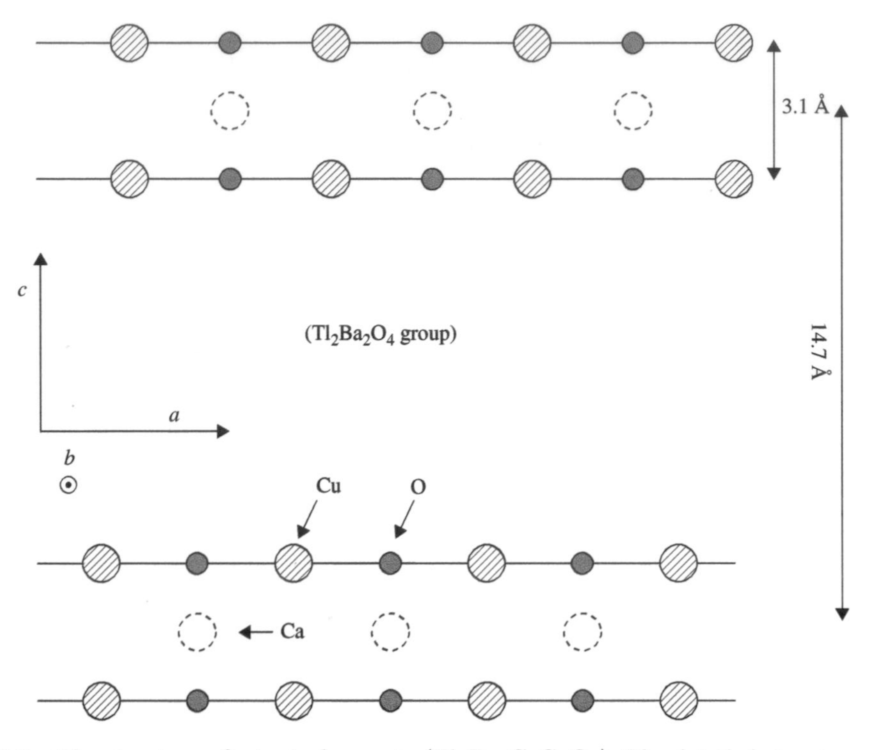

Fig. 7.1 The structure of a typical cuprate ($Tl_2Ba_2CaCuO_8$). The detailed structure of the "charge reservoir" group $Tl_2Ba_2O_4$ is not shown; see e.g. Shaked et al. 1994, p. 46.

Table 7.2 Some important superconducting cuprates.[4]

Common name	Formula	n, A, X	Remarks
LSCO	$La_{2-x}Sr_xCuO_4$	1, –, $La_{2-x}Sr_xO_2$	Original "high-T_c" material
YBCO	$YBa_2Cu_3O_{6+\delta}$	2, Y, $Ba_2Cu_{2+\delta}$	Probably most studied
BSCCO	$Bi_2Sr_2CaCu_2O_8$	2, Ca, $Bi_2Sr_2O_4$	Best for ARPES, EELS
HgBCO	$HgBa_2Ca_2Cu_3O_8$	3, Ca, $HgBa_2O_2$	Highest T_c to date
NCCO	$Nd_{2-x}Ce_xCuO_4$	1, –, $Nd_{2-x}Ce_xO_2$	"Electron-doped"
∞-layer	$Sr_xCa_{1-x}CuO_2$	∞, Ca(Sr),–	No charge-reservoir group

are listed in Table 7.2. A very useful source for the structure of many cuprates is the "pocket handbook" by Shaked et al. (1994).

I should mention at this point an alternative point of view on the structure of the cuprates, which comes naturally to chemists: A single-phase cuprate such as La_2CuO_4 may be thought of as a slight variant of the perovskite structure, ABO_3, in which

[4]The structures can be found in Shaked et al. 1994, except for BSCCO and HgBCO which are isostructural to Tl-2212 and Tl-1223 respectively.

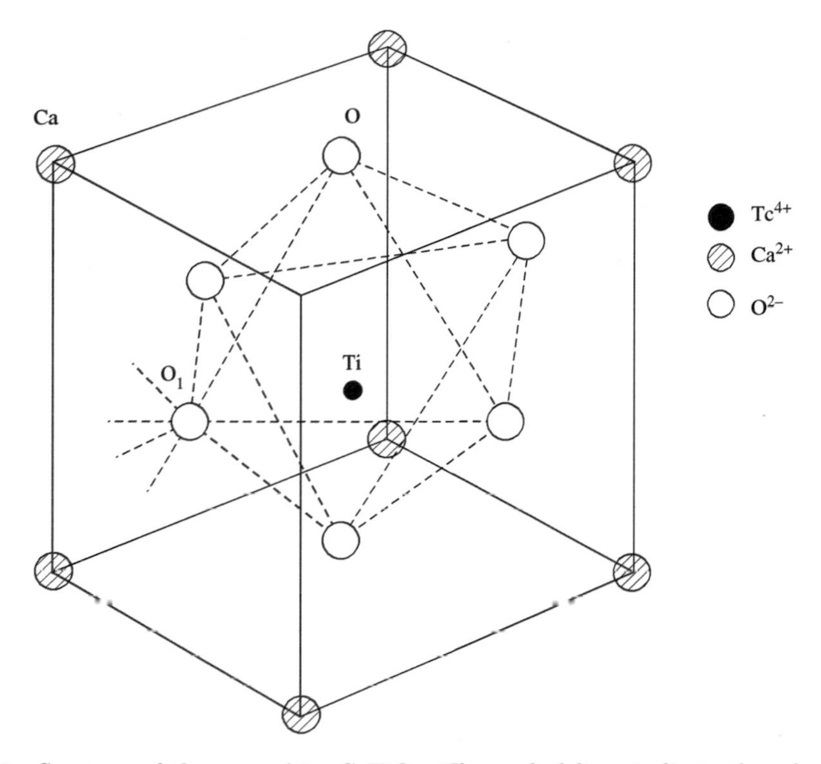

Fig. 7.2 Stucture of the perovskite $CaTiO_3$. The pecked lines indicate the edges of the oxygen octahedron. The dashed-dotted lines indicate that the oxygen atom labelled O_1 is shared with the octahedron in the next cube to the left.

the element B (Ti in the parent material, Ca Ti O_3) sits at the center of an oxygen octahedron, while the element A sits at the corners of the corresponding cube (see Fig. 7.2). In this (cubic) structure, each oxygen atom is shared between two octahedra. In La_2CuO_4 the role of B is played by Cu, and while the oxygens are still shared in the planar direction, those in the vertical direction (the so-called "apical" oxygens) are staggered relative to one another as in Fig. 7.3; thus we have an extra La and O in the formula unit. The same picture applies to a multilayer cuprate except that the oxygen octahedra are now as it were elongated in the vertical direction.

Let's now turn to the structure and chemistry of the CuO_2 planes,[5] which are almost universally believed to be the principal seat of superconductivity in the cuprates (see Section 7.8). To a first approximation, the structure of a CuO_2 plane is that of a simple square lattice with side approximately 3.84 Å, with Cu atoms at the corners and O atoms at the midpoints of the sides: see Fig. 7.4a. In the multilayer case, the lattices in the different CuO_2 planes within a multilayer are stacked on top of one another (i.e. not staggered[6]) and the intercalant element A is situated midway between the planes

[5]The raw data on which the statements below are based are mostly from standard x-ray diffraction (for the structure) and spectroscopy, augmented by standard valence considerations.

[6]However, as already stated, neighboring multilayers may be staggered relative to one another.

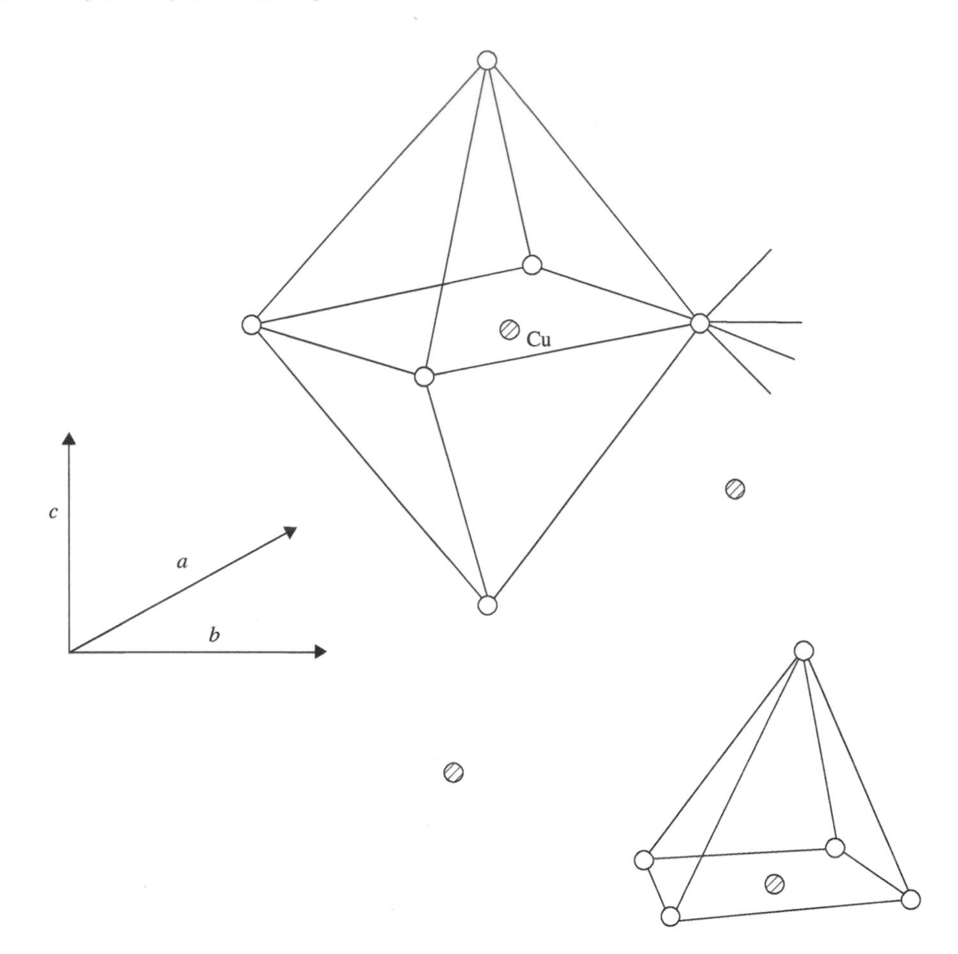

Fig. 7.3 The structure of La_2CuO_4. The sides of the oxygen octahedra are now indicted by solid lines.

in the center of the "unit cell", as indicated in the side view shown in Fig. 7.4b. Thus, to a first approximation the structure of a single CuO_2 plane is completely universal, independently of the particular cuprate compound in which it occurs.

On closer inspection the picture gets a little more complicated. In the first place, the lattice is often not quite square, and both the absolute values and the ratios of the a- and b-dimensions of the unit cell vary somewhat from compound to compound (the range of the former is 3.79 to 3.95 Å, and of the latter from 1 to 1.02 (for the extreme case of YBCO). Secondly, in most compounds (and all bilayer compounds) the planes are slightly "buckled", that is the Cu–O bonds do not all lie exactly parallel to the a- and b-axes; the degree of buckling varies from less than 0.5° (for $HgBa_2CuO_4$) to $\sim$14° (for YBCO). A further complication is that in the Bi compounds, probably because of the distorting influence of the BiO planes, there is a "superlattice" structure in the CuO_2 planes with a period $\sim$5 formula units.

As to the chemistry of the CuO_2 units and the associated spacer elements, the almost universal belief is that "at stoichiometry" (e.g. for La_2CuO_4 or $YBa_2Cu_2O_6$)

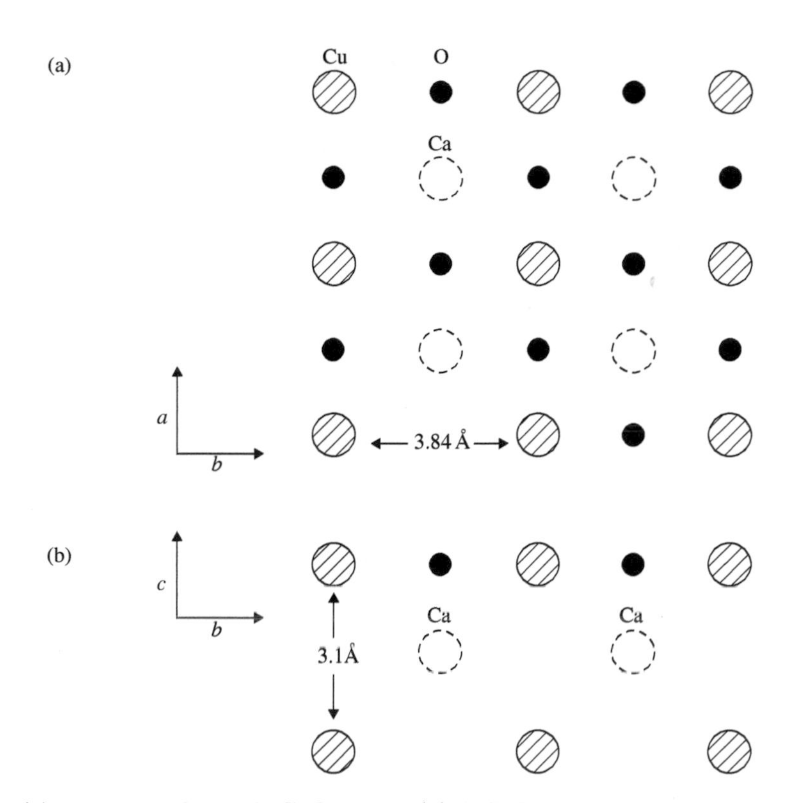

Fig. 7.4 (a) Structure of a single CuO_2 plane. (b) A CuO_2 bilayer viewed in the a-direction.

the valence state of the oxygen atoms is (to a first approximation) 2−, that of the alkaline-earth spacers 2+ (or for the rare earths or Y 3+), and that of the Cu atoms 2−; thus the oxygen and spacer atoms have closed electronic shells,[7] and the electronic configuration of the Cu atoms is $3d^9$, with the singly occupied orbital being the $d_{x^2-y^2}$ one. (A more accurate approximation allows hybridization between this orbital and the O p_x and p_y orbitals.) I return below to the effect of doping away from stoichiometry.

While the CuO_2 planes are the only structural units common to all cuprates, and are generally believed to be the main seat of superconductivity, other units may play a nonnegligible role. A particularly important case is the "chains" which are present in YBCO and isostructural RE compounds and run (by convention) parallel to the b-axis; while their exact role in superconductivity is controversial, the fact that the "number of superconducting electrons" as inferred from the London penetration depths is a factor of 2.5 greater for the b-direction than for the a-direction, even though the anisotropy of the CuO_2 planes themselves is only a factor of 1.02, would seem to imply a substantial role, whether direct or indirect, for the electrons in the chains. Again, in the heavy-metal cuprates such as the Bi series, which typically have BiO (or similar) planes, it

[7]Except of course for the tightly-bound $4f$ shells of the rare earth elements, which are not usually believed to play a role in superconductivity.

is not entirely clear whether these planes are metallic and whether, if so, they play any substantial role in the superconductivity.

In the cuprates superconductivity almost never occurs at exact stoichiometry; it is almost always necessary to "dope" the system by adjusting the properties of one or more of the "charge-reservoir" element so as to have a nonintegral average number for formula unit. Thus, for example, the compounds $La_{2-x}Sr_xCuO_4$ and $YBa_2Cu_3O_{6+\delta}$ show superconductivity for a range of fractional positive values of x and δ, respectively. This effect has apparently little direct connection with the presence of the extra (or substituted) atoms themselves; rather, it is a consequence of the change in valency of the formula unit. In fact, since the valency of La, Sr and O is respectively -2, -3 and $+2$, we would expect that the average number of valence electrons per formula unit is decreased by x and 2δ in the two cases respectively, or equivalently that we inject this number of holes. However, it is not always clear a priori where these holes go (e.g. in the case of YBCO, whether they migrate to the CuO_2 planes or to the chains or both). This question leads us to our next subject, the phase diagram of the cuprates.

Of the variables which control the phase diagram of the cuprates (temperature, pressure, magnetic field, chemical composition, etc.) the most significant are temperature and chemical composition ("doping") which it is conventional to plot along the vertical and horizontal axes respectively. Temperature needs no comment, but there is some ambiguity in the definition of doping. Clearly, for any given material we could plot, along the horizontal axis, the actual chemical stoichiometries (e.g. x, δ) themselves, or alternatively the total number of injected holes (x or 2δ). However, were we to do so we would find that the phase diagrams we obtain for different cuprates are displaced horizontally and/or squashed relative to one another. An alternative strategy consists essentially in *assuming*, for any given compound, that the relation between the chemical stoichiometry (x, δ, etc.) and the number of holes injected into a CuO_2 plane, which we shall call p, is such that the p-dependence of the phase diagram is universal; the absolute scale of p can then be obtained from those cases, such as LSCO, where one can be reasonably confident on chemical grounds that all the injected holes go into the planes (i.e. that in this case $p = x$). Although this fitting procedure might at first sight seem rather arbitrary and dangerous, the results appear to be consistent with the known chemical constraints, and it gains some support from the observation that when plotted in terms of p not only the phase diagram itself[8] but many experimental properties appear to possess a reasonable (though not total) degree of universality. From now on I will adopt this strategy without further comment.

Even with the above procedure, it is not really possible to obtain the complete phase diagram from a single system, since (at least to my knowledge) there is none which is chemically and structurally stable over the whole range of p which is of interest; rather, one has to use several different systems whose regions of stability in p overlap. The result of such a procedure is the by now standard phase diagram which is shown, for $0 < p < 0.4$, in Fig. 7.5, and on whose main features I will now comment.

[8] In a few cases the situation is complicated by the occurrence of one or more crystallographic phase transitions which are specific to the system in question. I ignore this complication in what follows.

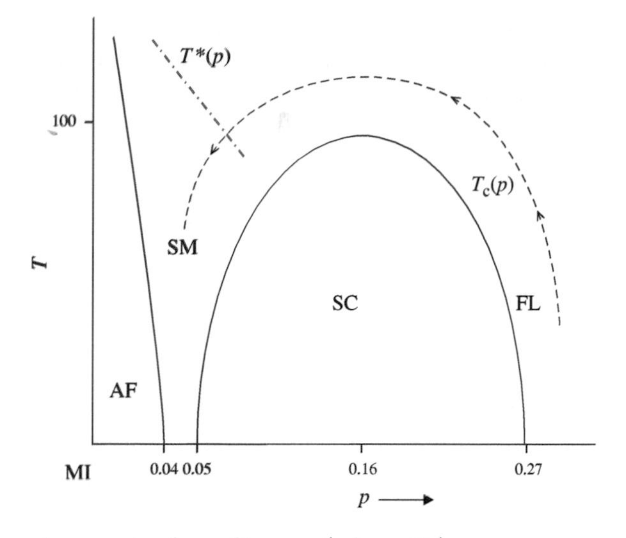

Fig. 7.5 The generic cuprate phase diagram (schematic).

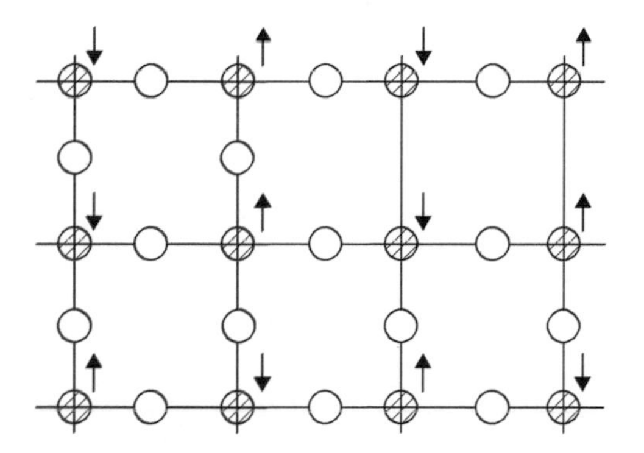

Fig. 7.6 The Mott-insulator state of a parent cuprate, e.g. La_2CuO_4. The view is of the CuO_2 plane from above.

At the limit of perfect "stoichiometry", $p = 0$, and at low temperature, the system is an electrical insulator, and neutron diffraction studies show that it is an antiferromagnet, with the magnetic unit cell twice the size of the crystal unit cell. Thus, the system is a classic *Mott insulator*: the electrons (or equivalently the holes) in the Cu $d_{x^2-y^2}$ orbital order so that there is exactly one localized electron per unit cell (or per Cu atom) with spins ordered antiferromagnetically; see Fig. 7.6. This Mott-insulator ("MI") phase persists, at $p = 0$, to a temperature of a few hundred K which does not vary much from material to material.

As we increase p away from zero at low temperature, the degree of antiferromagnetic order decreases and eventually tends to zero at a second-order phase transition which intersects the horizontal axis at approximately $p = 0.04$, more or less independently of material. I postpone for a moment a discussion of the nature of the phase to the right of this transition, but note that even at $T = 0$ it definitely seems to occupy a nonzero region of the horizontal axis, and to be (probably) electrically insulating, with magnetic behavior of the "spin-glass" type. At approximately $p = 0.05$, at $T = 0$, we find the onset of the superconducting (SC) phase, which occupies a "dome" of variable height in the phase diagram; despite this, the maximum of $T_c(p)$ always seems to occur at a p-value of approximately 0.16, and superconductivity disappears, at $T = 0$, at approximately $p = 0.27$. To the right of this point, the ("overdoped") phase which is realized at low temperatures appears to correspond roughly to a textbook "Fermi liquid" (FL) normal metal (though it should be said that the data in this regime are less extensive than in other parts of the phase diagram).

If now we start in this "FL" regime and trace a path in the phase diagram towards lower values of p which goes around (above) the superconducting "dome" (indicated by the dashed line in Fig. 7.5), we find that while the system remains essentially metallic, its properties diverge more and more from the textbook FL behavior. In fact, at $p = 0.16$ ("optimal doping") most of the experimental properties, when plotted as a function of temperature, bear little resemblance to the standard metallic behavior: see next section. In the region of the phase diagram to the left of the superconducting dome, at temperatures $\lesssim$200 K, the properties are even more anomalous, and for this reason this regime is often called the "strange metal" (SM) regime (also, for reasons we will see below, the "pseudogap" (PG) regime). In particular, if we stay to the left of the maximum T_c and decrease the temperature as shown by the dashed line in Fig. 7.5, the resistivity increases dramatically, apparently tending to infinity as $T \to 0$.

Although the experimental properties of the system in the SM regime are qualitatively quite different from those in the FL regime, the transition between them appears to be continuous (provided we avoid the SC region as on the dashed path). Nevertheless, while there is (at least to my knowledge) no evidence in any experiment to date for a phase transition, there is considerable evidence that the line marked $T^*(p)$ in Fig. 7.5 corresponds to a relatively sharp "crossover" between two rather different patterns of quantitative behavior. The reason that the dotted line does not extend beyond $p = 0.13$ in Fig. 7.5 is that at the time of writing it is controversial whether it should be extended so as to join the curve $T_c(p)$ as in Fig. 7.7.b, or rather dives down so as to enter the SC phase and, perhaps, intersect the horizontal axis around $p = 0.19$ (Fig. 7.7a); in the present author's opinion, various experiments in which $T_c(p)$ has been suppressed by doping or a magnetic field (see below) tell in favor of the latter hypothesis (cf. also Tallon and Loram 2001), but the matter is not yet entirely settled. Irrespective of this point, it is not entirely clear whether the line $T^*(p)$ is indeed merely a "crossover" line or whether it rather is evidence for a hidden phase transition whose presence is obscured by the effects of disorder. If the latter is the case, the nature of the order which sets in on the low-temperature side is a matter of conjecture.

While the "width" of the SC region in the p-T plane, i.e. the values of p between which it occurs, appears to be more or less universal,[9] the height $T_{c\,\max}$ (i.e. the maximum value of $T_c(p)$) varies widely from system to system and in many cases even from sample to sample within a single system, from a value below 10 K for some samples

[9]Although this is of course to some extent a consequence of the ansatz used to convert the stoichiometric quantities x, δ to p, see above.

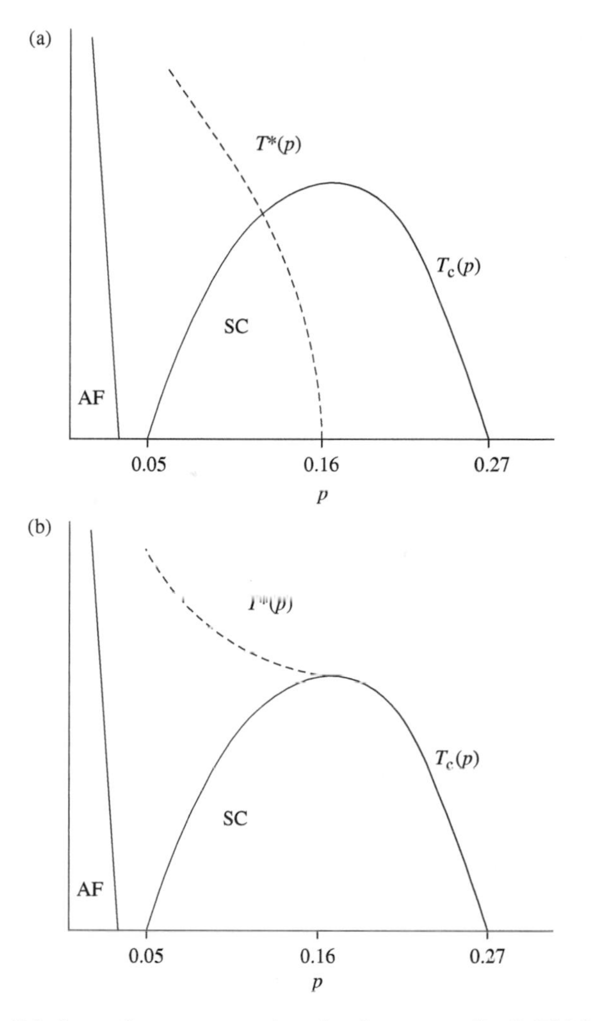

Fig. 7.7 Two possible hypotheses concerning the "crossover line" $T^*(p)$ (after Tallon and Loram 2001).

of single-layer BSCCO to ~160 K for trilayer HgBCO at high pressures. (The *shape* of the curve of $T_c(p)$, when scaled by $T_{c\,max}$, however appears to be approximately universal: $T_c(p)/T_{c\,max} \cong 1 - 82.6(p - 0.16)^2$.) If for a moment we consider a single system at a given value of p, then a general rule of thumb is that $T_c(p)$ increases with pressure at least up to a few GPa, and is depressed by a magnetic field and by the presence of impurities in the CuO_2 planes, especially by Zn substituting for Cu.

Even at zero field and ambient pressure, and with the "best" samples available,[10] $T_{c\,max}$ still varies widely between systems. Numerous attempts have been made to

[10] "Best" in this context refers to the degree of crystallographic purity, and to the absence of impurities other than the intended dopants. Generally speaking, the "best" (or "optimally doped") samples as so defined have the highest values of $T_{c\,max}$.

Table 7.3 Transition temperatures in various homologous series.[11]

Common name	X group	T_c for: $n=1$	2	3	4	5
Hg-12, $n-1$, n	$HgBa_2O_{3+\delta}$	98	126	135	125	110
Tl-22, $n-1$, n	$Tl_2Ba_2O_4$	95	118	125	112	105
Tl-12, $n-1$, n	$TlBa_2O_{3-\delta}$	70(?)	103	123	112	107
Pb-12, $n-1$, n	$PbSr_2O_{3+\delta}$	~40	97(?)	122	107	
Bi-22, $n-1$, n	$Bi_2Sr_2O_4$	10–20	89	107		

correlate $T_{c\max}$ to the various parameters which vary from system to system, such as the degree of orthorhombic anisotropy and/or of buckling of the CuO_2 planes, the chemical environment of the planes or the distance to the apical oxygens; in no case does the correlation seem so striking as to exclude other possibilities. However, there is one correlation which is both striking and universally accepted, namely that of $T_{c\max}$ with n in homologous series spaced with Ca. From Table 7.3 it is clear that in such series $T_{c\max}$ always increases when we go from $n = 1$ to $n = 2$, and again from $n = 2$ to $n = 3$, the increase in the latter case being approximately $\frac{1}{3}$ of that in the former when they can be compared. It also appears probable (though not certain, because of uncertainties regarding the uniformity of the doping, etc.) that $T_{c\max}$ decreases slightly for $n \gtrsim 4$. This striking behavior has motivated a number of theories of cuprate superconductivity, in particular the "interlayer tunneling model" of P. W. Anderson, see Section 5.

In this context I would like to draw attention to a curious and little-noticed fact concerning the class of nonsuperconducting cuprates mentioned above: To the best of my knowledge, a necessary condition to be a member of this class is to be a bilayer cuprate intercalated ("spaced") entirely with either Sr or Ba. Moreover, until the recent demonstration by Di Luccio et al. (2003) of superconductivity in Sr-intercalated bilayer BSCCO, the condition seems also to have been sufficient. A clue to the origin of this curious regularity may come from the studies of Shaked et al. (1993) on bilayer LSCO ($La_{2-x}Sr_xCaCu_2O_6$), in which substitution of the intercalated Ca by Sr drives T_c to zero; they found that the correlation of T_c was not with the concentration of intercalated Sr itself, but rather with that of the "intruder" oxygen atoms which, owing to the slightly larger size of the Sr^{++} ion relative to the Ca^{++} one, were able to sneak in between the two CuO_2 planes of the bilayer. But why these "intruder" oxygens should have such a drastic effect on superconductivity remains a mystery.

A final note: although the vast majority of superconducting cuprates have a phase diagram of the type shown in Fig. 7.5, there is at least one system, $Nd_{2-x}Ce_xCuO_4$ or "NCCO", in which superconductivity occurs at *negative* values of p. For these "electron-doped" superconductors the phase diagram is, crudely speaking, a reflection

[11]The Tl-12 and Pb-12 series have been studies less intensively than the other three, so the data on them may be subject to more uncertainty.

(though not an exact one) of the conventional one in the temperature axis. Most of the experimental evidence on these materials is consistent with the assumption that they do not differ qualitatively from the hole-doped majority.

7.3 The cuprates: principal experimental properties

The experimental data base on the cuprates is enormous, comprising more than 20,000 papers in the literature over the last 20 years. Thus, to get any kind of overall perspective on the experimental properties it is necessary to rely on the secondary literature (books or review articles). In doing this some caveats should be borne in mind: First, these materials are an order of magnitude more complicated in their crystallographic, chemical and materials properties than the classic superconductors, and since their superconductivity appears to be very sensitive to all these factors, not all conclusions drawn on the basis of early experiments have stood the test of time (though many have). Secondly, it needs to be borne in mind that because of practical experimental constraints, a particular kind of experiment may be feasible only on one or a few cuprates: for example, neutron scattering and NMR experiments, which require large single-crystal samples, are mostly done on YBCO and LSCO, while transmission EELS (electron energy loss spectroscopy), which optimally requires samples in the form of thin "flakes", is done almost exclusively on BSCCO; this is also true to a lesser extent of ARPES. Also, it needs to be remembered that it is very much easier to vary T at constant p than vice versa, and very few experiments have done the latter. Thirdly, the diversity both of the materials and of the experiments conducted on them is so great that virtually any theoretical proposal concerning cuprate superconductivity can be defended by being sufficiently selective in one's choice of the evidence (and, if challenged, dismissing the "unwelcome" part of it as due to bad samples and/or experimental error); thus, any review of the experimental situation written by theorists or others with an axe to grind should be read with an appropriate degree of caution. In Appendix 7.A I list a few general references which seem to me fairly even-handed; in the present section I will try as far as possible to confine myself to statements which I believe to be relatively uncontroversial.

At the start, I note that the elastic properties[12] of the cuprates (density, compressibility, thermal expansion, ultrasound velocity, Grüneisen parameter, etc.) are not particularly remarkable in any part of the phase diagram (except in some cases close to a crystallographic phase transition), being crudely comparable to those of a typical cubic perovskite such as $SrTiO_3$; the only difference is that, as we might perhaps expect, quantities such as the compressibility and thermal expansion are anisotropic, particularly for the bi- and trilayer compounds, being typically a factor $\sim$3 larger along the c-axis than in the ab-plane. An interesting observation is that under c-axis compression the shift of the apical oxygens towards the CuO_2 planes is appreciably more than proportional to the contraction of the overall lattice parameter. In some compounds, e.g. $REBa_2Cu_3O_3$, an orthorhombic-to-tetragonal crystallographic transformation occurs under a pressure $\sim$20 GPa.

[12]See Allen et al. (1989), Schilling and Klotz (1992).

From now on I will treat successively the experimental properties in (A) the normal state at optimal doping,[13] (B) the "pseudogap" regime, and (C) the superconducting state. Also, to prevent the recitation sounding too much like the proverbial laundry list, I shall usually remind the reader, in regimes (A) and (C), of the "textbook" results, i.e. those which follow from the Landau Fermi-liquid theory and the BCS theory respectively.

7.4 Normal state at optimal doping

7.4.1 Electronic specific heat

Because of the high T_c's of the cuprates relative to the Debye temperature, the experimental specific heat tends to be dominated even for $T \sim T_c$ by the phonon contribution; Loram and co-workers (1994) obtain the electronic specific heat by subtracting the experimental specific heat of a "reference" compound, usually the "parent" (e.g. $YBa_2Cu_3O_6$) which is an antiferromagnetic insulator and hence presumed to have negligible electronic specific heat up to room temperature.

The textbook result is, for $T \ll T_F$,

$$c_V^{el} = \gamma T, \qquad \gamma = \frac{2\pi^2}{3} k_B^2 N(0) \tag{7.4.1}$$

where $N(0)$ is the single-electron density of states at the Fermi surface per unit energy and spin. Loram and co-workers find that their results fit Eqn. (7.4.1) rather well for optimally doped YBCO, LSCO and Tl-2201, with γ-values which they actually quote in mJ/g atom K^2, but which when converted to "per mole planar CuO_2" is almost the same for the three materials, namely

$$\gamma = 6.5 \frac{\text{mJ}}{\text{mol}(\text{CuO}_2)\ \text{K}^2} \tag{7.4.2}$$

If we interpret this result in terms of a density of states $N(0)$ according to Eqn. (7.4.1), we get

$$N(0) \cong 1.4\ \text{eV}^{-1}\ \text{spin}^{-1}\ (\text{CuO}_2\ \text{unit})^{-1} \tag{7.4.3}$$

This is a factor ~4 times the value $(ma^2/2\pi\hbar^2)$ which would be predicted from a free-electron model.

Thus the specific heat data at optimal doping (and for the overdoped case) for $T \lesssim$ RT seem prima facie consistent with a Fermi-liquid model of the normal state with a modest enhancement of the electron effective mass.

7.4.2 Magnetic properties (Pennington and Slichter 1992: Kastner et al. 1998)

The textbook prediction, for $T \ll T_F$, is that the total spin susceptibility χ should be temperature-independent, with the "Wilson ratio" $((\pi k_B/\mu_B)^2/3) \cdot (\chi/\gamma) \sim 1$: the Knight shift K_s should be proportional to χ and thus also temperature-independent; and the NMR relaxation rate T_1^{-1} should be of the form $T_1^{-1} = cT$ (the Korringa law) with the constant c proportional to K_s^2.

[13] In regime (A) I will be most interested in the properties below and around room temperature (RT), since these seem most likely to give clues concerning the superconducting state.

Most magnetic data are on YBCO. According to Loram (op. cit., Fig. 5) the total spin susceptibility is indeed nearly temperature-independent, with a Wilson ratio close to 1 (the result expected for a Fermi liquid with the Landau parameter F_0^{a} equal to zero). As regards the NMR data, the situation is quite complicated; I refer to Pennington and Slichter (1992) for details and give only a summary here. Briefly, the NMR data for all the "interesting" nuclei (the in-plane Cu's and O's, and the intercalant Y's) appear to conform to the FL predictions, with the exception of the nuclear relaxation rate of the in-plane Cu's, which seems to be better fitted by the formula $T_1^{-1} = aT + b$. However, the *magnitude* of the Cu Knight shift relative to the Y and O ones is considerably larger than expected from the known structure of the electron–nuclear interaction, and this may give an important clue to the structure of the electronic spin fluctuations in the N state.

Recently, the picture has been considerably complicated by the results obtained by the Slichter group using a more sophisticated ("SEDOR") technique (Haase and Slichter 2000). The results appear to indicate that the magnetic and electric-quadrupole fields seen by different (crystallographically equivalent) nuclei are not identical, rather there is a distribution of both: it is significant, however, that the deviations of the two fields from the mean appear to be connected, as if there is a single parameter controlling both. The magnitude of the effect increases with decreasing temperature. This phenomenon is not well understood at present.

7.4.3 Transport properties (Ong 1990, Iye 1992, Uher 1992, Cooper and Gray 1994: Kastner et al. 1998)

The transport coefficients are second-rank tensors, which from symmetry have their eigenvectors parallel to the three crystal axes. Unless explicitly otherwise stated I will restrict myself to the *ab*-plane eigenvalues and assume them to be degenerate (i.e. ignore the effects of the possible orthorhombic in-plane anisotropy).

The textbook predictions, for a Fermi-liquid-like description of the N phase at $T \ll T_{\mathrm{F}}$, are as follows: The d.c. resistivity $\rho(T)$ is given by the standard Bloch–Grüneisen formula, and is thus proportional to T for $T \gtrsim \theta_{\mathrm{D}}$ (Debye temperature), complicated for intermediate temperature, and in the limit $T \ll \theta_{\mathrm{D}}$ is of the form $a + bT^5$. The a.c. conductivity has, for ω small compared to both the band gap and the plasma frequency ω_{p}, the form

$$\begin{gathered}\sigma(\omega, T) = \frac{\sigma(0)}{1 + i\omega\tau} \equiv \frac{\omega_{\mathrm{p}}^2 \tau}{(1 + i\omega\tau)} \\ \left(\sigma(0) \equiv \rho^{-1}(T),\ \tau^{-1}(T) \equiv \frac{\sigma(0)}{\omega_{\mathrm{p}}^2}\right)\end{gathered} \tag{7.4.4}$$

The thermal conductivity (mostly due to phonons for not too low T) is approximately temperature-independent; the thermoelectric power is proportional to T. Finally, the Hall coefficient should be independent of temperature, and in the high-field limit ($\omega_{\mathrm{c}}\tau \gg 1$, $\omega_{\mathrm{c}} \equiv eB/m$) should be given in a free-electron model by $1/ne$, where n is the (volume) number density of carriers and e their charge. The Hall *angle* is defined as $\tan^{-1}[\sigma_{xy}(B,0)/\sigma_{xx}(0)]$, and thus from the above considerations should be proportional to τ^{-1} and hence, for $T \gtrsim \theta_{\mathrm{D}}$, to T.

The (in-plane) d.c. resistivity $\rho(T)$ is one of the most widely measured properties of the cuprates, and has a very striking and apparently universal behavior: Throughout the whole of the N-state region of the phase diagram, with the exception of the "pseudogap" regime (on which see below) $\rho(T)$ appears to be very well fitted by a power-law behavior, $\rho(T) \propto T^{\alpha}$, where the exponent α increases monotonically with the doping x, reaching a value of 2 in the "FL" regime to the right of the superconducting dome. At optimal doping α appears to be 1 within experimental error for all cuprates so far measured, that is, *the resistivity is strictly linear in temperature for all* $T > T_c$.[14] It should be emphasized that this behavior persists not only in low-T_c compounds such as Bi-2201 where the temperature range in question extends far below the temperature (typically $\sim\theta_D/3$) where the Bloch–Grüneisen law deviates strongly from a linear form, but even down to $T \sim 0$ in experiments where the superconductivity is suppressed by a strong magnetic field (Boebinger et al. 2000, and earlier references cited therein). Thus, the behavior appears to have nothing to do with the traditional Bloch–Grüneisen picture.[15]

An obvious question is whether, if different compounds are compared at optimal doping, the (2D) resistance per CuO_2 plane is universal? If this is so, then at any given temperature the ratio of resistivities should be the ratio of the mean spacing of CuO_2 planes along the c-axis. Inspection of e.g. Fig. 5 of Iye (1992) indicates that at first sight this does not work very well: e.g. LSCO has a much higher resistivity than YBCO, even though the average interplane spacing is only slightly greater[16] (6.5 Å vs. 5.9 Å). However, if we restrict ourselves to the higher-T_c (say $T_c \gtrsim 80$ K) compounds it appears to work reasonably well, with a 2D resistance $R_\square$ which at RT is $\sim 3k\Omega$, i.e. about 0.12 of the "quantum unit of resistance" h/e^2.

The a.c. conductivity does not appear to have a simple Drude form. It is possible to fit it to a Drude formula, $\sigma(\omega) \sim \omega_p^2\tau/(1+i\omega\tau)$, but only if τ is allowed to be itself a function of ω, with $\tau^{-1}(\omega) \sim \max[\omega, kT/\hbar]$ (thus giving the d.c. result $R \sim T$). See also below on the optical properties.

The *Hall effect* is also anomalous. In YBCO (pure or Zn-doped) R_H^{-1} is closely proportional to $a + bT$, but in other superconducting cuprates where it has been measured the dependence is considerably weaker[17] (and approaches a constant as we overdope). The sign is usually positive.

The *Hall angle* shows a very characteristic behavior, at least in (pure or Zn-doped) YBCO: at $B = 8$ T,

$$\cot\theta_H = aT^2 + b \qquad b = f(\text{doping}),\ a = \text{independent of doping} \tag{7.4.5}$$

The *thermoelectric power S* usually has a positive value at T_c and a constant negative slope. A very intriguing observation is that as p is varied, the RT value of

[14] There is sometimes in addition a small constant term, but it appears to vanish in the limit of high sample purity.

[15] It is worth noting that in compounds such as YBCO which possess a strong crystalline anisotropy, the eigenvalues $\rho_a(T)$ and $\rho_b(T)$, though different (in YBCO by a factor ~ 2) are still each strictly linear in T.

[16] However, the case of YBCO is complicated because it is not clear whether, when considering the b-axis resitivity, the chains should be counted as an "extra plane".

[17] Note that Iye's Figs. 14 and 15 plot R_H while Fig. 17 plots R_H^{-1}.

S crosses zero at almost exactly the point where $T_c(p)$ has its maximum, $p \cong 0.16$ (Obertelli et al. 1992).

Finally, the thermal conductivity is the one transport property which behaves reasonably "normally": it is usually either $\sim$ const. or weakly decreasing as a function of T.

I now turn more briefly to the c-axis transport coefficients (Cooper and Gray 1994, Takagi 1998). In contrast to the ab-plane ones, these are very far from universal (at least at first sight!). The *d.c. c-axis resistivity* is usually relatively well fitted (at optimal doping) by a power law, $\rho_c(T) \sim T^\alpha$, but the exponent α can range from $\sim$+1 (e.g. YBCO, Tl-2201) through 0 (Hg-1201, LSCO) to $\sim$−1 (Bi-2212). The absolute magnitude of ρ_c (at $T \sim T_c$, say) is always much larger than that of ρ_{ab}, by a factor which varies from $\sim$30 for optimally doped YBCO to $\sim 10^5$ for Bi-2212. The optical conductivity is featureless as a function of ω right up to frequencies $\sim$1 eV, except for isolated peaks which can be correlated to known phonons. (What is usually measured directly is the reflectance, which depends on the *complex* dielectric constant, but barring pathologies a constant reflectance as measured (see e.g. Cooper and Gray (loc. cit. Fig. 13) implies constant ϵ_1 and ϵ_2, hence $\sigma(\omega) \sim \omega^{-1}$.)

7.4.4 Spectroscopic probes[18]

Since for technical reasons a given spectroscopy is often feasible on only a subset of cuprates, I shall indicate in this subsection the systems which have been principally investigated by each technique.

It is convenient to divide spectroscopic probes into those which probe the bulk material (neutron scattering, transmission EELS, and to a lesser extent optical reflectivity and Raman spectroscopy) and those which are sensitive only to conditions within a few atomic layers of the surface (tunneling, photoemission). The volume probes tend to be sensitive to the motion of the ions (i.e. to phonons) as well as that of the electrons; in the ensuing discussion I shall assume that these phonon effects can be reliably identified, and will concentrate on the remaining (electronic) effects.

Neutron scattering (Mook et al. 1997: Norman and Pepin 2003) (LSCO, some YBCO)

Neutrons couple negligibly to the electronic charge degree of freedom: the principal coupling to the electrons is via their spin, so the neutron scattering cross-section $\sigma(\boldsymbol{q}, \omega)$ measures (after subtraction of phonons) the spectrum of spin fluctuations of wave vector $\boldsymbol{q}$ and frequency ω. Neutron scattering experiments need large crystals and even then are very time-consuming, so that the error bars are often comparable to the real data.

If for optimally doped YBCO in the normal phase we plot $\sigma(\boldsymbol{q}, \omega)$ as a function of $\boldsymbol{q}$, then there is a marked maximum close to the "commensurate" value (in units of π/a) (0.5, 0.5); this is exactly the point at which in the AF phase of the parent compound we get magnetic "superlattice" scattering. It is debated whether at optimal doping the peak is a single, exactly commensurate one (i.e. exactly at (0.5, 0.5)) or whether it is really four peaks at $(0.5 \pm \delta, 0.5 \pm \delta)$ where δ is small (for *underdoped*

[18]A good general reference is the proceedings of the series of conferences on Spectroscopies of Novel Superconductors, published in *J. Phys. Chem. Solids.*

YBCO it seems almost certain that the latter assignment is correct). At RT the cross-section at (0.5, 0.5) has very little energy-dependence, but as T is lowered there are some indications of a broad peak centered at 34 meV.

Raman scattering (Blumberg et al. 1998b, Rübhausen 1998)

In Raman scattering, one shines on the system light of frequency ω_i and definite polarization $\hat{\epsilon}_i$, and detects the light scattered with frequency ω_f and polarization $\hat{\epsilon}_f$, in general different from $\hat{\epsilon}_i$. If the *difference* in wave vector of the incident and scattered light is $\boldsymbol{q}$ (note that $\boldsymbol{q}$ is almost invariably very small on the scale of the reciprocal lattice, etc.) and ω the difference $\omega_i - \omega_f$, and if we assume that the intermediate state (wave vector $\boldsymbol{q_i}$, energy ω_i) is not too close to a resonance, then what Raman scattering essentially measures is the fluctuations of the dielectric constant tensor, $\langle \epsilon_{\alpha\beta} : \epsilon_{\gamma\delta} \rangle (\boldsymbol{q}, \omega)$ where the indices $\alpha, \beta, \gamma, \delta$ depend on the polarizations. Very often in the literature, it is assumed that particular choices of $\alpha, \beta, \gamma, \delta$ (e.g. the so-called B_{1g} geometry) correspond to particular mechanisms of scattering (e.g. the so-called "two-magnon" scattering), but this may be dangerous. Like neutrons, Raman scattering is sensitive to phonons as well as electrons.

The electronic Raman spectra of optimally doped YBCO and BSCCO appears to be almost totally featureless for ω up to at least 2 eV, and moreover at RT to be essentially identical in the A_{1g} and B_{1g} channels; however, for $T \lesssim 2T_c$ some difference appears (and the intensity increases somewhat overall). Note that this is far from what would be prima facie expected for the "textbook" model of a metal, where the fluctuations of $\epsilon(\boldsymbol{q}, \omega)$ should depend on those of physical quantities such as the charge and spin density, and thus prima facie be limited to a frequency regime $\sim v_F q \ll 2$ eV.

Optical reflectivity (many cuprates) (Timusk and Tanner 1989, Tanner and Timusk 1992: Norman and Pepin 2003)

Optical reflectance experiments measure the real and imaginary parts of the complex $q \cong 0$ dielectric constant $\epsilon(\omega)$, either indirectly from the reflection coefficient $R(\omega)$ via use of the Kramers–Kronig relation or directly by ellipsometric techniques. In comparing the reflectance data on different cuprates, it is essential to bear in mind that $\epsilon(\omega)$ is a "3D" quantity; thus, even if at (say) optimal doping the behavior of the individual CuO_2 planes is identical in different materials, the fact that the density of planes differs from material to material means that $\epsilon(\omega)$ will for that reason alone be different, and this will be reflected in $R(\omega)$ in a complicated way. A further real-life complication which is easy to forget is that while the (mostly insulating) "reservoir" layers between the CuO_2 planes are unlikely to contribute much to the conductivity at frequencies $\lesssim$3–4 eV (i.e. to the imaginary part of ϵ) they will contribute very importantly to the *real* part and hence to the reflectance.[19] In view of these complications it is perhaps at first sight surprising that the *ab*-plane optical properties of the cuprates show any "universality" at all.

[19]However, it may be possible to take this effect into account semiquantitatively for members of the same homologous series.

The most dramatic "universal" property comes out most clearly if we plot not the reflectance or the real part of the conductivity, but the so-called loss function $L(\omega) \equiv -\text{Im}\ \epsilon^{-1}(\omega)$. If we consider a simple textbook model of a metal and for the moment neglect band-structure effects, then $\epsilon(\omega)$ has the simple form

$$\epsilon(\omega) = 1 - \frac{\omega_{\text{p}}^2}{\omega(\omega + i/\tau)} \tag{7.4.6}$$

where $\omega_{\text{p}} \equiv (ne^2/m\epsilon_0\epsilon_\infty)^{-1/2}$ is the (3D) plasma frequency, τ is the Drude relaxation time and ϵ_∞ the "high-frequency" dielectric constant. Thus in the usual limit $\omega_{\text{p}}\tau \gg 1$, $L(\omega)$ would consist of a sharp peak of width $\sim 1/\tau$ at the plasma frequency, plus a small background. By contrast, all the superconducting cuprates (indeed, to my knowledge, all the metallic ones even when nonsuperconducting) have a strong and broad spectrum of $L(\omega)$, typically extending from $\sim$0.1 eV to a fairly sharp upper cutoff at a value of ω which varies from $\sim$1 to 2 eV depending on the cuprate. This "mid-infrared (MIR) peak" is one of the most striking generic properties of the cuprates, and attracted attention from an early stage. Also characteristic is the near-zero value of $L(\omega)$ which occurs at the upper edge of the MIR peak; at higher energies $L(\omega)$ again has some weight, but it is not so spectacular and the detailed form tends to be material-specific as one might expect. The temperature dependence of the MIR optical properties is surprisingly large, see e.g. Holcomb et al. 1996.

Electron-energy-loss spectroscopy (EELS) (Nücker et al. 1989)

The simplest kind of EELS experiments, "transmission EELS", measures the fluctuations of the charge density in the scattering system, and thus should be closely related to the optical experiments. In fact, for an isotropic 3D system in the limit $\boldsymbol{q} \to 0$ the EELS cross-section $\sigma(\boldsymbol{q}, \omega)$ should be directly proportional to $L(\omega)$. In a strongly layered system this equivalence holds only for $qd \ll 1$ where d is the interlayer spacing, so one must employ caution in applying it to the cuprates. However, it is reassuring that experiments in BSCCO (and, to a large extent, in YBCO) with $\boldsymbol{q} \parallel ab$, $|q| \sim$0.05–0.1 Å^{-1} (so that $qd \lesssim 1$) do show a spectrum which appears to be consistent with the optical data, in particular they show a strong MIR peak. For larger values of q (up to $\sim$0.5 Å^{-1}, the upper limit for practical reasons) the peak persists but is somewhat attenuated relative to the background.

I now turn to "surface" probes. For such experiments the system of choice is often BSCCO, and in this context it is worth noting that this material almost always cleaves preferentially along the "weak link" between the two BiO planes, so that the 'nearest" CuO_2 plane is two layers down (but cf. below, on the S phase).

Tunneling (I–V characteristics) (Hasegawa et al. 1992: Hussey 2002) (LSCO, YBCO, BSCCO, etc.)

Tunneling experiments on the cuprates are mostly done using not macroscopic tunnel oxide junctions but STM (point-contact) techniques, and there is some evidence that the structure of the measured I–V characteristic may be sensitive to the tip-to-surface distance. The measured differential conductance $G_n(V) \equiv \partial I/\partial V$ should measure the product of an appropriate squared matrix element times the density of singe-particle states at energy eV. The textbook prediction is that $G_n(V)$ should be flat on the scale of the Fermi energy (or possibly some characteristic energy associated with the junction, if smaller) and should be temperature-dependent only on this scale. The actual result is that in all experiments to date $G_n(V)$ appears to be well fitted by the formula

$$G_n(V) = a + b\,|V| \tag{7.4.7}$$

with b virtually independent of T and a decreasing weakly with decreasing T. The absolute values of a and b are of course junction-specific, but the ratio a/b is typically $\sim$50–100 mV. This behavior is unusual but not without parallel; something rather similar is found, e.g. in BKBO, which is generally believed to be a member of the "classic" class of superconductors.

ARPES (mostly BSCCO, some YBCO and LSCO)

The full theory of angularly resolved photoemission (ARPES) experiments is quite complicated, but in strongly layered materials such as the cuprates they are generally believed to measure the "spectral function" $A(\boldsymbol{k}, \epsilon)$, i.e. the probability that an electron has in-plane momentum $\boldsymbol{k}$ and energy ϵ; if this is so, then by integrating the result over ϵ we obtain the average occupation number $\langle n(\boldsymbol{k})\rangle$ of the state $\boldsymbol{k}$. If this function shows a discontinuity as $|\boldsymbol{k}|$ is varied for fixed direction $\hat{\boldsymbol{n}}$, we identify this value of $|\boldsymbol{k}|$ with the Fermi surface. Thus we plot out the Fermi surface (actually line) as a function of direction $\hat{\boldsymbol{n}}$.

When this procedure is applied to the cuprates in the N phase, the results are complicated but two things stand out. First, in contrast to the textbook prediction that $A(\boldsymbol{k}, \epsilon)$ should be sharply peaked near a "quasiparticle energy" $\epsilon = \epsilon(\boldsymbol{k})$, with a width which tends to zero (at $T = 0$) at least as k approaches k_{F} (cf. Eqn. 5.A.14), the spectral function of the cuprates is very smeared-out in ϵ, with at least 90% of the "weight" in the background for all values of $\boldsymbol{k}$. Secondly, if we integrate over ϵ to find $\langle n_{\boldsymbol{k}}\rangle$, we indeed get a sharp drop as a function of $|\boldsymbol{k}|$ for fixed $\hat{\boldsymbol{n}}$, though it is again at most $\sim$10% of the mean value. If we use this drop to define a Fermi surface as above, we find that, crudely speaking,[20] we get one of the form shown in Fig. 7.8, so that the "Fermi-sea" is hole-like and centered at (π, π), with a total area that corresponds roughly to $(1 + p)$ ($\cong$1.19 for optimal doping) holes per CuO_2 unit (note not 0.19!). It is somewhat reassuring that this is just what is predicted by a phenomenological band-structure calculation based on a tight-binding model with reasonable values of the hopping parameters.

7.5 The "pseudogap" regime (Tallon and Loram 2001)

A simple and hopefully not too misleading way of summarizing the experimental properties of the pseudogap or strange metal regime ($p < 0.16$) is to say that most of them behave as if, below the "crossover" line $T^*(p)$, the electronic density of states close to the Fermi surface is much attenuated. For example, if we write the *specific heat* (Loram et al. 1994) $c_V(x, T)$ in the conventional form $\gamma(x, T) \cdot T$, then the coefficient

[20]There is actually more structure than this, and in particular there is some evidence in the YBCO data for (at least) two Fermi surfaces, corresponding to the expected "even" and "odd" two-layer bands.

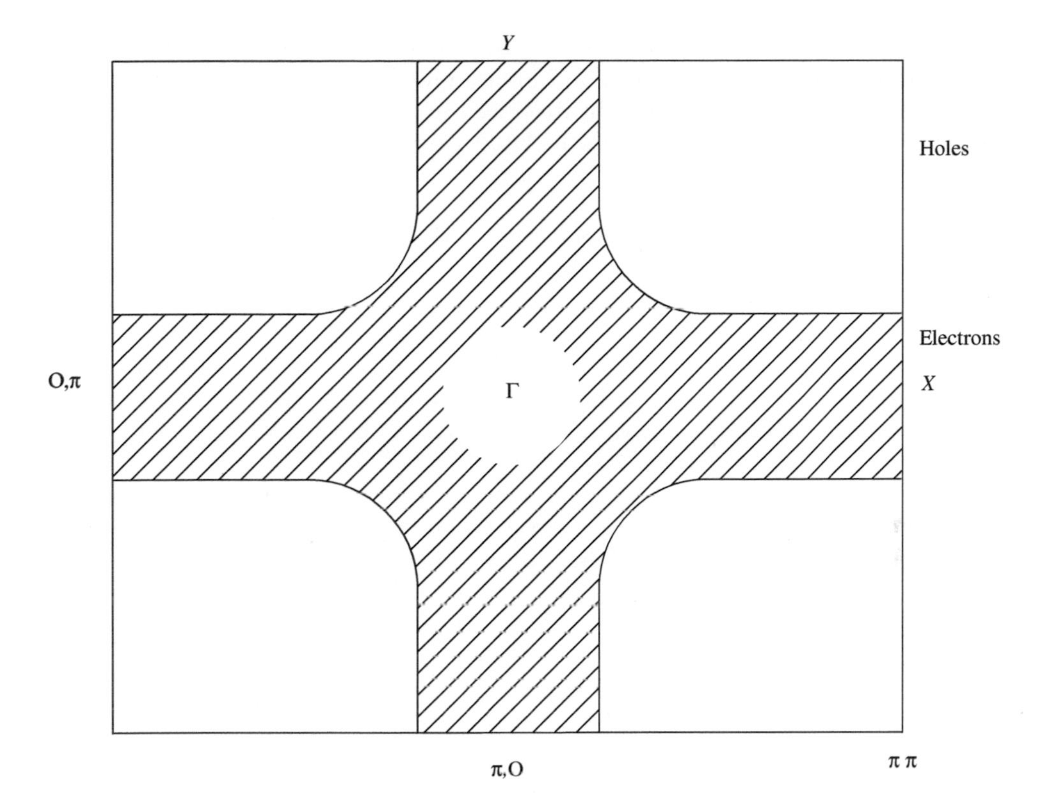

Fig. 7.8 The Fermi "surface" (line) of a typical cuprate in the *ab*-plane. The "Fermi sea" is hole-like and obtained by reconstituting the four blank areas, cf. fig. 7.12.

γ is almost independent of T for $T \gg T^*(p)$, but for $T \lesssim T^*(p)$ drops rapidly; this behavior is largely compensated by an increased anomaly in c_V at the superconducting transition, so that the high-temperature entropy is not strongly x-dependent. A similar behavior is seen in the Pauli *spin susceptibility* χ as deduced from the Knight shift: again we find a substantial drop from the (constant) high-temperature value for $T \lesssim T^*(p)$ (see Pennington and Slichter 1992). It is interesting that the "Wilson ratio", when written in terms of $S/\chi T$ rather than $c_V/\chi T$, is almost independent of x and T at a value close to 1. The nuclear relaxation constant $(T_1T)^{-1}$ also falls below $T^*(p)$.

The *ab-plane d.c. resistivity* $\rho_{ab}(T)$, which is decreasing with decreasing T roughly as T^α $(0 < \alpha < 1)$ for $T \gg T^*(p)$, falls more sharply than this below $T^*(p)$ but has a minimum at a lower temperature; in the small interval of p $(0.04 < p < 0.05)$ where there is no phase transition down to zero temperature, $\rho_{ab}(T)$ appears to diverge as $T \to 0$. If one tries to fit the a.c. data by a Drude formula with frequency-dependent scattering rate $\Gamma(\omega)$ and effective mass $m^*(\omega)$, then for $\omega < 500$ cm^{-1} $\Gamma(\omega)$ must decrease with temperature. On the other hand, in this same frequency regime the *c-axis resistivity* $\rho_c(\omega)$ increases with decreasing T for $T \lesssim T^*(p)$. These observations are qualitatively consistent with a picture in which the main mechanism of resistivity in the *ab*-plane is electron–electron scattering, while the *c*-axis conductivity is by an incoherent hopping process (cf. Section 7.8, below): in such a picture an attenuation of the electronic density of states will by itself lead to decreased conductivity both along the *c*-axis and in the *ab*-plane, but in the latter case may be over-compensated, in the appropriate regime, by a decrease in the electron–electron scattering rate (cf. below, on the superconducting phase).

More direct evidence for the attenuation of the electronic density of states close to the Fermi surface comes from *tunneling* and *ARPES* measurements; both these techniques show evidence for an energy gap similar to that seen in the superconducting state (see below), but persisting up to a temperature $\sim T^*(p)$ (Renner et al. 1998, Matsuda et al. 1999). The symmetry of this "gap", as measured by ARPES, seems to be the same as normally assumed for the superconducting-state energy gap, namely the so-called $d_{x^2-y^2}$ state (crudely, $|\Delta|(\boldsymbol{\theta}) \sim |\cos 2\theta|$ where θ is the angle with respect to the crystal a-axis, see Section 7.8).

Finally, I note that the electronic *Raman scattering* spectrum, essentially featureless in all channels for optimal doping, in the pseudogap regime develops a broad peak at $\sim$500 cm^{-1} in the so-called B_{1g} channel; and that the magnetic neutron scattering peak which is seen around $(0.5, 0.5)$ is definitely split in the pseudogap regime into four peaks at $(0.5 \pm \delta, 0.5 \pm \delta)$, where δ is small and x-dependent.

Overall, it seems possible to regard many of the properties of the pseudogap regime, including some not mentioned in this subsection, as qualitatively similar to those found in the superconducting state. Whether or not this resemblance is accidental is currently a matter of spirited debate (cf. Section 7.9).

7.6 Superconducting state

Most experiments on the superconducting state of the cuprates have been done at optimal doping, and I shall assume this condition unless otherwise stated. What evidence there is on the p-dependence tends to suggest that, as in the normal state, the behavior becomes more "conventional" with increasing p. In the discussion of this subsection I shall take as "reference" behavior that of a classic (s-wave) BCS superconductor.

7.6.1 Specific heat (Tallon and Loram 2001)

The general pattern is similar to that for the BCS case, with the differences that (a) in some systems there is a "precursor" contribution for $T > T_c$, which becomes more pronounced and more smeared-out as we approach underdoping, and (b) at low T it is certainly not exponential, but is probably of power-law form, $c_v \sim aT^n$ with $n \cong 2$, and (c) there is evidence for the effect of critical fluctuations close to T_c. If we integrate the specific heat to get the zero-temperature condensation energy of the S state relative to the (notional) N ground state, we find (in YBCO) a sharp peak at $p = 0.19$ (note this is somewhat on the overdoped side of optimal) at a value of 33 J/mole, corresponding to a condensation energy of $\sim$2 K per CuO_2 unit.

7.6.2 NMR (Pennington and Slichter 1992, Tallon et al. 1998)

The Pauli *spin susceptibility* as inferred from the Knight shift appears to follow approximately the BCS-Yosida behavior for $T/T_c \gtrsim 0.5$ (the c-axis Knight shift is temperature-independent, but this is believed to be due to the fact that because of a fortuitous cancellation the Pauli effect makes no contribution). At low temperatures the T-dependence is apparently linear in T. The *NMR relaxation rate* T_1^{-1} drops precipitously below T_c and appears to be proportional to T^3 at low T.

7.6.3 Macroscopic d.c. electromagnetic properties (Basov et al. 2005)

These seem understandable within a GL description as applied to the classic Type-II superconductors, with two provisos: First, as we might anticipate from their strongly layered structure, the relevant parameters are strongly anisotropic. In fact, if we extract the pair radius $\xi(0)$ from the (extrapolated[21]) value of $H_{c2}(0)$, then $\xi_{ab}(0)$ is about 30 Å and $\xi_c(0)$ is 2–3 Å (since this latter number is less than the intermultilayer spacing, it is not clear that it has any direct physical meaning). Similarly, if the London penetration depth is measured as described below, we find that a typical value of $\lambda_{ab}(0)$ is 1000–4000 Å, while $\lambda_c(0)$ can be as large as 100 μ (for BSCCO). A further difference from the classic Type-II superconductors is the large value of the ratio $(a/\xi_{ab}(0))$ (a = spacing between conduction electrons) which is at most 0.01 and often much less for a typical classic superconductor but is ~0.25 for the cuprates at optimal doping; as a result, even within the GL theory we expect fluctuation effects to be many orders of magnitude more important than for a classic superconductor, and this indeed seems to be the case, see e.g. Malozemoff (1989), and Section 7.10 below.

The quantity which is in some sense the most direct measure of the superconducting properties, and which I shall therefore discuss in some detail, is the penetration depth $\lambda(T)$; since the cuprates are always in the extreme Type-II limit, it is not usually necessary to distinguish between the measured quantity and the (notional) London penetration depth, so that $\lambda(T)$ can be directly related to the superfluid density $\rho_s(T)$. As noted above, the penetration depth is highly anisotropic. While numerous experiments in the literature have claimed to measure both the absolute value and the temperature-dependence of $\lambda(T)$, the most reliable method is probably the measurement of surface impedance; this is described in detail in Bonn and Hardy (1996), from which much of the ensuing discussion is taken. Where necessary, the surface-impedance data can be supplemented with the results of muon spin resonance (μSR) experiments, which should also be fairly reliable.

Typical values of $\lambda_{ab}(0)$ range from ~800 Å for the b-axis of YBCO 1248 to ~4000 Å for optimally doped LSCO. The temperature-dependence is more naturally plotted in terms of the relative superfluid density $\rho_s(T)/\rho_s(0) = \lambda^2(0)/\lambda^2(T)$; generally speaking, for optimally doped pure materials, the curve of $\rho_s(T)/\rho_s(0)$ is roughly as indicated in Fig. 7.9; it is linear at low T and near T_c appears to behave as $(T_c - T)^\zeta$ where ζ is close to $\frac{2}{3}$, the behavior expected in the critical regime for a 3D XY-model (single complex order parameter). Obviously, this behavior is quite different from that found in a classic superconductor.

It is a very interesting question whether the ab-plane penetration depth data are compatible with the hypothesis of universal behavior in the CuO_2 planes at the

[21] For most cuprate superconductors the value of $H_{c2}(0)$ obtained by extrapolation from higher-temperature measurements is too large (~100 T) to be measured directly.

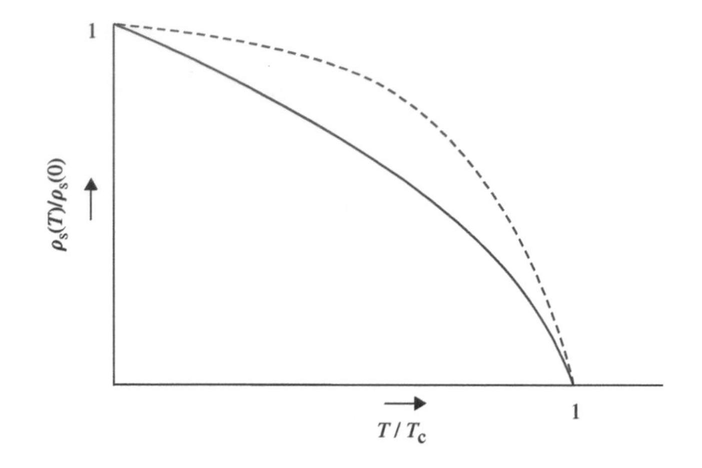

Fig. 7.9 The form of the relative superfluid density as a function of temperature for a BCS superconductor (pecked line) and for a typical cuprate (solid line) (qualitative).

same level of doping. Recall that λ_{ab}^{-2} measures the 3D superfluid density; thus if the hypothesis of universality is correct, one would expect the relation

$$\frac{\lambda_{ab}^2(0)}{\bar{d}} = \text{const.} \tag{7.6.1}$$

to hold, where $\bar{d}$ is the *average* distance between CuO_2 planes. While the microwave data alone are hardly sufficient to test this hypothesis, we can try to compare the values inferred from μSR (Uemura et al. 1989); ratios may be hoped to be given by this technique more reliably than absolute values. The data of Uemura et al. (op. cit.) appear compatible with the hypothesis as regards the higher-T_c materials, i.e. the ratio is the same[22] within the error bars for optimally doped *Tl*-2223 and (near)-optimally doped YBCO, and if we take the *a*-axis value for the latter from the microwave data the constant comes out to be near 5×10^5 Å. For LSCO the number is quite different, about a factor of 2 larger.

The data of Uemura et al. were actually presented as evidence of an intriguing correlation between $\lambda_{ab}^{-2}(0)$ and the transition temperature T_c; for doping *below* optimal the relationship, for the nine different systems measured, appears to be rather convincingly linear. However, their Fig. 2 also shows that the increase of $\lambda_{ab}^{-2}(0)$ with doping persists *beyond* the maximum in T_c.

One may ask how well the data fit a naïve picture, in which the superfluid density per plane is simply expressed as $ne^2\mu_0/m^*$, where n is the number of carriers per unit area and m^* ($\sim$4 m) the effective mass inferred from the specific heat measurements, so that the quantity $\lambda_{ab}^{-2}(0)$ is $n_{3D}e^2\mu_0/m^*$. For optimally doped YBCO, $n_{3D} \cong p_{\text{eff}} \times 1.1 \times 10^{22}$ cm^{-3}, where p_{eff} is the effective number of carriers per CuO_2 unit (see below), and the quantity $\lambda_{ab}^{-2}(0)$ is therefore approximately $1.5p_{\text{eff}}(m/m^*)\ 10^{-6}$ Å^{-2}.

[22] Actually, the values of $\bar{d}$ and $\lambda_{ab}^{-2}(0)$ separately are closely similar for the two materials, but this is not particularly significant since the multilayering structure is different.

We can therefore fit the a-axis data ($\lambda_{ab}^{-2}(0) \cong 0.4 \times 10^{-6}$ Å^{-2}) with an m/m^* ratio of 0.25 as indicated by the normal-state specific heat, *provided* we take p_{eff} at optimal doping to be not p but rather $(1+p)$, i.e. all the holes in the Cu $3d^9$ band and not just the excess ones over the parent compound contribute. Note that to get the b-axis value right (assuming the behavior of the planes themselves is nearly isotropic) we need to assume that the chains contribute $\frac{3}{2}$ as much as *both* planes! That this may indeed be so is indicated by the even smaller λ_b value for YBCO 1248 (~800 Å) (and by the anisotropy of the normal-state resistivity).

Now let's turn to the c-axis penetration depth $\lambda_c(T)$. In addition to the methods (microwave, μSR, powder magnetization, etc.) used for λ_{ab}, there is now the possibility of measuring this, at least within ~10–20% accuracy, by STM studies of the structure of vortices (the reason this technique does not work for λ_{ab} is insufficient resolution). In general, measured values of $\lambda_c(0)$ are much larger than those of $\lambda_{ab}(0)$, ranging from ~11,000 Å for slightly overdoped YBCO to the enormous value of ~100 μm (0.1 mm!) for Bi-2212. Not surprisingly, as YBCO is underdoped λ_c increases much faster than λ_{ab}; this presumably reflects the fact that the chain O's, which are the ones removed when the sample is underdoped, play an important role in the contact between neighboring bilayers. Although the general shape of $\lambda_c(T)$ on a scale $T \sim T_c$ is qualitatively similar to that of $\lambda_{ab}(T)$, the first corrections to the $T = 0$ value are much smaller, and in fact are often approximated by a T^n form with $n \sim 2.5$ (however, they do *not* seem well fitted by an exponential).

An obvious question, given the very weak contact between neighboring multilayers in most cuprates, is how well the value of $\lambda_c(0)$ is reproduced by modeling each inter-multilayer link as an independent Josephson junction and applying to it the standard formulae, in particular the Ambegaokar–Baratoff relation (see Chapter 5, Section 5.10)

$$I_c(0)R_N = \frac{\pi\Delta(0)}{2e} \tag{7.6.2}$$

Since the c-axis resistivity ρ_c is $R_N A/d_{\text{int}}$ (d_{int} = mean spacing between multilayers) and in a simple Josephson model $\rho_s^{(c)}(0) \sim \lambda_c^{-2}(0)$ is proportional to $I_c(0)$, the predicted relation is

$$\frac{\lambda_c^{-2}(0)\rho_c d_{\text{int}}}{\Delta(0)} \sim \text{const.} \tag{7.6.3}$$

Since ρ_c is often strongly temperature-dependent, a quantitative test of this prediction is somewhat ambiguous. However, Basov et al. (1994) have shown that the values of $\lambda_c^{-2}\rho_c(T_c)$ for eight different systems measured fit the prediction within a factor ~2–3.

The low-frequency ($\omega \lesssim kT_c/\hbar$) *a.c. conductivity* in the superconducting state is a complicated function of frequency and temperature, both for the classic superconductors and for the cuprates. The most salient difference between these two classes is that in the cuprates the (real part of the) a.c. conductivity $\sigma_1(\omega)$ is appreciable at frequencies well below $3.5kT_c/\hbar$ (which in BCS theory is 2Δ, the lowest allowed excitation frequency), although it is smaller than in the normal state and does show signs of dropping towards zero at the lowest frequency (note that the "data" are obtained by Kramers–Kronig transformation of the reflectance). Equally striking is the fact that if we sit at a fixed frequency $<2\Delta(0)$ and lower the temperature through T_c, $\sigma_1(\omega)$ actually *increases* initially before eventually dropping well below the N-state value.

This suggests strongly that the effective scattering rate which enters the (pseudo-) Drude formula is primarily due to electrons, which are gapped below T_c.

The question of whether one can ever actually "see" the superconducting gap in the optical conductivity of the cuprates at all is a vexed one, see Tanner et al., op. cit.

Turning to the *c*-axis conductivity, this typically *decreases*, in the superconducting state, typically in the frequency region below $\sim$150 cm^{-1} for optimally doped samples but up to $\sim$500 cm^{-1} in underdoped ones. A peculiar characteristic of the *c*-axis electrodynamics of the cuprates is that the plasma frequency (which is proportional to $I_c^{1/2}$) is so low that it may actually be $<2\Delta$; thus, while the "plasmon" is strongly damped and hence invisible in the normal state, in the superconducting state it may sharpen up enough to be seen. Indeed, evidence for such *c*-axis plasmons inferred from the reflectivity has been reported for LSCO and BSCCO (see, e.g. Van der Marel 2004).

7.6.4 Thermal conductivity

In a typical metal, the heat is carried both by electrons and by phonons. To estimate their relative contribution to the thermal conductivity κ, it is adequate to use the "classical" formula

$$\kappa \sim \tfrac{1}{3} c_v \bar{v} \ell \tag{7.6.4}$$

where c_v = specific heat of carriers, $\bar{v}$ = mean velocity, ℓ = mean free path. Since for $T \ll \theta_D$ and $\sim$1 free electron per unit cell, we have $c_v^{el} \sim nk_B(T/T_F), c_v^{ph} \sim nk_B(T/\theta_D)^3$, $\bar{v}_{el} = v_F$, $\bar{v}_{ph} = c_s$, and $c_s/v_F \sim \theta_D/T_F$, the ratio is of order

$$K_{ph}/K_{el} \sim (T/\theta_D)^2(\ell_{ph}/\ell_{el})$$

For conventional superconductors this ratio is usually $\ll$1 by the time T_c is reached, but for the cuprates, which simultaneously have much higher T_c's, lower electron densities and shorter mean free paths, the phonon contribution may be dominant in the N phase, or at least comparable to the electronic one. Since in a conventional superconductor the heat transport for $T \lesssim T_c$ is overwhelmingly electronic (carried by the normal component), and the electronic mean free path is usually predominately due to impurity scattering, we expect that below $T_c\, K(T)$ would drop off approximately as $\rho_n(T)/\rho$, and this seems consistent with most of the data on such systems.

In the superconducting cuprates, almost universally, $K(T)$ *rises* below T_c, peaking in the cleanest samples at a value $\sim 2K(T_c)$ at a temperature $\sim 0.5T_c$ and thereafter falling to zero as $T \to 0$. In many cases the low-temperature $K(T)$ appears to be linear in T, though in BSCCO it looks like T^2 without any visible T région. The rise below T_c indicates a decrease in scattering mechanisms, but does not by itself distinguish between electron and phonon contributions. However, the fact that in conventional YBCO crystals K_a and K_b are different, with a ratio close to the normal-state ratio of R_a^{-1} and R_b^{-1}, indicates that at least in this material the mechanism may be primarily electronic, and this is consistent with the decrease of K in a magnetic field.

The *c*-axis thermal conductivity has not been much studied but appears to drop off smoothly below T_c (Uher 1992, Figs. 20 and 21). Presumably this is almost entirely due to phonons, and the scattering mechanism seems likely to be primarily by impurities, although this is not completely clear.

Ong and co-workers have reported a large increase in the Nernst coefficient (transverse thermopower), that is the xy-component of the Seebeck tensor S_{ij} in a z-axis magnetic field, of pure crystal YBCO (Krishana et al. 1995). As the fields in question were $\gg H_{c1}$, this may be a "vortex" effect; at any rate, it is currently a major puzzle (see e.g. Ong et al. 2004).

7.6.5 Spectroscopic probes

Tunnelling (Renner et al. 1998, Matsuda et al. 1999: Hussey 2002)

We recall (a) that whereas in a classic superconductor in the normal state the differential tunneling conductance $G_{nn} \equiv \partial I/\partial V$ is essentially flat or at most weakly parabolic, in a typical cuprate it often (though not invariably) has the form $a + b|V|$; (b) that for a classic (s-wave) superconductor in the superconducting state, the ratio of the differential conductance G_{ns} to its normal state value G_{nn} is given by the ratio of the density of quasiparticle states and is thus of the form (at $T = 0$)

$$\frac{G_{ns}(E)}{G_{nn}(E)} = \frac{E}{\sqrt{E^2 - \Delta^2}}\theta(E - \delta) \tag{7.6.5}$$

Although in practice gap inhomogeneity over the FS, lifetime effects etc., tend to smooth out the singularity, the general pattern is as shown in Fig. 7.10a, with a peak to peak difference which is close to 2Δ as estimated independently; see, e.g. Tinkham (1996), Section 3.8.

In the cuprates, ab-plane tunneling measurements yield a $G(V)$ characteristic qualitatively similar to Fig. 10a, see Fig. 10b. The principal differences are that (a) the peak-to-peak splitting is considerably larger than $3.5k_BT_c$, typically 7–8 k_BT_c, though it decreases towards the BCS value with overdoping; (b) the conductance is certainly not zero in the low-energy region, and in fact may behave as a power of $|V|$; (c) there is a noticeable "dip" below the normal state curve *above* the peak. In addition to these differences, which appear to be approximately independent of the direction of tunneling in the ab-plane, for some directions, in particular [011], there appears a "zero-bias anomaly" (peak in $G(V)$ at zero voltage, not shown in figure); this is of great inererest in the context of symmetry of the order parameter, see Greene et al. (2004). In addition, some measurements in the [100] direction with planar tunnel oxide junctions show a very considerable asymmetry in the $G(V)$ characteristic.

Most existing STM measurements in the c-direction show $G(V)$ characteristics which, while qualitatively similar to Fig. 10b, differ in some respects (in particular, the peak-to-peak splitting is smaller and the (reduced) low-voltage conductance higher). However, these experiments were done on samples where the exposed surface is a BiO plane; recent experiments which exposed one of the CuO_2 planes themselves found characteristics closer to Fig. 10a, in fact intermediate between that and the behavior shown in Fig. 10b.

ARPES (Campuzano et al. 2002, Damascelli et al. 2003, Norman and Pepin 2003)

Recall that the vast majority of ARPES experiments have been done on BSCCO with the surface in the c-direction (but the surface itself is a BiO layer!) In principle, subject to the usual assumptions, the photocurrent measures the single-particle spectral function $A(\boldsymbol{k}, \omega)$ for $\boldsymbol{k}$ in the ab-plane (see above). In practice, the accuracy obtainable

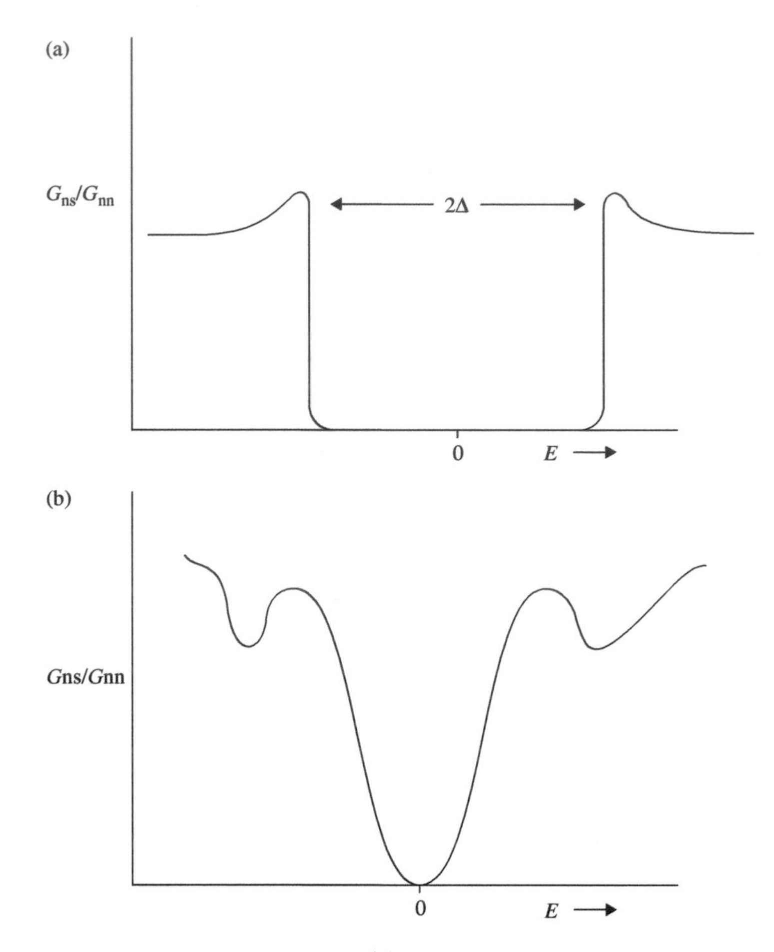

Fig. 7.10 Tunnelling density of states. (a) for a BCS superconductor. (b) for a typical cuprate (*ab*-plane).

for $\boldsymbol{k}$ (i.e. the "width" of the resolution function in $\boldsymbol{k}$-space) is $\sim$0.01–0.04 Å^{-1}, and the energy accuracy is $\sim$15–35 meV, now reduced to $\sim$2 meV (see Damascelli et al., op. cit.): note that these are smaller than the expected momentum and energy scales of the pairs, though in the momentum case still not by a large margin.

In the normal state, for given $\boldsymbol{k}$ near the (inferred) Fermi momentum, the ARPES data as a function of energy have a steep drop at the (inferred) Fermi energy but are otherwise featureless. In the superconducting state, if we consider for definiteness the $(\pi, 0)$ direction, the curve develops a sharp peak whose leading edge is "pulled back" from the Fermi energy by an amount $\Delta\epsilon$, and below the peak the curve dips beyond the N-state value (cf. the tunneling data). Beyond this point, there is a weak "hump". At optimal doping both the height of the peak and its displacement $\Delta\epsilon$ from the Fermi energy increases smoothly as T falls below T_c and at low temperature $\Delta\epsilon(\pi, 0) \cong$ 35 meV, corresponding to $\sim$4–5$k_B T_c$. The shift is also seen in the underdoped materials though it is less pronounced and persists above T_c (cf. below).

The most interesting feature of the peak in the superconducting state is its dependence on the angle on the Fermi surface. Both the height and $\Delta\epsilon$ are a maximum in the $(\pi, 0)$ direction (i.e. along the crystal axes) and decrease towards zero as we move towards the (π, π) direction; it is now generally believed that both quantities are zero when one sits precisely at (π, π) (though one cannot absolutely rule out the possibility that the zeros are split, on either side of this).

Neutron scattering (Mook et al. 1998, 1999: Fong et al. 1999: Norman and Pepin 2003)

We noted above that the N-state neutron scattering data on all cuprates so far examined are fairly strongly peaked as a function of $\boldsymbol{q}$, with either single or multiple peaks close to $(0.5, 0.5)$ (the AF superlattice), but featureless as a function of energy. If we restrict ourselves to the energy range below $\sim$30 meV, then this statement remains qualitatively true for all systems in the S phase in the "even" channel, although as a function of $\boldsymbol{q}$ the peaks sharpen somewhat. However, in YBCO a new feature appears in the "odd" channel close to $(1/2, 1/2, 1)$ (i.e. $q_x \cong q_y \cong \pi/2a$, $q_z = \pi/d$, where d is the intrabilayer spacing $\cong$3.1 Å): we get a striking peak at an energy which ranges from $\sim$41 meV for optimally doped YBCO to $\sim$34 meV for $x = 0.6$ with an energy width $\sim$5–10 meV (narrow by neutron standards) and a momentum width $\sim$–0.2 RLU (Mook et al. 1993).[23] The position of the peak appears to be independent of temperature, but its amplitude is strongly temperature-dependent and in optimally doped samples it cannot be seen above T_c. A peak with essentially the same properties has been seen in Bi-2212 (Fong et al., ref. cit.), but searches in LSCO have failed to detect anything remotely similar (note that there is no trace, in YBCO, of this resonance in the "even" $(\pi, \pi, 0)$ channel). As a function of doping, the energy of the peak seems to scale with T_c both for UD and OD regimes (but in all cases is independent of temperature); the weight scales with temperature roughly as $\Delta^2(T)$, where Δ is the "gap" inferred from the photoemission spectrum (thus, in the UD regime, the peak persists above T_c and roughly up to some $T^*(p)$).

Both the origin of this "41 meV peak", and its possible relation to the anomalies in the tunneling and ARPES spectra, are currently a topic of great interest.

Raman Scattering (Strohm and Cardona 1998, Blumberg et al. 1997, Rübhausen 1997)

We noted above that in the N state of optimally doped cuprates the electronic Raman spectrum is essentially channel-independent and featureless for $\omega < 2$ eV. In the S state things are much more interesting. In the first place, both for YBCO and for Bi-2212, a broad peak develops at a frequency of the order of 2Δ. Unfortunately, although it is tempting to identify this peak as corresponding to a bound pair of magnon-like excitations seen at 41 meV (Blumberg et al., ref. cit.), its isotope- and doping-dependence (Hewitt et al. 1999) appear to indicate that this is at least as likely to be a vibrational mode (which was presumably too overdamped in the N phase to

[23] Note that (at least roughly) the energy width is resolution-limited but the momentum width is not.

be visible there.) To the best of my knowledge this peak has not been seen (maybe not looked for) in LSCO.

The second point of interest in the Raman behavior is the low-frequency dependence of the intensity. In YBCO (at least) this is somewhat similar for the A_{1g} and B_{1g} channels; in each case there is a term linear in ω, and in B_{1g} channel also a term of comparable magnitude proportional to ω^3. In the OD regime the linear term in B_{1g} is reduced relative to the ω^3 one, and this behavior is also seen in overdoped Tl-2201 and (probably) also in Bi-2212. The question of the difference between the asymptotic behavior of the Raman intensities in the two geometries (A_{1g} and B_{1g}) is of considerable significance in the context of the question of gap symmetry.

Optics (Holcomb et al. 1996, Molegraaf et al. 2002, Rübhausen et al. 2001)

At least until quite recently, the investigation of the possible changes in the optical (visible-region) behavior of the cuprates when the system goes superconducting has not been a particularly fashionable subject. Probably the main reason for this lack of interest has been that any simple BCS-like theory would predict that any such relative changes are likely to be of the order or $(k_B T_c/\hbar\omega)^2$, which for ω in the optical region (say $\hbar\omega \sim 1$ eV) would be $\sim 10^{-4}$, probably too small to see in most experiments.

In fact, measurements of the real and imaginary parts of the dielectric constant, whether indirect (via reflectance measurements and the use of a Kramers–Kronig relation), or direct (ellipsometric), find changes in these quantities of order of 1–5%, right up to energies ~3.5 eV. It is interesting that the formula for the change $\delta\epsilon(\omega)$ from the N-state value $\epsilon_n(\omega)$ which would follow from the "naive" hypothesis of a simple (average) shift down by δ of the relevant initial-state energy levels, namely

$$\delta\epsilon(\omega) = -\delta\left(\frac{\partial\epsilon_n(\omega)}{\partial\omega}\right) \tag{7.6.6}$$

does not appear to fit the data at all well. My own belief is that these unexpectedly large effects must be an important input into any successful theory of cuprate superconductivity.

EELS (Phelps et al. 1994, Mills et al. 1994)

In view of the surprisingly large effects of the onset of superconductivity on the visible-region (1–3 eV) optical properties, it would be very interesting to know whether the transmission EELS cross-section, which should be a direct measure of the charge fluctuations and thus the Coulomb energy, undergoes significant changes in this region of the spectrum. Unfortunately, there is at present no published data on the differential EELS cross-section (i.e. $\sigma_s(q\omega) - \sigma_n(q\omega)$) in this region. EELS experiments have been conducted (in the reflection geometry) in the superconducting state, but at considerably lower energies (~100 meV) and with disappointing results (see Mills et al., ref. cit.); in particular, there is no feature in the data which can be plausibly identified with the superconducting gap. Quite apart from the fact that the cross-section in this regime is in practice dominated by surface optical phonons, which may well mask any small changes, it is not clear to me that we should even expect to see anything interesting in this energy region, since theoretically the normal-state loss function is expected to be very small for $\omega < 100$ meV.

7.7 Some preliminary comments on the experimental data

7.7.1 Universality

If one excepts phenomena such as crystallographic phase transitions or the weak ferromagnetism of the RE elements in $ReBa_2Cu_3O_{6+\delta}$, which do not obviously have much to do with the CuO_2 planes, the only group of N-state properties of the cuprates which seems to lack *qualitative* universality is the c-axis transport.[24] However, it should be emphasized that not only the transition temperature T_c, but the very existence of the superconducting state is nonuniversal within the class of cuprates (cf. Section 7.2).

The question of quantitative universality needs to be made a bit more precise. Let us start with the normal state and first ask: Given a particular (superconducting) cuprate at the values of hole concentration p, impurity concentration (zero) etc. which maximize T_c, and at given $T > T_c$, are the properties "per CuO_2 plane" uniquely determined, i.e. the same for all cuprates under these conditions? Within the error bars, the answer seems to be yes for at least two quantities, the Sommerfeld coefficient γ in the electronic specific heat, which seems to be $\cong$6.5 mJ/mol (CuO_2) K^2 for all compounds on which it has been reliably measured, and the thermoelectric power S at room temperature (but not, interestingly, more generally). On the other hand, while the in-plane resistivity $\rho_{ab}(T)$ is fairly universally proportional to T and hence "qualitatively" universal, the coefficient of proportionality is fairly clearly different for LSCO than for the higher-T_c materials (YBCO, Tl-2201, Bi-2212 etc.) and it is not entirely clear that it is quantitatively universal even within the latter group. This difference may reflect in part the fact that transport properties, unlike thermal ones such as the specific heat, are likely to be sensitive to the degree of localization ("trapping") of the in-plane carriers by near-plane disorder, such as naturally occurs in $La_{2-x}Sr_xCu_2O_4$.

The (infrared and visible region) ab-plane optical properties are evidently qualitatively universal (all known cuprates show the characteristic MIR peak); the question of quantitative universality is a bit more complicated, since as noted above the charge-reservoir material may contribute appreciably. However, it seems probable that the data below say 2 eV are consistent with quantitative universality of the "per-plane" response, the differences in the directly measured quantity (reflectance) arising primarily from the different density of planes in the different materials.

In the superconducting phase the natural question would seem to be: At a given value of p (e.g. the "optimal" value $\sim$0.16) and a given value of the reduced temperature $T/T_c(p)$ (e.g. zero), are the per-plane properties the same for different cuprates? This does not seem to be true for thermal properties such as condensation energy (cf. Table II of Munzar et al. (2001)); for transport-type properties such as the penetration depth $\lambda_{ab}(0)$ the universality is quantitative[25] only among the higher-T_c materials (YBCO (a-axis), Tl-2223, probably Bi-2212 etc.). A more systematic experimental investigation of this question would seem to be of great value.

When we come to spectroscopic probes of the superconducting state, a very interesting situation arises: while the Raman and optical behavior appears to be at least qualitatively universal, the neutron scattering shows a striking feature, namely

[24] Which however might be regarded as "qualitatively universal" in the sense that (e.g.) $\rho_c(T) \sim T^\alpha$ with α material-dependent.

[25] Except perhaps at the lowest temperatures (where it is very sensitive to impurity levels, etc.)

the "41 meV peak" which has been clearly observed in YBCO and BSCCO but equally reliably observed not to occur in LSCO. Further, the "peak-dip" structure in the ARPES characteristics appears in the former two compound but not in the last (at least so far). This clear lack of even qualitative universality raises several obvious questions:

1. Is the origin of the 41 meV peak intrinsically associated with the bilayer structure of YBCO and BSCCO, or is its absence in LSCO a consequence of other factors (possibly the same ones as are responsible for the lack of quantitative universality of $\rho_{ab}, \lambda_{ab}(0)$ etc.)? The appearance of a similar peak at 47 meV in single-layer Tl-2201 (He et al. 2002) seems to favor the latter hypothesis.
2. Is there a causal connection between the 41 meV peak and the peak-dip structure of the ARPES spectrum (and possibly that of the tunneling spectrum)?
3. Does either phenomenon play a role in the mechanism of superconductivity?

7.7.2 Electron–electron umklapp scattering

So long as one uses a "band" picture of the electronic states and ignores umklapp (U) processes, it is easy to show that (a) the imaginary part not only of the density correlation function but of the transverse current function (i.e. in effect, the conductivity $\sigma(\boldsymbol{q}, \omega)$) should be essentially zero for $\omega \gg qv_\mathrm{F}, \Gamma_0$ (where Γ_0 is the scattering rate due to phonons and impurities), i.e. there should be no weight in the MIR region; (b) electron–electron scattering processes cannot by themselves give rise to a finite d.c. resistivity. Now, both the imaginary part of the density correlation function (as measured by EELS) and the optical conductivity show a strong MIR peak; moreover, both the temperature-dependence of the d.c. *ab*-plane resitivity and the drop in the a.c. resistivity below T_c suggest strongly that the main scattering mechanism is electronic rather than by phonons or impurities. Consequently, we can conclude that within a "band" picture U-processes must be vastly more important than in the classic superconductors (or more generally in "textbook" metals). Of course, one might equally well take the view that the whole idea of starting from a "band" picture is misguided.

7.7.3 The superconducting state

One (negative) conclusion one can draw more or less directly from the experimental data is that, whatever may be the nature of the superconducting state, the electronic excitation spectrum is definitely not of the BCS type with isotropic gap; such a hypothesis could not explain (e.g.) the nonzero value of $G(V)$ at low voltages or the T^3 behavior of T_1^{-1}, and moreover even if generalized to an anisotropic gap with nodes would apparently not explain the "peak-and-dip" structure observed in tunneling and ARPES experiments.

7.8 What do we know *for sure* about cuprate superconductivity?

In this section I shall address the question:

How much, if anything, can we infer about the general nature of the normal and/or superconducting states of the cuprates without recourse to a specific microscopic model?

Probably the single most important piece of experimental information we have on cuprate superconductivity concerns flux quantization and the Josephson effect.

These experiments were done on YBCO at an early stage, and gave the results which are standard for classic superconductors, i.e. the unit of flux quantization is $h/2e$ (not, e.g. h/e or $h/4e$) and the Josephson frequency-voltage relation is $\omega = 2eV/\hbar$. However, there is one subtle point which is often overlooked: the circuits used in the experiments were without exception such that the "paths" with respect to which the flux is quantized (etc.) lie entirely in the *ab*-plane. It is theoretically conceivable (though to my mind improbable, in view of the considerations below) that a direct experiment using an "all *c*-axis" circuit, should it be possible, would give a different result.

The significance of these results is that, according to the arguments of Chapter 1, Section 1.4 and Chapter 5, Section 5.6, they provide very strong evidence that the superconducting state of the cuprates possesses long-range order in the two-particle correlation function (and does not have it in the one-particle one), which is, crudely speaking, equivalent to the statement that the "topology" of the wave function corresponds to formation of Cooper pairs just as in the classic superconductors. Thus, we reach the conclusion:

1 Superconductivity in the cuprates is due to formation of Cooper pairs

A corollary of statement (1) is that if we assume, as is almost universally done, that this result holds for the *c*-axis as well as for the *ab*-plane, then this knowledge is sufficient for us to set up a Ginzburg–Landau description in terms of an order parameter which, just as in the classic superconductors, will have the physical significance of the center-of-mass wave function of the Cooper pairs. However, in distinction to the case of a classic (isotropic) superconductor the parameters of the theory will evidently distinguish between *ab*-plane and *c*-axis.

At this point, anticipating the conclusions to be obtained below, we might ask whether the fact that the *internal* state of the Cooper pairs will turn out, almost certainly, to be "exotic", that is to have a symmetry lower than that of the lattice, will effect the validity of the GL description? The answer is no, at least so long as it corresponds to a single nondegenerate irreducible representation of the crystal symmetry group (see below), but the reason is quite subtle: Although the OP does in a sense possess an "orientation", that orientation is *not free to adjust itself arbitrarily*, but is pinned to the original crystal lattice, and therefore does not constitute a real "degree of freedom" which needs to be explicitly taken into account. Were the orientation free to adjust, as for example in the case of the l-vector in superfluid ^{3}He-A, it would have to be incorporated explicitly and the description would become more complicated.

Thus, we proceed just as in the classic case but with allowance for the anisotropy: we treat the coordinate $\boldsymbol{r}$ for the moment as a continuous variable and define, just as there, a complex scalar order parameter $\Psi(\boldsymbol{r})$ and write the usual terms proportional to $|\Psi(\boldsymbol{r})|^2$ and $|\Psi(\boldsymbol{r})|^4$ in the free energy. The gradient term, however, must now be treated as a *tensor* quantity γ_{ab} with eigenvalues $\gamma_{\|}$, $\gamma_{\perp}$ corresponding to in-plane[26]

[26] We implicitly assume isotropy within the *ab*-plane; where this is not present (as in YBCO) the appropriate generalization is obvious.

and c-axis variation. Thus, the relevant form of the free energy density is

$$F\{\Psi(\boldsymbol{r})\} = -\alpha(T)|\Psi(\boldsymbol{r})|^2 + \frac{1}{2}\beta(T)|\Psi(\boldsymbol{r})|^4 + \sum_{\alpha\beta} \gamma_{\alpha\beta}(T)\{(\boldsymbol{\nabla}_\alpha + 2ie\boldsymbol{A}_\alpha(\boldsymbol{r}))\Psi^*(\boldsymbol{r})(\boldsymbol{\nabla}_\beta - 2ie\boldsymbol{A}_\beta(\boldsymbol{r}))\Psi(\boldsymbol{r})\} \quad (7.8.1)$$

where as usual in the limit $T \to T_c$ we assume the temperature and magnetic field dependence $\beta(T) \sim \text{const.} \equiv \beta$, $\gamma_{\alpha\beta}(T) \sim \text{const.} \equiv \gamma_{\alpha\beta}$, $\alpha(T) = \alpha_0(T_c - T)/T_c$ ($\alpha_0 =$ const.).

It is worth taking a moment to discuss the limits of validity of Eqn (7.8.1). Strictly speaking, it is valid only in the limits $T \to T_c$ and infinitely slow spatial variation. A generalization to arbitrary T can (as in the classic case) be simply achieved by replacing the first two terms in F by a more general function $F_{\text{loc}}\{|\Psi(\boldsymbol{r})|^2, T\}$, and usually does not change things qualitatively. The question of the spatial variation, however, is more tricky. We recall that for a given eigenvalue γ of $\gamma_{\alpha\beta}$, the GL healing length $\xi(T)$ is given by $\xi(T) \equiv (\gamma(T)/\alpha(T))^{1/2} = \xi_0(1 - T/T_c)^{-1/2}$ where $\xi_0 \equiv (\gamma_0/\alpha_0)^{1/2}$. Crudely speaking, $\xi(T)$ is the distance over which the order parameter has to bend appreciably either in amplitude or in phase before the bending energy exceeds the original condensation energy; thus, the maximum gradient of the OP which is physically realistic is of order $\xi^{-1}(T)$. The GL description will therefore be a generally valid description, at given T, if $\xi(T)$ exceeds by an appreciable margin any "microscopic" lengths in the problem (since correction terms, e.g. of the form $|\boldsymbol{\nabla}\Psi|^4$, may be expected to become appreciable when the bending is over such a microscopic length). We recall for orientation that in the standard BCS case the longest such microscopic length is the (nearly temperature-independent) pair radius ξ_p, which in BCS theory is of the same order as the prefactor ξ_0 in $\xi(T)$; thus, for $t = 1 - T/T_c \ll 1$ the GL description is generally valid. In the case of a layered system like the cuprates, this argument goes over unchanged as regards the ab-plane behavior. However, in the case or c-axis bending it may turn out that the prefactor $\xi_0^{(c)}$ of $\xi^{(c)}(T)$ is only of the order of a few Å or even less, and thus, in particular, smaller than the characteristic microscopic scale of the lattice structure (i.e. the (effective) c-axis cell dimension, ~6–15 Å). In this case, Eqn. (7.8.1) will still be a valid description in the limit $T \to T_c$, but for T appreciably away from T_c it may need to be replaced by a more microscopic description; to formulate the latter we need to know a little about the "locus" of superconductivity in the cuprates, and it is to this question that I now turn.

The 200-odd known superconducting cuprates span an enormous variety of chemical compositions and crystal structures; they have one and only one[27] thing in common, namely the CuO_2 planes. When this observation is combined with the qualitative and often quantitative universality of the S-state properties across the known cuprates, the following conclusion seems irresistible.

[27]Even the "apical oxygens", while a very ubiquitous structural feature of the cuprates, are missing in the "infinite-layer" material.

2 *The principal locus of superconductivity in the cuprates is the CuO_2 planes*

It should be cautioned that (2) does not imply that all cuprates are superconducting (as we saw in Section 7.2, they are not) nor that structural elements other than the CuO_2 planes (e.g. the chains in YBCO or the BiO planes in BSCCO) may not play a role; however, it assures us that a reasonable zeroth-order description of superconductivity may be obtained by focusing on the behavior of the order parameter within the CuO_2 planes.

This observation allows us to extend the GL formalism to the case where the formally calculated $\xi_c(T)$ is of order of the c-axis lattice constant or smaller. Let's start by considering a "single-plane" material such as Tl-2201. Then it is natural to define a discrete order parameter $\Psi_n(\boldsymbol{r}_\parallel)$ for the nth CuO_2 plane, where $\boldsymbol{r}_\parallel$ is the in-plane (xy-) coordinate. As a function of this order parameter the GL free energy would be expected to have the usual "bulk" terms, and the terms corresponding to in-plane bending should also be essentially identical to those for the 3D case. However, we should expect that for "bending" along the c-axis the continuous gradient terms would be replaced by an expression proportional to the square of the difference of the discrete quantities $\Psi_n(\boldsymbol{r}_\parallel)$ and $\Psi_{n+1}(\boldsymbol{r}_\parallel)$. The lowest-order expression which is compatible with gauge invariance, etc., is

$$\begin{aligned} F^n_{\text{band}}(\boldsymbol{r}_\parallel) &= \left|\Psi_{n+1}(\boldsymbol{r}_\parallel) - \Psi_n(\boldsymbol{r}_\parallel)\right|^2 \\ &\equiv |\Psi_{n+1}(\boldsymbol{r}_\parallel)|^2 + |\Psi_n(\boldsymbol{r}_\parallel)|^2 - 2|\Psi_n(\boldsymbol{r}_\parallel)||\Psi_{n+1}(\boldsymbol{r}_\parallel)|\cos\Delta\phi_n(\boldsymbol{r}_\parallel) \qquad (7.8.2) \end{aligned}$$

where $\Delta\phi(\boldsymbol{r}_\parallel) = \phi_{n+1}(\boldsymbol{r}_\parallel) - \phi_n(\boldsymbol{r}_\parallel)$, ϕ_n being the phase of the complex quantity Ψ_n. Thus, this term has the characteristic form of a *Josephson coupling* between neighboring CuO_2 planes.

The total free energy obtained in this way has the form

$$F = \sum_n \int \mathrm{d}\boldsymbol{r}_\parallel F_n\{\Psi_n(\boldsymbol{r}_\parallel)\} \qquad (7.8.3)$$

where

$$\begin{aligned} F_n\{\Psi_n(\boldsymbol{r}_\parallel)\} \equiv & -\alpha(T)|\Psi_n(\boldsymbol{r}_\parallel)|^2 + \tfrac{1}{2}\beta(T)|\Psi_n(\boldsymbol{r}_\parallel)|^4 \\ & + \gamma_\parallel(T)|(\boldsymbol{\nabla}_\parallel - 2ie\boldsymbol{A}_\parallel(\boldsymbol{r})/\hbar)\Psi_n(\boldsymbol{r}_\parallel)|^2 \\ & + K|\Psi_{n+1}(\boldsymbol{r}_\parallel) - \Psi_n(\boldsymbol{r}_\parallel)|^2 \qquad (7.8.4) \end{aligned}$$

Here we chose for simplicity a gauge in which $A_\perp(\boldsymbol{r}) \equiv 0$ (this is always possible); if for any reason we wish to deviate from this condition, then to maintain gauge invariance we should replace the last term, most generally, by an expression of the form[28]

$$K\left|\Psi_{n+1}\exp\left[i\int_{z_n}^{z_{n+1}} 2e\boldsymbol{A}\cdot \mathrm{d}\boldsymbol{l}/\hbar\right] - \Psi_n\right|^2 \qquad (7.8.5)$$

The model described by Eqn. (7.8.5) is known as the *Lawrence–Doniach model*; we see that it is equivalent to regarding the CuO_2 planes as a set of Josephson junctions

[28]Tinkham's Eqn. (9.3) is consistent with this in the limit that A_z is slowly varying over the interlayer spacing.

in series. An important point to notice is that while it would be formally possible to incorporate the terms in $|\Psi_n(\boldsymbol{r}_\parallel)|^2$ arising from the last (K-) term in the term $-\alpha(T)|\Psi(\boldsymbol{r}_\parallel)|^2$, *this is not a natural thing to do*; we expect physically that in equilibrium $\Psi_{n+1} = \Psi_n$ and thus the last term as a whole is zero, so that the equilibrium value $\Psi(T)$ of Ψ_n is (in the mean-field approximation) the same as in a 3D case with the same parameters α and β. Thus, in any simple model of the Lawrence–Doniach type, *we do not expect the interplane Josephson coupling to raise* T_c.[29]

It is clear that under conditions where $\Delta\phi_n$ is small compared to 1, i.e. where the bending of the order parameter over the interplane distance c is small, the LD model reduces to a GL theory, with the correspondence

$$\gamma_\perp = Kd \tag{7.8.6}$$

It seems that for YBCO this regime is already reached at $1 - T/T_c \sim 0.1$, while for the much more anisotropic BSSCO-2212 compound it occurs only at $1 - T/T_c \sim 10^{-3}$ and hence is barely visible: BSCCO is almost always in the "true LD" limit.

One may ask whether, apart from its a priori plausibility, there is any direct experimental evidence for the picture of the CuO_2 planes as a series of Josephson junctions in series? If this view is correct, then one would expect that under appropriate conditions the *nonlinear* current–voltage characteristics would show the typical Josephson features, and indeed this seems to be the case in Bi-2212 (Kleiner and Müller 1994). Note that in the true LD limit, in strong distinction to the GL case, the critical current can be exceeded without heating the sample into the normal phase.

So far, we assumed we are dealing with a single-plane material. What about multiplane materials such as Bi-2212? The most obvious assumption (which I have implicitly used a couple of times above, when referring to experimental data on this compound) is that the CuO_2 planes within a single multilayer are coupled together so strongly that it is legitimate in the present context to treat each multilayer as a single plane; then the above analysis goes through unchanged. It is conceivable that this assumption is wrong, in which case the LD model would have to be generalized, but at present there seems no strong evidence for this.

The LD model, or its limiting form, the 3D GL theory, has been very widely applied in theoretical papers to calculate the macroscopic electromagnetic properties of the cuprates, and the agreement of its predictions with experiment, while not perfect, seems as satisfactory as one could reasonably hope. An excellent account of this subject can be found in Chapter 9 of Tinkham's (1996) book, and I shall not spend time on it here;[30] I merely want to emphasize that the validity of the LD (or 3D GL) models depends only on the truth of the assertions (1) and (2) above, and is completely independent[31] of any microscopic model of the superconducting state.

[29]This point has been widely misunderstood in the literature, where one can find numerous attempts to invoke this coupling to explain the increase of T_c with n in homologous series. Of course, once one goes beyond the standard mean-field approximation, the situation becomes more complicated; see Section 7.9 below.

[30]However, I return to some of the idiosyncrasies of the macroscopic EM behavior of the cuprates in Section 7.10 below.

[31]Though the definition of the GL order parameter in terms of the pair wave function F defined below may be sensitive to some of the symmetries of the latter.

The most general definition of the "Cooper pairing" postulated in conclusion (1) above is that the two-particle reduced density matrix

$$\rho(\mathbf{r}_1\sigma_1 t_1 : \mathbf{r}_2\sigma_2 t_2 : \mathbf{r}_3\sigma_3 t_3 : \mathbf{r}_4\sigma_4 t_4) \equiv \langle \psi^\dagger_{\sigma_1}(\mathbf{r}_1 t_1)\psi^\dagger_{\sigma_2}(\mathbf{r}_2 t_2)\psi_{\sigma_3}(\mathbf{r}_3 t_3)\psi_{\sigma_4}(\mathbf{r}_4 t_4)\rangle \tag{7.8.7}$$

has one and only one eigenvalue which is of order N in the thermodynamic limit. Given this statement, we can define the "Cooper-pair wave function" as the corresponding eigenfunction; with the standard convention for the meaning of the average $\langle\rangle$ (see Chapter 5, Section 5.4) this can then be written

$$F(\mathbf{r}_1\alpha_1 t_1 : \mathbf{r}_2\beta_2 t_2) \equiv \langle \psi_\alpha(\mathbf{r}_1 t_1)\psi_\beta(\mathbf{r}_2 t_2)\rangle \tag{7.8.8}$$

What can we say about the structure of the pair wave function (order parameter) F?

The Pauli principle implies that $F(\mathbf{r}_1\alpha_1 t_1 : \mathbf{r}_2\beta_2 t_2)$ is odd under interchange of all three indices ($\mathbf{r}_1 \rightleftharpoons \mathbf{r}_2$, $\alpha \rightleftharpoons \beta$, $t_1 \rightleftharpoons t_2$). In principle this gives two possibilities (or a "mixture" of them): (a) F is odd under exchange of the coordinate and spin indices and even under $t_1 \rightleftharpoons t_2$; (b) F is even under the first exchange and odd under $t_1 \rightleftharpoons t_2$. The latter possibility (known in the literature as "odd-frequency" pairing, see Abrahams et al. 1995) implies that F is zero at $t_1 = t_2$; there seems no positive experimental evidence in favor of it, and as we shall see below some evidence against it, so from now on I shall make the "default" assumption (a). Since apart from possible pathologies F is then nonzero for $t_1 = t_2$, we may simplify the discussion by considering the quantity (for which I use the same symbol F)

$$F(\mathbf{r}_1\alpha : \mathbf{r}_2\beta) \equiv \langle \psi_\alpha(\mathbf{r}_1)\psi_\beta(\mathbf{r}_2)\rangle \tag{7.8.9}$$

where the time arguments of the ψ's are now taken equal. The quantity (7.8.9) must then be odd under the simultaneous exchange of coordinates and spins.

In general, the group G of exact symmetry of the Hamiltonian of the cuprates is of the form

$$G = U(1) \otimes T_l \otimes H \tag{7.8.10}$$

where $U(1)$ is the usual gauge group, T_l is the (abelian) group of lattice translations and H is the point group of the crystal in question (which we note for most cuprates includes spatial inversion, P). If we take the BCS "particle-nonconserving" convention (see Chapter 2, Section 2.4 and Appendix 5A) then the existence of ODLRO automatically implies "spontaneous" breaking of the $U(1)$ symmetry (in the alternative convention mainly used in this book, the latter is unbroken). What of the other elements in (7.8.10)?

Although there are theoretical papers in the literature which postulate spontaneous breaking of the lattice translation symmetry along the c-axis and/or in the ab-plane, to my mind there is no experimental evidence for any such phenomena in the generic case, and to simplify the argument I shall therefore again make the "default" assumption that the translational symmetry is unbroken. The question of the "symmetry of the order parameter" then reduces to its behavior under the crystal point group H.

Because of the presence of spin–orbit coupling, H does not in general contain the spin rotation group as a subgroup. Nevertheless, conservation of parity forbids the mixing[32] of spin singlet and triplet components in the order parameter (7.8.9), so that the natural assumption is that this quantity corresponds either to a spin singlet or to a triplet (in the latter case, in general with nontrivial spin–orbit correlations). Which is it? Here we appeal to the experimental observation, by now made in many different cuprate superconductors, that the Knight shift (or more accurately its "Pauli" part) tends to zero in the superconducting state in the limit $T \to 0$, independently of the field direction. In any simple weak-coupling theory of the BCS type, this behavior would be unambiguous evidence of spin singlet pairing. It may perhaps be argued that in a more general theory we cannot be absolutely sure that a triplet-paired (or perhaps mixed singlet-triplet) state would not behave in the same way, but to the best of any knowledge no concrete theory having this property has been proposed, and while effects beyond BCS might lead to some reduction in χ_s (cf. also the behavior of ^{3}He-B) it seems a priori very implausible that they would lead to its complete vanishing.

Thus, we can draw our third conclusion:

3 The order parameter is a spin singlet

In other words, it must have the form

$$F(\boldsymbol{r}_1\alpha, \boldsymbol{r}_2\beta) = (i\sigma_y)_{\alpha\beta} F(\boldsymbol{r}_1, \boldsymbol{r}_2) \tag{7.8.11}$$

where to respect the Fermi antisymmetry $F(\boldsymbol{r}_1, \boldsymbol{r}_2)$ must be even under the exchange of its arguments, or equivalently, for fixed COM coordinate $\boldsymbol{R} \equiv (\boldsymbol{r}_1 + \boldsymbol{r}_2)/2$ the parity of $F(\boldsymbol{r}_1 - \boldsymbol{r}_2)$ is even.

It remains to discuss the symmetry of the spatial part of the pair wave function, $F(\boldsymbol{r}_1 \boldsymbol{r}_2)$ or as we shall refer to it for brevity, the order parameter (OP), under the operations of the "orbital" part of the crystal point group (i.e. those operations which provided the spin degrees of freedom have the form (7.8.11) leave the Hamiltonian unchanged). Let us for the moment neglect the complications associated with the c-axis layering structure and thus restrict $\boldsymbol{r}_1$ and $\boldsymbol{r}_2$ to lie within a single CuO_2 plane. Although the ensuing argument is quite general, it helps one's intuition if one also fixes the center-of-mass coordinate $(\boldsymbol{r}_1 + \boldsymbol{r}_2)/2$ to lie at a point of high symmetry in the lattice, let us say the position of a Cu atom, and studies the dependence of $F(\boldsymbol{r}_1, \boldsymbol{r}_2)$ on the relative coordinate $\boldsymbol{r}_1 - \boldsymbol{r}_2$.

If we neglect for the moment yet another complication, namely that of orthorhombic anisotropy (which is small or zero in most materials other than YBCO), then the symmetry group of the CuO_2 planes is that of the square, namely C_{4v} in the standard group-theoretical terminology. This is a rather simple group: the primitive operations are: (a) rotation through $\pi/2$ about the (001)(z-)axis ($\boldsymbol{R}_{\pi/2}$); (b) reflection in a crystal axis, e.g. (100)($\boldsymbol{I}_{\text{axis}}$); and (c) reflection in a 45°(110) axis, $\boldsymbol{I}_{\pi/4}$; any one of these three can be represented as a product of the other two, i.e. $\boldsymbol{I}_{\text{axis}} I_{\pi/4} R_{\pi/2} = 1$. In view of the above conclusion about the spin singlet nature of the pairing and the consequent even parity of the state, all states we are interested in here satisfy $\boldsymbol{R}^2_{\pi/2} = +1$, and moreover of course $\boldsymbol{I}^2_{\text{axis}} \equiv \boldsymbol{I}^2_{\pi/4} = +1$. Thus, all the even-parity irreducible representations can be labeled by the possible eigenvalues, ± 1, of (e.g.) the operators $\boldsymbol{R}_{\pi/2}$ and

[32] As in the case of the orbital symmetry to be discussed below, one cannot absolutely exclude the possibility that both components are present, but there seems no positive reason for this.

Table 7.4 The four singlet pairing states of a single plane with square symmetry.

Informal name	Group-theoretic notation	$R_{\pi/2}$	I_{axis}	$I_{n/4}$	Representative state
s (or s^+)	A_{1g}	$+1$	$+1$	$+1$	const.
g (or s^-)	A_{2g}	$+1$	-1	-1	$xy(x^2-y^2)$
$d_{x^2-y^2}$	B_{1g}	-1	$+1$	-1	x^2-y^2
d_{xy}	B_{2g}	-1	-1	$+1$	xy

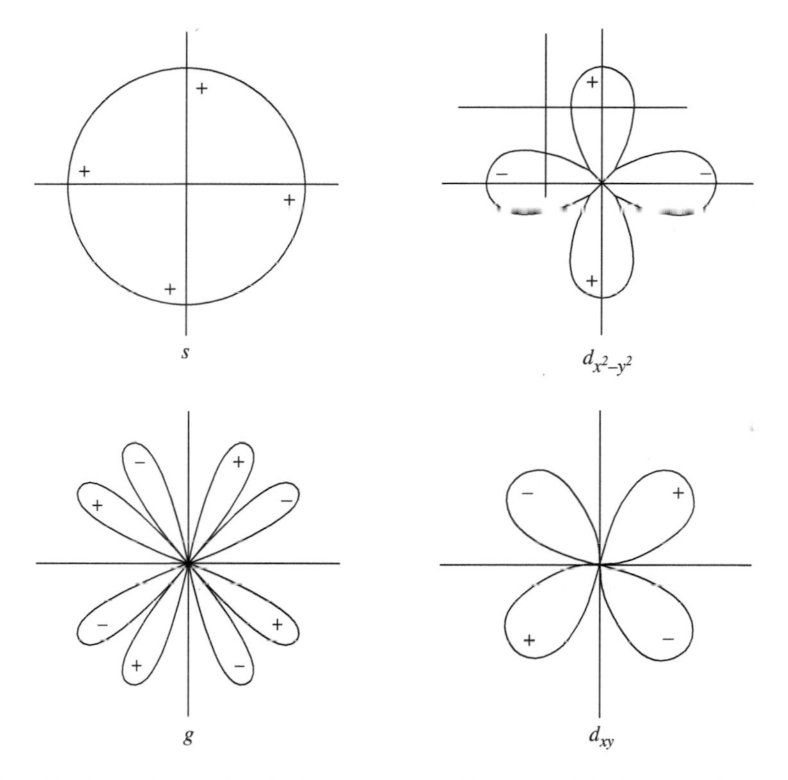

Fig. 7.11 The four singlet irreducible representations possible in a single square CuO_2 plane.

$\boldsymbol{I}_{\text{axis}}$. These four possible representations, which are all one-dimensional, are denoted A_{1g}, A_{2g}, B_{1g} and B_{2g} in the standard group-theoretic notation, or informally respectively as s (or $s^{(+)}$), g (or s^-), $d_{x^2-y^2}$ and d_{xy}; their symmetry properties are shown in Table 7.4, and correspond to the "representative" functions shown; A_{1g} is the "identity" (or "trivial" representation). Representative forms of the corresponding functions are shown in Fig. 7.11; however, it should be emphasized that the actual form may look quite different, provided only that it transforms in the correct way. For example, a possible form of s-wave OP as a function of the angle θ in the CuO_2 plane is $A + B\cos 4\theta$, which for $|B| > |A|$ has eight nodes (such a form of OP is sometimes called an "extended s-wave" state).

In general, the OP of a superconducting cuprate may correspond to a single irreducible representation ("irrep") of the group C_{4v}, or to a superposition of irreducible representations (the latter possibility is often labeled in shorthand (e.g.) "$s + id_{x^2-y^2}$" to indicate the representations involved). However, the latter possibility appears to be ruled out in bulk, barring pathology, by the failure to date to observe any kind of further phase transition within the superconducting state: for the details of the argument, see Annett et al. (1996), Section 3. Thus, we can conclude that the (bulk) order parameter of the cuprates corresponds to a single irrep of the group C_{4v}.

Before embarking on a discussion of experimental tests, we should say a word about the complications due to orthorhombicity and c-axis structure. While the members of the Hg series are strictly tetragonal and those of the two Tl series very nearly so, most other cuprates have some orthorhombic asymmetry. In particular, in YBCO (both 1237 and 1248) the chains break the tetragonal symmetry, so that while $\boldsymbol{I}_{\text{axis}}$ remains a good symmetry $\boldsymbol{R}_{\pi/2}$ and $\boldsymbol{I}_{\pi/4}$ are no longer so. In LSCO and the Bi series, on the other hand, it is the two orthogonal 45° axes which are inequivalent, while remaining mirror planes; thus in this case $\boldsymbol{I}_{\pi/4}$, but not $\boldsymbol{R}_{\pi/2}$ or $\boldsymbol{I}_{\text{axis}}$, remains a good symmetry. As a result of the orthorhombicity, the states cannot be classified rigorously as (e.g.) s or $d_{x^2-y^2}$; however, we can still classify them as "s-like" or "d-like" according as the OP preserves or changes its sign under a $\pi/2$ rotation (The magnitude will in general change, e.g. in YBCO it seems plausible that the OP along the b-axis will be greater in magnitude than that along the a-axis.) One further note concerns possible complications associated with the c-axis layering structure: for these we refer the reader to Annett et al. (ref. cit.) Section 4.3.

In the simple BCS theory of anisotropic pairing as presented in Chapter 6, and presumably in any theory not too qualitatively different from it, the angular anisotropy of the order parameter (which is identical in $\boldsymbol{r}$-space or in $\boldsymbol{k}$-space) is reflected in that of the energy gap Δ_k in $\boldsymbol{k}$-space; in particular, in the s^-, d_{xy} and $d_{x^2-y^2}$ states the energy gap must have nodes on the Fermi surface, while the $s^+(\equiv s)$ state need not. There exists by now a vast body of experiments, both spectroscopic and on the asymptotic temperature-dependence of quantities such as the penetration depth as $T \to 0$, which give apparently incontrovertible evidence that the single-particle density of states in a typical cuprate superconductor is nonzero down to an energy of at most of the order of $0.01T_c$. Unfortunately, this does not by itself rule out the s-wave scenario: all it tends to show is that the single-particle energy gap has nodes (or at least very low minima), and this is entirely consistent with an "extended s-wave scenario". A further difficulty in interpreting such experiments as giving information on the order parameter is that the very existence of the pseudogap phenomenon shows that we can get a thinning-out of the low-energy DOS even when the superconducting OP is zero, so that there is no one-one correspondence between the two quantities (although it is admittedly difficult to envisage a scenario where the OP is nonzero in all directions but the DOS at low or zero energy is nevertheless also nonzero).

A much more direct and, at least in the opinion of most workers in the field, reliable way of determining the symmetry of the order parameter is via phase-sensitive experiments which exploit the Josephson effect. I postpone a discussion of the general nature of these experiments to Section 7.10, and for the details of the theoretical analysis and an up-to-date statement of the experimental results refer the reader to Annett et al. (1996) and Tsuei (2000, 2004) respectively; here I just state my conclusion. Namely, of the 30-odd experiments of this type conducted over the last twelve years the vast majority appear to be consistent with, and when taken together uniquely to establish, a bulk order parameter with pure $d_{x^2-y^2}$ symmetry. To be sure, there exists one series of experiments whose results appear to be inconsistent with such an assignment (see Klemm 2005), but it is not clear to me that they are consistent with *any* reasonable assignment, and in view of the various uncertainties involved I would regard them as substantially outweighed by the rest of the experimental evidence. Consequently, I believe we can draw our fourth conclusion, namely that (in bulk).

4 *The orbital symmetry of the order parameter is* $d_{x^2-y^2}$

It should be emphasized: (a) that this statement refers only to the behavior under the operations of C_{4v}, and certainly does *not* imply that the dependence of the OP (or the energy gap) on the Fermi surface is specifically $k_x^2 - k_y^2$; (b) that it does not exclude a different or more complicated behavior near some or all surfaces; indeed, there is considerable circumstantial evidence (see e.g. Greene et al. 2003) that near a (110) surface the form may be $d_{x^2-y^2} + is$.

Let's next consider the question of how far we can regard the formation of the superconducting state as taking place independently within the individual multilayers, without appreciable effect from intermultilayer coupling. The crucial input here is the order of magnitude of the c-axis resistivity $\rho_c(T)$. If we interpret the latter provisionally in terms of an incoherent hopping process between multiplanes, then we can extract a "hopping time" $\tau(T)$, and it turns out that in all cases except (marginally) that of optimally doped YBCO we have the order-of-magnitude estimate $\hbar/\tau(T_c) \ll k_B T_c$. If this is the case, it is at least highly plausible that the intermultilayer hopping contact, while no doubt essential to stabilize true ODLRO in the superconducting state, must be irrelevant to a first approximation to the process of forming the pairs within each multilayer. It might be asked whether we could avoid this conclusion by postulating that the c-axis transport is, rather, of the conventional Bloch-wave type? Although the argument is then a little more complicated, the answer is no, since the matrix element t_c for coherent intermultilayer transport also turns out to be $\ll k_B T_c/\hbar$. Thus, if we exclude the possibility[33] that intermultilayer tunneling is somehow orders of magnitude more efficient in the S state than in the N state, we can presumably conclude that

5 *The formation of Cooper pairs takes place independently within different multilayers*

A further question relates to the role of the electron–phonon interaction. Let us ask the specific question: By how much is the kinetic energy of the ions lowered (or increased) in the S ground state with respect to the N ground state? We can obtain information

[33] A theory of cuprate superconductivity having in effect this character was proposed soon after the original discovery by P.W. Anderson, but does not at the time of writing seem to command widespread support, see Section 7.9.

on this from the isotope effect: Suppose that the change in ionic KE is a fraction η of the total condensation energy (note that η is not a priori inevitably positive). Then we have directly from the Feynman–Hellman theorem

$$\frac{\partial(\ln E_c)}{\partial(\ln M)} = -\eta \tag{7.8.12}$$

Ideally, we should measure the isotope effect on E_c directly. However, if we assume that the transition temperature T_c scales as E_c^β, and define the conventional "isotope exponent" $\alpha \equiv -\partial(\ln T_c)/\partial(\ln M)$, then we evidently have $\eta = \alpha/\beta$. Thus, within a BCS theory, which gives $\beta = \frac{1}{2}$, an observed value of α equal to $\frac{1}{2}$ (as in most classic superconductors) implies that $\eta = 1$, i.e. the superconducting condensation energy is accounted for entirely by the change in ionic kinetic energy. In the case of the cuprates, the isotope exponent is found experimentally to be small (typically only $\sim$0.01 for both Cu and O) except for the lower-T_c systems and, sometimes, close to a crystallographic phase transition, and this would indicate that the saving in ionic kinetic energy, expressed as a function of the maximum achievable condensation energy, is generally small. To be sure, this does not establish rigorously that phonons are not playing an important role in the formation of Cooper pairs,[34] and there is some evidence from recent ARPES data that a particular phonon mode involving the apical O's may be more strongly coupled to the electrons than was previously suspected. However, barring pathologies it would seem safe to draw the conclusion that:

6 *The electron–phonon interaction is not the principal mechanism of the formation of Cooper pairs*

In arguing for conclusions (1)–(6), I have not assumed that a BCS-type theory is even qualitatively applicable to the Cooper pairs formed in the cuprates. However, if we are prepared to postulate that at least some qualitative features of this theory apply, we can draw a couple more conclusions. The first concerns the size of the Cooper pairs in the *ab*-plane[35] (let us say the r.m.s. radius of the function $|F(\boldsymbol{r}_1 - \boldsymbol{r}_2)|^2$ for some reference value of the COM coordinate). From the GL description above, without any reliance on BCS theory, we can derive from the experimentally measured critical field $H_{c\perp}(T)$ the in-plane GL healing length $\xi_{ab}(T)$ for T close to T_c,[36] and in this region it appears to scale as $(1-T/T_c)^{-1/2}$. This is, of course, precisely the behavior predicted in BCS theory, and in this theory the prefactor is, apart from a constant of order unity, the size ξ_0 of the Cooper pair. A second way of estimating the latter within BCS theory is to use the relation $\xi_0 \sim \hbar v_F/k_B T_c$; we can try to apply this result to the cuprates if we take $v_F \sim \hbar k_F/m^*$, where $k_F \sim 1$ Å^{-1} is the radius of the Fermi surface apparently observed in ARPES experiments and $m^* \sim 4$ m the effective mass needed to fit the specific heat to a Fermi-liquid picture (see Sections 7.3–7.7). The two methods give the same estimate for the pair radius ξ_0 within a factor of 2–3, and this might give us sufficient confidence to conclude that (at least for the higher-T_c systems).

[34] A few cases are known of classic superconductors with α close to zero.

[35] In view of conclusions (2) and (3), the concept of the "size" of the pairs in the c-direction is not really well defined.

[36] As noted earlier, for lower values of T the critical field is usually too large to be measured.

7 The size of the Cooper pairs in the ab-plane lies in the range 10–30 Å

It is possible that in the future it will become possible to confirm (or refute) this estimate by measuring the quantity $|F(\boldsymbol{r}_1 - \boldsymbol{r}_2)|^2$ directly, e.g. by a two-photon spectroscopic technique. Note that if (7) is indeed valid, then since at optimal doping the average distance a between conduction electrons (or rather holes) in the ab-plane is approximately 9 Å, this means that the ratio ξ_0/a is of order 1, in strong contrast to a classic superconductor where it is typically $\sim 10^4$ (cf. above).

Finally, let us ask the question: Are the Cooper pairs formed, as in standard BCS theory, by pairing electrons in time-reversed states? This is actually not totally obvious a priori: in particular, it has been pointed out (Tahir-Kheli 1998) that if one adopts a picture of the N state in which more than one band intersects the Fermi surface, then it is not obviously excluded that the Fourier-transformed orbital pair wave function $F(\boldsymbol{k})$ has the form

$$F(\boldsymbol{k}) = \langle a_{\boldsymbol{k}i} a_{-\boldsymbol{k}j} \rangle \tag{7.8.13}$$

where the suffixes i and j refer to different bands. However, at least within the band picture, there is a strong argument (Leggett 1999a) that, barring pathology, such a hypothesis would predict a strong absorption of electromagnetic radiation just above the gap edge, something which certainly does not seem to be seen experimentally. To the extent that one is willing to assume that this result generalizes beyond the simple band picture, one can thus conclude that just as in traditional BCS theory:

8 The Cooper pairs are formed from time-reversed states

While it is certainly not true that all theoretical papers on cuprate superconductivity accept conclusions (1)–(8), it is probably fair to say that the vast majority assume all of them, either explicitly or (more often) implicitly. If we do so, then we can divide them into two groups. Conclusions (2), (5) and (6) can effectively be combined into the following statement, where to avoid inessential complications I refer to a single-plane cuprate: For the purpose of discussing cuprate superconductivity, a sufficient zeroth-order approximation to the Hamiltonian is a sum of independent single-plane Hamiltonians of the form

$$\hat{H} = \hat{T}_{\parallel} + \hat{U} + \hat{V} \tag{7.8.14}$$

where $\hat{T}_{\parallel}$ denotes the in-plane kinetic energy of the conduction electrons, $\hat{U}$ the single-electron potential energy due to the field of the static CuO_2 lattice, and $\hat{V}$ the interconduction-electron Coulomb interaction, appropriately screened by the in-plane and off-plane atomic cores. It cannot be over-emphasized that the Hamiltonian (7.8.14) should be in principle a *complete* zeroth-order description of the problem; any considerations concerning spin fluctuations, excitons, anyons or any other exotic entities should be *derived* from it, not added to it!

The other conclusions (1), (3), (4), (7) and (8) describe the output we must require from a viable theory: the formation of Cooper pairs in a spin singlet, $d_{x^2-y^2}$ orbital state, with radius $\sim$10–30 Å and formed from time-reversed single-electron states. This paragraph and the last define, in a nutshell, the theoretical problem posed by cuprate superconductivity.

7.9 The cuprates: questions and ideas

At the time of writing there exists no "theory" of superconductivity in the cuprates which has received anything approaching the degree of assent in the relevant community which was accorded to the BCS theory of classic superconductivity already within a few years of its publication. Indeed, it is not even clear what a successful "theory" of cuprate superconductivity would be like: we will know it when we see it! For this reason I shall not attempt in this section to describe any of the many current theoretical approaches in detail; rather I will outline some obvious questions and sketch, at a qualitative level, a selection of current ideas which may be relevant to them.

First, then, the questions:

1. Is the normal state of the cuprates well described by (some variant of) the traditional Landau Fermi-liquid theory? This question is actually not quite as clear-cut as it looks: the Fermi-liquid hypothesis is essentially a statement about the asymptotic low-energy (hence low-temperature) behavior of the system, and the fact that the onset of superconductivity often occurs at values of T_c/T_F which are much greater than in the classic superconductors or in ^{3}He means that this limit is not always accessible. Presumably, a relatively definitive experimental answer to the question could in principle be obtained by a sufficient range of experiments on samples in which superconductivity is suppressed by doping and/or magnetic fields, or on the class of cuprates which never becomes superconducting at all (see Section 7.1). However, this raises the next question:
2. Should the anomalous properties of the S state be regarded as reflecting those of the N state, or rather vice versa, i.e. are the N-state properties anomalous because of the occurrence of high-temperature superconductivity? In particular, should the pseudogap regime be regarded as an extension of the S regime?
3. Related to the above, is there a "hidden" phase transition somewhere in the neighborhood of the "crossover" line $T^*(p)$? If so, what is the nature of the phase which occurs below this line? Does it correspond to the spontaneous breaking of some symmetry (other than the $U(1)$ symmetry which is often said to be "broken" by the onset of Cooper pairing), and if so which one?
4. Is the two-dimensional layered structure of the cuprates an essential ingredient in their high-temperature superconductivity?
5. Related to the last, why does T_c increase with n in homologous series up to $n = 3$ (and thereafter apparently decline somewhat)? As pointed out in Section 7.8, the "naive" explanation which attributes the increase simply to the Josephson (or LD) interplane coupling certainly does not work, at least without considerable modifications (cf. below, on the ILT model).
6. Does the long-range part of the Coulomb interaction play a significant role in cuprate superconductivity?
7. Why do cuprates having very similar CuO_2-plane characteristics (e.g. the Hg and Tl series) nevertheless have substantially different (maximum) values of T_c?
8. Why, despite the fact that the *ab*-plane properties of the cuprates appear to be at least qualitatively universal (see Sections 7.3–7.7), does the *c*-axis N-state resistivity behave qualitatively differently for different systems?

Finally, two rather more wide-ranging questions:

9. Most current theoretical approaches to the problem of cuprate superconductivity start from a "model", that is, they postulate some form of effective Hamiltonian which, it is hoped, describes the "low-energy" states of the many-body system in at least a qualitatively correct way. Typically, "low-energy" means with energy per electron of order $k_B T_c$, or at least small compared to the (perturbative) Fermi energy, say $\sim$1 eV for the cuprates. Such a procedure was of course spectacularly successful in the case of the BCS theory (with the effective Hamiltonian being the electron-phonon one). Is it obvious that *any* approach of this type will work for the cuprates?
10. Is the current maximum value of T_c (150–160 K, depending on the exact definition) the theoretical upper limit for any material of the cuprate class? Irrespective of this, are there other classes of materials (presumably not existing in nature) which would show even higher T_c's, perhaps up to or above room temperature?

I now sketch a subset of the bewildering variety of theoretical ideas which have been proposed in order to answer some of these questions. For a first pass, it is convenient to classify these into two main groups, depending on whether they start from something resembling a traditional "band" model or from a more localized picture. (However, not all of the ideas to be discussed are easily classified in this way.) Among the first group we may note:

1. *van Hove singularities* (Markiewicz 1997). This is probably the most conservative approach to the problem of the high T_c's of the cuprates; it is often, though not always, combined with invocation of the traditional electron-phonon interaction as the mechanism of formation of Cooper pairs. One starts from the BCS formula

$$T_c \sim \omega_D \exp -(N(0)V_0)^{-1} \tag{7.9.1}$$

and notes that for fixed V_0 an increase in the single-electron density of states at the Fermi surface, $N(0)$, may lead to a dramatic increase in T_c. (Such an effect is undoubtedly seen in the alkali-doped fullerides, see Chapter 8, Section 8.1.) It is known that in a 2D (or 3D) crystalline lattice potential the energy band structure $\epsilon(\boldsymbol{k})$ can have saddle points, and that those will than give rise to "van Hove" singularities in the DOS $N(\epsilon)$. Should such a singularity occur close to the Fermi surface of the system in question, $N(0)$ should be much enhanced, thus possibly leading to an enhancement in T_c. Such a hypothesis would appear to offer a natural qualitative explanation of the sensitivity of T_c to doping (since the latter shifts the Fermi energy), and some of the ARPES data indeed seems consistent with a substantially enhanced DOS in the relevant region; however, if the V_0 of (7.5.1) is attributed to the electron–phonon interaction, we seem to encounter the difficulties mentioned under point (6) of Section 7.8, and more generally it is not clear whether this approach pushes the BCS scheme beyond its regime of validity.

2. *Marginal Fermi liquid* (Varma et al. 1989: Varma et al. 2002, Section 7). This is a phenomenology of the N state which was developed early in the history of the subject. It is assumed that the general structure of the energy levels in the N state conforms to the canonical Fermi-liquid pattern, (in the sense that there is, or at least

may be, a one-one correspondence to those of the noninteracting Fermi gas), but that for some reason the energy-dependence of the quasiparticle decay rate $\Gamma(\epsilon)$ is linearly proportional to the difference $|\epsilon|$ from the Fermi surface (rather than to ϵ^2 as in the standard Fermi-liquid picture, cf. Appendix 5A). Such a hypothesis explains in a natural way many of the anomalous temperature-and frequency-dependences of the N-state properties, but does not in itself explain the origin of the linear form of $\Gamma(\epsilon)$. For the latter various proposals involving interactions of the quasiparticles with different boson-like collective excitations of the electron system (e.g. excitons) have been made; of those the most quantitatively developed is associated with ...

3. *Spin fluctuations* (Pines 1998, Scalapino 1999). It is noted that in the AF Mott-insulator regime of the phase diagram, the *static* spin density structure factor $S(\boldsymbol{q})$ has a pole at the "superlattice" points $\boldsymbol{q} = (\pm\pi/2a, \pm\pi/2a)$ which can be interpreted as saying that the frequency of spin excitations tends to zero at these points. It is then a natural conjecture that in regions of the phase diagram which border the AF phase the spin fluctuation spectrum for $\boldsymbol{q}$ near these superlattice points will have a strongly enhanced weight at low frequencies (energies). Interaction of the single-particle-like quasiparticle excitations with these low-energy spin fluctuations should then give rise to a marginal-Fermi-liquid-like behavior of the N phase; in addition exchange of virtual spin fluctuations could play a role analogous to the exchange of virtual phonons in BCS theory and thus give rise to an effective electron–electron attraction which could lead to formation of Cooper pairs and thus to superconductivity.

While there are a number of variants of the "spin-fluctuation" hypothesis in the literature, they have in common one striking prediction, as follows. If we assume a Fermi surface of roughly the structure indicated by the ARPES experiments (see Fig. 7.8), then a simple geometrical construction indicates that the pairs of points on the Fermi surface which are connected by superlattice vectors $\boldsymbol{q}_0$ lie as shown in Fig. 7.12, that is, each lies close to the crystal (a and b) axis direction and they are approximately related by a $\pi/2$ rotation. If scattering by exchange of a virtual spin fluctuation with $\boldsymbol{q}$ close to $\boldsymbol{q}_0$ is indeed important in the formation of Cooper pairs, this would suggest that the pair amplitude ought to be large in directions close to the crystal axes. However, a more interesting conclusion relates to the relative sign at points A and B. As is shown in Appendix 7B, in a general anisotropic BCS-like state scattering by static impurities is pair-breaking (and so depresses $T_{\rm c}$, possibly to zero) unless it possesses the "correct" sign (or more generally phase) of the order parameter. Although it is not perhaps immediately obvious, it turns out (Millis et al. 1988) that a similar consideration applies to pairing induced by low-energy virtual excitations: the structure of the pair wave function should be such that *real* scattering processes involving such excitations should preserve the correct symmetry. Now, when an electron scatters from a spin fluctuation (with $\boldsymbol{q}$ close to $\boldsymbol{q}_0$), it flips its spin and thus changes the sign of the pair wave function[37] (cf. the magnetic-impurity case discussed in Chapter 5, Section 5.9), and hence the sign of the "orbital" part of the wave function must be opposite at A and B. Thus, the spin-fluctuation hypothesis predicts that the order parameter of the cuprates must (a) be large in the region of the $a-$ and $b-$ crystal axis directions, and (b) change sign under $\pi/2$ rotations in the

[37] Assumed to be a spin singlet for the reasons discussed under (4) above.

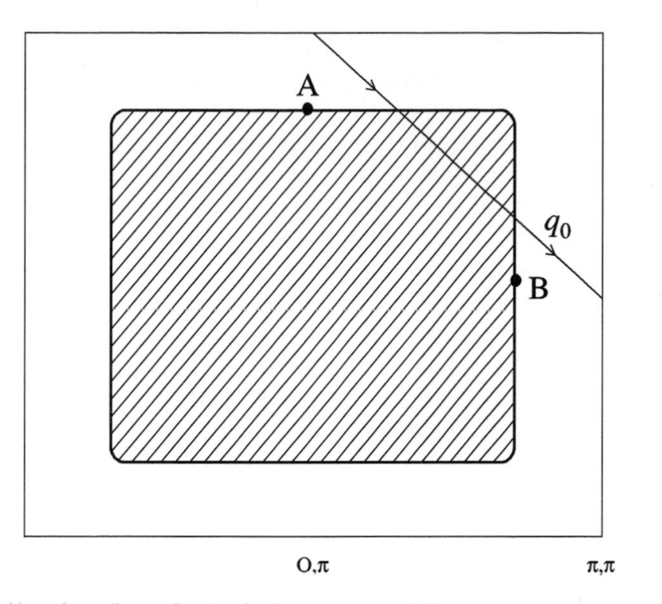

Fig. 7.12 The Fermi surface of a typical cuprate and the superlattice vector $\boldsymbol{q}_0$ (see text).

ab-plane. On inspection of Table 7.3, we see that these requirements uniquely fix the symmetry of the OP as $d_{x^2-y^2}$. This prediction, which was made before the $d_{x^2-y^2}$ symmetry was definitely established by the Josephson experiments, is arguably the only substantive example in the cuprates to date of theory anticipating experiment.

Finally in this group of ideas I include somewhat arbitrarily

4. *"Pre-formed pairs" and phase fluctuations* (Uemura 1997, Emery and Kivelson 1995). This complex of ideas refers mainly to the "pseudogap" or "strange metal" region of the cuprate phase diagram and postulates that in this regime we have in one sense or the other "local" pairing of fermions without ODLRO. In the most extreme form of this scenario (Uemura 1997), pairs of electrons literally form real bosons, which are assumed to have undergone BEC (and hence produced ODLRO) in the S region of the phase diagram; the pseudogap regime then corresponds to the region where the bosons are formed but do not undergo BEC. At the other end of the spectrum is the "phase-fluctuation" scenario proposed by Emery and Kivelson (1995); while the general structure of the S phase is assumed in some sense BCS-like (in particular, no real bosons are formed), it is pointed out that as a result of the very much reduced superfluid density (relative to a typical 3D classic superconductor) phase fluctuations cost much less energy and may, rather than vanishing of the energy gap, be responsible for the destruction of superconductivity. Various scenarios intermediate between those two extreme versions have been explored, and recently interesting attempts have been made to relate them to the "BEC-BCS crossover" behavior of the Fermi alkali gases (see Chapter 8, Section 8.4). In most scenarios of this general class, the line (region) of anomalies $T^*(p)$ in the cuprate phase diagram (see Sections 7.3–7.7) is identified with the region in which the pairs disassociate (or, in the Emery–Kivelson version, with the "ideal" ("mean-field") T_c which one would get could phase fluctuations be

entirely neglected); advocates of such a scenario often postulate that the "line" $T^*(p)$ merges smoothly into the part of the curve $T_c(p)$ which lies in the overdoped regime, as in Fig. 7.7b.

I now turn to the second class of theoretical ideas, which starts from the observation that the S region of the cuprate phase diagram lies close to a regime where the electrons are not merely antiferromagnetic but localized, that is, form a Mott insulator, and thus regards the problem of cuprate superconductivity as either identical with or closely related to, that of a Mott insulator doped with a relatively small number p of holes. In this connection the following observation is quite suggestive: Consider a single hole in an otherwise perfect antiferromagnetic Mott-insulator lattice (Fig. 13). Assume, as seems plausible, that the matrix element for hopping of the hole (i.e. of the electron it replaces) is largest along a crystal axis direction. It is clear that such hopping processes have the effect of disrupting the AF order (see Fig. 13) and thus cost an energy which is linear in the number of sites transversed. ("Diagonal"

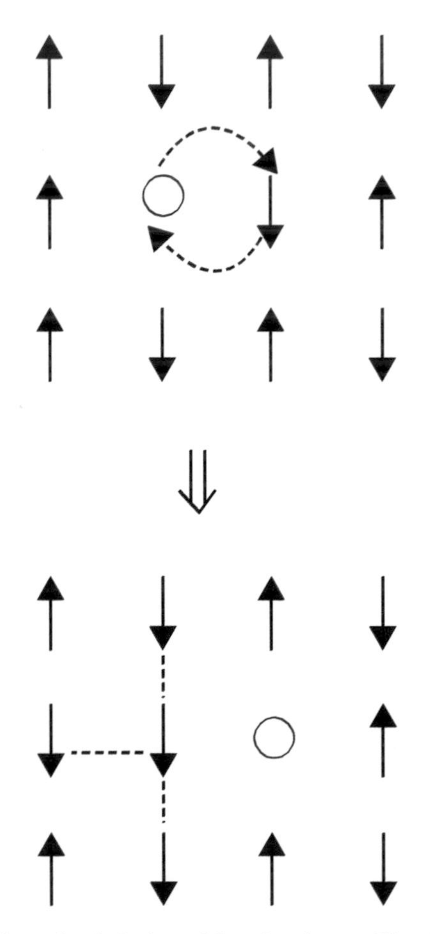

Fig. 7.13 Effect of motion of a hole in a Mott insulator. The "bad" bonds are indicated by straight pecked lines.

hopping processes do not incur this energy cost, but are assumed less probable.) Now consider *two* holes separated by n lattice spacings, and hopping along the same path. It is clear that the second hole restores the AF order which the first has destroyed: consequently, the cost in energy of the hopping is now simply proportional to n and independent of the total number of sites transversed by the pair. Consequently, the pair of holes acts like a bosonic object whose COM motion is determined by an effective mass related to the single-electron tunneling matrix element, and with an effective binding energy linearly proportional to the pair size – a situation analogous to the phenomenon of "confinement" in quantum chromodynamics. If one could assume that those considerations extend, at least qualitatively, to the case of many pairs, then at first sight it seems that the above observation might hold the key to the mechanism of Cooper pairing and hence of superconductivity in the cuprates.

Unfortunately, there is a serious snag with this scenario: The "confinement" mechanism requires, at least at first sight, a nonzero degree of AF order in the many-body state, whereas experimentally it is found quite unambiguously that antiferromagnetism (as distinct from localization) disappears at a value of $p(\sim 0.04)$ considerably less then that (~ 0.05) needed for the onset of superconductivity. Nevertheless, the idea is sufficiently appealing that it seems attractive to try to derive the occurrence of superconductivity within a model which at least permits (as the band model does not) the existence of a Mott-insulator state for appropriately small values of p and in particular for $p = 0$. The very simplest example of such a model is the

5. *Hubbard model.* This model, first introduced by Hubbard in 1964, is probably the best-known low-energy effective Hamiltonian in all of condensed matter physics. In its very simplest ("one-band, nearest-neighbor") form it refers to a single set of atomic orbitals $\varphi(\boldsymbol{r} - \boldsymbol{R}_i^0)$ localized near lattice sites $\boldsymbol{R}_i^0$ and the associated electron creation operators $a_{i\sigma}^+$ (where the suffix $\sigma = \pm 1$ refers to the spin), and in terms of the latter reads (where δ runs over the bonds connecting to i to its nearest neighbors).

$$\hat{H}_{\text{Hubb}} = -t \sum_{i,\delta,\sigma} a_{i+\delta,\,\sigma}^+ a_{i\sigma} + \frac{1}{2} U \sum_{\sigma} a_{i\sigma}^+ a_{i\sigma} a_{i,-\sigma}^+ a_{i,-\sigma} \tag{7.9.2}$$

where the operators $a_{i\sigma}$ satisfies the standard anticommutation relations

$$\{a_{i\sigma}, a_{j\sigma'}\} = 0, \quad \{a_{i\sigma}, a_{j\sigma'}^+\} = \delta_{ij}\delta_{\sigma\sigma'} \tag{7.9.3}$$

and the "filling" of the lattice is specified by the condition (recall that p is conventionally defined as the number of holes per CuO_2 unit relative to the state with one electron per unit)

$$N^{-1} \sum_{i\sigma} \langle a_{i\sigma}^+ a_{i\sigma} \rangle = 1 - p \tag{7.9.4}$$

(N = total number of sites). The first term in (7.9.2) describes hopping between nearest-neighbor sites in the tight-binding approximation, while the second takes account of the cost in Coulomb energy of double occupation of the orbital on site i by electrons with opposite spins (occupation by two or more electrons of the same spin is of course automatically forbidden by the Pauli principle which follows from the anticommutation relations (7.9.3)). The Hamiltonian (7.9.2), supplemented with the conditions (7.9.3) and (7.9.4), is believed by many[38] to be qualitatively adequate as a starting point for the discussion of both the N and S state properties of the cuprates.

Despite its apparent simplicity, the problem defined by Eqn. (7.9.2)–(7.9.4) is, as far as is presently known, analytically soluble in 3D, or even in the 2D case presumably relevant to the cuprates,[39] only in certain limits: in the limit $U/t \to 0$ it fairly obviously reduces to the problem of noninteracting electrons described by a tight-binding band structure, while in the limit $U/t \to \infty$ with $p = 0$ ("half-filling") it is reliably believed to produce a Mott-insulator state (in which each lattice point is occupied by exactly one electron) which for t/U very small but nonzero is antiferromagnetic (cf. below, on the $t - J$ model). Away from these simple limits one must at present resort either to (digital) numerical computation or to approximate analytic schemes. An intriguing possibility which has opened up in the last couple of years is that it may be possible to use fermion alkali gases in optical lattices (cf. Chapter 4, Section 4.6) to do reliable analog simulations of the problem posed by (7.9.2)–(7.9.4) which much exceed in power the present digital techniques; if such simulations indeed turn out to be feasible, they should answer once and for all the question whether a simple Hamiltonian of the type (7.9.2) is capable of producing (inter alia) a Cooper-paired state which resembles that found experimentally in the cuprates.

In the limit of small but nonzero t/U, it is plausible that one can trust the hopping term as a perturbation, and thereby derive from the Hubbard model the so-called

6. *t–J model*: The effective Hamiltonian is given in terms of the operators $a^+_{i\sigma}$, $a_{j\sigma'}$ by

$$\hat{H}_{t-J} = -t\sum_{i,\delta\sigma} a^+_{i\sigma}a_{j\sigma} + J\sum_{i,\delta} \boldsymbol{S}_i \cdot \boldsymbol{S}_{i+\delta}, \quad \boldsymbol{S}_i \equiv \sum_{\alpha\beta} a^+_{i\alpha}\boldsymbol{\sigma}_{\alpha\beta}a_{i\beta} \tag{7.9.5}$$

with an explicit prohibition of double occupancy. Here the antiferromagnetic exchange coupling constant J is given by $J = 4t^2/U$, and determines the spin-wave excitation spectrum. By measuring the latter by inelastic neutron scattering, one deduces that at half filling (i.e. in the "parent" compound, $YBa_2Cu_30_6$, etc.) the value J is approximately 1500 K (0.13 eV), and since tight-binding band structure calculations give a value of t of the order of 0.5 eV, this would imply that the on-site Coulomb repulsion is ~8 eV, which for a pair of hybridized $2p - 3d_{x^2}$ orbitals is probably not unreasonable. The t–J Hamiltonian has been used as the starting point for many calculations on the cuprates. However, one should be worried that it (along with its parent, the Hubbard model) leaves out at least one effect which may be important, namely the long-range part of the Coulomb interaction (see Leggett 1999b).

7. *Spin-charge separation* (Lee, 1999). One of the main attractions of the $t - J$ model in the context of a theory of the (N and S states of the) cuprates is that it does not assume a priori that the ground state of the system is anything like a traditional Fermi liquid. Indeed, currently there is a major industry of postulating various kinds of exotic groundstate (usually with one or more symmetries other than $U(1)$ spontaneously broken) and exploring their properties. A'very persistent theme

[38]Not including the present author.

[39]It is analytically soluble in 1D, but the solutions to 1D problems are notoriously unreliable as guides to behavior in higher dimensions.

in this context is the notion of spin-charge separation. In many 1D models this idea is almost trivial. For example, consider a 1D chain of fermions with exactly one fermion per site. Suppose initially the spins are oriented in some particular configuration. Then if there is any spin–spin coupling (irrespective of sign) neighboring spins can flip and this will result in transfer of spin: e.g.

$$\uparrow\uparrow\uparrow\uparrow\downarrow\downarrow\downarrow\downarrow\rightarrow\uparrow\uparrow\uparrow\downarrow\uparrow\downarrow\downarrow\downarrow\rightarrow\uparrow\uparrow\downarrow\uparrow\uparrow\downarrow\downarrow\downarrow\ \dots$$

Evidently, a "down" spin moves from R to L without any charge transfer. If now we introduce a hole, then if the lattice is initially disordered it will propagate without (on average) spin transfer:

$$\uparrow\downarrow\downarrow\uparrow\uparrow\downarrow\ \bigcirc\ \uparrow\rightarrow\uparrow\downarrow\downarrow\uparrow\uparrow\ \bigcirc\ \downarrow\uparrow\rightarrow\uparrow\downarrow\downarrow\uparrow\ \bigcirc\ \uparrow\downarrow\uparrow\ \dots$$

In the case of two or more dimensions, the generalization of this idea is not so clear. What is usually done is the following: one splits the electron creation operator $c_{i\sigma}^{+}$ for site i and spin σ into a "spinon" and a "holon" operator, $f_{i\sigma}^{+}$ and b_i^{+} respectively:

$$a_{i\sigma}^{+} = f_{i\sigma}^{+} b_i \tag{7.9.6}$$

The operator $f_{i\sigma}^{+}$ obeys Fermi statistics and b_i^{+} obeys Bose statistics. The constraint of no double occupancy is replaced by the condition

$$f_{i\sigma}^{+} f_{i\sigma} + b_i^{+} b_i = 1 \tag{7.9.7}$$

which is enforced by introducing an appropriate Langrange multiplier.

While it is almost trivial to write the electron operators in terms of spinon and holon operators, it is not at all obvious how to invert the process and express the spinons and holons in terms of real electrons. Thus it is not altogether clear (at least to me) that the procedure described by (7.9.6) is well defined. Nevertheless, let us see what consequences follow.

The most obvious guess is that the holons, which obey Bose statistics with (presumably) conserved particle number, undergo BEC at a temperature $T_{\text{BEC}} = 2\pi pt$ (where p is the doping). On the other hand, the exchange term can be written in terms of the spinon operators alone:

$$\begin{aligned} J\boldsymbol{S}_i \cdot \boldsymbol{S}_j &= -J|f_{i\alpha}^{+} f_{j\alpha}|^2 \\ &= -J(f_{i\uparrow}^{+} f_{j\downarrow}^{+} - f_{i\downarrow}^{+} f_{j\uparrow}^{+})(f_{i\downarrow} f_{j\uparrow} - f_{i\uparrow} f_{j\downarrow}) \end{aligned} \tag{7.9.8}$$

The last expression is an identity, but it suggests making various mean-field types of decoupling. First, it is natural to suppose that at low temperatures $\langle f_i^{+} f_j \rangle \neq 0$, which when Fourier transformed implies the existence of an energy band and a Fermi surface. Secondly, it is tempting to suppose that at sufficiently low temperature the quantity $f_{i\uparrow}^{+} f_{j\downarrow}^{+} - f_{i\downarrow}^{+} f_{j\uparrow}^{+}$ also acquires an expectation value (in the usual BCS sense):

$$\langle f_{i\uparrow}^{+} f_{j\downarrow}^{+} - f_{i\downarrow}^{+} f_{j\uparrow}^{+} \rangle = \Delta_{ij} \tag{7.9.9}$$

Since the sites i and j involved here are different, the "gap" Δ_{ij} cannot have s-wave symmetry, and since it is confined (prima facie at least) to nearest neighbors and is a spin singlet, the only possibility is d-wave (in fact, in a square lattice with only nearest-neighbor pairing, $d_{x^2-y^2}$). The resulting phase diagram is predicted to look as in Fig. 14, which is reproduced from Lee, Fig. 3.

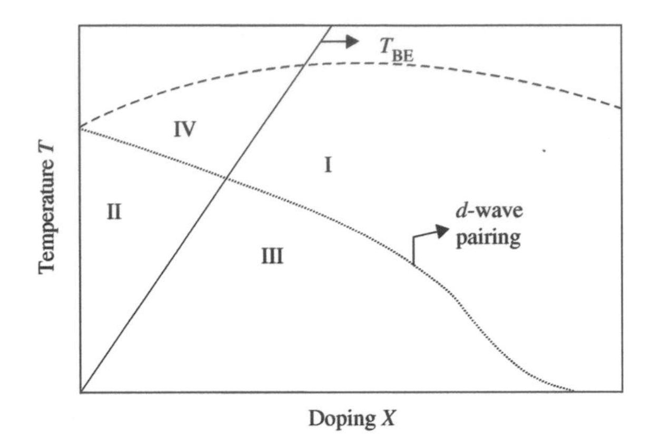

Fig. 7.14 Schematic mean-field phase diagram of the $t-J$ model (Reprinted from *Physica C*, Vol. 317, P. A. Lee, Pseudogaps in Underdoped Cuprates, pp. 194–204, Copyright (2000) with permission from Elsevier.).

To the right of the solid line we have BEC of the holes, and below the lower dashed line $\Delta_{ij} \neq 0$ for the spinons. Superconductivity, that is a finite $\langle c^+_{i\sigma} c^+_{j-\sigma} \rangle$ where the c_i's are true electron operators, requires simultaneous BEC and pairing ($\Delta_{ij} \neq 0$), thus it corresponds to region III. Region I is a FL-like phase (despite the BEC of holons!), region II is the "spin-gap" (pseudogap) phase and region IV is said to be a "strange-metal" phase. For further discussion, see Lee, 1999.

8. *Quantum critical point* (Sachdev 2003). As already mentioned, another recurrent theme which arises in some of the conjectured solutions of the Hubbard or $t-J$ models (although it is of course of more general significance) is that in addition to the $U(1)$ symmetry which is (said to be) "broken" in the S phase (and the $SU(2)$ spin symmetry which is broken in the Mott-insulator phase), some other symmetry related to the spin and/or orbital motion is broken in some part of the phase diagram, usually below the $T^*(p)$ line. In connection with this idea it is sometimes conjectured that this line (assumed to correspond to a "hidden" second-order phase transition) extends into the S region and meets the $T = 0$ axis at a p-value of about 0.18, thus defining at this point a "quantum critical point" (QCP) (see Sachdev, 2003). Such a hypothesis holds out some promise of explaining some of the anomalous N-state properties of the cuprates, such as the fact that at optimal doping ($p = 0.16$, close though not identical to the p-value of the postulated QCP) the resistivity appears to be fairly accurately linear in temperature. A somewhat related though not identical idea (Zhang 1997) is that of "$SO(5)$ symmetry": the basic notion is that the symmetries $U(1)$ and $SU(2)$, which are broken in the S and AF phases respectively, may be regarded as subgroups of a larger group, $SO(5)$, and that the pseudogap regime thus in some sense interpolates between the S and AF phases.

9. *Phase segregation* (Kivelson and Fradkin 2005). Another theme which has attracted much attention over the last few years is that the translational symmetry of the Hamiltonian may be spontaneously broken in parts of cuprate phase diagram, leading to alternating one-dimensional regions of high and low charge density ("stripes"), two-dimensional regions or perhaps something even more complicated. There is by now fairly convincing experimental evidence that such inhomogeneities do occur in

some specific cuprates, but it is far from obvious whether the phenomenon is in any sense universal, and even if it is, whether it has much to do with superconductivity. This whole issue is complicated by the fact that recent STM measurements (Hoffman et al. 2002) seem to indicate that at least some of the cuprates may be inhomogeneous in an irregular and unsystematic way.

I would finally like to mention one idea, not easily classifiable in the above scheme, which has, historically, played an important role in the theory of cuprate superconductivity and which, while it is no longer believed even by its original proponents to solve the main part of the problem, is of considerable interest in its own right, namely

10. *Interlayer tunneling.* This general idea was proposed by P. W. Anderson in 1987 and developed by him and various collaborators over the next few years; it forms the core of his book '*The Theory of High-Temperature Superconductivity*'. Let us consider for definiteness a single-layer cuprate such as *Tl*-2201, and imagine the process of c-axis tunneling between neighboring CuO_2 planes to be analogous to tunneling across an oxide junction between two bulk superconductors. We have already seen (Section 7.8, point 2) that the conventional way of taking this process into account in the S phase is via the Lawrence–Doniach (LD) model, which treats the interplane barrier as a Josephson junction with all the usual properties. Within the framework of this model, the difference ΔE between the energies of the superconducting groundstate, measured relative to the (putative) normal groundstate, with and without interlayer tunneling is given by the expression

$$\Delta E = J - K \cos \Delta\varphi \tag{7.9.10}$$

where $\Delta\varphi$ is the phase difference of the Cooper pairs in the two layers, and (at least in the simplest models) $K \approx J$. Thus, as already remarked in Section 7.8, point 2, the interlayer tunneling process, while constraining $\Delta\varphi$ to be zero in equilibrium, does not appreciably increase the condensation energy of the S state. The basic reason is that the lowering of the kinetic energy associated with c-axis tunneling is already present in the N state and is merely "shuffled around" in the S state.

However, this argument rests on the implicit assumption that (single particle) tunneling is indeed already present in the N state. Anderson pointed out that if the nature of the in-plane N-state many-electron wave function is very different from the standard Fermi-liquid one, this may not be so; the c-axis N-state tunneling may be blocked. On the other hand, it is not altogether implausible that in the S (Cooper-paired) state the tunneling of Cooper pairs might not be so inhibited. Were this true, the effect would be to make term J in (7.9.10) essentially zero, so that one would obtain a superconducting condensation energy per plane of order K; thus it would become energetically advantageous for the electrons within a single CuO_2 plane to form Cooper pairs, not because of any attractive interaction but because by doing so they are able to reduce the c-axis kinetic energy. (Compare the mechanism of "confinement" of two holes in the Mott-insulating phase discussed above.)

Whatever its other merits or demerits, the ILT theory of cuprate superconductivity possesses one very attractive feature which makes it almost[40] unique among existing theories of cuprate superconductivity: it is "falsifiable" in the Popperian sense, that

[40]But see above, on the prediction by SF model of $d_{x^2-y^2}$ symmetry.

is, it makes at least one specific prediction with such confidence that should that prediction be experimentally refuted, it is recognized that the theory has to be given up. The specific prediction relates in this case to the zero-temperature c-axis London penetration depth $\lambda_c(0)$ of a single-phase cuprate which within the simplest[41] version of the ILT theory can be uniquely related to the superconducting condensation energy. In the case of Tl-2201 the prediction is that $\lambda_c(0)$ should be about 1 μ, whereas an experimental measurement gave a value of about 20 μ; note that since the physically significant quantity is the superfluid density $\rho_s \propto \lambda_c^{-2}$, the true discrepancy is a factor of 400 (similar results were found for Hg-1201). Thus the ILT theory had to be (and has been) discarded as a general theory of cuprate superconductivity; it remains an open question whether it is a valid explanation of the behavior of $T_c(n)$ in homologous series.

I would like to emphasize that this section can in no sense be regarded as an adequate review of the current state of the theory of cuprate superconductivity; not only does it (partly for reasons of space) fail to do justice to the subtlety of some of the ideas listed, but it omits completely others which may well play a role in a future comprehensive theory.

7.10 Novel consequences of Cooper pairing in the cuprates

As should be clear from previous sections of this chapter, not only is the structure and chemistry of the cuprates very different from that of most classical superconductors, but it is very likely that the nature of the superconducting (Cooper-paired) phase is also substantially different. One might therefore expect that the consequences of pairing in the cuprates might include phenomena not seen in more traditional materials. In this section I shall briefly review three such phenomena which are associated respectively with the extreme anisotropy of the cuprates, with their very small coherence lengths and with the "exotic" symmetry of the order parameter. My guess is that if and when we attain a generally agreed understanding of the nature of the pairing in those materials we will discover more.

7.10.1 Consequences of extreme anisotropy;[42] "pancake" vortices

For pedagogical convenience I assume that we are in the "3DGL" regime (although for the most anisotropic cuprates such as the Bi series this regime exists only very close to T_c): the results obtained within the LD model are not qualitatively different. We first define the GL healing lengths $\xi_c(T), \xi_{ab}(T)$ in the obvious way ($\xi_c(T) \equiv (\gamma_\perp/\alpha(T))^{1/2}$, etc.) and also the penetration depths $\lambda_c(T), \lambda_{ab}(T)$; here it is essential to appreciate that $\lambda_c(T)$ means the penetration depth *which screens currents flowing along the c-axis*, for which the direction of penetration actually lies in the ab-plane. Now, by deriving an expression for the electric current from Eqn. (7.8.1) in the standard way and inserting it into the London equation, we see that for given $\Psi(\mathbf{r})$ the eigenvalue of λ is proportional to $\gamma^{-\frac{1}{2}}$, while as we have seen the eigenvalue of ξ is proportional to $\gamma^{1/2}$. Thus we have the important relation, for any given T,

$$\xi_{ab}\lambda_{ab} = \xi_c\lambda_c = f(T) \sim (1 - T/T_c)^{-1} \tag{7.10.1}$$

[41] For some possible subtleties, see Leggett (1998).

[42] A good general reference on the material of the subsection is Tinkham (1996), Chapter 9.

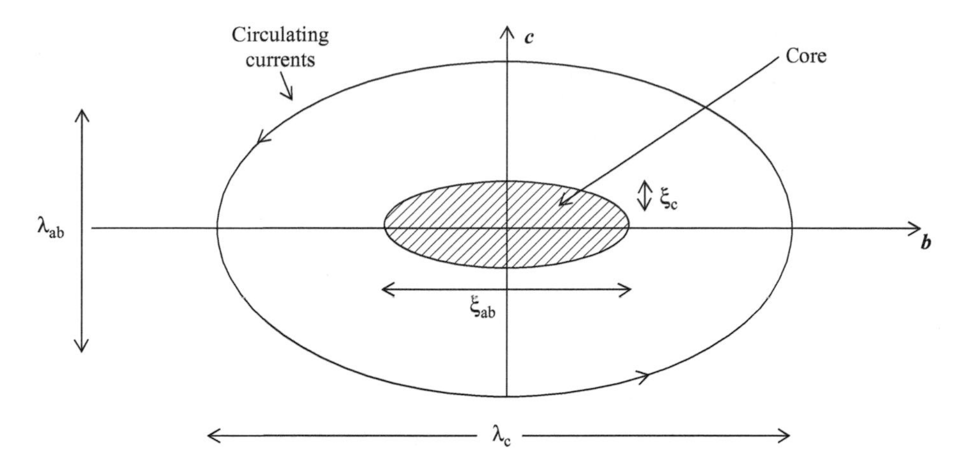

Fig. 7.15 An anisotropic vortex. The magnetic field is out of the paper.

Thus, while the healing (coherence) length along the c-axis is much smaller than that in the ab-plane, the penetration depth for currents flowing along the c-axis is much *larger*.

Consider now the structure of the vortex in a cuprate superconductor. If $\boldsymbol{H}$ is parallel to the c-axis (normal to the ab-plane), then all currents flow in the plane and the anisotropy has no effect; the theory is identical to that for an isotropic superconductor, provided that for ξ and λ we use the ab-plane values ξ_{ab} and λ_{ab}, respectively. In particular, the upper critical field in the z-direction ($H_{c2}^{\perp}$) is given by

$$H_{c2}^{\perp}(T) = \varphi_0/2\pi\xi_{ab}^2(T) \tag{7.10.2}$$

For a field lying in the ab-plane, we can obtain the form of the vortex by scaling the coordinate axes appropriately. Both the core and the overall size of the vortex are strongly elongated in the direction in the ab-plane perpendicular to the field, see Fig. 7.15; because of the relation (7.7.1) and the meaning of λ_c, we see that the *shape* of the vortex is not a function of temperature (provided we stay in the region where (7.10.1) remains valid). If we introduce, as is conventional, the anisotropy ratio $\eta \equiv \xi_{ab}/\xi_c$, then this gives the ellipticity of the vortex. The upper critical field in the ab-plane ($H_{c2}^{\parallel}$) is given by the expression

$$H_{c2}^{\parallel}(T) = \Phi_0/2\pi\xi_{ab}(T)\xi_c(T) \tag{7.10.3}$$

so that

$$H_{c2}^{\parallel}(T)/H_{c2}^{\perp}(T) = \eta \tag{7.10.4}$$

and since η is often large compared to 1, the critical field in the ab-plane is generally much larger than the (already large!) one along the c-axis.

A particularly interesting situation occurs when the external magnetic field is neither parallel nor perpendicular to the ab-plane. Under these conditions one expects to produce vortices with are *on average* parallel to the field. However, it is easily verified that to produce a given current in the c-direction costs an energy $\gamma_{\parallel}/\gamma_{\perp}$ times that necessary to produce the same current in the ab-plane, and thus the currents much prefer to flow in the planes. The result is a set of so-called "pancake" vortices which are staggered from one plane to the next, and the magnetization is not parallel to the field but oriented more along the *c*-axis: see Tinkham (1998), Section 9.3. The critical value of a field making an arbitrary angle θ with the *ab*-plane turns out to be given by

$$H_{c2}(\theta) = \frac{H_{c2}^{\perp}\eta}{(\cos^2\theta + \eta^2\sin^2\theta)^{1/2}} \tag{7.10.5}$$

Note that if $\eta \gg 1$, this means that for all but the smallest values of θ the critical field is determined by the condition $H_c(\theta)\sin\theta = H_{c2}^{\perp}$, i.e. the *c*-axis component of the field is equal to the critical field in this direction. The physical reason for this result is that almost all the energy of the vortices is associated with currents flowing in the *ab*-plane, very little with those flowing along the *c*-axis.

7.10.2 Consequences of the small coherence length: fluctuations

The existence of fluctuations in the superconducting order parameter, and of consequences thereof for the stability of supercurrents, etc., is not unique to the cuprate superconductors: in fact the experimental and theoretical investigations of such phenomena in the classic superconductors goes back to the 1960s. However, the importance of such effects is enormously enhanced in the cuprates, because of their very short coherence lengths (a secondary cause of enhancement is the much higher temperatures to which the cuprates remain superconducting); we recall from Section 7.9, point 4 that such fluctuation effects have even been invoked to explain the gross structure of the cuprate phase diagram, and it is universally agreed that they have major effects on the behavior of the resistivity in a magnetic field.

To investigate the importance of fluctuations qualitatively, let us first follow an argument originally given by Ginzburg and Pitaevskii in the context of the classic superconductors. We assume that we are in zero field and close enough to T_c to use the GL description, so that the free energy F is given as a functional of the order parameter $\Psi(r)$ by Eqn. (5.7.2), namely

$$F\{\Psi(r)\} = F_0 + \int dr\{\alpha|\Psi(r)|^2 + \frac{1}{2}\beta|\Psi(r)|^4 + \gamma|\nabla\Psi(\boldsymbol{r})|^2\} \tag{7.10.6}$$

when α, β, γ are functions of temperature. The *mean* value of $\Psi(r)$ is given by minimizing F, and for the simple translation-invariant case described by (7.10.6) is evidently a position-independent constant $\Psi(= (|\alpha|/\beta)^{1/2})$. However, there will be thermal fluctuations around this "mean-field" value, with a probability given by the expression

$$P\{\Psi(r)\} = Z^{-1}\exp -\Delta F\{\Psi(r)\}/k_BT \tag{7.10.7}$$

where ΔF is $F\{\Psi(r)\}$ minus its value for the mean-field solution and Z is an appropriate normalization constant chosen so that the sum of the probabilities of all possible configurations $\Psi(\boldsymbol{r})$ is unity.

Let us try to formulate an intuitive criterion for the fluctuations to be "important". A plausible guess is that a region with dimension of the order of the healing length $\xi(T)$ should have a probability of order 1 of fluctuating into the normal phase; in other words (assuming $T \sim T_c$) we should have

$$F_{\text{cond}}(T)\xi^3(T) \lesssim k_BT_c \tag{7.10.8}$$

(where F_{cond} is the value of the first and second terms in (7.10.6)). A more quantitative investigation supports the order-of-magnitude estimate (7.10.8); for example, when this condition is fulfilled one can no longer safely use the simple "mean-field" calculation of the specific heat.

Now, within the framework of a GL description, with the standard temperature-dependences of α, β and γ (see Chapter 5, Section 5.7) we have

$$\xi(T) = \xi_0(1 - T/T_c)^{-1/2} \tag{7.10.9a}$$

$$F_{\mathrm{cond}}(T) = E_0(1 - T/T_c)^2 \tag{7.10.9b}$$

Equation (7.10.9a) is a definition of the "coherence length" ξ_0 (which within a BCS-type theory is of the order of the pair radius, and quite generally may be obtained from the upper critical field $H_{c2}(T)$), while in Eqn. (7.10.9b) the order of magnitude of the constant E_0 may be reasonably taken to have the BCS value $n(k_B T_c)^2/\epsilon_F$ where n is the density. Substituting (7.10.9a) and (7.10.9b) in (7.10.8), we thus find that the condition on $\Delta T/T_c \equiv 1 - T/T_c$ is

$$\frac{\Delta T}{T_c} \lesssim (E_0\xi_0^3)^{-2} \sim \left(\frac{\epsilon_F}{k_B T_c} \cdot \frac{1}{n\xi_0^3}\right)^2 \equiv \zeta \tag{7.10.10}$$

(For a BCS-like system this expression can be further simplified, since then $\epsilon_F/k_B T_c \sim n^{1/3}\xi_0$.) If we put numbers into Eqn. (7.10.11), we see that for a clean classic superconductor the range of $\Delta T/T_c$ so defined is $\sim 10^{-16}$, whereas for a typical cuprate (regarded as a 3D system) it is of general order of magnitude unity. In other words, even in zero magnetic field it is never really legitimate to neglect fluctuations in the cuprates! Indeed, it is an experimental fact that quantities such as the specific heat or the resistivity, which in classic superconductors have a sharp discontinuity at T_c, are in the cuprates much more rounded.

As a second illustration, consider the problem of the metastability of supercurrents in a ring of cross-section A, which we take (somewhat arbitrarily) to be of order ξ_0^2. One mechanism of decay of such supercurrent, which actually turn out to be the physically dominant one under appropriate conditions, is for a cross-section of the ring of length $\sim\xi(T)$ to fluctuate into the normal phase, thereby destroying the "topological" conservation of the winding number which we have seen (Chapter 2, Section 2.5) to be the origin of the metastability. The free energy of such a fluctuation is of order $E_{\mathrm{cond}}\xi(T)A \sim E_{\mathrm{cond}}\xi(T)\xi_0^2$; while the condition for this to be $\lesssim k_B T_c$ (and thus for such a fluctuation to have appreciable probability) is less stringent than (7.10.8), it turns out that in terms of the parameter ζ of Eqn. (7.10.10), it can be written in the form

$$\Delta T/T_c \lesssim \zeta^{1/3} \tag{7.10.11}$$

and thus, here again, large values of ζ imply a substantial effect of fluctuations.

In a nonzero external magnetic field the effects of fluctuations are even more pronounced. The two basic qualitative points to appreciate are (1) that "superconductivity" in the sense of zero resistivity cannot be maintained in the presence of vortices, unless these are pinned, and (2) that because of the very different orders of magnitude of the relevant parameters, in particular temperature, it is far more difficult to pin vortices than in a classic superconductor. As a result, the question "are cuprate superconductors in a magnetic field really superconducting?" does not have a trivial answer.

To take point (1) first, a vortex of circulation $\hat{\boldsymbol{n}} \oint \boldsymbol{v}_\mathrm{s} \cdot d\boldsymbol{\ell} = \boldsymbol{\kappa}$ ($\hat{\boldsymbol{n}}$ = direction of axis) placed in a flow field such that the flow velocity at ∞ is $\boldsymbol{v}$ will feel a so-called Magnus force of magnitude

$$F_\mathrm{M} = \rho \boldsymbol{v} \times \boldsymbol{\kappa} \tag{7.10.12}$$

where ρ is the density of the fluid forming the vortex. The Magnus force has nothing to do with quantum mechanics (it was originally discovered in *classical* fluids, cf. Chapter 3, Section 3.5); for a neutral system it is straightforward to obtain it by considering a tube of finite width and calculating the total kinetic energy as a function of vortex position. For a charged system where the vortex is effectively of finite extent, $\sim\lambda$, this argument in its simple form does not work, but more sophisticated arguments give the same result (cf. Tinkham (1998), Section 5.2). In the case of a superconductor, $\boldsymbol{\kappa}$ is equal to $\hat{\boldsymbol{n}}(h/2m)$, and if we assume that the bulk velocity $\boldsymbol{v}$ is associated with the same "density" ρ as appears in Eqn. (5.6.4) (i.e. the superfluid density ρ_s) then we can rewrite (7.6.12) in terms of the electric current density $\boldsymbol{J}(\boldsymbol{r})$

$$\boldsymbol{F}_\mathrm{M} = \boldsymbol{J} \times \hat{\boldsymbol{n}} \cdot \Phi_0 \qquad (\Phi_0 \equiv h/2e)$$

Note that this relation is independent of the value of ρ_s and hence of T. If the Magnus force $\boldsymbol{F}_\mathrm{M}$ is not balanced by some "pinning" force which tends to keep the vortex close to a given impurity (etc.), then its effect will be to accelerate the vortex transverse to the current $\boldsymbol{J}$; eventually its effect will be balanced by some frictional force, and the vortex will reach a terminal (steady-state) velocity $\boldsymbol{u}$, which is the simplest case would be expected to be proportional to $\boldsymbol{v}$.

Now consider the total phase difference $\Delta\varphi_{12} \equiv \int_1^2 \boldsymbol{\nabla}\varphi \cdot d\boldsymbol{\ell}$ between two points in the system separated along the direction of the current flow: for definiteness we choose a straight contour to connect them. Whenever a vortex moves across the contour, the integral decreases by an amount 2π (cf. Chapter 3, Section 3.5). But according to the Josephson relation, we expect the voltage difference V_{12} between the points 1 and 2 to be proportional to the rate of change of $\Delta\varphi_{12}$ (cf. Chapter 3, Section 3.4):

$$\frac{d}{dt}\Delta\varphi_{12} = \frac{2e}{\hbar}V_{12} \qquad (\equiv 2\Delta\mu_{12}/\hbar)$$

Consequently, the *average* voltage $\bar{V}$ is

$$\bar{V}_{12} = \frac{\hbar}{2e}\frac{d}{dt}\Delta\varphi_{12} = \frac{\hbar}{2e}2\pi n_v u s_{12} \equiv \Phi_o n_v u s_{12}$$

where n_v is the number of vortices per unit area and s_{12} the distance between 1 and 2. Thus if u is linearly proportional to $\boldsymbol{J}$ as in the simplest case, $\bar{V}_{12}$ is proportional to s_{12} and to $\boldsymbol{J}$, i.e. the system displays a simple ohmic resistive behavior (In the more general case, the I–V characteristic is nonlinear.) This is known as the *flux-flow mechanism of resistivity*. To calculate an actual value for the resistivity ρ, one needs a theory of the frictional force acting on a moving vortex: the simplest (Bardeen–Stephen) theory yields a linear friction coefficient $\mu_v \approx \Phi_0 H_{c2}/\rho_\mathrm{n}$ where ρ_n is the normal-state resistivity, and this then gives the remarkable result that in the limit $H \to H_{c2}$ the flux-flow resistivity approaches the normal-state value.

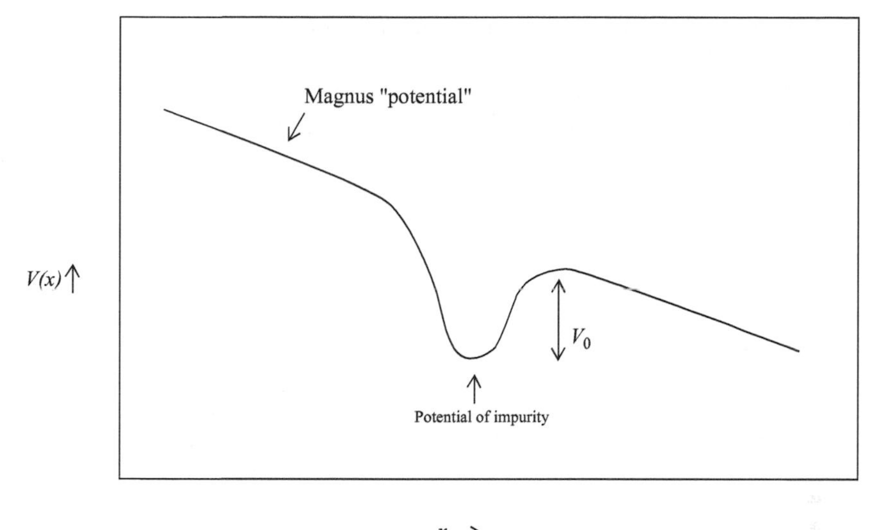

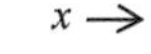

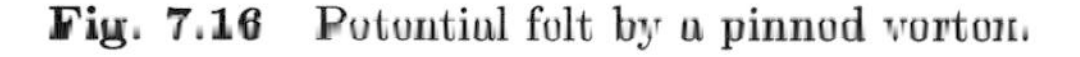

Fig. 7.16 Potential felt by a pinned vortex.

The above discussion rested essentially on the assumption that the vortices are free to move. In real life, they tend to be pinned by impurities, lattice defects, etc., which is why many classic Type-II superconductors have essentially zero resistivity even in fields comparable to H_{c2}. As a result of the pinning plus the Magnus force, the effective potential seen by a vortex is of the form shown in Fig. 7.16; at zero temperature the vortex will be pinned in the metastable well, but at finite temperature there will be the chance of thermal activation over the barrier, with a rate proportional to the Arrhenius factor $\exp(-V_0/kT)$ (note that V_0 is a strongly decreasing function of increasing J). Now it turns out (not obviously!) that typical (single-defect) pinning energies in the cuprates are not very different in order of magnitude from those in classic superconductors, while T_c is much larger; consequently, it is very difficult to avoid depinning. The situation is further complicated by the fact that at lower temperatures interactions between vortices may lead to *collective* pinning ("glass transition"): see Tinkham (1996), Section 9.5.

7.10.3 Consequences of OP symmetry: "exotic" Josephson circuits

Although the symmetry of the Cooper pairs in the cuprates is believed to be "exotic", i.e. the order parameter transforms in a nontrivial way under the symmetry group of the crystal, the associated orientational anisotropy is pinned to the crystal axes: for example, in the very simplest interpretation of "$d_{x^2-y^2}$" symmetry we have

$$\Psi(\boldsymbol{r}) \propto \cos 2\theta$$

where θ is the angle with either the a- or the b-crystal axis. We see moreover (also in the case of a more general $d_{x^2-y^2}$ OP) that interchanging the role of the a- and b-axis merely multiplies the OP (pair wave function) by an overall factor of -1 and therefore has no physical consequences. Thus, one would not at first sight expect to see any analog of the rich variety of effects which arise in superfluid ^{3}He from spatial variation of the OP; in particular, there is no obvious analog of the topological singularities which occur in that system.

Nevertheless, the cuprates do display a class of effects, associated with the exotic symmetry of the OP, which are in some sense the analog of the topological defects in superfluid ^{3}He. Historically, the principal significance of this class of experiments has been that they were the first to demonstrate unambiguously the by now standard assignment of $d_{x^2-y^2}$ symmetry of the OP; a detailed discussion of the logic of the argument is given in (Annett et al. (1996), and an up-to-date review of the relevant experiments in (Tsuei et al., 2000, 2004). In the present subsection I shall as it were approach the problem from the opposite end, by assuming from the start that the OP is indeed $d_{x^2-y^2}$ and asking what novel types of behavior might then be expected; I will not spend time on the many technical details which need to be discussed if one wishes to use the experiments as evidence for the standard assignment.

To orient the ensuing discussion, let's start by considering a thought-experiment which would be very difficult, indeed probably impossible, to realize in practice: Imagine a toroidal strip (of cross-section $\gg \lambda_L^2$) of a tetragonal cuprate, say Tl-2201, which is bent so that as we go once around the torus the *ab*-phase crystal axes are rotated through $\pi/2$ (see Fig. 7.17). Then, if we make the very natural assumption that the orientation of the order parameter "follows" that of the crystal axes adiabatically, it will have been itself rotated through $\pi/2$ and, assuming it is of $d_{x^2-y^2}$ symmetry, will thus acquire a phase of π. However, the single-valuedness boundary condition (cf. Chapter 2, Section 2.2) requires that the pair wave function, when brought around the ring in this fashion, should "come back to itself", including phase. Thus there must be a compensating gradual phase change in the pair wave function which when integrated around the loop makes up the extra required π phase difference. However, such a phase gradient, if uncompensated by a magnetic vector potential $\mathbf{A}$ will give rise to an electric current, and we knew that on paths in the interior of the superconductor, such as the path C shown in Fig. 7.17, such currents must have been screened out by the Meissner effect (which, we recall, is a direct consequence of the GL description and is thus (qualititatively) independent of the OP symmetry). Thus the only possible conclusion is that the phase bending on the contour C must be compensated by a vector potential so as to give total current zero; the line integral of the vector potential, i.e. the total flux trapped through C, must be just $\Phi_0/2$ where as in Chapter 5, $\Phi_0 \equiv h/2e$ is the superconducting flux quantum. Thus we reach the conclusion that *in thermodynamic equilibrium the system shown in Fig. 17 traps half a quantum of flux.* It is clear that this result is reminiscent of the "Berry-phase" effects mentioned in Chapter 2, Section 2.2; note that the half-flux quantum can have either sign, i.e. the groundstate is twofold degenerate.

Consider now a circuit composed of a set of N bulk superconductors, the OP's of which may be of conventional and/or "exotic" type, connected by N Josephson junctions (Fig. 7.18); even when two or more bulk superconductors are of identical material (e.g. YBCO) they may have different orientations of the crystal axes. The only property of the Josephson junctions which we need to invoke is that they are sufficiently "similar" that the preferred relative orientation of the OP's on the two sides is identical in all cases *when viewed from the reference frame of the junction.* For example, in the case of a junction between a (100) surface of (say) Tl-2201 and

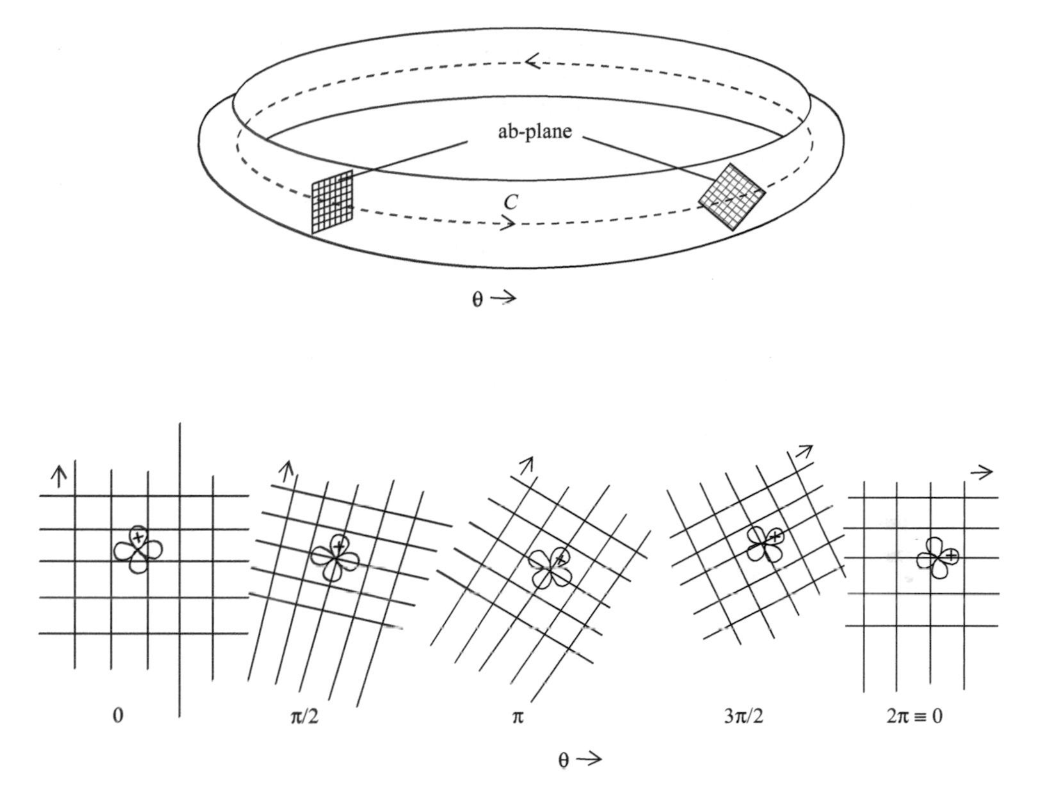

Fig. 7.17 A circuit where *d*-wave pairing will generate half a quantum of flux. The row of figures illustrates the *ab*-plane as viewed from the negative *c*-direction, as a function of θ. The path C is distant by $\gg \lambda_L$ from any surface of the toroid.

a conventional *s*-wave superconductor such as Al, it is natural to assume that the "preferred" orientation is one in which the sign of the lobe of the Tl OP which points along the *a*-axis (i.e. is normal to the junction) is the same as that of the *s*-wave Al OP; but we could equally well make the opposite assumption, provided that we make it consistently for all "sufficiently similar" junctions of the circuit.

It now follows immediately that if the various junctions j of the circuit (meaning the tunnel-oxide barriers themselves *plus* the pair of bulk crystals involved) are related to one another by some sort of symmetry operations $\hat{S}_j$, then when we traverse the circuits once we effectively perform on the pair wave function the transformation $\hat{S}_{\text{tot}} \equiv \Pi_{j=1}^{N-1} \hat{S}_j$ (note that in certain circumstances it may be important to keep the order of the operations corresponding to the physical process). If the OP has no particular symmetry under the operation $\hat{S}_{\text{tot}}$, then we cannot draw any further conclusions without more detailed microscopic assumptions about the junctions, etc.; and if the OP corresponds to an eigenfunction of $\hat{S}_{\text{tot}}$ with eigenvalue $+1$ (as would trivially be the case, for example, for an arbitrary circuit containing only *s*-wave

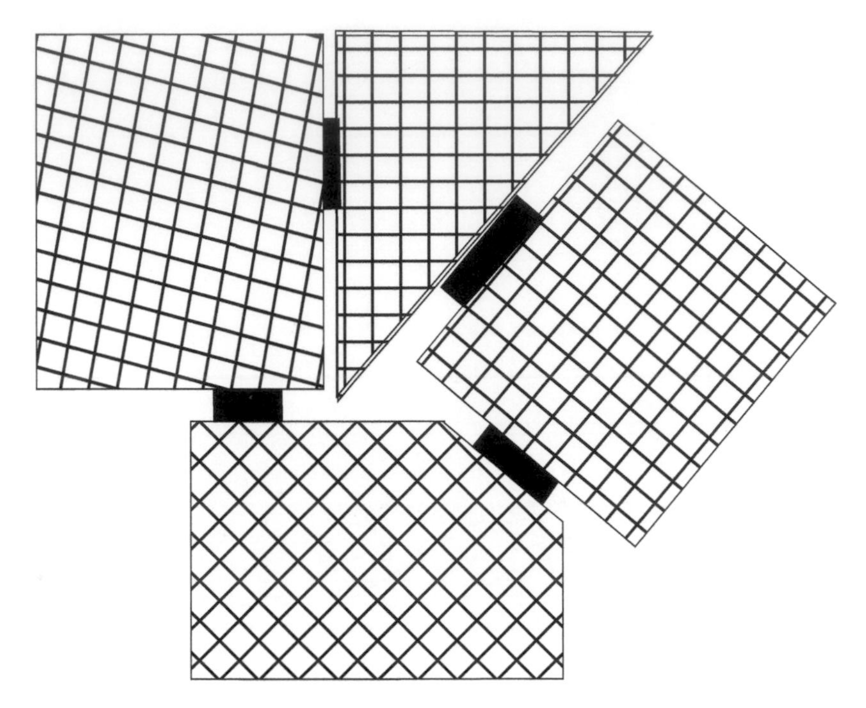

Fig. 7.18 A generic Josephson circuit with $N = 4$.

bulk superconductors and only "normal" junctions), then we find nothing particularly interesting. If, however, the OP of one or more of the bulk superconductors is an eigenfunction of $\hat{S}_{\text{tot}}$ with eigenvalue -1, a more interesting result can occur: If the product of the pair wave functions of all the bulk superconductors in the circuit is an eigenfunction of $\hat{S}_{\text{tot}}$ with eigenvalue -1, then we have an "intrinsic" phase change of π; this must then be made up, just as in the toroidal thought-experiment discussed above, by a phase gradient, and provided the bulk superconductors are large on the scale of the London penetration depth this must be compensated by a vector potential, again leading to trapping by the circuit of half a quantum of flux.

To put some flesh on the rather abstract-seeming considerations of the last paragraph, it may be helpful to consider two specific examples, which are in effect somewhat idealized versions of two actual experiments. In the first, which is an idealization of the experiment of Wollman et al. (1993), we consider a circuit composed of a tetragonal cuprate and a conventional superconductor such as Nb which is known to have s-wave symmetry, with the two junctions rotated through an angle $\pi/2$ relative to one another; see Fig. 7.19. Thus, $\hat{S}_{\text{tot}}$ in this case is $\hat{R}_{\pi/2}$, a rotation in the ab-plane through $\pi/2$. Since the s-wave order parameter is invariant under this operation and the d-wave parameter assumed to characterize the cuprate changes sign, we predict that the "intrinsic" phase change should be π and thus the circuit should spontaneously trap a half quantum of flux, as indeed is in effect[43] observed. Note that

[43]A subsequent experiment using a similar geometry (Mathai et al. 1995) explicitly observed half-integral flux.

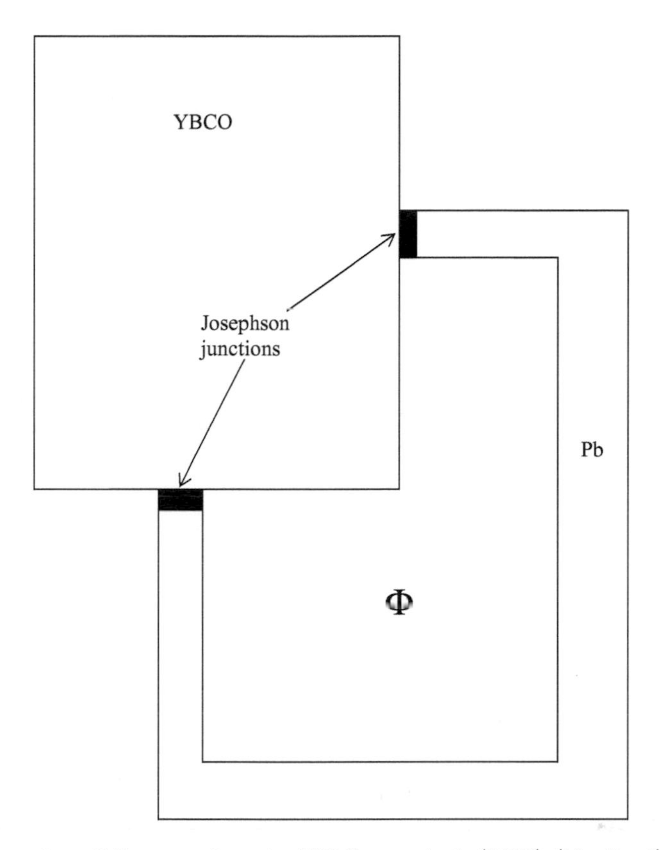

Fig. 7.19 Geometry of the experiment of Wollman et al. (1993) (idealized).

the argument as given would work equally well for a d_{xy} state, which also changes sign under a $\pi/2$ rotation; for any s-state (of the cuprate) there would be no flux trapping.

A somewhat more sophisticated application of the principle is to the experiment sketched in Fig. 7.20, which an idealization (in this case a relatively slight one) of the experiment of Tsuei et al. (1994). It is clear that in this case the junction A is related to junction B by inversion in the crystal axis ($\hat{I}_{\text{axis}}$), while B is related to C by a $\pi/2$ rotation ($\hat{R}_{\pi/2}$). Consequently, $\hat{S}_{\text{tot}} = \hat{I}_{\text{axis}}\hat{R}_{\pi/2} \equiv \hat{I}_{45}$, an inversion in the 45° axis. Since we have three bulk cuprate crystals and the $d_{x^2-y^2}$ state is odd under this operation, the total intrinsic phase change is π and we again predict trapping of a half-quantum of flux (in this case, in the real-life experiment, this prediction was directly checked by a scanning SQUID microscope technique). Again, note that a similar result would follow on the assumption that the symmetry of the cuprate OP irreducible representation ("irrep") is s^- (but not if it is s^+ or d_{xy}).

The above discussion rests explicitly on the assumption that the OP corresponds to a single irrep of the symmetry group of the lattice. If one allows the possibility of superpositions of different irreps, either in bulk or (more plausibly) near certain kinds of surface (cf. Greene et al. 2003), the analysis becomes considerably more complicated

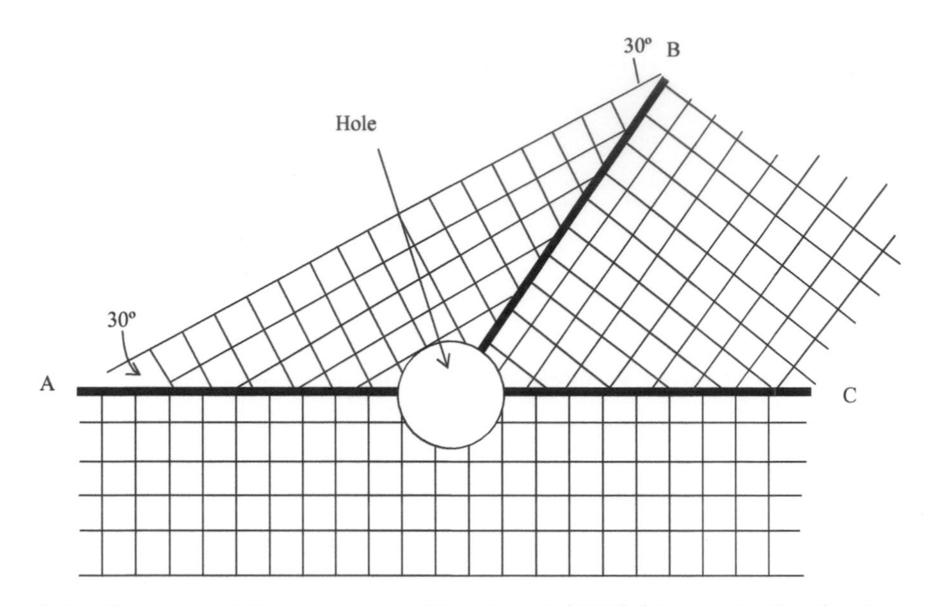

Fig. 7.20 Geometry of the experiment of Tsuei et al. (1994) (slightly idealized). All crystals are YBCO (twinned). The heavy lines indicate grain boundaries, which are believed to act as Josephson junctions.

and is to my knowledge not yet complete: for a partial discussion see Annett et al. (1996).

Appendices

7A Some reviews of various experimental properties of the cuprates

A number of good reviews of experimental work up to 1996 on specific aspects of cuprate superconductivity may be found in the series *Physical Properties of High-Temperature Superconductors*, Vols. I–V, ed. D. M. Ginsberg, Singapore, World Scientific 1989–1996. I list below a few more recent reviews; the list is not intended to be exhaustive.

B. Batlogg, Solid State Comm. 107, 639 (1998) (experimental overview).
M. Kastner et al., Revs. Mod. Phys. 70, 897 (1998) (magnetic, transport, optical).
C. C. Tsuei and J. R. Kirtley, Revs. Mod. Phys. 72, 969 (2000) (Josephson experiments).
J. L. Tallon and J. W. Loram, Physica 349, 53 (2001) (phase diagram).
N. E. Hussey, Adv. Phys. 51, 1685 (2002) (thermodynamics, transport, tunnelling).
J. C. Phillips et al., Reps. Prog. Phys. 66, 2111 (STM and other probes of disorder).
M. R. Norman et al., Reps. Prog. Phys. 66, 1547 (2003) (ARPES, neutron scattering, optics).
A. Damascelli et al., Revs. Mod. Phys. 75, 403 (2003) (ARPES).
D. N. Basov and T. Timusk, Revs. Mod. Phys. 77, 721 (2005) (electrodynamics).

7B Pair-breaking in non-s-wave states[44]

To avoid irrelevant complications, let's assume the Hamiltonian of our system is invariant under time reversal but that the potential term favors pairing in an anisotropic state, that is, one which is not invariant under the operations of the symmetry group of the "pure" system; for definiteness let us suppose this state is a spin singlet and hence has nonzero even "angular momentum". Examples include the hypothetical d-wave state of a system such as liquid ^{3}He, or the $d_{x^2-y^2}$ state believed to be actually realized in the cuprate superconductors. It is intuitively plausible that any impurities, even nonmagnetic ones, will fail to commute with the pair-creation operator $\hat{\Omega}$, and that (unlike in the case of nonmagnetic impurity scattering in an s-wave paired system) this cannot in general be remedied by choosing a different basis for the pairing. Thus, as in the case of magnetic impurity scattering in an s-wave state discussed in Chapter 5, Section 5.8, we have the qualitative choice between pairing in the actual single-particle eigenstates and thereby driving the pairing terms to zero, or "reconstituting" the Fermi sea so as to be able to pair in plane-wave states, and the latter choice becomes unviable as soon as $\Gamma_\Lambda \gtrsim \Delta_0$, where Δ_0 is the BCS gap in the pure system and Γ_Λ is now, crudely speaking, the energy width of a plane-wave state due to impurity scattering (see below for a more accurate definition). Thus we reach the important conclusion that (barring pathological cases, cf. below) *any* impurities, whether magnetic or not, will tend to suppress non-s-wave pairing.

We can attempt to make this argument a little more quantitative as follows.[45] Let us define an operator $\hat{\Lambda}$ such that it includes all (relevant) operations of the symmetry group of the pure system and when applied to the favored pairing state gives eigenvalue $+1$. It is simpler to illustrate this by examples: for an ABM-like state oriented along the z-axis we would have

$$\hat{\Lambda} = \frac{1}{2\pi}\int_0^{2\pi} e^{-i\phi}\hat{R}(\varphi)d\varphi \tag{7.B.1}$$

and for the $d_{x^2-y^2}$ state of a cuprate superconductor

$$\hat{\Lambda} = \frac{1}{4}\left(1 + \hat{R}(\pi) - \hat{R}\left(\frac{\pi}{2}\right) - \hat{R}\left(\frac{3\pi}{2}\right)\right) \tag{7.B.2}$$

where in each case $\hat{R}(\phi)$ indicates a rotation of ϕ around the z-axis. We now construct, as in Section 5.8, a pair operator of the form

$$\hat{\Omega} = \sum_{mn} c_{nm} a_n^\dagger a_m^\dagger \tag{7.B.3}$$

and it is then intuitively plausible that the expectation value of the pairing term in the potential energy will be given by an expression analogous to (5.9.16), namely

$$\langle V \rangle = -V_0 |\Psi|^2, \qquad \Psi \equiv \sum_{mn} c_{nm}\Lambda_{mn} \tag{7.B.4}$$

[44]While the effect of impurities, both magnetic and nonmagnetic, on Cooper pair formation in exotic superconductors is an important and interesting topic in its own right, this appendix is included here primarily because of the application to superfluid 3-He in aerogel (Chapter 8, Section 8.2).

[45]I should emphasize that the argument which follows is intended to be plausible rather than rigorous (cf. Preface). For a much more careful discussion see Balatsky et al. (2006).

where V_0 is the coefficient of the pairing potential in the "favored" channel. Once we accept (7.B.4), we can proceed in exact analogy to the calculation of Section 5.8, and if we assume the relaxation of Λ is exponential with lifetime Γ_Λ^{-1}, then we recover a result completely analogous to (5.9.26), namely

$$\frac{1}{\pi}\int_0^{\epsilon_c} d\epsilon \int_0^{\epsilon_c} d\epsilon' \frac{(\tanh(\beta_c\epsilon/2) + \tanh(\beta_c\epsilon'/2))\Gamma_\Lambda}{[(\epsilon-\epsilon')^2 + \Gamma_\Lambda^2](\epsilon+\epsilon')} = \int_0^{\epsilon_c} \frac{\tanh(1/2)\beta_{c0}\epsilon}{\epsilon} d\epsilon \tag{7.B.5}$$

or as it is more usually written (cf. 5.9.27)

$$\ln \frac{T_{c0}}{T_c(\Gamma_\Lambda)} = \psi\left(\frac{1}{2} + \frac{\hbar\Gamma_\Lambda}{4\pi k_B T}\right) - \psi\left(\frac{1}{2}\right) \tag{7.B.6}$$

Thus, unless by some pathology the impurity scattering conserves $\hat{\Lambda}$, it is indeed deleterious to any non-s-wave state. For practical application of formula (7.B.6) it is usually reasonable to approximate Γ_Λ by the measured inverse transport relaxation time, although of course they in general differ by a numerical factor.

8

Miscellaneous topics

8.1 Noncuprate "exotic" superconductors

Under this heading I classify a variety of noncuprate materials which have been found to be superconducting over the last 30 years, but which lack one or more of the properties, listed at the beginning of chapter 7, which are characteristic of the classic superconductors. I shall not attempt a detailed description either of the experimental properties of these materials or of our current theoretical understanding of them, but will focus primarily on the ways in which they differ qualitatively from the classic type, and on what they have in common with the cuprates.

8.1.1 Alkali fullerides

These materials are crystals of fullerene (usually C_{60}) molecules doped with various alkali atoms, typically with a ratio of close to 3 to 1 (e.g. K_3C_{60}). In the C_{60} molecule the 60 C atoms arrange themselves approximately over the surface of a sphere of radius ~5 Å in a regular pattern of hexagons and pentagons; of the four hybridized $2s$–$2p$ orbitals characteristic of the C atom, three are used up in binding nearest neighbors, leaving a fourth "p_z-like" orbital whose amplitude is maximum in the direction perpendicular to the local surface of the sphere. Linear combinations of these p_z-like states and of higher atomic states can then be used to construct molecular orbitals (MOs), which have different degeneracies depending on their symmetries; it turns out that the 60 electrons (one per C atom) so far unused exactly fill the lowest seven MOs, and that the lowest unoccupied molecular orbital (LUMO) is three-fold orbitally degenerate and lies about 0.6 eV above the highest occupied one (HOMO).

The crystalline structure of solid C_{60} is fcc, with (cubic) lattice parameter 14.2 Å corresponding to close packing of the molecules; the large compressibility of the lattice (about 12 times that of diamond) indicates that the intermolecular bonding is of van der Waals type. In the crystal the molecular orbitals of the isolated C_{60} molecule evolve into bands which are reasonably well described by a tight-binding approximation, with a HOMO-LUMO band gap of approximately 1.9 eV. Thus, in the crystalline states of pure C_{60} the lowest seven bands are completely filled and the LUMO band is completely empty, so that the system is an insulator[1] as predicted by naive band theory. The LUMO band is of width approximately 0.4 eV, and contains six states

[1]Because of its relatively small band gap, it is often classed as a wide-gap semiconductor, but the difference in the present context is merely one of name.

per C_{60} molecule (or per primitive unit cell); as we shall see, its properties are crucial for the superconductivity of the alkali fullerides.

The fcc structure of crystalline C_{60} allows 12 "voids" per cubic unit cell (three per C_{60} molecule), which can thus accommodate up to three small atoms per C_{60}. The most interesting case is where these "intercalated atoms" are alkali atoms of a single type, e.g. $K_{3-x}C_{60}$. (We will use the generic notation A_3C_{60} for compounds of this type.) The *ns* ground state of the alkali always lies well above the bottom of the LUMO band of the host C_{60}, so that the single alkali valence electron is donated to this band. The role of the alkali donors thus appears to be similar to that frequently played by elements of the "charge reservoir" layers of the cuprates: having surrendered their valence electrons to the host, they then essentially play the role of passive spectators.

If this picture is correct, the normal state of the A_3C_{60} compounds should be reasonably well described by a free-electron-like (actually tight-binding) model of the LUMO band, with occupation $\frac{1}{2}$. The density of states at the Fermi surface, $N(0)$, is calculated to lie in the range 6–10 states/(eV.spin.C_{60}) for a cubic lattice constant of 14.24 Å (corresponding to K_3C_{60}) and 15-25% higher for the spacing of 14.44 Å which corresponds to Rb_3C_{60} (because the larger Rb atom expands the lattice somewhat). The Fermi velocity is calculated to be around 1.8×10^7 cm/sec and the plasma frequency around 1.2 eV. Apart from the photoemission data, which is difficult to interpret, most of the experimentally measured properties of the normal state of the A_3C_{60} compounds seem at least qualitatively consistent with this simple picture.[2]

Superconductivity occurs in the $A_{3-x}C_{60}$ compounds over a narrow range of x close to 0 (i.e. for a very nearly half-filled band); the transition temperature is a steeply increasing function of the lattice spacing a, with dT_c/da ~33 K/Å in the usual (merohedrally disordered) case, and reaches a value of ~40 K for Cs_3C_{60} under pressure. This behavior can be understood qualitatively if we assume that T_c is, as in BCS theory, proportional to $e^{-1/N(0)V_0}$, and that the effective coupling constant comes mainly from intramolecular effects and is therefore almost independent of a, while $N(0)$ depends on a as is found experimentally; then $N(0)V_0$ increases with a, and the increase of T_c with a is consistent.

The A_3C_{60} superconductors are strongly Type-II, with an (extrapolated) $H_{c2}(T = 0)$ of ~30 T, implying an extrapolated GL healing length ξ_0 of ~26 Å (<2 lattice spacings!); note that this is an order of magnitude smaller than the estimate following from T_c (~500 Å). The penetration depth is a few thousand Å, and decreases with increasing a, a behavior which is inconsistent with "clean-limit" predictions but consistent with the "dirty-limit" case.

Turning to the pairing state, we find that both T_1 and infrared reflectivity measurements indicate that the quasiparticle density of states is very small below a gap Δ which in the limit $T \to 0$ is approximately equal to the BCS value $1.76k_BT_c$. This indicates strongly that the pairing state is *s*-wave as in the classic superconductors, and at the time of writing there appears to be no evidence against this hypothesis.

[2] I have ignored, here, the question of "merohedral" disorder (disorder in the orientation of the individual C_{60} molecules; see e.g. Lu and Gelfand 1995), which may be important in interpreting the electrical resistivity.

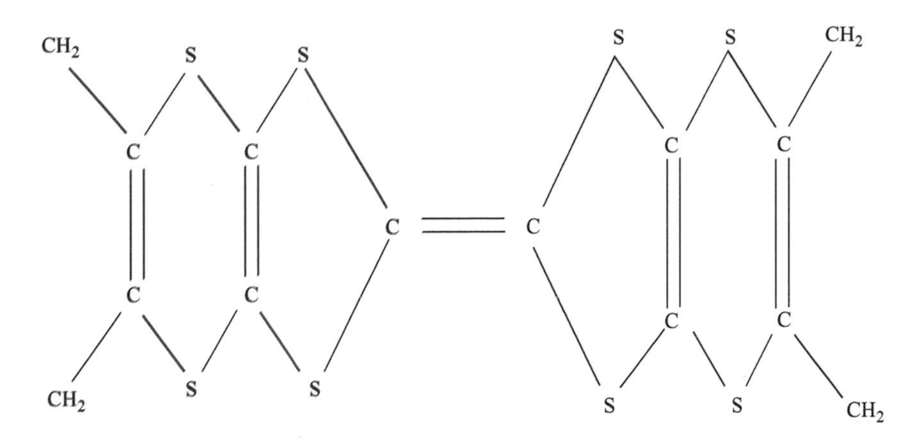

Fig. 8.1 Structure of the molecule bis(ethylene-dithio-)tetrathiofulvalene (BEDT-TTF, ET).

As to the pairing mechanism, we note that even if this is of the traditional phonon-exchange type the situation is rather different from that in the classic superconductors, since in the latter the ratio of the Debye energy $\hbar\omega$, to the Fermi energy ϵ_F is typically $\sim 10^{-2}$, while in the alkali fullerides, because of the lightness of the C atom combined with the narrowness of the electronic bands, this ratio is $\sim$0.3–0.6. One would therefore not necessarily expect the BCS formulae for (e.g.) the isotope effect to be quantitatively valid. Nevertheless, the experimentally measured isotope effect is $\sim$0.4, and this would suggest rather strongly that the basic mechanism is indeed phonon exchange (although an alternative, purely electronic mechanism has been suggested (Chakravarty et al. 1991) which is claimed to accommodate this data).

In sum, there appears at present no compelling evidence against the "default" hypothesis, namely that the alkali fulleride superconductors are well described by standard BCS theory but, like MgB_2, manage (in their case probably because of their strongly molecular structure) to avoid the traditional upper limit on T_c.

8.1.2 Organics

The organic superconductors are mostly[3] quasi-2D crystals based on the planar molecule bis(ethylene-dithio)-tetrathiofulvalene (usually abbreviated BEDT-TTF or ET), the structure of which is shown in Fig. 8.1. The chemical formula is $(ET)_2X$, where X is a monovalent anion, e.g. I_3^-, $Cu(NCS)_2^-$; the (positively charged) $(ET)_2$ complexes form the conducting layers, while the anions form "blocking" layers between them. The conduction electron density is typically in the range of 10^{21} cm^{-3}, comparable to that of the cuprates; the anisotropy is small in the *ab*-plane but very strong along the *c*-axis, e.g. the ratio of ρ_c to ρ_{ab} is $\sim 10^2$–10^3, comparable to that of a typical cuprate. The superconducting transition temperature is typically in the range 10–12 K; the materials are extreme Type-II, with $\xi(0) \sim 50$ Å for the *ab*-plane and ~ 5 Å for the *c*-axis (smaller than the interplane spacing, so that one might think

[3]But a few are quasi-1D.

a LD model would apply away from T_c). The magnetic properties show a considerable degree of irreversibility, as one might expect.

As to the mechanism of superconductivity, while the effect of deuteration is complicated and varies from compound to compound, other substitutions give a "normal" isotope effect, which tends to suggest that the mechanism is the standard phonon one; this is supported by observation of (small) effects of superconductivity on the phonon spectrum.

Regarding the symmetry of the order parameter, the evidence appears at present to be not entirely consistent: the specific heat measured in most experiments appears to be roughly exponential at low temperatures, but the nuclear spin relaxation rate T_1^{-1} and the London penetration depth show clear (but different) power-law behaviors. In fact, the normal density usually appears to behave as $(T/T_c)^{3/2}$, a behavior which is inconsistent with any simple BCS-like model; this is not currently understood.

8.1.3 Heavy fermions

The name "heavy-fermion (HF) system" is applied to a wide class of compounds containing rare-earth (usually Ce) or actinide (usually U) elements; it comes from the fact that the electronic specific heat at low temperatures, while apparently associated as in a textbook metal with *mobile* electrons (cf. below) exceeds that of such a metal by a factor ~100–1000, indicating that the electrons have a very large effective mass. These systems have different crystal structures (usually simple cubic or hcp) but are all 3D rather than layered. Among the different HF compounds, some show no transition at all at low temperatures, some become antiferromagnetic (only), some ($CeCu_2Si_2$, UBe_{13}, UPt_3) become superconducting (only), while a fourth class (URu_2Si_2, UNi_2Al_3, UPd_2Al_3, UGe_2) exhibit superconductivity in coexistence with antiferromagnetism; in the last case the AF transition occurs at temperatures ~5–20 K, while superconductivity sets in only at a considerably lower temperature, ~0.5–2 K (as it does also in the third class). Historically, the HF superconductors were the first metallic systems to show superconductivity which is fairly clearly not of the BCS type, and work on them has strongly affected thinking about the cuprates.

The normal state behavior of the superconducting (and nonsuperconducting) HF systems is quite different from that of a textbook metal, and is not entirely universal. A reasonable qualitative understanding may be obtained from the following model: We separate the valence electrons into two classes, namely the fairly tightly bound "core" electrons ($4f_1$ for Ce^{3+}, $5f_2$ for U^{4+}) which have a very small (say $\lesssim 1$ K) interatomic hopping matrix element, and a set of much more mobile ($s-$ or $d-$) electrons with a very much larger matrix element, perhaps of the order of 0.1 eV (10^3 K). Since any band associated with the core electrons is very narrow (~10 K), for temperatures of the order of room temperature we can equally well regard these electrons as localized; they will thus make a contribution to the magnetic susceptibility χ proportional to $1/T$ (the Curie law). Meanwhile the mobile electrons, while contributing little to the susceptibility, will dominate transport properties such as the DC resistivity; the two kinds of electrons are coupled only by relatively weak collisions. As the temperature falls to a value of the order of the core-electron bandwidth, it is necessary to describe these electrons also by band theory, with a very large effective mass; moreover, Kondo-type effects of the mobile-core interactions may become important. As a result, for those HF systems which do not undergo an AF transition in this regime, a Fermi-liquid-type description of the normal state is not qualitatively misleading. That the "heavy" electrons which contribute the huge low-temperature specific heat are indeed mobile is confirmed by the fact that the relative specific heat jump $\Delta c_V/\gamma T_c$ at the superconducting transition is of the order of the BCS value of 1.42, indicating that the electrons which give rise to the large value of γ are forming Cooper pairs.[4] This rather qualitative picture appears to be consistent with the bulk of the normal-state behavior of the HF superconductors, although the quantitative details are often far from clear.

In considering the nature of the superconducting state of the HF systems, it is necessary to bear in mind that because of the strong spin–orbit scattering the spin of a Cooper pair is not in general a good quantum number; in fact, the most useful classification[5] of possible pairing states is their behavior under the operations of spatial inversion (P) and time reversal (T), and since all the states actually realized appear to be invariant under T, the most important question is whether they are even or odd under P ("even-parity" or "odd-parity"), with a further classification, in the odd-parity case, according to the behavior under other lattice symmetry operations such as $\pi/2$ rotations (for a cubic crystal). Crudely speaking, the simplest even-parity state is analogous to the s-wave spin singlet state realized in the absence of spin-orbit coupling, while the odd-parity states are the analogs of the triplet states. It should be borne in mind that, just as in the cuprates, the question of gap nodes is not identical to that of symmetry (an "extended s-wave" state can have nodes!).

Evidence for the existence or not of gap nodes can be obtained from the low-temperature behavior of various properties (specific heat, nuclear spin relaxation, (electronic) thermal conductivity, etc.) associated exclusively with the normal component, while evidence concerning the even/odd nature may in principle be obtained from a variety of experiments (Knight shift, H_{c2}, sensitivity to nonmagnetic scatterers, etc.). Without going into the details, I just list here the most plausible assignments of the pairing state in some of the best-known HF superconductors.

UPt_3 (hcp, paramagnetic, $T_c = 0.56$ K): almost certainly odd-parity, possibly E_{2u}, gap nodes.
$CeCu_2Si_2$ (simple cubic, paramagnetic, $T_c = 0.65$ K): very probably even-parity, but has gap nodes.
UBe_{13} (simple cubic, paramagnetic, $T_c = 0.9$ K): probably odd-parity, gap nodes.
UPd_2Al_3 (hcp, AF ($T_N = 14.5$ K), $T_c = 2$ K): very probably even-parity, but has gap nodes.
$CeColn_s$ (tetragonal, paramagnetic, $T_c = 2 \cdot 3$ K): probably $d_{x^2-y^2}$.
UNi_2Al_3 (tetragonal, AF($T_N = 4.6$ K, $T_c = 1$K): probably odd-parity.
URu_2Si_2 (tetragonal, AF($T_N = 17.5$ K, $T_c = 0.8$K): most likely odd-parity.

[4]A skeptic might perhaps argue that this fact is insufficient to prove that they are mobile in the normal state (cf. the phenomenon of "localized superconductivity" in an Anderson insulator, see Chapter 5, Section 5.8). However, I know of no current model which instantiates such an assumption.

[5]The classic discussion is that of Gor′kov and Volovik (1985).

Thus the pairing state of the HF superconductors appears not to be universal, even within compounds of the same crystal symmetry.

As regards the mechanism of pairing, none of the HF superconductors shows any appreciable isotope effect, nor is there evidence of much effect of superconductivity on the phonon properties. This tells strongly against the phonon mechanism, and one is left by default with an all-electronic mechanism, the most obvious candidate being the exchange of antiferromagnetic spin fluctuations. This whole topic is at the time of writing still controversial.

8.1.4 Ruthenates (Sr_2RuO_4)

This is the newest of the exotic superconductors; indeed one reason for its intensive investigation over the last few years is that it is the only known superconducting layered perovskite not containing Cu (it is in fact isostructural to the parent compound of LSCO, La_2CuO_4). A good review is Mackenzie and Maeno (2003).

At temperatures ~RT, Sr_2RuO_4 is not typically metallic in its behavior, but for $T \lesssim 25$ K in the N phase, it appears to behave as a highly anisotropic Fermi liquid. The specific heat is $\gamma T + O(T^3)$, with $\gamma \sim 375$ mJ/mol K^2 (intermediate between conventional metals and HF's), χ is $\sim$ const. $\sim 9 \times 10^{-3}$ emu/mol, giving an (average, see below) Wilson ratio of ~1–2. The electrical resistivity is ~T^2 both in and out of plane: $\rho_{ab}/T^2 \sim 4.5$–7.5 nΩ K^{-2}, $\rho_c/T^2 \sim 4$–7 $\mu\Omega$K^{-2}, so $\rho_c/\rho_{ab} \sim 10^3$ (similar to cuprates in magnitude though not in T-dependence).

DHvA measurements show three peaks in the amplitude spectrum as $f(B)$, which have been assigned to three nearly cylindrical F surfaces, two (α, β) electron-like and one (γ) hole-like; these are thought to be hybridized Ru($4d$)–O($2p$) bands. The m^*/m ratio is respectively 3.4(α), 7.5(β) and 14.6(γ) (in agreement with the specific heat data).[6]

Thus the normal state at $T \gtrsim T_c$ appears rather well understood.

T_c for the present samples is 1.5 K; it seems almost certain that the pairing state is non-s-wave, and very probable that it is spin triplet (odd-parity). Evidence:

1. T_c is extremely sensitive to nonmagnetic impurities such as Al (a mean free path as long as 10^3 Å is sufficient to destroy superconductivity altogether), which by the arguments of Appendix 7B would imply that it is not s-wave.
2. The Knight shift for H in the ab-plane is unchanged from the normal state (at least down to 15 mK), which suggests a spin triplet state.
3. T_1^{-1} shows no Hebel–Slichter peak, and below 0.7 K $1/(T_1T) =$ const.; again this suggests a spin triplet.
4. The specific heat shows a large residual DOS for $T < T_c$.
5. The strongest evidence for odd-parity pairing comes from the observations in μSR of a *spontaneously generated magnetic field* in the superconducting state (Uemura et al. 1998).

Very recently, Josephson experiments similar to those done at UIUC on YBCO (see Chapter 7, Section 7.10) have been carried out on Sr_2RuO_4, with however the junctions attached on opposite faces of the crystal rather than mutually perpendicular.

[6]The Wilson ratio is said to be 2.2 for α, β, 1.2 for γ (Mackenzie and Maeno 2003).

(Nelson et al. 2004). In the three samples examined, a maximum in the critical current of the device was reported at $\Phi = 1/2\Phi_0$ ("π-shift"). This is consistent with the odd-parity assignment provided that the two junctions are assumed similar enough that they break the symmetry in the same way; however, at the time of writing it is not entirely clear why this should be so, and further experiments along these lines would seem likely to be revealing.

As to the mechanism, this is very probably as in the HF systems purely electronic, but beyond that is controversial.

8.2 Liquid ^{3}He in aerogel

As we saw in Chapter 7, Appendix 7.B, one of the most interesting theoretical predictions concerning an anisotropic Fermi superfluid is that the transition should be strongly depressed by any kind of impurity scattering, whether it is time-reversal breaking or not. In the cuprates this prediction, while apparently qualitatively consistent with the data, is for a number of reasons difficult to verify quantitatively; so that it is a natural question whether one could test the theory of "anisotropic pair-breaking" quantitatively in ^{3}He by deliberately introducing impurities into the bulk liquid. Alas, this turns out to be impractical: just about any impurity one tries to introduce promptly precipitates out.[7]

However, in the last 10 years it has been realized that one may achieve what is in effect a pair-breaking situation by introducing liquid ^{3}He into the pores of the material known as aerogel. Aerogel[8] is an assembly of small ($\sim$3 nm diameter) particles of amorphous silicate (SiO_2) into a fractal structure which can be crudely thought of as a sort of dilute spaghetti; it has the remarkable property that a very large fraction of the total volume ($\sim$97–99.5% in the samples used in the experiments) is free space ("open volume"), with the solid silicate occupying only the residual 0.5–3%. It is possible to introduce either ^{4}He or ^{3}He (or of course a mixture of the two) into this structure, at pressures below the bulk melting pressure. In the case of ^{4}He, most of the properties shown by the system in the aerogel matrix are not very different from those in bulk;[9] this is not too surprising, since the characteristic length scales of bulk liquid ^{4}He (such as the $T = 0$ healing length, in so far as it can be defined) are of the order of a few Å, while as we shall see the smallest length scale of the aerogel which is likely to be relevant is a few hundred Å. In the case of liquid ^{3}He, where the characteristic bulk length scale (the coherence length ξ_0, see below) is precisely of this latter order of magnitude, we might expect to see more interesting behavior, and indeed we do. Before discussing this in detail, we need to say a little more about the structure of aerogel.

A good discussion of this subject, based mainly on x-ray scattering experiments, is given by Porto and Parpia (1999). For our purposes the salient results are as

[7] Even ^{4}He is not stable in bulk liquid ^{3}He, at least down to a very tiny concentration.

[8] The ensuing statements refer to "base-catalyzed" aerogel, which is the type which has been used in most experiments involving ^{3}He.

[9] Exceptions include the propagation of ultrasound (which is clearly sensitive to the presence of boundaries) and the critical behavior (since this involves a divergence of the characteristic length scale of the bulk liquid).

follows: (a) the particles agglomerate into fractal, somewhat rod-like structures, with diameter $\sim$ a few times 3 nm and a correlation length $\sim$1000 Å; the mean separation clearly depends on the free volume fraction, but is typically $\sim$200–300 Å; (b) if one tries to calculate a mean free path l by the standard geometric arguments, it ranges from $\sim$500 Å for 95% open volume fraction to $\sim$3000 Å for 99%. We note for comparison that the coherence length ξ_0 of superfluid ^{3}He as conventionally defined ($\equiv \hbar v_F / 2\pi k T_c$) ranges from $\sim$150 Å at melting pressure to $\sim$800 Å at SVP; thus, the ratio ξ_0/l is typically of order unity (and can be varied by changing either pressure or open volume fraction or both). However, it should be emphasized that theoretical estimates of l are not particularly reliable; indeed, it is possible that in the context of superfluidity in this kind of geometry the whole concept of a "mean free path" is not really meaningful.

When ^{3}He at pressures below the bulk melting pressure is introduced into 97–99.5% open volume aerogel and cooled below 3 mK, we find some surprises. The overall phase diagram appears to be qualitatively similar to that of bulk liquid ^{3}He, with two different phases[10] stable in zero magnetic field: see Fig. 8.2[11] (and compare with Fig. 6.1); for reasons which will be obvious from Fig. 8.2 itself and from the discussion below, these are conventionally referred to as the "A-like" and "B-like" phases respectively. However, we see that not only is the transition temperature from the N phase depressed relative to the bulk value (as we should expect, cf. below) but the region of stability of the A-like phase at high pressures is considerably smaller than that of the bulk A phase. Moreover, not only does the A-like phase supercool appreciably (as in bulk), there is a very high degree of superheating of the B-like phase[12] and a considerable regime of temperature where the two phases apparently co-exist; note that this behavior is quite different from that of the bulk (where the B phase does not superheat, and coexistence is impossible under homogeneous conditions).

What is the nature of the "A-like" and "B-like" phases? I will make the obvious default assumption that they are both Cooper-paired phases, but what is the nature of this pairing? In particular, are they identical to the bulk A and B phases respectively? Some important clues come from the NMR behavior. First, the static susceptibility is the same as the N-phase one (within experimental error) in the A-like phase, and reduced in the B-like one; according to the arguments of Chapter 6, Section 6.2, this indicates that the A-like phase is an ESP phase and the B-like phase is not. Secondly, the shifts in the NMR frequency away from the Larmor value are always positive in the B-like phase but can have either sign in the A-like one. This behavior is characteristic of the bulk B and A phases respectively; however, in the latter case there exist other known phases, even in the unitary class,[13] which can give the same behavior.

Clearly, the default option is to identify the A-like phase with the ABM (bulk A) phase and the B-like one with the BW (bulk B) phase, and to suppose that in both

[10]As in the bulk case, a third "A_1-like" phase appears in high magnetic field as a sliver between the A-(like) and N phases.

[11]This (thermodynamic) phase diagram is not measured directly but is inferred from measurements of the NMR shifts as described in Baumgardner and Osheroff (2004).

[12]Indeed, some experiments have failed to observe the A-like phase at all on heating.

[13]Within the class of unitary states the BW phase is unique in producing only positive (or zero) NMR shift independently of orientation.

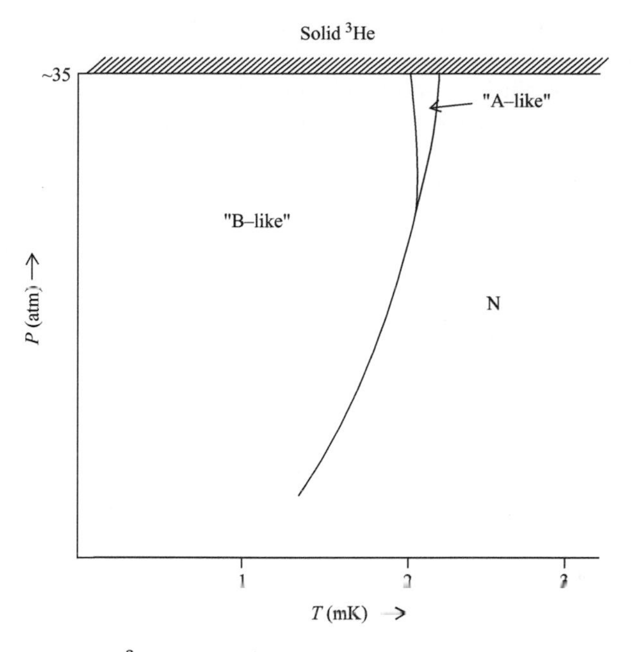

Fig. 8.2 Phase digram of ^{3}He in 99.3% aerogel in low magnetic field (approximate, after Baumgardner and Osheroff 2004).

cases the "orientation" (of the $\boldsymbol{l}$-vector) in the former and the $\boldsymbol{\omega}$-vector in the latter) deviates from its optimum (bulk) form because of the effects of pinning by the "walls" (i.e. the solid aerogel); this hypothesis is qualitatively consistent with all the NMR data, including the nonlinear data. At the time of writing there seem to be only two obvious objections to this identification. The first is that to yield the phase diagram shown in Fig. 8.2 the values of either the β_i's or the I_i's (Chapter 6, Section 6.3) must be substantially different than those for bulk ^{3}He; since the β_i are presumably mainly dominated by short-range effects which should be insensitive to the confinement, this suggests that it is the I_i's, i.e. the actual structure of the pairing, which is different for at least one of the two phases. However, it is not entirely clear that with a typical interstrand distance which may be less than the coherence length the β_i's could not be affected sufficiently to yield the observed behavior (cf. also below, on the superfluid density). The second counterargument against the "default" identification is due to Fomin (2005), who actually raises two different points. The first is the claim that fluctuations make other ESP phases (what he calls "robust" phases) more stable in aerogel than the A phase; whether or not this is true is a matter for detailed calculation. The second point, which is more interesting in the context of this book, is that if the A phase were indeed to form in aerogel, then because of the random orienting effect of the strands it would form a "superfluid glass", that is a configuration in which domains are formed in which the $\boldsymbol{l}$-vector points in different directions, and that this configuration would not be superfluid in the phenomenological sense (cf. Volovik 1996). My own belief is that the first part of the above statement is correct (i.e. $\boldsymbol{l}$-domains are indeed formed) but that the second is not, i.e. that a "superfluid glass" as defined above will in fact show all the usual signatures of superfluidity (cf. Mineev and Zhitomirsky 2005). Thus, I would regard the most plausible identification at the time of writing of the two phases of ^{3}He seen in aerogel as the "default" one (A-like = ABM, B-like = BW).

Irrespective of the nature of the order parameter, however, there are other features of the behavior of ^{3}He in aerogel which are quite puzzling. If one had an independent measure of the "mean free path" l due to the aerogel, one could plug this into the Abrikosov–Gor'kov formula (7.B.6) for the depression of T_c and compare with the experiments. In practice, it makes more sense to use the latter to fix l, in which case one finds that the naïve geometrical estimate is too small. Irrespective of this, having fixed l in this way one can then use the AG theory to compute the superfluid density ρ_s in the two phases, and compare the prediction with experiment. On doing this we get two surprises: First, the *absolute* magnitude of the superfluid fraction $\rho_s(T)/\rho$ is substantially less than the theoretical prediction (as is also the order parameter inferred from the magnitude of the NMR shifts); it has been suggested (Thuneberg et al. 1998, Harrison et al. 2003) that this may be due to the inhomogeneity of the scattering by the aerogel. Secondly, the ratio ρ_{sA}/ρ_{sB} for the A- and B-like phases when compared at the same temperature is smaller than 0.5, a situation which could occur in the bulk liquid only if the $\boldsymbol{l}$-vector is overwhelmingly oriented parallel to the direction of flow. This seems very implausible in the aerogel geometry (and is probably incompatible with the NMR data); at the same time, the suggestion (Nazaretski et al. 2004) that, unlike in the bulk liquid, the A-phase gap is much smaller than the B-phase one at the same temperature seems difficult to reconcile with energy considerations (cf. Baumgardner et al. 2004). All in all, the behavior of superfluid ^{3}He in aerogel seems to pose some rather nontrivial challenges to theory.

8.3 Supersolids

The word "supersolidity" or "supersolid behavior" has become conventional to describe a phenomenon which, while the subject of speculation for 35 years, has apparently become an experimental reality only in the last two years. At the time of writing the nature and significance of this phenomenon is still far from clear; however, unless the data should turn out simply to be attributable to some experimental error (something the present author would regard as unlikely) it seems certain to necessitate an substantial revision of some of our present concepts regarding superfluidity or the solid state or both, and for this reason I feel it would be a pity to omit all mention of it in this book. Briefly put, the most natural interpretation of the data seems to indicate that a crystalline solid can show at least some of the behavior which has been up to now regarded as unique to a superfluid liquid.

The experiments (Kim and Chan 2004b) which are most significant for the present discussion[14] report the rotational behavior of a torsional oscillator containing an annular cavity filled with ^{4}He pressurized to pressures ranging from 26 to 66 atm;

[14] An earlier round of experiments by the same group (Kim and Chan 2004a) was on ^{4}He contained in the pores of Vycor glass; because of uncertainties about the microscopic nature of the relevant phase, these allow more freedom in their interpretation, and I shall not attempt to discuss them here.

I note (and will return to this question below) that essentially all previous experimental work on ^{4}He in this range of pressures has been consistent with the standard view of this system as a regular crystalline solid. The basic experiment consists in simply driving the oscillatory angular motion with an external AC torque and measuring the amplitude and phase of the induced oscillation as a function of the frequency and amplitude of the drive and other variables such as temperature and ^{4}He pressure. In the following it is essential to distinguish between the AC angular frequency ω of the drive (and hence of the response) which is always of order $2\pi \times$ 1 kHz, and the peak amplitude Ω of the actual rotational motion, which is very much smaller and may be either small or large compared to the characteristic quantum unit of rotation $\Omega_c \equiv \hbar/mR^2$ (where R is, for definiteness, the mean radius of the cavity containing the ^{4}He and m the mass of the ^{4}He atom); in the actual experiment the numerical value of Ω_c is approximately 6×10^{-4} rad/sec[15]). It should be noted that even at the largest values of Ω at which an effect is seen the linear displacement corresponds to only ~70 atomic spacings.

Let us start by asking what we should expect if the whole of the torsional oscillator, including the ^{4}He in the cavity, behaves like a simple rigid body. Then the complex response $\chi(\omega)$ (that is, the ratio of the Fourier transform of the angular velocity $\Omega(t)$ to that of the drive amplitude $F(t)$) should be a Lorentzian, with the peak of the resonance frequency ω_0 given by

$$\omega_0 = \left(\frac{K}{I}\right)^{1/2} \tag{8.3.1}$$

where K is the torsional constant of the suspension and I the moment of inertia, and with a width $\Delta\omega = \omega_0/Q$ where Q is the quality factor of the suspension (in the actual experiment Q is of order 10^6). Precisely this behavior is observed at temperatures above about 230 mK, with a value of ω_0 which is well fitted by the independently measured[16] value of K and the calculated (temperature-independent) value of the moment of inertia I. Note that the contribution of the ^{4}He to the total moment of inertia is only ~0.6%, so that the resonance frequency of the "empty" cell (no ^{4}He in the cavity) differs from (8.3.1) only by ~0.3%.

Below 230 mK Kim and Chan find a totally unexpected pattern of behavior: for small drive amplitude the resonance remains relatively sharp but shifts to a higher frequency. The magnitude of the shift depends on temperature, pressure and drive amplitude,[17] disappearing by the time the peak velocity reaches 400 μsec; as a function of temperature it appears to set in smoothly and to saturate, for $T \lesssim 40$ mK, at a value of around 40 ppm.

From Eqn. (8.3.1), we see that such an increase in the resonant frequency could be a result of either an increase in the torsion constant K or a decrease in the moment of inertia I (or both). However, the first possibility appears to be ruled out by control

[15]The experimental paper reports values of the peak linear velocity v rather than angular velocity; the value of v corresponding to Ω_c is approximately 3 μ/sec.

[16]ω_0 has in fact a weak temperature-dependence which appears also for the "empty" cell (no ^{4}He) and is presumably a result of a weak T-dependence of K.

[17]Or, what is essentially equivalent since the effect is so small, on the amplitude of the induced velocity.

experiments with an empty cell or one filled with ^{3}He; in neither case is any shift observed, which seems to prove not only that the effect comes from a decrease in the moment of inertia I but that the latter is associated specifically with the ^{4}He. The maximum change in the ^{4}He moment of inertia which would be necessary to explain the observed frequency shift is approximately 2%.

Now, the mere fact of a change in the moment of inertia of the ^{4}He at temperatures below 200 mK is not in itself evidence for any kind of superfluidity; much more mundane explanations are possible, for example a redistribution (for reasons which are admittedly unknown) of the density as a function of the radial coordinate. However, this kind of explanation is made a great deal less plausible by an elegant control experiment, in which the annular cavity was blocked; under these conditions the frequency shift was absent.[18] Thus, the most natural interpretation of the data is that in these experiments *the* ^{4}*He has acquired a superfluid density* ρ_s *which at the lowest temperature is a fraction* $\sim$*2% of its total density.*

Before proceeding to explore the implications of this hypothesis, however, we should pause to make one important caveat: The experiment of Kim and Chan is not the analog of the Hess–Fairbank experiment on liquid ^{4}He (an experiment involving DC rotation) and does not in itself demonstrate the property of nonclassical rotational inertia (NCRI, see Chapter 2, Section 5). Rather, it is the analog of the original Andronikashvili experiment on liquid ^{4}He, and what it directly shows is that *over time scales of the order of the oscillation period* ($\sim$1 msec) the "effective" moment of inertia of the ^{4}He is reduced by up to $\sim$2%. In fact, if we look at the numbers in the Andronikashvili experiment, it turns out that the peak angular velocity Ω of the disks was always large compared to the relevant Ω_c, so that at least over the vast bulk of the oscillation cycle the state of the liquid ^{4}He could not possibly have been the one in thermodynamic equilibrium with the rotating "walls" (disks). The same is largely true in the Kim–Chan experiment, although for the lowest drive amplitudes Ω is only slightly greater than Ω_c, so that the behavior is consistent with the true equilibrium behavior being realized over an appreciable fraction of the cycle (that part for which the angular velocity is $<1/2\Omega_c$). However, it should be emphasized that we cannot rigorously conclude, on the basis of the existing experiments, that the system is realizing NCRI; it is impossible to totally exclude the hypothesis that what we are seeing is a drastic slowing-down below 200 mK (needless to say of unknown origin) of the process of rotational equilibration with the walls of the cavity. That said, just as in retrospect the Andronikashvili experiment can be regarded as the natural harbinger of the Hess–Fairbank experiment which was realized more than 20 years later, so the "default" interpretation of the Kim–Chan experiment is that it is indeed circumstantial evidence for NCRI, and that if and when a true Hess–Fairbank experiment is performed on ^{4}He in the relevant pressure range this phenomenon will be unambiguously found. From now on I shall make this assumption.

If, then, the phenomenon of NCRI really is occurring, with a superfluid fraction $\sim$2%, in bulk ^{4}He under pressures of 26–66 atm, what are the implications? We should first reiterate that not only are all previous experiments on ^{4}He in this pressure

[18]To be precise, it occurred but only at the level of $\sim$1% of the original effect; this is probably attributable to the fact that the blocking was not perfect.

range (including those done below 200 mK) apparently consistent with the "default" hypothesis that it is a regular crystalline solid (albeit one with an anomalously large zero-point motion of the atoms around their lattice sites) but several experiments, in particular by elastic x-ray or neutron scattering, have provided direct evidence that the density is a nontrivially periodic function of position, as this hypothesis would of course imply. Moreover, while to my knowledge it has not been checked explicitly that in the temperature regime below 200 mK the number of atoms is equal to the number of lattice sites (i.e. there are no detectable interstitials or vacancies), at higher temperatures the number of such interstitials/vacancies is found to follow an Arrhenius-type law with an activation energy in each case of a few K; thus at $T < 200$ mK the equilibrium number would be only of order 10^{-10} or less. Thus, there appear to be only two obvious possibilities: (a) the phase of ^{4}He realized in the Kim–Chan experiment is quite different from that normally realized under similar conditions (perhaps some kind of metastable phase); (b) it is the same as the normally realized phase, but this phase is actually quite different from our traditional theoretical conceptions, in a way so subtle that it has until now escaped detection.

Taking option (a) first, possible scenarios are, in increasing order of plausibility (or perhaps one should say in decreasing order of implausibility!), that the Kim–Chan system is an overpressurized liquid:[19] that it is a system of microcrystallites with interfaces in which a small fraction of liquid can form: or that it is essentially bulk crystalline solid but with an anomalously large number of topological defects such as dislocations which can accommodate mobile atoms. While none of these hypotheses can be rigorously excluded on the basis of the existing experiments, it should be possible to test all of them in the reasonably near future by x-ray and/or neutron scattering, provided the obvious experimental problems of using these probes "in situ" can be overcome. Should one of these explanations in fact prove correct, it would be very interesting but would not necessarily radically overthrow any of our current preconceptions concerning superfluidity. A more exotic scenario along these general lines is explored below.

Hypothesis (b) is more intriguing, as it suggests that the innocent-looking crystals of ^{4}He on which we have experimented for decades may actually be in one sense or another wolves in sheep's clothing. What can we say about this possibility?

It is first necessary to make an important remark about our traditional conception of a "quantum crystal." Let's approach this conception by stages. First, imagine a system of "classical" (in both senses, i.e. Newtonian and distinguishable) atoms interacting by some reasonably physical two-body potential, say the van der Waals potential. At zero temperature it is highly plausible that such a system could form a regular lattice (most likely close-packed, i.e. fcc or hcp) with a lattice constant such that nearest-neighbor atoms sit at the minimum of their mutual interaction potential. Next, let us introduce quantum-mechanical effects (so that the atoms satisfy Schrödinger's equation rather than Newton's), but are still distinguishable. The most naive modification of the classical ground state is a "quantum Einstein model" in which each atom is localized near a particular site which is specified by a vector $\boldsymbol{R}_{0i}$,

[19]But it would then be difficult to understand why the $T = 0$ superfluid fraction is not 100% as in the usual liquid (cf. Leggett 1998).

those vectors corresponding to the classical sites. In this model the ansatz for the ground state wave function would be

$$\Psi_0^{(\text{dist})}(\boldsymbol{r}_1\boldsymbol{r}_2\ldots\boldsymbol{r}_N) = \phi_0(\boldsymbol{r}_1 - \boldsymbol{R}_{01})\,\phi_0(\boldsymbol{r}_2 - \boldsymbol{R}_{02})\ldots\phi_0(\boldsymbol{r}_N - \boldsymbol{R}_{0N}) \qquad (8.3.2)$$

with the function ϕ_0 something like a 3D harmonic-oscillator ground state wave function. At the present stage it is a crucial assumption that the ϕ_0's based in neighboring wells do not overlap appreciably.

The ansatz (8.3.2) has the unphysical property that the atoms are localized in absolute space rather than relative to one another as one would expect the potential to require. On the other hand, it is plausible that unless the crystal is floating freely in a spaceship, the *overall* translational symmetry would be "spontaneously" broken (actually broken by, for example, rugosities in the walls of the cavity). Thus we need to localize the center of mass while still allowing *relative* (small) motion of the atoms, i.e. zero-point phonons. I do not believe this complication affects the points to be discussed qualitatively,[20] so shall omit this stage and use the "naive" ansatz (8.3.2).

The next stage is to take into account the indistinguishability of the atoms. Since we are interested in ^{4}He, we will assume Bose statistics, and thus must symmetrize the many-body wave function with respect to exchange of any pair of coordinates $\boldsymbol{r}_i$, $\boldsymbol{r}_j$. This gives the ansatz

$$\Psi_0(\boldsymbol{r}_1\boldsymbol{r}_2\ldots\boldsymbol{r}_N) = (N!)^{-1/2}\sum_{P\{\boldsymbol{r}_i\}}\phi_0(\boldsymbol{r}_1 - \boldsymbol{R}_1^0) \times \phi_0(\boldsymbol{r}_2 - \boldsymbol{R}_2^0)\ldots\phi_0(\boldsymbol{r}_N - \boldsymbol{R}_N^0) \qquad (8.3.3)$$

where $P\{\boldsymbol{r}_i\}$ denotes all possible permutations of the $\boldsymbol{r}_i$ (keeping the $\boldsymbol{R}_i^0$ fixed). With a caveat concerning nearest-neighbor exchange (see below) and the one already mentioned concerning zero-point phonons, Eqn. (8.3.3) is usually thought to be a reasonable schematic description of the ground state of a quantum crystal such as solid ^{4}He has been traditionally believed to be.

Let us consider the properties of the ground state ansatz (8.3.3) keeping in mind that for the moment we are assuming the overlap of the ϕ_0's based in neighboring wells (and hence a fortiori those based in nonneighboring ones) to be negligible. It is intuitively clear that each lattice site is occupied by exactly one atom, and thus that there is no BEC; indeed, explicit evaluation of the single-particle density matrix it to have the form

$$\rho_1(\boldsymbol{r}, \boldsymbol{r}') = \sum_{i=1}^{N}\phi_0^*(\boldsymbol{r} - \boldsymbol{R}_i^0)\,\phi_0(\boldsymbol{r}' - \boldsymbol{R}_i^0) \qquad (8.3.4)$$

so that it manifestly has N degenerate eigenvalues each equal to 1. Moreover, the state (8.3.3) cannot show NCRI (is not "superfluid"); to see this we note that in order to accommodate the necessary modification in the "single-valuedness boundary condition" required by the rotation[21] it is sufficient to consider each "branch"

[20]Though it may have to be analyzed further in the context of possible explanations of the Kim–Chan experiments which attribute them to kinetic effects.

[21]This refers to an alternative way of deriving the OF quantization rule (2.2.9), see Leggett (1998).

(particular permutation of the $\boldsymbol{r}_i$'s) and put in the necessary "kink" across a surface where $\phi_0(\boldsymbol{r}_i - \boldsymbol{R}_j^0)$ is negligibly small.

We now come to what is perhaps the trickiest point in this whole discussion. What happens if we allow in (8.3.3) for finite overlap of the function ϕ_0 based on neighboring sites? Intuitively, this corresponds to the possibility of nearest-neighbor atoms "changing places", and in solid ^{3}He this is known to lead to a real physical effect, namely the exchange coupling between the nuclear spins of nearest-neighbor atoms. In solid ^{4}He, which has no nuclear spin degree of freedom to act as a "tag", one might question whether the concept of "exchange" is really meaningful; but at least one can legitimately raise the question whether overlap of the ϕ_0's – or more generally a many-body wave function which allows neighbors to exchange – can invalidate the conclusions reached above.

Let me at this point briefly comment on one point which seems capable of causing confusion in this context. In a 1970 paper I showed that, quite generically, if a system which shows crystalline order (or indeed any nontrivial variation in its equilibrium density) were somehow to become superfluid, then one can put a rigorous upper limit on the superfluid fraction (as measured by NCRI) from a knowledge of the $\boldsymbol{r}$-dependence of the density $\rho(\boldsymbol{r})$: crudely speaking, the smaller the density becomes on the "interstitial" planes the lower the upper limit on ρ_s. Now, once the ϕ_0's overlap, the interstitial-plane density is in general nonzero, and hence the upper limit established on ρ_s is not zero. However, this emphatically does *not* mean that the system will automatically be superfluid! A nonzero interstitial value of $\rho(\boldsymbol{r})$ is a necessary, but by no means sufficient, condition for the latter phenomenon.

In fact, whatever the detailed nature of the many-body wave function, nearest-neighbor exchanges by themselves cannot lead either to BEC (in the sense of ODLRO) or to NCRI; what is necessary for either, crudely speaking, is that the probability amplitudes with $N-1$ of the atoms fixed, to find the last one at positions $\boldsymbol{r}$ and $\boldsymbol{r}'$ should be simultaneously nonzero in the limit $|\boldsymbol{r} - \boldsymbol{r}'| \to \infty$. In the case of ODLRO this is obvious, since the condition stated is essentially just the definition in words of that concept. If now we accept the conventional view the ODLRO is a necessary as well as a sufficient condition for NCRI ("superfluidity"), then it of course follows immediately that the condition stated is also necessary for NCRI. However, this is the kind of situation where the ODLRO-NCRI link is not in fact entirely obvious, and a more detailed consideration is necessary. While such a discussion can be given within the "wavefunction" language we are using, a more elegant version has been recently presented by Prokofiev and Svistunov (2005) using the language of path integrals, and I refer the reader to that: the conclusion is that the stated condition is indeed necessary for NCRI.

How might this condition (call it (A)) be fulfilled? One possibility is that in the ground state the number of atoms might not be equal to the number of lattice sites, i.e. one has a nonzero number either of "zero-point vacancies" or of "zero-point interstitials". The latter case is easier to visualize: if in the ground state there are present a nonzero number of interstitial atoms, then the various interstitial positions are evidently energetically equivalent, the atoms can tunnel between them and form Bloch waves, and if they undergo BEC in the $\boldsymbol{k} = 0$ Bloch wave state then condition (A) is automatically fulfilled. Such a hypothesis (and the analogous one for the vacancy case) was actually proposed as long ago as 1969 by Andreev and Lifshitz; we may call it the "incommensurate supersolid" scenario. The most obvious experimental test would evidently be to determine simultaneously the number of atoms (e.g. in principle, by simply weighing the sample!) and the number of sites (by x-ray diffraction).

Suppose however it turns out that the number of atoms is exactly equal to the number of sites. Is there any way in which condition (A) can be fulfilled? Indeed there is, namely what we may call the "commensurate supersolid" scenario. To explain this, let's digress for a moment to the problem discussed in Chapter 4, Section 4.6, namely a Bose gas confined in an optical lattice, with the number of atoms equal to the number of sites. We saw that depending on the U/t ratio, two qualitatively different states are realized: the "Mott-insulator" state, in which the many-body wave function is schematically of the form (8.3.3), and the "superfluid" state in which, to a zeroth approximation, the atoms undergo BEC into the lowest ($\boldsymbol{k} = 0$) Bloch wave state in the periodic potential. Now, were it not for the fact that the periodic potential is externally provided rather than self-consistently generated, the MI state of the atoms in the optical lattice would be the exact analog of the quantum-crystal ground state traditionally ascribed to solid ^{4}He; in fact, we can make the analogy exact if we imagine that as we gradually switch off the laser-generated potential the MI state for some reason remains stable. Let us similarly imagine that we start from the "superfluid" state (BEC in the (nontrivial) $\boldsymbol{k} = 0$ Bloch wave state) and again gradually switch off the laser-generated external potential, and that as we do so the original state remains stable: i.e. that when the external periodic potential is zero the system prefers to remain Bose-condensed in a *nontrivially periodic* Bloch wave state. Then, while the *average* occupation of a "site" is exactly 1, there are (unlike in the MI state) large fluctuations around this average, and it is moreover immediately obvious that condition (A) is fulfilled.

It is rather obvious that there are considerable difficulties with both the incommensurate and the commensurate supersolid scenarios. In the first place, while some very schematic ("lattice-gas") model Hamiltonians do appear to have ground states of this type for certain choices of the parameters (see Liu and Fisher 1973), it is not at all obvious that for a realistic (e.g. van der Waals) interatomic potential any such state could ever be stable simultaneously with respect to the Bose-condensed liquid state and to the conventional quantum-crystal ground state. Secondly, the nature of these states, particularly the commensurate one, is so qualitatively different from the conventional quantum-crystal state that it would be very surprising that no other qualitative differences from the behavior predicted for the latter have been noticed over the half-century or so for which intensive experimental work on solid ^{4}He has been done. Both these objections are of course somewhat mitigated if one assumes that the Kim–Chan samples are in some kind of metastable state not previously attained (possibility (a) above), and it seems certain that experimental work in the near future will try to resolve this question.

8.4 Fermi alkali gases: the BEC-BCS crossover

A central theme of this book has been that the Bose–Einstein condensation (BEC) which occurs in systems of simple bosons, and the Cooper pairing (CP) which is realized in degenerate Fermi systems, are in some sense just different aspects of the same basic phenomenon, namely the onset of a single macroscopic eigenvalue in the relevant reduced density matrix. In particular, one might imagine (cf. Chapter 1, Section 1.4) that by considering a two-species system of fermions and "tuning" the interaction between them appropriately, one might be able to achieve a continuous transition between a Bose condensate of diatomic molecules and Cooper pairing in a dilute Fermi gas – a transition known in the literature as the "BEC-BCS crossover". While theoretical speculations on this topic go back more than 35 years, it is only within the last two years that it has become an experimental subject. Although relevant experiments (and related theoretical developments) are now being reported almost weekly, so that anything written today may well be out of date within a few months, I feel that the centrality of this topic to the theme of the book is such that it is worthwhile to give, at least, a brief review of the experimental systems in question and the general theoretical framework within which they are usually considered.

The relevant system is a laser-trapped dilute ($\sim 10^{12}$–10^{13} atoms/cc) set of fermion alkali atoms (to date ^{6}Li or ^{40}K), prepared in an incoherent mixture[22] of two different hyperfine species in equal numbers; the most commonly used pairs of species are the $m_J = \frac{1}{2}$, $m_I = \frac{1}{2}$ and $-\frac{1}{2}$ states of ^{6}Li (in a field $B \gg B_{\text{hf}}$) and the $F = \frac{9}{2}$, $m_F = -\frac{9}{2}$ and $-\frac{7}{2}$ states of ^{40}K (in a field $B \ll B_{\text{hf}}$). In the following I shall label the two hyperfine species by a "pseudospin" index σ which takes values $\pm\frac{1}{2}$ (although in some contexts is may be necessary to remember that this does not represent a "real" spin).

The "tuning" of the interatomic interaction is achieved by sweeping through a Feshbach resonance (see Chapter 4, Section 4.2.2, and Appendix 4.A). For the two pairs of hyperfine states mentioned above, such Feshbach resonances occur at magnetic fields of about 822 G (for ^{6}Li) and about 220 G (for ^{40}K). A serendipitous aspect of such resonances is that since only two hyperfine species are present in the open channel, the Pauli principle forbids the strong three-body recombination which occurs close to the resonance in the bosonic case, so the system is relatively stable in the relevant regime.

A point which it is crucial to appreciate is that in the context of the BEC-BCS crossover the "interesting" regime of detuning δ (see Chapter 4, Section 4.2.3 and Appendix 4A) is such that $|k_{\text{F}}\, a_{\text{s}}(\delta)| \gtrsim 1$, where k_{F} is the Fermi wave vector of the free gas and $a_{\text{s}}(\delta)$ ($\sim$−const. δ^{-1}) the s-wave scattering length. Now, under the conditions of the existing experiments it turns out that the Fermi energy $\epsilon_{\text{F}}(\equiv \hbar^2 k_{\text{F}}^2/2m)$ is always very small[23] compared to the characteristic one-body energy δ_{c} characterizing the resonance (see Appendix 4A), and it then follows from the considerations of that appendix that *the "interesting" regime for the BEC-BCS crossover corresponds to* $\delta \ll \delta_{\text{c}}$. This has the consequences (cf. Appendix 4A) that (a) the Feshbach resonance can be treated exactly like a simple single-channel ("potential") resonance, with the

[22]Such a mixture can be prepared in various ways, e.g. by starting with all atoms in a single hyperfine state, applying a $\pi/2$ RF pulse and allowing the system to equilibrate in the presence of an inhomogeneous magnetic field, or by a rapid succession of pulses.

[23]In the literature this state of affairs is often called a "broad resonance"– to my mind a slightly unfortunate nomenclature, since its occurrence is a function not only of the two-body parameter δ_{c} but also of the many-body parameter ϵ_{F}.

s-wave scattering length $a_s(\delta)$ taken directly from experiment; (b) the "molecules" which are formed (in the two-body problem) for $\delta < 0$ are formed almost entirely in the open channel[24] and have a radius which is typically orders of magnitude greater than that of the tightly bound closed-channel molecules. In fact, for the purposes of considering the BEC-BCS crossover we can (as of now) forget about the existence of the closed channel entirely.[25] Of course, future experiments may not necessarily operate in this regime.

Let's now consider the behavior of the system at $T = 0$ as δ is swept from large positive to large negative values (but always within the range $|\delta| \ll \delta_c$, see above). Qualitatively, what we expect is the following: For large positive δ, where the quantity $a_s(\delta)$ and hence the effective interatomic interaction (see Chapter 4, Section 4.2) is negative and small, the system would tend to form Cooper pairs, corresponding to a macroscopic (but $\ll N$) eigenvalue N_0 of the two-particle density matrix, with a corresponding eigenfunction (the "pair wave function", see Chapter 5, Section 5.4) whose radius in the relative coordinate, ξ_0 is large compared to the interparticle separation. On the other hand, for large negative δ, where $a_s(\delta)$ is small and positive, we would expect that the fermions form molecules of a radius ($\sim a_s$) much less than the interparticle spacing, and that these molecules would then (at $T = 0$) undergo almost complete BEC. In such a state the two-fermion reduced density matrix again has a macroscopic eigenvalue N_0, this time nearly equal to N, with an eigenfunction whose radius in the relative coordinate ($\sim a_s$) is small compared to the interparticle spacing. In the light of the above, the obvious default hypothesis would seem to be that there is a continuous transition between these two limiting situations, with N_0 gradually increasing and ξ_0 gradually decreasing. We shall see that the experimental data appears to be for the most part consistent with this qualitative picture.

How to implement this scenario quantitatively? The very simplest approach[26] (which I will refer to below as the "naive ansatz") was originally written down by Eagles (1969) in the context of the problem of superconductivity in a degenerate semiconductor, and was applied explicitly to the Fermi alkali-gas problem by the present author some years later. It consists of simply writing down the generalized BCS ansatz (5.4.3) for *arbitrary* (positive or negative) values of δ, and optimizing the coefficients in the standard way, just as in Chapter 5, Section 5.4. The only real points of difference from the original BCS calculation are (a) that the chemical potential μ, which in the BCS limit could be safely set equal to the Fermi energy ϵ_F, must now be obtained self-consistently by setting the average number of particles equal to the specified value ($=k_F^3/3\pi^2$) and (b) that the full interatomic potential $V(r)$, rather than being renormalized to the BCS constant V_0, must now be eliminated in favor of the (two-body) s-wave scattering length $a_s(\delta)$.

There is, however, one point to which I should call explicit attention: The calculation of Chapter 5, Section 5.4 (which is effectively equivalent to the original BCS

[24]That this is so in the real-life ^{40}K experiments is confirmed by a measurement of the magnetic moment of the molecules.

[25]Of course if we sweep to large enough negative values of δ the molecules may switch to a predominately clos ed-channel configuration, but by that time their interaction is negligible ($k_F a_s(\delta) \ll 1$).

[26]In the following I shall consider a fictitious translationally invariant system, ignoring the complications associated with the real-life trap geometries.

calculation based on a reduced Hamiltonian) sets both the Hartree and the Fock terms in the expectation value of the energy equal to zero. For a dilute alkali gas, the neglect of the Fock term is almost certainly legitimate, since this operates only between parallel-spin atoms, which must be in a $l > 0$ relative state and hence have negligible probability of approaching one another to within the range of $V(r)$ (cf. chapter 4, Section 4.2). The neglect of the Hartree term is however more problematic, in particular since the direct evaluation of this term over the many-body wave function (5.4.3) using a realistic interatomic potential (with hard core) is liable to give a divergent result. The best one can hope for is that a calculation which ignores this term will give a reasonable result for the difference in energy between the true (paired) ground state and the notional "normal ground state" (and will also give reasonable results for $F(r)$, etc.). This point should be borne in mind where the output of the naive ansatz is compared with that of more sophisticated calculations, or with the experimental data.

Let us then postulate the ansatz (5.4.3) for the many-body wave function, with the function $\phi(\boldsymbol{r}_1\boldsymbol{r}_2\sigma_1\sigma_2)$ having the s-wave singlet form. Then, following through the analysis of Chapter 5, Section 5.4 word for word, we obtain the gap equation corresponding to (5.4.22):

$$\Delta_{\boldsymbol{k}} = -\sum_{\boldsymbol{k}'} V_{\boldsymbol{k}-\boldsymbol{k}'} \frac{\Delta_{\boldsymbol{k}'}}{2E_{\boldsymbol{k}'}} \quad (V_q \equiv \text{FT of } V(r)) \tag{8.4.1}$$

where the symbols have their standard meaning; note in particular the definition

$$E_{\boldsymbol{k}} \equiv \sqrt{(\xi_{\boldsymbol{k}} - \mu)^2 + |\Delta_{\boldsymbol{k}}|^2}, \quad \xi_{\boldsymbol{k}} \equiv \hbar^2 k^2/2m \tag{8.4.2}$$

We can eliminate explicit reference to the potential components $V_{\boldsymbol{k}-\boldsymbol{k}'}$ in favor of the scattering length $a_s(\delta)$ by using the two-body Schrödinger equation and a standard renormalization procedure (see Appendix 8.A). The result is

$$\sum_{\boldsymbol{k}} (\xi_{\boldsymbol{k}}^{-1} - E_{\boldsymbol{k}}^{-1}) = \frac{m}{2\pi\hbar^2 a_s} \tag{8.4.3}$$

where now the quantity $E_{\boldsymbol{k}}$ is defined by (8.4.2) with $\Delta_{\boldsymbol{k}}$ replaced by its $\boldsymbol{k} \to 0$ limit, which we denote by Δ. The sum is convergent in 3D, so the upper limit may be taken to infinity. Equation (8.4.3) is one equation for the two unknowns μ and Δ. To solve it we need a second equation, for which we use the condition of particle number conservation[27] (cf. Eqn. 5.4.28)

$$\sum_{\boldsymbol{k}} \left(1 - \frac{\xi_{\boldsymbol{k}} - \mu}{E_{\boldsymbol{k}}}\right) = \frac{k_F^3}{3\pi^2} \tag{8.4.4}$$

Equations (8.4.3) and (8.4.4) may be solved for the two unknowns Δ and μ; it is clear that we have

$$\Delta = \epsilon_F \, f(\zeta) \tag{8.4.5a}$$

$$\mu = \epsilon_F g(\zeta) \quad (\zeta \equiv -(k_F a_s(\delta))^{-1}) \tag{8.4.5b}$$

[27] The factor of $\frac{1}{2}$ in (5.4.28) is cancelled by the spin sum, which is omitted in (8.4.4). The sum over k is again convergent for large k, since according to (5.4.28) the summand falls off as k^{-4}.

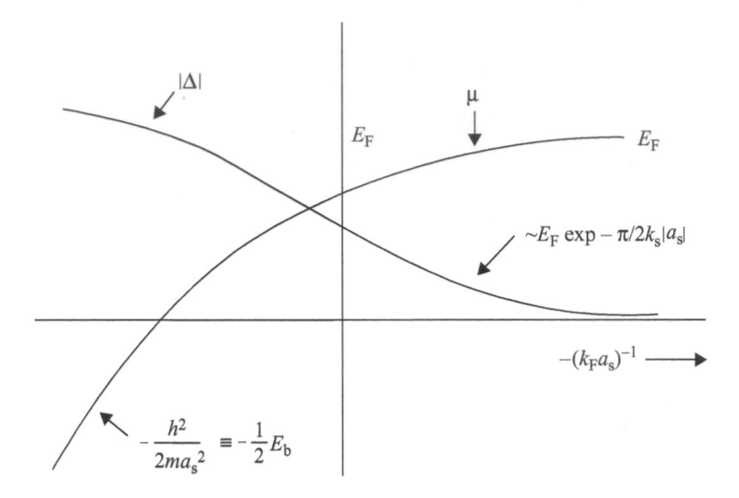

Fig. 8.3 Behavior of μ and $|\Delta|$ in the "naive" ansatz.

where $\epsilon_F \equiv \hbar^2 k_F^2/2m$ is the Fermi energy of the free gas, and the functions $f(\zeta)$ and $g(\zeta)$ must be obtained by numerical computation. The solutions are presented schematically in Fig. 8.3; I note the following points.

1. In the limit $\zeta \to +\infty$ (BCS limit) μ tends to ϵ_F and Δ tends to a constant[28] times $\exp(-\pi/2k_F|a_s|)$. Since the single-spin density of states at the Fermi energy, N_0, is $k_F/2\pi^2$, while the effective long-wavelength ($k \ll a_s^{-1}$) interaction is $4\pi\hbar^2 a_s/m$ (see Chapter 4, Section 4.2), the exponent can be written as $(-1/N_0V_0)$, so we recover the standard BCS results (to logarithmic accuracy).
2. In the limit $\zeta \to -\infty$ (BEC limit) we recover the theory of the free Bose gas: in this limit the chemical potential μ tends to $-\hbar^2/2ma_s^2$, i.e. to half the binding energy of the molecule, as it should (since dissociation of the molecule is equivalent to removing *two* fermions to infinity), while the quantity Δ (which actually has no great physical significance in this limit) is proportional to $n^{1/2}a_s^{-1/2}$.
3. There is no singularity of any kind at the "unitarity limit" $\delta = 0$ (the point at which the two-body state becomes bound in free space).[29].
4. However, the point at which $\mu = 0$, which occurs on the "BEC" side of the resonance, is special, in the sense that according to (8.4.2) the minimum value of the quantity E_k (which as in the usual BCS theory turns out to be the quasiparticle excitation energy) changes its dependence on μ and Δ from $|\Delta|$ to $(\mu^2+|\Delta|^2)^{1/2}$. While it seems unlikely that this has much effect at $T = 0$, it seems probable that at finite T it will produce higher-order discontinuities in thermodynamic quantities (though cf. below).

[28]The constant is of order ϵ_F, but is not correctly recovered in this approximation, see Gor'kov and Melik-Barkhudarov (1961); however the correction is only a factor of order unity.

[29]This result has long been known in the theory of nuclear matter: the equation of state of the latter is insensitive to whether or not the deuteron is bound.

5. On the basis of Eqns. (8.4.3) and (8.4.4) it is possible to obtain other quantities of interest besides μ and Δ, for example the pair radius ξ, as a function of ζ, see e.g. Randeria (1995).

It is clear that the naive ansatz, while possibly giving a good qualitative account of the crossover, cannot be expected to get things quantitatively right outside the two extreme limits, even at $T = 0$. Its most obvious defect is that it leaves out all "Bogoliubov-level" effects, that is, the effects of scattering of pairs of particles out of the condensate; one would expect these to be particularly important for small negative ζ. Indeed, it has been shown explicitly (Petrov et al. 2004) by solution of the relevant four-particle problem that it does not even get the ground state energy right at the GP level (the coefficient of the interaction term is a factor of 10/3 too large). For this reason, there have been many attempts in the literature to go beyond the naive ansatz: for a partial review, see Stajic et al. (2005). A particularly challenging problem is to obtain an accurate value of the ground state energy E at unitarity: since at this point the only characteristic length[30] in the problem is the interparticle spacing $n^{-1/3}$ (n = density), it is clear on dimensional grounds that we must have

$$E = A\frac{n^{2/3}\hbar^2}{m} \tag{8.4.6}$$

so the problem reduces to finding an accurate value for the dimensionless constant A. Despite its apparent simplicity, this problem is not as far as is currently known susceptible to analytic solution, so it appears that one must resort to numerical computation (see, for example, Carlson et al. 2003).

At nonzero temperatures is seems unlikely that the naive ansatz will retain even a qualitative validity. The reason is the following: In the BCS limit, as the temperature is lowered, the onset of Cooper pairing ("pseudo-BEC") and hence, presumably, of superfluidity, is an instability of the normal Fermi sea, and is thus extremely sensitive to the inter-fermion interaction (cf. Eqn. 5.5.18). Looking at it the other way around, as we *increase* the temperature the normal phase sets in as a result of (collective) "dissociation" of the Cooper pairs. By contrast, in the BEC limit the di-fermionic molecules are tightly bound and do not dissociate up to a temperature $T_0 \sim \hbar^2/ma_s^2$ much too high to be of interest;[31] the restoration of the normal state is due, rather, to the fact that the previously "condensed" molecules can now all be accommodated in nonzero-$\boldsymbol{k}$ states, a process which is entirely omitted in the naive ansatz. It is conceivable that one could extend this ansatz so as to include these "bound-but-not-condensed" pairs, but so far this has not turned out to be achievable in any rigorous way, and one must resort to approximate analytic techniques, to scaling arguments (which are particularly useful close to unitarity) or to numerical calculation: see e.g. Stajic et al. (2005). Without going into the details, I remark that there is one qualitative point on which almost all calculations seem to agree, and which can in fact be reasonably inferred by extrapolation of the known behavior in the $\zeta \to \pm\infty$ limits: If we plot, in the (T, ζ) plane, the locus $T_c(\zeta)$ of onset of "condensation" (i.e. of a macroscopic

[30] In the approximation that the range r_0 of the potential is negligible.

[31] The dissociation is expected to be a gradual process, not a phase transition.

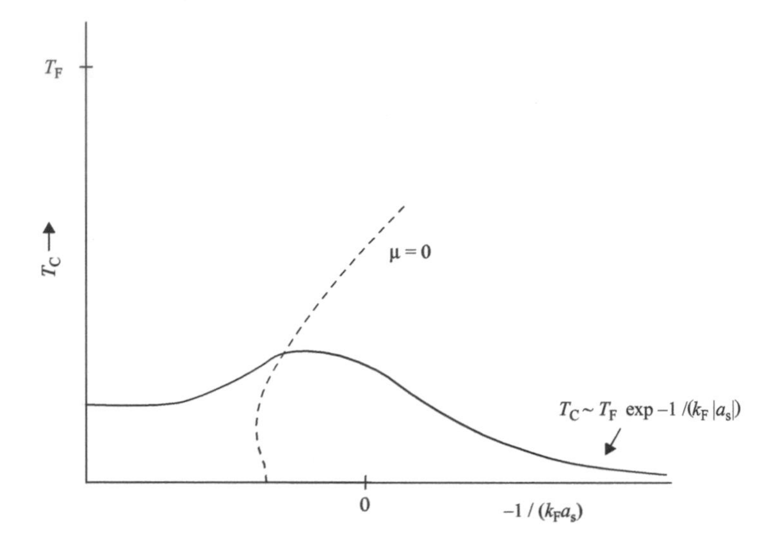

Fig. 8.4 The expected behavior of the temperature T_c for the onset of quantum condensation, as a function of $\zeta \equiv -1/(k_F a_s)$. The separatrix $\mu(T, \zeta) = 0$ is also shown.

eigenvalue of the two-fermion density matrix) then we know that as $\zeta \to +\infty$ (BCS limit) T_c should be given by the BCS formula $T_c \sim T_F \exp(-\pi/2k_F|a_s|)$, while for $\zeta \to -\infty$ (BEC) limit it should tend to the BEC transition temperature of a gas of bosons of mass $2m$ and density $n/2$ (where m and n are the *fermion* mass and density); this turns out to be equal to $0.21T_F$. Moreover, since the effective molecule–molecule interaction is repulsive (Petrov et al. 2004) and it has been convincingly demonstrated that repulsion *increases* the T_c of a dilute Bose gas (see e.g. Kashurnikov et al. 2001), we expect that the curve $T_c(\zeta)$ rises as ζ increases from a large negative value towards zero. Putting these considerations together, we might guess that $T_c(\zeta)$ would have the qualitative form shown in Fig. 8.4, rising to a maximum value of order T_F at a presumably small value of $|\zeta|$ (but not necessarily at unitarity). To calculate the curve of $T_c(\zeta)$ quantitatively is at the time of writing a major theoretical challenge. In Fig. 8.4 I have also sketched (qualitatively) the locus of the curve $\mu(T, \zeta) = 0$; while this appears not to be a separatrix between two differently (dis)ordered phases, it may for the reason given above correspond to a higher-order phase transition in the traditional thermodynamic sense, at least for $T < T_c(\zeta)$.

Over the last two years, following the pioneering work of Regal et al. (2004), there have been dozens of experiments which have probed the behavior of degenerate dilute Fermi alkali gases close to a Feshbach resonance; these have explored, inter alia, the behavior as a function of ζ of the density profile and collective excitation in a trap geometry, the NMR profiles (which can detect, in principle, both the pairing gap and the presence of a small closed-channel component) and the specific heat. A particularly interesting class of experiments (but one which is often quite difficult to interpret reliably) is those in which the system has been swept from positive to negative δ (i.e. ζ) and properties such as the condensate fraction or the persistence of vorticity measured on the BEC side. Because of the high risk of obsolescence of the discussion, I shall not attempt to describe or analyze these experiments in detail here; suffice it to say that at the time of writing (August 2005), while the existence of "condensation" (i.e. of a macroscopic eigenvalue of the two-particle density matrix) for positive ζ has not been directly established beyond reasonable doubt,[32] there appears to be convincing evidence for the persistence into this region of (a) a nonzero energy gap, (b) metastability of superflow (in the form of vortices), and (c) a nonzero closed-channel component.[33] Thus there is considerable circumstantial evidence for Cooper pairing on the "BCS side" of the transition. Further, the experiments on the density profile and collective excitations, which probe the energy of the system as a function of ζ, seem at least qualitatively consistent with the general picture sketched above on the basis of the naive ansatz, and the specific-heat experiments may be consistent with an approximate analytic scheme (Stajic et al. 2005) which is in some sense a nonzero-temperature generalization of it. In particular, with the possible exception of some puzzling data on a particular collective excitation, all experiments to date seem consistent with the "default" hypothesis of a continuous crossover from "BCS" to "BEC" behavior postulated earlier in this section.

There seems little doubt that the behavior of dilute Fermi alkali-gas systems in the neighborhood of a Feshbach resonance will continue to be an active area of research for the next few years. While all experiments to date have been on resonances in which the open channel corresponds to particles in a relative s-state, an intriguing prospect, which though not yet an experimental reality has begun to be explored by theorists, is that it may be possible to explore also $l \neq 0$ resonances. If so, this would open up to experimental investigation a whole host of new questions, including the perennial one of the spontaneous angular momentum of the condensate (see Appendix 6A). The field is still in its infancy!

Appendix

8A Renormalization of the gap equation for a dilute two-species Fermi gas

The procedure used is closely similar to that employed in Appendix 5F for a general Cooper-paired system in the limit $\Delta/\epsilon_F \to 0$; the difference is that in the present case we no longer require this limit, and moreover can express the result of the renormalization in terms of the s-wave scattering length which characterizes the two-particle problem (for which see Chapter 4, Section 4.2).

We consider a two-species Fermi gas in the "dilute" limit, meaning that the interparticle spacing $n^{-1/3}(\sim k_F^{-1})$ is much greater than the range r_0 of the interatomic potential. Since, as we see in Section 8.4, the interesting regime for many-body effects is $|k_F a_s| \gtrsim 1$, this condition means that we also have $a_s \gg r_0$; and since, as we see in that section, the pair radius ξ_0 is at least of the order of the smaller of a_s and k_F^{-1}, it too is large compared to r_0. The crucial point is that the pairing potential $V_{\boldsymbol{kk'}} \equiv V_{\boldsymbol{k}-\boldsymbol{k'}}$ has significant structure only on the scale of r_0^{-1}, so that it is effectively constant for $|\boldsymbol{k} - \boldsymbol{k'}|$ of the order of the inverse lengths k_F, a_s^{-1} important for the

[32] I would regard the interpretation of the experiments of Regal et al. (2004) and Zwierlein et al. (2004) as controversial at the time of writing.

[33] Which is indirect evidence for condensation in a two-particle bound state, see Partridge et al. (2005).

many-body problem. Consider then the many-body system at $T = 0$; we can follow the argument of Chapter 4, Section 4.4 word for word and derive the standard BCS gap equation (5.4.22), which it is convenient to rewrite explicitly in terms of the pair wave function $F_{\boldsymbol{k}}$:

$$F_{\boldsymbol{k}} = -\frac{1}{2E_{\boldsymbol{k}}}\sum_{\boldsymbol{k}'} V_{\boldsymbol{k}\boldsymbol{k}'}F_{\boldsymbol{k}'}, \quad E_{\boldsymbol{k}} \equiv ((\xi_{\boldsymbol{k}} - \mu)^2 + |\Delta_{\boldsymbol{k}}|^2)^{1/2},$$

$$\xi_{\boldsymbol{k}} \equiv \frac{\hbar^2 k^2}{2m} \tag{8.A.1}$$

We may compare (8.A.1) with the zero-energy two-body Schrödinger equation

$$\psi_{\boldsymbol{k}} = -\frac{1}{2\xi_{\boldsymbol{k}}}\sum_{\boldsymbol{k}'} V_{\boldsymbol{k}\boldsymbol{k}'}\psi_{\boldsymbol{k}'} \tag{8.A.2}$$

From this comparison it is clear that for k large compared to the "many-body" wave vectors $(2m|\mu|/\hbar^2)^{1/2}$, $(2m|\Delta|/\hbar^2)^{1/2}$ the functional form of $F_{\boldsymbol{k}}$ is identical to that of $\psi_{\boldsymbol{k}}$. Transforming this result into coordinate space, we reach the important conclusion that *for r small compared to k_{F}^{-1}, a_{s} (and thus in particular for $r \lesssim r_0$) the real-space form of the pair wavefunction $F(\boldsymbol{r})$ is identical to that of the zero-energy two-body scattering state $\psi(\boldsymbol{r})$*; the only difference lies in the normalization. This result holds over the whole "crossover" region.

To express the gap equation (8.A.1) in terms of the two-body s-wave scattering length a_s, we select a cutoff wave vector k_c large compared to k_{F}, a_{s}^{-1} but small compared to $\hbar/mr_0$, and proceed exactly as in Appendix 5F, obtaining the renormalized gap equation

$$\sum\nolimits'_{k<k_c} (2E_k)^{-1} = t_0^{-1}(\epsilon_c) \tag{8.A.3}$$

where $\epsilon_{\mathrm{c}} \equiv \hbar^2 k_{\mathrm{c}}^2/2m$ is the cutoff energy corresponding to k_{c}, and $t_0(\epsilon_{\mathrm{c}})$ is the constant matrix

$$\hat{t}_0(\epsilon_{\mathrm{c}}) \equiv (1 + \hat{V}\hat{P}_2\hat{K})^{-1}\hat{V} \tag{8.A.4}$$

where the symbols $\hat{P}_2$ and $\hat{K}$ have the same meaning as in Appendix 5F. In (8.A.3) the "gap" $\Delta_{\boldsymbol{k}}$ occurring in $E_{\boldsymbol{k}}$ is a $\boldsymbol{k}$-independent constant Δ. Because the matrix $\hat{P}_2\hat{K}$ is independent of $|\Delta|$ and μ, $t_0(\epsilon_{\mathrm{c}})$ is identical to the corresponding quantity occurring in the two-body problem. Now the latter quantity may be rescaled for arbitrary small values of the cutoff by a further renormalization procedure analogous to that used in Appendix 5F, so that

$$t_0^{-1}(\epsilon'_{\mathrm{c}}) = t_0^{-1}(\epsilon_{\mathrm{c}}) + \sum_{k'_{\mathrm{c}}<k<k_{\mathrm{c}}} (2\xi_k)^{-1} \tag{8.A.5}$$

But by the arguments of Chapter 4, Section 4.2 the quantity $\lim_{\epsilon'_{\mathrm{c}}\to 0} t(\epsilon'_{\mathrm{c}})$, the effective interaction in the long-wavelength limit, is just equal to $4\pi\hbar^2 a_s/m$, so that

$$t_0^{-1}(\epsilon_{\mathrm{c}}) = \left(\frac{4\pi\hbar^2 a_{\mathrm{s}}}{m}\right)^{-1} + \sum_{k<k_{\mathrm{c}}} (2\xi_k)^{-1} \tag{8.A.6}$$

Substituting (8.A.4) in (8.A.3) and extending the (convergent) sum over $\boldsymbol{k}$ to infinity, we obtain Eqn. (8.4.3) of the text.

Bibliography

Abo-Shaeer, J.R., et al., 2001, *Science* 292, 476.
Abrahams, E., et al., 1995, *Phys. Rev.* 52, 1271.
Acheson, D.J., 1990, *Elementary Fluid Dynamics*, Clarendon Press, Oxford.
Albiez, M., et al., 2005, *Phys. Rev. Lett.* 95, 010402.
Allen, P.B., et al., 1989, in Ginsberg, vol. I.
Allum, D.R., et al., 1976, *Phys. Rev. Lett.* 36, 1313.
Amit, D.J., 1978, *Field Theory, the Renormalization Group and Critical Phenomena*, McGraw-Hill, New York.
Anderson, B.P. and Kasevich, M.A., 1998, *Science* 282, 1686.
Anderson, M.H., et al., 1995, *Science* 269, 198.
Anderson, P.W., 1966, *Revs. Mod. Phys.* 38, 298.
Anderson, P.W. and Morel, P., 1961, *Phys. Rev.* 123, 1911.
Andrews, M.R., et al., 1996, *Science* 273, 84.
Andrews, M.R., et al., 1997, *Science* 275, 637.
Annett, J.F., 2004, *Superconductivity, Superfluidity and Condensates*, Oxford University Press, New York.
Annett, J.F., et al., 1996, in Ginsberg, vol. V.
Ashcroft, N.W. and Mermin, N.D., 1976, *Solid State Physics*, Holt, Rinehart and Winston, New York.
Ashhab, S., 2002, Ph.D. thesis, University of Illinois.
Ashhab, S. and Leggett, A.J., 2003, *Phys. Rev. A* 68, 63612.
Awschalom, D.D. and Schwarz, K.W., 1984, *Phys. Rev. Lett.* 52, 49.
Balatshy, A.V., et al., 2006, *Revs. Mod. Phys.* 78, 373.
Basov, D.N. and Timusk, T., 2005, *Revs. Mod. Phys.* 77, 721.
Basov, D.N., et al., 1994, *Phys. Rev. B* 50, 3511.
Batlogg, B., 1998, *Solid State Comm.* 107, 639.
Baüerle, C., et al., 1996, *Nature* 382, 332.
Baumgardner, J.E. and Osheroff, D.D., 2004, *Phys. Rev. Lett.* 93, 155301.
Baym, G. and Pethick, C.J., 1991, *Landau Fermi-Liquid Theory: Concepts and Applications*, Wiley, New York.
Baym, G. and Pethick, C.J., 1996, *Phys. Rev. Lett.* 76, 6.
Berry, M.V. and Robbins, J.M., 1997, *Proc. Roy. Soc.* (London), series A, 453, 1771.
Berry, M.V. and Robbins, J.M., 2000, *J. Phys. A* 33, L 207.
Bloch, I., et al., 2000, *Nature* 403, 166.

Blumberg, G., et al., 1997, *Science* 278, 1427.
Blumberg, G., et al., 1998, *J. Phys. Chem. Solids* 59, 1932.
Boebinger, G., et al., 2000, *Phys. Rev. Lett.* 85, 638.
Bonn, D.A. and Hardy, W.N., 1996, in Ginsberg, vol. V.
Boronat, J., et al., 1995, *Phys. Rev. B* 52, 1236.
Brinkman, W.F., et al., 1974, *Phys. Rev. A* 10, 2386.
Burger, S., et al., 2001, *Phys. Rev. Lett.* 86, 4447.
Campuzano, J.C., et al., 2002, cond-mat/0209476.
Carlson, J., et al., 2003, *Phys. Rev. Lett.* 91, 050401.
Castin, Y. and Dalibard, J., 1997, *Phys. Rev. A* 55, 4330.
Cataliotti, F.S., et al., 2001, *Science* 293, 843.
Cataliotti, F.S., et al., 2002, *Phys. Rev. Lett.* 89, 088902.
Ceperley, D.M., 1995, *Revs. Mod. Phys.* 67, 279.
Ceperley, D.M. and Pollock, E.L., 1986, *Phys. Rev. Lett.* 56, 351.
Chakravarty, S., et al., 1991, *Science* 254, 970.
Chevy, F., et al., 2000, *Phys. Rev. Lett.* 85, 2223.
Chikkatur, A.P., et al., 2000,*Phys. Rev. Lett.* 45, 483.
Chossat, P. and Iooss, G., 1994, *The Couette–Taylor Problem*, Springer, New York.
Combescot, R., 1978, *Phys. Rev. B* 18, 6071.
Cooper, S.L. and Gray, K.E., 1994, in Ginsberg, vol. IV.
Corwin, K.L., et al., 1999, *Phys. Rev. Lett.* 83, 1311.
Damascelli, A., et al., 2003, *Revs. Mod. Phys.* 75, 473.
De Gennes, P.-G., 1966, *Superconductivity of Metals and Alloys*, Benjamin, New York (reprinted, Perseus Books, Reading, MA 1999).
Di Luccio, T., et al., 2003, *Phys. Rev. B* 67, 92504.
Donnelly, R.J., 1967, *Experimental Superfluidity*, University of Chicago Press, Chicago.
Donnelly, R.J, 1991, *Quantized Vortices in Helium-II*, Cambridge University Press, Cambridge.
Draeger, E.W. and Ceperley, D.M., 2002, *Phys. Rev. Lett.* 89, 015301.
Duck, I. and Sudarshan, E.C.G., 1998, *Am. J. Phys.* 66, 284.
Dunningham, J.A. and Burnett, K., 2000, *J. Phys. B* 33, 3807.
Eagles, D.M., 1969, *Phys. Rev.* 186, 456.
Emery, V.J. and Kivelson, S.A., 1995, *Nature* 374. 434.
Feenberg, E., 1969, *Theory of Quantum Fluids*, Academic, New York.
Ferrell, R.A., 1959, *Phys. Rev. Lett.* 3, 262.
Fetter, A.L., 1972, *Ann. Phys.* 70, 67.
Feynman, R.P. and Hibbs, A.R., 1965, *Quantum Mechanics and Path Integrals*, McGraw-Hill, New York.
Fisher, M.P.A., et al., 1989, *Phys. Rev. B* 40, 546.
Fomin, I.A., 2005, *JETP Lett.* 81, 298.
Fong, H.F., et al., 1999, *Nature* 398, 588.
Fujita, T. and Quader, K., 1987, *Phys. Rev. B* 36, 5152.
Gati, R., et al., 2006, *Phys. Rev. Lett.* 96, 130404

Gavoret, J. and Nozières, P., 1964, *Ann. Phys.* 28, 349.
Ginsberg, D.M., ed., 1989–1996, *Physical Properties of High Temperature Superconductors*, vols. I–V, World Scientific, Singapore. (Referred to throughout the Bibliography as "Ginsberg.")
Glyde, H.R., 1994, *Phys. Rev. B* 50, 6726.
Glyde, H.R., et al., 2000, *Phys. Rev. B* 62, 14337.
Goldenfeld, N., 1992, *Lectures on Phase Transitions and the Renormalization Group*, Addison-Wesley, Reading, MA.
Goldstein, H., 1980, *Classical Mechanics*, 2nd edn, Addison-Wesley, Reading, MA.
Gor'kov, L.P. and Melik-Barkhudarov, T.K., 1961, *Soviet Phys. JETP* 13, 1018.
Gor'kov, L.P. and Volovik, G.E., 1985, *Soviet Phys. JETP* 61, 843.
Greene, L.H., et al., 2003, *Physica C* 387, 162.
Greiner, M., et al., 2002a, *Nature* 415, 39.
Greiner, M., et al., 2002b, *Nature* 419, 51.
Gribakin, A. and Flambaum, V.V., 1993, *Phys. Rev. A* 48, 546.
Griffin, A., 1993, *Excitations in a Bose-Condensed Liquid*, Cambridge University Press, Cambridge, UK.
Griffin, A. and Stringari, S., 1996, *Phys. Rev. Lett. 76*, 259.
Haard, T.M., et al., 1994, *Phys. Rev. Lett.* 72, 860.
Haase, J. and Slichter, C.P., 2000, *J. Sup.* 13, 723.
Hagley, E.W., et al., 1999, *Phys. Rev. Lett.* 83, 3112.
Hall, D.S., et al., 1998, *Phys. Rev. Lett.* 81, 1543.
Harrison, W.A., 1970, *Solid State Theory*, McGraw-Hill, New York.
Hasegawa, T., et al., 1992, in Ginsberg, vol. III.
He, H., et al., *Science* 295, 1045 (2002).
Hewitt, K.C., et al., 1999, *Phys. Rev. B* 60, R9943.
Ho, T.-L., 1982, *Phys. Rev. Lett.* 49, 1837.
Ho, T.-L. and Shenoy, V.B., 1996, *Phys. Rev. Lett.* 77, 2592.
Ho, T.-L. and Yip, S.-K., 2000, *Phys. Rev. Lett.* 84, 4031.
Hoffman, J.E., et al., 2002, *Science* 297, 1148.
Hohenberg, P.C. and Platzman, P.M., 1966, *Phys. Rev.* 152, 198.
Holcomb, M.J., et al., 1996, *Phys. Rev. B* 53, 6734.
Huang, K., 1987, *Statistical Mechanics*, 2nd edn, Wiley. New York.
Hugenholtz, N.M. and Pines, D., 1959, *Phys. Rev.* 116, 489.
Huse, D.A. amd Siggia, E.D., 1982, *J. Low Temp. Phys.* 46, 137.
Hussey, N., 2002, *Adv. Phys.* 51, 1685.
Inguscio, M., et al., eds., 1999, *Bose–Einstein Condensation in Atomic Gases, Proc. Enrico Fermi International School of Physics*, IOS Press, Amsterdam.
Inouye, S., et al., 2001, *Phys. Rev. Lett.* 87, 080402.
Iye, Y., 1992, in Ginsberg, vol. III.
Jackson, H.W. and Feenberg, E., 1962, *Revs. Mod. Phys.* 34, 686.
Kalos, M.H., et al., 2005, *J. Low Temp. Phys.* 138, 747.
Kashurnikov, V.A., et al., 2001, *Phys. Rev. Lett.* 87, 120402.
Kastner, M., et al., 1998, *Revs. Mod. Phys.* 70, 897.

Kennedy, T., et al., 1988, *Phys. Rev. Lett.* 61, 2582.
Ketterle, W., et al., 1999, in M. Inguscio et al., eds., *Bose–Einstein Condensation in Atomic Gases*, IOS Press, Amsterdam.
Kim, E. and Chan, M.W.H., 2004a, *Nature* 427, 225.
Kim, E. and Chan, M.W.H., 2004b, *Science* 305, 1941.
Kivelson, R. A. and Fradkin, E., 2005, cond-mat/0507459 (to be published in *Treatise of High-Temperature Superconductivity*, ed. J.R. Schrieffer).
Kleiner, R. and Mueller, P., 1994, *Phys. Rev. B* 49, 1327.
Klemm, R A., 2005, *Phil. Mag.* 85, 801.
Kohl, M., et al., 2005, *Phys. Rev. Lett.* 94, 080403.
Kosztin, I., et al., 1998, *Phys. Rev. B* 58, 9365.
Krishana, K., et al., 1995, *Phys. Rev. Lett.* 75, 3529.
Kuklov, A.B. and Svistunov, B.V., 2002, *Phys. Rev. Lett.* 89, 170403.
Landau, L.D. and Lifshitz, E.M., 1958, *Statistical Physics*, trans. E. Peierls and R.F. Peierls, Pergamon, London.
Landau, L.D. and Lifshitz, E.M., 1977, *Quantum Mechanics (Non-relativistic Theory)*, trans. J.B. Sykes and J.S. Bell, Pergamon, Oxford.
Landau, L.D. and Lifshitz, E.M., 1987, *Fluid Mechanics*, trans. J.B. Sykes and W.H. Read, Pergamon, Oxford.
Law, C.K., et al., 1998, *Phys. Rev. Lett.* 81, 5257.
Leanhardt, A.E., et al., 2002, *Phys. Rev. Lett.* 89, 190403.
Lee, P.A., 2000, *Physica C* 317, 194.
Lee, T.D. and Yang, C.N., 1957, *Phys. Rev.* 115, 1119.
Lee, T.D., et al., 1957, *Phys. Rev.* 106, 1135.
Leggett, A.J., 1965, *Phys. Rev. Lett.* 14, 536.
Leggett, A.J., 1972, *Phys. Rev. Lett.* 29, 1227.
Leggett, A.J., 1974, *Ann. Phys.* 85, 11.
Leggett, A.J., 1975, *Revs. Mod. Phys.* 47, 331.
Leggett, A.J., 1992, in *Physical Phenomena at High Magnetic Fields*, ed. E. Manousakis et al., Addison-Wesley, Redwood City, CA.
Leggett, A.J., 1995, *Found. Phys.* 25, 113.
Leggett, A.J., 1996, *Science* 274, 587.
Leggett, A.J., 1998, *J. Stat. Phys.* 93, 927.
Leggett, A.J., 1999a, unpublished.
Leggett, A.J., 1999b, *Proc. Nat. Acad. Sci.* 96, 8635.
Leggett, A.J., 2000, *Mod. Phys. Lett. (suppl. issue) B* 14, 1.
Leggett, A.J., 2001, *Revs. Mod. Phys.* 73, 307.
Leggett, A.J., 2003, *New J. Phys.* 5, 103.
Leggett, A.J., 2004, *Synthetic Metals* 141, 51.
Leinaas, J.M. and Myrheim, J., 1977, *Nuovo Cimento B* 37, 1.
Lindenau, T., et al., 2002, *J. Low Temp. Phys.* 129, 143.
Liu, K-S. and Fisher, M.E., 1973, *J. Low Temp. Phys.* 10, 655.
Loram, J.W., et al., 1994, *Physica C* 235–240, 134.
Loram, J.W. and Tollan, J.R., 1998, *J. Phys. Chem. Solids* 59, 2091.
Lu, J.P. and Gelfand, M.P., 1995, *Phys. Rev. B* 51, 16115.

Mackenzie, A.P. and Maeno, Y., 2003, *Revs. Mod. Phys.* 75, 657.
Madison, K., et al., 2000, *Phys. Rev. Lett.* 84, 806.
Makhlin, Yu., et al., 2001, *Revs. Mod. Phys.* 73, 357.
Maki, K., 1969, in Parks, 1969.
Maki, K. and Tsuneto, T., 1964, *Prog. Theor. Phys.* 31, 945.
Malozemoff, A.P., 1989, in Ginsberg, vol. I.
Mandel, O., et al., 2003, *Phys. Rev. Lett.* 91, 010407.
Markiewicz, R.S., 1997, *J. Phys. Chem. Solids* 58, 1179.
Mathai, A., et al., 1995, *Phys. Rev. Lett.* 74, 4523.
Matsuda, A., et al., 1999, *Phys. Rev. B* 60, 1377.
Matthews, M.R., et al., 1999a, *Phys. Rev. Lett.* 83, 2498.
Matthews, M.R., et al., 1999b, *Phys. Rev. Lett.* 83, 3358.
McMillan, W.L. and Rowell, J.M., 1969, in Parks, 1969.
Mermin, N.D., 1979, *Revs. Mod. Phys.* 51, 591.
Mermin, N.D., 1990, *Boojums All the Way Through: Communicating Science in a Prosaic Age*, Cambridge University Press, Cambridge, UK.
Millis, A.J., et al., 1988, *Phys. Rev.* 37, 4975.
Mills, D.L., et al., 1994, *Phys. Rev. B* 50, 6394.
Mineev, V.P. and Zhitomirsky, M.E., 2005, *JETP Lett.* 81, 296.
Molegraaf, H.J.A., et al., 2002, *Science* 295, 2239.
Mook, H., et al., 1993, *Phys. Rev. Lett.* 70, 3490.
Mook, H., et al., 1997, in SNS97.
Mook, H., et al., 1998, *Nature* 395, 580.
Mook, H., et al., 1999, *Science* 289, 1344.
Moroni, S., et al., 1997, *Phys. Rev. B* 55, 1040.
Munzar, D., et al., 2001, *Phys. Rev. B* 64, 024523.
Nazaretski., E., et al., 2004, *J. Low Temp. Phys.* 134, 763.
Nelson, K.D., et al., 2004, *Science* 306, 1151.
Norman, M.R. and Pepin, C., 2003, *Reps. Prog. Phys.* 66, 1547.
Noziéres, P., 2004, *J. Low Temp. Phys.* 137, 45.
Noziéres, P. and Pines, D., 1989, *Theory of Quantum Liquids*, vol. II, Addison-Wesley, Redwood City, CA.
Nücker, N., et al., 1989, *Phys. Rev. B* 39, 12379.
Obertelli, S.D., et al., 1992, *Phys. Rev. B* 46, 14928.
Ong, N.P., 1990, in Ginsberg, vol. II.
Ong, N.P., et al., 2004, *Annalen der Physik* 13, 9.
Onofrio, R., et al., 2000, *Phys. Rev. Lett.* 85, 2228.
Orzel, C., et al., 2001, *Science* 291, 2386.
Osheroff, D.D., et al., 1980, *Phys. Rev. Lett.* 44, 792.
Parks, R.D., ed., 1969, *Superconductivity*, Marcel Dekker, New York.
Partridge, R.B., et al., 2005, *Phys. Rev. Letters* 95, 020404.
Pennington, C. and Slichter, C.P., 1992, in Ginsberg II.
Penrose, O. and Onsager, L., 1956, *Phys. Rev.* 104, 576.
Pethick, C.J. and Smith, H., 2002, *Bose Einstein Condensation in Dilute Gases*, Cambridge University Press, Cambridge.
Petrov, D.S., et al., 2004, *Phys. Rev. Lett.* 93, 090404.

Phelps, R.B., et al., 1994, *Phys. Rev. B* 50, 6526.
Phillips, J.C., et al., 2003, *Reps. Prog. Phys.* 66, 2111.
Pines, D., 1998, in *Proc. Conf. on Gap Symmetry in High-T_c Superconductors*, Plenum, New York.
Pines, D. and Noziéres, P., 1966, *Theory of Quantum Liquids*, vol. I., Benjamin, New York.
Pitaevskii, L.P. and Stringari, S., 2003, *Bose–Einstein Condensation*, Oxford University Press, Oxford.
Politzer, H.D., 1996, *Phys. Rev. A* 54, 5048.
Porto, J.V. and Parpia, J.M., 1999, *Phys. Rev. B* 59, 14583.
Prange, R.E. and Girvin, S.M., eds., 1990, *The Quantum Hall Effect*, 2nd edn., Springer, New York.
Prokofiev, N.V. and Svistunov, B.V., 2005, *Phys. Rev. Lett.* 94, 155302.
Putterman, S.J., 1974, *Superfluid Hydrodynamics*, North-Holland, Amsterdam.
Raman, C., et al., 1999, *Phys. Rev. Letters* 83, 2502.
Randeria, M., 1995, in ed. A. Griffin et al., *Bose–Einstein Condensation*, Cambridge University Press, Cambridge, UK.
Rayfield, G.W. and Reif, F., 1964, *Phys. Rev. A* 136, 1194.
Regal, C.A., et al., 2004, *Phys. Rev. Lett.* 92, 040407.
Renner, C.H., et al., 1998, *Phys. Rev. Lett.* 80, 149.
Rübhausen, M., 1998, in *J. Phys. Chem. Solids* 59, 2009.
Rübhausen, M., et al., 2001, *Phys. Rev. B* 63, 224514.
Ruutu, V., et al., 1996, *Nature* 382, 334 (1996).
Sachdev, S., 2003, *Revs. Mod. Phys.* 75, 913.
Sadovskii, M.V., 1997, *Phys. Repts.* 282.
Scalapino, D.J., 1969, in Parks, 1969.
Scalapino, D.J., 1999, *J. Low Temp. Phys.* 117, 179.
Schilling, J.S. and Klotz, S., 1992, in Ginsberg, vol. III.
Schrenk, R., et al., 1996, *Phys. Rev. Lett.* 76, 2945.
Schwabl, F., 1992, *Quantum Mechanics*, Springer-Verlag, Berlin.
Shaked, H., et al., 1993, *Phys. Rev. B* 48, 14921.
Shaked, H., et al., 1994, *Crystal Structures of the High-T_c Copper Oxides*, Elsevier, Amsterdam.
Sheshadri, K., et al., 1993, *Europhys. Lett.* 22, 257.
Slichter, C.P., 1992, *Principles of Magnetic Resonance*, Springer-Verlag, Berlin.
Snow, W.M. and Sokol, P.E., 1995, *J. Low Temp. Phys.* 101, 881.
Sokol, P.E., 1987, *Can. J. Phys.* 65, 1393.
Stajic, J., et al., 2005, *Phys. Repts.* 412, 1.
Steel, M.J., et al., 1999, *Phys. Rev. A* 58, 4824.
Steglich, F., et al., 1996, *Physica* C263, 498.
Stenger, J., et al., 1999, *Phys. Rev. Lett.* 82, 4569.
Stone, M. and Roy, R., 2004, *Phys. Rev. B* 69, 184511.
Streater, R.F. and Wightman, A.S., 1964, *PCT, Spin and Statistics and All That*, New York, Benjamin.
Strohm, T. and Cardona, M., 1997, *Phys. Rev. B* 55, 12725.
Sunakawa, S., et al., 1962, *Prog. Theor. Phys.* 27, 589.

Tahir-Kheli, J., 1998, *Phys. Rev. b* 58, 12307.
Takagi, H., and Hussey, N. 1998, *Proc. Intl. School of Physics, course CXXXVI*, IOS Press, Amsterdam.
Tallon, J.R. and Loram, J.W., 2001, *Physica C* 349, 53.
Tallon, J.R., et al., 1997, in SNS97.
Tallon, J.R., et al., 1998, *J. Phys. Chem. Solids* 59, 2145.
Tanner, D.B. and Timusk, T., 1992, in Ginsberg, vol. III.
Thuneberg, E.V., et al., 1998, *Phys. Rev. Lett.* 80, 2861.
Tiesinga, E., et al., 1996, *J. Res. Natl. Inst. Stand. Technol.* 101, 505.
Timusk, T. and Tanner, D.B., 1989, in Ginsberg, vol. I.
Tinkham, M.W., 1996, *Introduction to Superconductivity*, 2nd edn., McGraw-Hill, New York.
Tsuei, C.C. and Kirtley, J.R., 2000, *Revs. Mod. Phys.* 72, 969.
Tsuei, C.C., et al., 1994, *Phys. Rev. Lett.* 73, 593.
Tsuei, C.C., et al., 2004, *Phys. Rev. Lett.* 93, 187004.
Tsuneto, T., *Superconductivity and Superfluidity*, 1998, Cambridge University Press, Cambridge, UK.
Uemura, Y.J., 1997, *Physica* 282, 194.
Uemura, Y.J., et al., 1989, *Phys. Rev. Lett.* 62, 2317.
Uemura, Y.J., et al., 1998, *Nature* 398, 558.
Uher, C., 1992, in Ginsberg, vol. III.
Van der Marel, D., 2004, *J. Sup.* 17, 559.
Varma, C.M., et al., 1989, *Phys. Rev. Lett.* 63, 1196.
Varma, C.M., et al., 2002, *Phys. Repts.* 361, 267.
Vollhardt, D. and Wölfle, P., 1990, *The Superfluid Phases of Helium-3*, Taylor and Francis, London.
Volovik, G.E., 1992, *Exotic Properties of Superfluid 3-He*, World Scientific, Singapore.
Volovik, G.E., 1996, *JETP Lett.* 63, 301.
Vuorio, M., 1974, *J. Phys. C* 7L5.
Weinberg, S., 1972, *Gravitation and Cosmology: Principles and Applications of the General Theory of Relativity*, Wiley, New York.
Werthamer, N.R., 1969, in Parks, 1969.
Wheatley, J.C., 1978, *Prog. in Low Temp. Phys.* VIIA, 1.
Wilks, J., 1967, *The Properties of Liquid and Solid Helium*, Clarendon Press, Oxford.
Wollman, D.A., et al., 1993, *Phys. Rev. Lett.* 71, 2134.
Woodgate, G.K., 1970, *Elementary Atomic Structure*, McGraw-Hill, London.
Wu, B. and Niu, Q., 2002, *Phys. Rev. Lett.* 89, 088901.
Wyatt, A.F.G., 1998, *Nature* 391, 56.
Yang, C.N., 1962, *Revs. Mod. Phys.* 34, 694.
Yarmchuk, E.J., et al., 1979, *Phys. Rev. Lett.* 43, 214.
Zambelli, F. and Stringari, S., 1998, *Phys. Rev. Lett.* 81, 1754.
Zhang, S.C., 1997, *Science* 275, 1089.
Ziman, J., 1964, *Principles of the Theory of Solids*, Cambridge University Press, Cambridge, UK.
Zwierlein, M.W., et al., 2004, *Phys. Rev. Letters* 92, 120403.

Index

Entries in boldface refer to pages where an extended discussion of the topic is given.

*(individual properties are not indexed.)